# FERROELECTRIC CRYSTALS

## FRANCO JONA

*Department of Materials Science and Engineering,*
*State University of New York at Stony Brook*

## G. SHIRANE

*Department of Physics,*
*Brookhaven National Laboratory*

DOVER PUBLICATIONS, INC.

NEW YORK

This Dover edition, first published in 1993, is an unabridged and unaltered republication of the work first published by the Pergamon Press, Oxford, England, and The Macmillan Company, New York, in 1962 as Volume I in the *International Series of Monographs on Solid State Physics*.

Manufactured in the United States of America
Dover Publications, Inc., 31 East 2nd Street, Mineola, N.Y. 11501

*Library of Congress Cataloging-in-Publication Data*

Jona, Franco, 1922–
    Ferroelectric crystals / Franco Jona, G. Shirane.
        p.    cm.
    Originally published: Oxford ; New York : Pergamon Press, 1962, in series: International series of monographs on solid state physics.
    Includes bibliographical references and indexes.
    ISBN 0-486-67386-3 (pbk.)
    1. Ferroelectric crystals. I. Shirane, G. II. Title.
QD931.J6 1993
548'.85—dc20                                                    92-31174
                                                                     CIP

# CONTENTS

CHAPTER X

# STRUCTURAL PROBLEMS AND RECENT DEVELOPMENTS

# PREFACE

Forty years have elapsed since the discovery of the dielectric phenomenon known today under the name of ferroelectricity. Most of our knowledge in this branch of Solid State Physics has been acquired in the years prior to 1950 through extensive investigations of three ferroelectric materials: Rochelle salt, potassium dihydrogen phosphate, and barium titanate. In the last decade the major progress in the field stems from the discovery of a greater number of ferroelectric crystals. Unfortunately our knowledge of the basic phenomenon has not been appreciably enhanced. Hence, a comprehensive treatment of ferroelectricity is very difficult, if not impossible, at the present stage.

In contrast to some other branches of Solid State Physics, the understanding of the physical properties of ferroelectric crystals requires a wide range of experimental and theoretical approaches. The large reversible polarizations of these materials have both electronic and ionic origins. The latter indicates the very close interdependence between ferroelectric activity and crystal structure. This interdependence makes the most refined structural analysis a very essential tool for the investigation of the materials. On the other hand, the existence of linear electromechanical effects, which require the consideration of interactions between electrical and mechanical quantities, makes the treatment of the ferroelectric phenomenon much more complex than, for example, that of the ferromagnetic analogue.

Several review articles and monographs have already been written by a number of authors on the subject of ferroelectricity. All these publications have been of invaluable help to us at various stages of the preparation of the present monograph. In particular, we have been repeatedly enlightened by the excellent review article published by Devonshire in 1954, which is, we believe, the best presentation of the ferroelectric phenomenon given to date. The most recent contributions are Megaw's book and Känzig's article, both published in 1957. A large number of ferroelectric materials have been discovered since, and are discussed in the present monograph to the extent to which pertinent information is available.

This monograph purports to present the state of the art in the spring of 1960, with the hope that it serves a dual purpose. First, it may be of help to graduate students and research scientists unfamiliar with the subject as an introduction to the problems in the field of ferroelectricity. To attain this goal we have included several sections of an introductory nature. As the second purpose this monograph should be useful as a reference work for researchers in the field. It is mainly with this idea in mind that we have adopted a presentation in terms of chapters and sections devoted to one compound at a time and we have tried to quote as many numerical results as we think may be desirable. In a presentation of this kind repetitions are often unavoidable; nevertheless, we have tried to introduce a number of cross references to help the reader interested in the continuity of a given phenomenon rather than the characteristics of a given compound.

The science of ferroelectricity has been expanding very rapidly and it is difficult to predict what its future will be. This high rate of expansion can be explained in part by the attractiveness of the bistable characteristics of ferroelectric materials and their resultant device applications. This device potentiality has not been realized, at least until now, but it has stimulated a number of excellent experimental and theoretical investigations of far-reaching scientific interest.

The volume of literature published on this subject is so large that a claim of completeness for our bibliography would be ill advised. While exercising some discrimination, we have tried to compile as complete a list of references as possible up to May 1960. Some of the papers published after this date have also been referred to, although they could not be properly co-ordinated. Lists of literature references are given at the end of each chapter and are arranged alphabetically according to the first letter of the name of the principal author. Researchers in the field of ferroelectricity will have noticed a marked increase in Russian literature in the most recent years. We have also tried to include the Russian contributions in our list of references, but unfortunately linguistic problems did not permit us to digest all the pertinent papers. Our quotations refer to the original Russian periodicals and not to existing translations published by the American Institute of Physics.

We are very grateful to many colleagues who have made the results of their investigations available to us prior to publication: R. C. Miller, A. Savage, A. G. Chynoweth and J. P. Remeika of the Bell Telephone Laboratories; B. C. Frazer and T. Mitsui of the Brookhaven National Laboratory; S. Triebwasser, R. Landauer, M. Drougard, G. Burns and T. Dunne of the IBM Research Center; E. Fatuzzo and W. J. Merz of the RCA Laboratories; E. C. Subbarao of the Westinghouse Research Laboratory; H. H. Wieder of the O.N.R. Laboratory. We are particularly indebted to Prof. Y. Takagi of the Atomic Energy Research Institute of Japan and to Dr. R. Landauer of the IBM Research Center for comments, criticism and illuminating discussions. Thanks are also due to E. C. Subbarao and D. E. Cox of the Westinghouse Research Laboratory; E. Giess, F. Morehead, and G. Cheroff of the IBM Research Center and B. C. Frazer of the Brookhaven National Laboratory for careful reading of parts of the manuscript and constructive criticism. The invaluable co-operation and suggestions of Prof. R. Smoluchowski are also gratefully acknowledged.

F. JONA

December, 1960                                              G. SHIRANE

# CHAPTER I

# INTRODUCTION

## 1. General Features of Ferroelectric Crystals

THE MAIN problems which arise in the theory of dielectric crystals are concerned with the polarization that can be induced in such non-conducting materials by means of an externally applied electric field. The polarization values which can be measured in normal dielectrics upon application of experimentally attainable fields are usually small. Consequently, the effects that the polarization is expected to have on a number of physical properties of the crystals, such as the elastic, optical, thermal behavior, etc., are too small to be observed. Fortunately, a rather large, but yet limited, number of crystals exhibit polarization values which are many orders of magnitude larger than those observed in most dielectrics. The detailed study of some of these crystals has revealed many peculiar effects which are interesting not only from the point of view of dielectric theory, but also from that of crystallography, crystal-chemistry, thermodynamics, and, last but not least, practical applications in the field of electrical engineering.

It has become customary to call "ferroelectricity" the phenomenon exhibited by these crystals, and "ferroelectric" the crystals themselves. The reason for this denomination is historical and is due to a formal similarity of the ferroelectric phenomenon with that of ferromagnetism. The similarity is mainly phenomenological: just as ferromagnetic materials exhibit a spontaneous magnetization and hysteresis effects in the relationship between magnetization and magnetic field, ferroelectric crystals show a spontaneous electric polarization and hysteresis effects in the relation between dielectric displacement and electric field. This behavior is mostly observed in certain temperature regions delimited by transition (or Curie) points above which the crystals are no longer ferroelectric and show normal dielectric behavior.

Examples of typical ferroelectrics are: potassium di-hydrogen phosphate, $KH_2PO_4$, and a number of isomorphous phosphates and arsenates; barium titanate, $BaTiO_3$, and other isomorphous double oxides; Rochelle salt (sodium potassium tartrate tetrahydrate, $NaKC_4H_4O_6 \cdot 4H_2O$), and a few isomorphous crystals. The properties of these and other ferroelectrics will be discussed in detail in the following chapters. Reviews of these properties and the problems involved in the field of ferroelectricity are already available in the articles of Känzig (K 1) and Forsbergh (F 1) and in the book of Megaw (M 1). These authors follow different approaches in order to accomplish the task. Both Känzig and Forsbergh treat the problem of ferroelectricity in terms of the *properties* which characterize the phenomenon. Megaw, on the other hand, presents her description of ferro-electricity in terms of *compounds*, rather than properties, and is particularly concerned with the crystallographic aspect of the problem. Our own approach, in the present treatment of the same problem, is to describe the properties of the various ferroelectric crystals individually, and to emphasize the *dielectric* character of the ferroelectric phenomenon as judged from the viewpoint of the solid-state physicist.

In order to describe the essential features of the ferroelectric phenomenon, let us introduce a model of a hypothetical ferroelectric crystal and see what kind

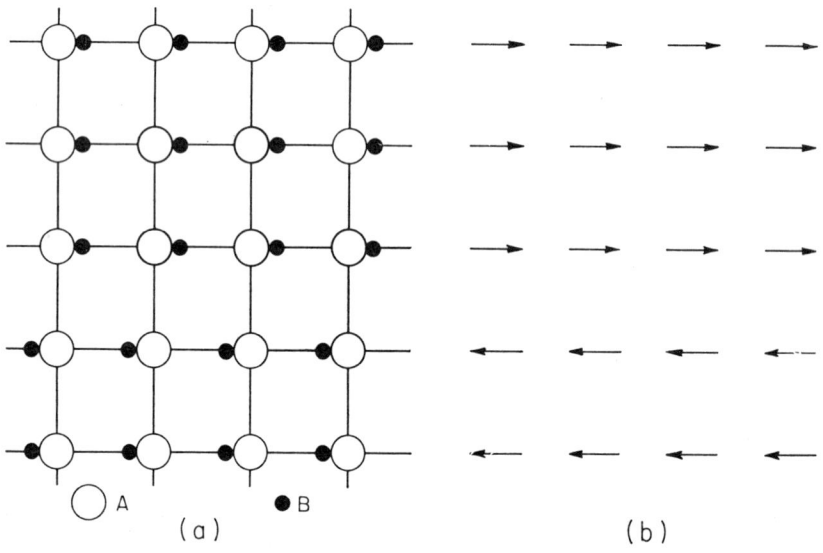

FIG. I-1. Schematic structure of a fictitious ferroelectric crystal.

of dielectric behavior it would show. The model, which has, of course, no general validity and is extremely oversimplified, is that of a two-dimensional crystal having the chemical formula $AB$ and a fictitious structure depicted in Fig. I-1 (a).

The $A$ ions, which we will assume to carry a negative charge, are located on the lattice points of a simple square net. The $B$ ions, carrying a positive charge, are located on the horizontal lines joining the $A$ ions; their equilibrium positions are such that they always lie closer to one of the two adjacent $A$ ions than to the other. Such a situation is possible if the potential between two adjacent $A$ ions is of the type sketched in Fig. I-2 (a): there are two equilibrium positions, corre-sponding to the same minimum value of the energy, for a $B$ ion on the line joining two $A$ ions. The $B$ ions can jump from one equilibrium position to the other but in order to do so they must be provided with the energy necessary to over-come the energy barrier $\Delta E$.

Suppose now that, at a given temperature $T$, all $B$ ions are closer to their $A$ partners on the left. We can visualize every group $AB$ as an electric dipole and the structure can then be schematically represented by an assembly of dipoles pointing all in the same direction, as in the three upper rows of Fig. I-1 (b). We say that the crystal is *spontaneously polarized*: the *spontaneous polarization* is measured in terms of dipole moment per unit volume, or, with reference to the charges induced on the surfaces perpendicular to the polarization, in terms

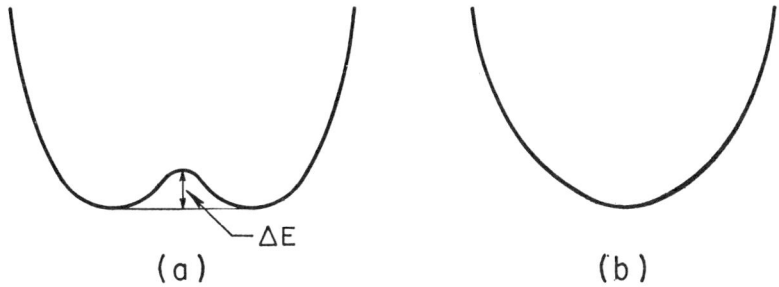

( a )   ( b )

FIG. I-2. Schematic potential wells.

of charge per unit area. The crystals having a spontaneous polarization are called *pyroelectric* (see Section 3) and the direction of the spontaneous polarization is called the *polar axis*.

Alignment of the electric dipoles may extend only over a region of the crystal, while in another region the direction of the spontaneous polarization may be reversed, as in the lower portion of Fig. I-1 (b). Such regions of uniform polarization are called *domains*, a term borrowed again from ferromagnetism. We are going to see in the following that energy considerations require the formation of domains.

Suppose now that we apply an electric d.c. field in the horizontal direction of Fig. I-1. The dipoles which are already oriented in the direction of the field will remain so aligned, but those which are oriented in the direction opposite (antiparallel) to the field will show a tendency to reverse their orientation. If the applied field is sufficiently large, the $B$ ions of our model will be able to overcome the barrier $\Delta E$, and in so doing will cause the corresponding dipole to flip (or switch) over into the direction of the field. This phenomenon of polarization reversal takes place by way of a nucleation process and domain–wall motions. The question which immediately arises is this: how does the process of polarization reversal affect the relation between polarization $P$ and applied electric field $E$?

Suppose that our crystal is initially composite of an equal number of positive and negative domains (i.e. domains oriented to the right and domains oriented to the left), which means that the overall polarization of the crystal is equal to zero. If we first apply a small electric field directed say, in the positive direction, we will have only a linear relationship between $P$ and $E$ because the field is not large enough to switch any of the domains and the crystal will behave like a normal dielectric. In the plot of $P$ vs. $E$, shown schematically in Fig. I-3, we obtain the portion $OA$ of the curve. If we increase the electric field strength, a number of the negative domains will switch over in the positive direction and the polarization will increase rapidly (portion $AB$), until we reach a state in which *all* the domains are aligned in the positive direction: this is a state of saturation (portion $BC$) and the crystal consists now of a single domain.

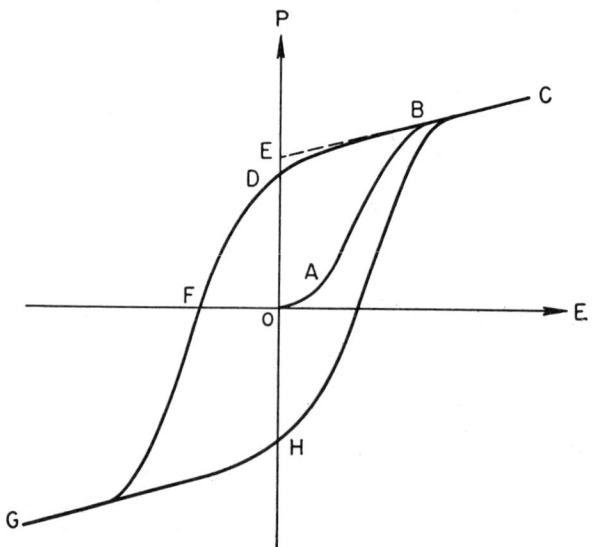

FIG. I-3. Ferroelectric hysteresis loop (schematic).

If we now decrease the field strength, the polarization will generally not return to zero but rather follow the path $CD$ of Fig. I-3. When the field is reduced to zero, some of the domains will remain aligned in the positive direction and the crystal will exhibit a *remanent polarization* $P_r(OD)$. The extrapolation of the linear portion $BC$ of the curve back to the polarization axis represents the value of the *spontaneous polarization* $P_s(OE)$.

In order to annihilate the overall polarization of the crystal, we find it necessary to apply an electric field in the opposite (negative) direction. The value of the field required to reduce $P$ to zero $(OF)$ is called the *coercive field* $E_c$. Further increase of the field in the negative direction will, of course, cause complete alignment of the dipoles in this direction $(FG)$, and the cycle can be completed by reversing the field direction once again $(GHC)$.

The relation between $P$ and $E$ is thus represented by a *hysteresis loop* $(CDGHC)$, which is the most important characteristic of a ferroelectric crystal. The essential feature of a ferroelectric is thus *not* the fact that it has a spontaneous polarization, but rather the fact that this spontaneous polarization can be reversed by means of an electric field. It may be noted that, owing to the relation between dielectric displacement $D$, electric field $E$, and polarization $P$,

$$D = E + 4\pi P,$$

the relation between $D$ and $E$ is also characterized by a hysteresis curve. The analogy with the well-known ferromagnetic hysteresis loop representing the relation between magnetic induction $B$ and magnetic field $H$ is obvious.

Ferroelectric hysteresis loops can be observed very easily on the screen of an oscilloscope by inserting the crystal in a simple circuit first described by Sawyer and Tower (S 1) and using an a.c. field (generally 60 c/s). The circuit is depicted schematically in Fig. I-4: the voltage lying across the crystal $C_x$ is put on the horizontal plates of the oscilloscope, thus plotting on the horizontal axis a quantity which is proportional to the field lying across the crystal. The linear capacitor

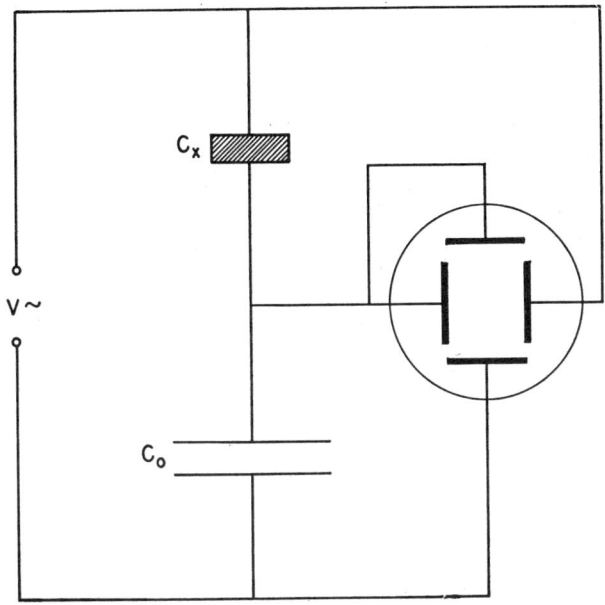

FIG. I-4. Schematic circuit for the observation of ferroelectric hysteresis loops (according to Sawyer and Tower (S 1) ).

$C_0$ is connected in series with the crystal $C_x$ and the voltage lying across $C_0$ is therefore proportional to the polarization of the crystal $C_x$. This voltage is laid across the vertical plates of the oscilloscope. The Sawyer and Tower circuit allows not only the display of the hysteresis loop on the scope screen, but also the measurement of important quantities such as the spontaneous polarization $P_s$ and the coercive field $E_c$.

We have mentioned above that the most important property of a ferroelectric is the reversibility of its spontaneous polarization. The fact that an applied field can cause the polarization to alter its orientation has some important consequences. In the first place, it means that the shift of the $B$ ions is small, and this small relative shift turns the crystal into its electrical twin. It is not illogical to expect, then, that if the energy barrier $\Delta E$ (Fig. I-2a) is so low as to be accessible with the help of an electric field, then it may also be affected by other factors, such as temperature changes.

With increasing temperature, the thermal motion of the atoms in the lattice may increase to such an extent that the $B$ ions of our fictitious crystal may be able to overcome the energy barrier $\Delta E$ without the help of an external field, and may therefore jump from one equilibrium position to the other.

Another possibility is that the shape of the potential curve between adjacent $A$ ions changes to the point of becoming as indicated schematically in Fig. I-2(b). In the latter case, there is only one equilibrium position possible for the $B$ ions, i.e. midway between two $A$ ions. In the former case, the statistical distribution of the $B$ ions will be symmetrical with respect to the $A$ ions. The result is the same: our crystal is no longer polar, and will behave rather as a normal dielectric material.

The temperature at which such a transition from the polar into the non-polar state occurs is often referred to as *Curie temperature* (or *Curie point*), again in analogy with ferromagnetism. Almost all ferroelectric crystals known to date have a Curie point, and those which do not have it simply decompose before such a temperature is reached.*

Another consequence of the small atomic shifts occurring in ferroelectric crystals is the behaviour of the dielectric constant of these crystals as a function of temperature. The dielectric constant is normally defined as the derivative of the dielectric displacement $D$ with respect to the field $E$. Similarly, the derivative of polarization $P$ with respect to field $E$ is usually defined as the dielectric susceptibility $k$. Owing to the well-known relationship between $D$, $E$ and $P$:

$$D = E + 4\pi P,$$

the dielectric constant $\varepsilon$ is related to the dielectric susceptibility $k$, in the c.g.s. system, through the equation

$$\varepsilon = 1 + 4\pi k.$$

This definition of the dielectric constant is significant and consistent for normal dielectrics in which the relationship between $D$ and $E$ is linear up to large values of $E$. In the case of ferroelectric crystals, $\varepsilon$ must be defined more precisely, owing to the more complicated relation between $D$ and $E$ (Fig. I-3). For our purposes, we are going to define the dielectric constant $\varepsilon$ as the slope of the $D$–$E$ curve at the origin:

$$\left(\frac{\partial D}{\partial E}\right)_{E=0}$$

and we are going to measure it with very small a.c. fields, so as not to reverse any domains.

As the potential curve between two adjacent $A$ ions of our crystal changes with temperature, the polarizability of the $B$ ions moving within such a curve will change accordingly. It is thus to be expected that the dielectric constant $\varepsilon$

---

* It has become customary, in many publications, to refer to the phase above the Curie point as to the "paraelectric" phase; ferroelectric crystals are, therefore, often said to become "paraelectric" above their Curie temperature. This term originates again from the formal similarity with the behaviour of a ferromagnetic crystal becoming paramagnetic above its transition point. We prefer, however, to refrain from usage of such a term, and we will refer to the state of a ferroelectric crystal above its Curie temperature as the "non-polar" state.

will exhibit some kind of anomaly at the transition temperature. Generally, any kind of phase transition in dielectrics is accompanied by an anomaly of the dielectric constant; a ferroelectric transition, being generally a transition from a polar into a non-polar phase, is no exception to this rule. In some ferroelectrics, the temperature dependence of the dielectric constant above the transition temperature can be described fairly accurately by a simple law, called the *Curie–Weiss law* from its ferromagnetic analog:

$$\varepsilon = \varepsilon_0 + \frac{C}{T - T_0}$$

where the temperature-independent part $\varepsilon_0$ can often be neglected, $C$ is the *Curie constant* and $T_0$ the *Curie–Weiss temperature*. It is advisable, in general, to distinguish between this Curie–Weiss temperature $T_0$ and the Curie (or transition) temperature $T_c$.

In the vicinity of the Curie–Weiss temperature, the dielectric constant becomes very large, and the relationship between dielectric constant and susceptibility can be approximated by

$$\frac{\varepsilon}{4\pi} \cong k \, .$$

We are going to see in the following that in certain ferroelectrics (which undergo a transition of the second order) the Curie–Weiss temperature $T_0$ practically coincides with the transition temperature $T_c$, but in others (which undergo a transition of the first order) it does not. It must be emphasized that a hyperbolic dependence of the dielectric constant on temperature, such as the Curie–Weiss law, is by no means a necessary condition for a ferroelectric transition, as it was customary to believe in the early stages of ferroelectric research. The fact that in some ferroelectrics the anomaly of the dielectric constant is large and the Curie–Weiss law is obeyed may simply mean that the co-operative phenomenon leading to ferroelectricity has a dielectric origin in these crystals, while in others it may have a different character. As we are going to see in the following, a number of characteristic quantities other than the dielectric constant may exhibit anomalies at the transition temperature. This point will be better discussed when we treat specific cases.

## 2. Spontaneous Polarization

There are a number of fundamental inaccuracies in the description of the model which we have introduced in the preceding section, in spite of the fact that it helps to account for the general features of ferroelectric crystals.

In the first place, we have arbitrarily picked out pairs of atoms and attributed to each of them a dipole moment, defined, as is customary, as the product of charge and separation. This procedure is definitely not legitimate in a real structure. It is true that in a ferroelectric crystal every unit cell carries a dipole moment, but we cannot generally attribute this dipole moment to a specific pair of atoms in the unit cell without making arbitrary assumptions about the interatomic forces.

All we can say is that, given a charge distribution $\varrho(r)$ within a given crystal, the electric moment with respect to an *arbitrary* origin,

$$\iiint \varrho(r) \cdot r \cdot dv,$$

is different from zero and the value of this volume integral is independent of the choice of the origin. To attribute this electric moment to a given pair of atoms would imply that this atomic pair is held together by forces which are stronger than those which bound it to the rest of the structure. This implication is incorrect, because the two atoms which we have picked out are bound to their neighbors by the same forces as those that bind them to each other.

A second important point which does not appear clear in the simple model introduced before is that the dipole moment of each unit cell does not consist only of ionic charges. Eventually, charge separation may occur through the distortion of the electronic cloud of a given atom. In other words, the charge distribution $\varrho(r)$ within the crystal involves not only the ionic but also the electronic charge. Although a distinction between these two contributions to the electric moment is experimentally quite difficult, the concepts of ionic and electronic polarization have been introduced often in the literature. It may be pointed out that these concepts constitute a very important difference between ferroelectricity and ferromagnetism. In ferroelectric crystals, the dipoles are imbedded in a highly polarizable medium ("background"), whose polarizability is temperature dependent, while in ferromagnetic crystals the magnetic moments due to the spin cannot be influenced by the "background".

It should also be pointed out that the simple model introduced in Section 1 is mainly concerned with phenomenological consequences of a spontaneous polarization which is somehow assumed to be already existing within the crystal. We have, in other words, approached the problem of ferroelectricity from the point of view which assumes an already *polarized* state. What we should also do, however, is to follow the opposite approach, starting from the *unpolarized* state, and inquire about the interatomic forces which are at work to bring about spontaneous polarization in a crystal lattice. This problem is of fundamental importance in the general theory of dielectrics, and the reader may be referred, for the details of the following arguments, to the textbooks of Böttcher (B 1), Brown (B 2), Dekker (D 1), Fröhlich (F 2), Kittel (K 2), or Smyth (S 2).

Given a particular type of atom in a substance, we define the polarizability $\alpha$ of such an atom as the ratio between the dipole moment (permanent or induced) of the atom and the electric field acting upon it:

$$\alpha = \mu/F.$$

The field $F$ is generally given by Lorentz's formula for the internal field acting on a dipole located in a lattice of *equal* dipoles with *cubic* symmetry:

$$F = E + \frac{4\pi}{3} P.$$

Straightforward use of this formula leads to the well-known Clausius–Mosotti equation for the dielectric constant $\varepsilon$:

$$\frac{\varepsilon - 1}{\varepsilon + 2} = \frac{4\pi}{3} N\alpha,$$

where $N$ is the number of the dipoles per unit volume. The above equation can be re-written in the following form:

$$\varepsilon = \frac{1 + (8\pi/3)N\alpha}{1 - (4\pi/3)N\alpha}.$$

From this, it follows that the dielectric constant $\varepsilon$ becomes infinite, corresponding to a finite polarization for zero applied field, when

$$N\alpha = \frac{3}{4\pi}.$$

Putting, for example,

$$1 - \frac{4\pi}{3} N\alpha = \frac{3}{C} (T - T_0),$$

we immediately obtain a Curie–Weiss law for the dielectric constant

$$\varepsilon \cong \frac{C}{T - T_0}$$

in agreement with the observed temperature dependence of the dielectric constant of some ferroelectrics above the Curie temperature.

Thus, it is seen that a simple assumption about the temperature dependence of the product $N\alpha$, combined with Lorentz's internal-field formula leads easily to a ferroelectric transition and a Curie–Weiss law. However, the shortcomings of this theory appear especially obvious when it is applied to polar liquids: when numerical values are inserted, it turns out that liquids such as water should be spontaneously polarized at room temperature, an obviously absurd result which has been called the "$4\pi/3$ catastrophe". This catastrophe occurs in the case of fields produced by permanent dipoles, and can only be avoided by using modified internal-field formulae derived by Onsager, Kirkwood, Oster and Fröhlich (see e.g., Refs. (B 1), (B 2), (F 2), (J 1), (S 1)). In the case of induced moments, the theory can be used, provided the proper correction of the Lorentz's formula for the internal field is applied, as was shown by Slater in the case of ferroelectric barium titanate (see Chapter IV).

It is, however, generally true that consideration of the long-range forces characterized by Lorentz's formula is insufficient to explain the onset of spontaneous polarization in a given crystal lattice. This is not only due to the fact that the Lorentz formula was derived for a lattice of equal dipoles with cubic symmetry and, as such, represents merely a first approximation. In some cases, the consideration of short-range coupling is necessary; in fact it may be sufficient to explain the occurrence of a ferroelectric transition, although it may not suffice to reproduce the exact character of the transition. A typical example of this is Slater's theory for $KH_2PO_4$. (See Chapter III.)

### 3. Crystallographic Considerations and Definition of a Ferroelectric

It is well known that any one crystal can be classified in one or another of thirty-two crystal classes (point groups) according to the symmetry elements which it possesses. A study of these thirty-two classes reveals that eleven of them are characterized by the existence of a center of symmetry: they are thus called *centrosymmetric*. A centrosymmetric crystal can of course possess no polar properties. If, for example, we apply a uniform stress to such a crystal, we will indeed cause a small displacement of the charges within the lattice, but the existence of a center of symmetry will bring about a compensation of the relative displacements. If we apply an electric field to a centrosymmetric crystal, we will indeed change its shape, but the strain will remain unchanged if we reverse the direction of the electric field. In other words, the strain is proportional to the square of the applied field: the effect is quadratic. This is the effect called *electrostriction*, which occurs actually in *all* substances, whether crystalline or amorphous, solid or fluid.

The remaining twenty-one crystal classes do not have a center of symmetry; they are non-centric. The absence of a center of symmetry makes it possible for crystals in these classes to have one or more polar axes and thus to show vectorial or tensorial properties. With one exception, all classes devoid of a center of symmetry exhibit the *piezoelectric effect*. The single exception is the cubic class 432 which, although without a center of symmetry, nevertheless has other symmetry elements that combine to exclude the piezoelectric activity. Piezoelectricity is the property of a crystal to exhibit electric polarity when subject to stress. The piezoelectric effect is a *linear* effect; application of pressure to a piezoelectric crystal plate between two electrodes causes a charge to flow in a certain direction through a measuring circuit. If the pressure is replaced by a tension, the charge will flow in the opposite direction. Also, if we apply an electric field to the crystal plate, it will be stretched; if we reverse the field direction, it will be compressed. This is the converse piezoelectric effect.

Out of the twenty piezoelectric classes, ten are characterized by the fact that they have a *unique* polar axis, i.e. an axis which shows properties at one end different from those at the other. Crystals in these classes are called *polar* because they are spontaneously polarized. Generally, this spontaneous polarization cannot be detected by charges on the surface of the crystal because these charges have been compensated through external or internal conductivity, or by twinning. The value of the spontaneous polarization, however, is dependent on temperature; thus, if the temperature of the crystal is altered a change in the polarization occurs and electric charges can be observed on those crystal faces which are perpendicular to the polar axis. This is the *pyroelectric* effect. The ten crystal classes which have a unique polar axis are also called pyroelectric classes.

It follows from the discussion in Section 1 that the ferroelectric crystals belong to the pyroelectric family. They constitute only that part of it, however, for which the direction of the spontaneous polarization can be reversed by application of an electric field. It is thus a *necessary* condition for a ferroelectric crystal to belong to any one of the ten polar classes (in its ferroelectric phase) but not a sufficient condition, as reversibility of the polarity must also occur. We can thus *define* a *ferroelectric* crystal as a *pyroelectric crystal with reversible polarization*.

This definition brings about an interesting difference between ferroelectricity and ferromagnetism. In the case of ferromagnetism, whenever the moments are spontaneously aligned they can also be reversed by means of an external field. Thus, the magnetic phenomenon is interesting *because* the moments are spontaneously aligned. The phenomenon of ferroelectricity, on the other hand, is interesting only when the moments are loosely aligned, i.e. when the interactions are so delicately balanced as to allow reversal of the spontaneous polarization.

It should be noted that the existence of a unique polar axis in the point group symmetry of a given crystal can, in principle, be established, e.g. by means of $X$-rays, but its reversibility can be established only by dielectric measurements.

## 4. Classification of Ferroelectrics

For a number of years, the phenomenon of ferroelectricity was known to occur in a rather limited number of crystals, viz. Rochelle salt, potassium di-hydrogen phosphate, barium titanate and a number of compounds isomorphous with these crystals. At this stage, various classifications of the ferroelectric materials were proposed, in order to facilitate the treatment of their properties. In recent years, however, owing particularly to the efforts of Matthias, Pepinsky, Smolenskii and co-workers, the number of ferroelectric crystals known has increased to a point where ferroelectricity appears to be a somewhat more common phenomenon than had been thought for a long time, and a satisfactory classification of all the materials has become very difficult.

Table I-1 provides a partial list of ferroelectric crystals arranged approximately in chronological order of their discovery. It is seen that, from the chemical point of view, representative ferroelectrics can be found among the tartrates, phosphates, arsenates, double oxides, sulfates, borates, propionates, nitrates, nitrites, etc. The symmetries of the non-polar phases vary from cubic to tetragonal, orthorhombic and monoclinic, and the Curie temperatures range from approximately 10 °K, for potassium tantalate, to about 840 °K for lead metaniobate. The values of the spontaneous polarization vary from the order of $10^{-7}$ C/cm² to about $10^{-4}$ C/cm². It may be interesting to point out in this respect that in normal dielectrics, such as, for example, the alkali halides, such high values of the polarization could theoretically be attained only with electric fields of the order of from $10^5$ to $10^8$ V/cm.

The following is a summary of the criteria according to which various classifications of the ferroelectric materials have been or could be proposed:

(i) Crystal–chemical classification. According to this classification, the ferroelectric compounds may be divided into two groups. The first group comprises hydrogen-bonded crystals, such as $KH_2PO_4$, Rochelle salt, tri-glycine sulfate, etc. The second group includes the double oxides, such as $BaTiO_3$, $KNbO_3$, $Cd_2Nb_2O_7$, $PbNb_2O_6$, $PbTa_2O_6$, etc.

(ii) Classification according to the number of directions allowed to the spontaneous polarization. Again, ferroelectric materials can be divided into two groups; a group involving a single axis of the spontaneous polarization, in which

TABLE I-1. PARTIAL LIST OF FERROELECTRIC CRYSTALS

(The values of the spontaneous polarization reported in the third column are either the room-temperature or the maximum values)

| Name and chemical formula | Curie temperature (°C) | Spontaneous polarization ($10^{-6}$ C/cm²) | Year in which reported |
|---|---|---|---|
| Rochelle salt<br>$NaKC_4H_4O_6 \cdot 4H_2O$ | + 23 | 0.25 | 1921 |
| Lithium ammonium tartrate<br>$Li(NH_4)C_4H_4O_6 \cdot H_2O$ | − 170 | 0.20 | 1951 |
| Potassium di-hydrogen phosphate<br>$KH_2PO_4$ | − 150 | 4. | 1935 |
| Potassium di-deuterium phosphate<br>$KD_2PO_4$ | − 60 | 5.5 | 1942 |
| Potassium di-hydrogen arsenate<br>$KH_2AsO_4$ | − 177 | 5.0 | 1938 |
| Barium titanate<br>$BaTiO_3$ | + 120 | 26.0 | 1945 |
| Lead titanate<br>$PbTiO_3$ | + 490 | > 50 | 1950 |
| Potassium niobate<br>$KNbO_3$ | + 415 | 30 | 1951 |
| Potassium tantalate<br>$KTaO_3$ | − 260 | ? | 1951 |
| Cadmium (pyro) niobate<br>$Cd_2Nb_2O_7$ | − 85 | ∼10 | 1952 |
| Lead (meta) niobate<br>$PbNb_2O_6$ | + 570 | ? | 1953 |
| Guanidinium aluminium sulfate hexahydrate<br>$C(NH_2)_3Al(SO_4)_2 \cdot 6H_2O$ | None | 0.35 | 1955 |
| Methylammonium aluminium alum<br>$CH_3NH_3Al(SO_4)_2 \cdot 12H_2O$ | − 96 | 1.0 | 1956 |
| Ammonium sulfate<br>$(NH_4)_2SO_4$ | − 50 | 0.25 | 1956 |
| Tri-glycine sulfate<br>$(NH_2CH_2COOH)_3 \cdot H_2SO_4$ | + 49 | 2.8 | 1956 |
| Colemanite<br>$CaB_3O_4(OH)_3 \ H_2O$ | − 7 | 0.65 | 1956 |
| Dicalcium strontium propionate<br>$Ca_2Sr(CH_3CH_2COO)_6$ | + 8 | 0.3 | 1957 |
| Lithium acid selenite<br>$LiH_3(SeO_3)_2$ | None | 10.0 | 1959 |
| Sodium nitrite<br>$NaNO_2$ | + 160 | 7.0 | 1958 |

we put the ferroelectrics which can polarize only along one axis, such as Rochelle salt, $KH_2PO_4$, colemanite, $PbTa_2O_6$, etc., and a group comprising crystals which can polarize along several axes that are crystallographically equivalent in the non-polar phase, such as $BaTiO_3$, $Cd_2Nb_2O_7$, the ferroelectric alums, etc. This classification may be particularly useful for the study of ferroelectric domains.

(iii) Classification according to the existence or lack of a center of symmetry in the point group of the non-polar phase. A first group of ferroelectrics is characterized by a non-polar phase which is piezoelectric (non-centrosymmetrical), such as Rochelle salt, $KH_2PO_4$, and isomorphous compounds. A second group of ferroelectrics is characterized by a centrosymmetrical non-polar phase, such as $BaTiO_3$, $Cd_2Nb_2O_7$, tri-glycine sulfate, etc. This classification may be particularly useful for the thermodynamic treatment of the ferroelectric transitions.

(iv) Classification according to the nature of the phase change occurring at the Curie point. A first group of ferroelectrics undergo a transition of the order–disorder type, as $KH_2PO_4$, tri-glycine sulfate and probably some of the alums. A second group of compounds undergo a transition of the displacive type, such as that of barium titanate and most of the double oxide ferroelectrics. This classification is practically equivalent to that done on the basis of the existence of permanent or induced dipoles in the non-polar phases of the crystals. A characterization of the nature of the phase change in the sense mentioned above (order–disorder vs. displacive) can, in principle, be met on the basis of accurate structural investigations. In some cases, however, this information is already available from the results of dielectric investigations.

If we examine the temperature dependence of the dielectric constant, or in other words, the value of the Curie constant $C$ appearing in the Curie–Weiss law, we see that the various ferroelectrics can be divided into two main groups (see Table I-2). Compounds in the first group have Curie constants of the order of $10^3$. It can be shown that this order of magnitude is to be expected for a substance containing a number of similar dipoles, each of which has two positions of equilibrium corresponding to opposite orientations of the dipole. This dipole model is practically equivalent to that of an ion moving in the type of potential shown in Fig. I-2(a). The dielectric constant $\varepsilon$ of such a model assumes the following form (D 2):

$$\varepsilon = \frac{4\pi}{\gamma} \frac{T_0}{T-T_0},$$

where $\gamma$ is the Lorentz factor appearing in the Lorentz internal-field formula $F = E + \gamma P$, and $T_0$ is the Curie–Weiss temperature. The factor $\gamma$ will depend on the particular atomic structure but will be of the order of $4\pi/3$. Thus, we can write for the dielectric constant:

$$\varepsilon \simeq \frac{3T_0}{T - T_0},$$

TABLE I-2. CURIE CONSTANTS $C$, DEFINED FROM THE CURIE–WEISS

LAW $\varepsilon = \dfrac{C}{T - T_0}$, FOR VARIOUS FERROELECTRICS

| Substance | $C$ (degrees) |
|---|---|
| Rochelle salt | $2.2 \times 10^3$ |
| $KH_2PO_4$ | $3.3 \times 10^3$ |
| Colemanite | $0.5 \times 10^3$ |
| Tri-glycine sulfate | $3.2 \times 10^3$ |
| Methylammonium aluminium alum | $1.0 \times 10^3$ |
| $NaNO_2$ | $5.0 \times 10^3$ |
| $BaTiO_3$ | $1.5 \times 10^5$ |
| $KNbO_3$ | $2.0 \times 10^5$ |
| $Cd_2Nb_2O_7$ | $1 \ \times 10^5$ |
| $PbNb_2O_6$ | $3 \ \times 10^5$ |

implying that the Curie constant $C$ is of the order of $3T_0$, i.e. for $T_0 \sim 300$ °K, of the order of $10^3$.

The second group of ferroelectrics shown in Table I-2 exhibit Curie constants of the order of $10^5$. This can be explained by assuming that the substances in this group contain, say, $N$ oscillating ions per unit volume and that each ion produces an electric moment $ex$ when it is displaced a distance $x$ from its normal equilibrium position. Expressing the energy of an ion relative to its equilibrium position in the form $ax^2 + bx^4$, Devonshire (D 2) has shown that the dielectric constant assumes the form:

$$\varepsilon = \frac{4\pi Ne^2}{6k} \frac{a}{b} \frac{1}{T - T_0},$$

where $k$ is Boltzmann's constant. Thus, the temperature dependence of the dielectric constant depends on the anharmonic term $bx^4$ in the energy of an ion. This corresponds to the model of an ion moving in a potential of the type shown in Fig. I-2(b). This equation shows that if the coefficient $b$ is small, the temperature dependence of $\varepsilon$ is small and hence the Curie constant will be large.

Clearly, any intermediate type of potential field between the two extreme cases shown in Fig. I-2 is possible for the non-polar phase of the ferroelectrics, but it is significant that most of the ferroelectrics known belong rather to either one of these extreme models as shown in Table I-2. A classification made on this basis may, therefore, be useful in discussing the types of ferroelectric transitions.

It is evident that the four classifications described above do not coincide with each other. Each one is useful only when discussing a particular aspect of the ferroelectric phenomenon, but a consistent classification of all ferroelectrics appears hardly possible at the present stage. It is thus more appropriate to restrain from making any such classification at this stage, but to proceed with the detailed description of the phenomenological and structural characteristics of the cases known. Accordingly, the phenomenological and model theories will be discussed, for several of the most typical ferroelectrics, in the corresponding chapters devoted to the particular crystals for which the theories were developed.

In order to supplement this approach we have provided in Appendix II a concise analytical index in table form. In this table we have listed the numbers of chapters and sections where the individual topics are treated somewhat extensively. This table is by no means complete; it is hoped, however, that it may help the reader interested in comparisons among various ferroelectrics.

## 5. Introduction to the Thermodynamic Theory of Crystals

Because the general features of the phenomenological behavior of most ferroelectrics are the same, it is convenient to present first an introduction to the thermodynamic theory of the crystalline state. The value of such a theory is significant in disclosing what thermodynamic properties are needed in order to account for the experimental facts, and therefore what features are to be looked for in some future successful model. The development of the following thermodynamic treatment is due to Mueller (M 2), Cady (C 1), and Devonshire (D 2).

A crystal, whether ferroelectric or not, is a thermodynamic system whose equilibrium state can be completely specified by the values of a number of variables. The internal energy of the system can, in fact, be expressed as a function of the mechanical and electrical stresses $X$ and $E$, respectively, or the mechanical and electrical strains $x$ and $P$, respectively, in addition to temperature and entropy. Of the four variables $X$, $E$, $x$ and $P$ we can choose two as independent, the other two as dependent variables, depending on what we have in mind. Experimentally, for example, it is generally easier to vary the external stresses $X$ and the applied electric fields $E$, so that it is logical to assume the strains $x$ and the polarization $P$ as dependent variables. For theoretical considerations, on the other hand, the opposite is true. Whatever the choice, we must realize that, as $P$ and $E$ are vectors (three components), and $x$ and $X$ tensors (six components), the energy state of our crystal is specified when we know the values of ten variables (these nine components plus the temperature). It is a basic postulate of thermodynamics that a unique function of these ten variables exists, which is called the free energy function; we are attempting to present a simple analytical expression of this function.

Suppose that we consider the strains $x$ and the polarization $P$ as independent variables. We can always expand the (unknown) free energy function in powers and products of the components of strain and polarization. We can then see whether this power series converges and determine how many terms we need in order to describe the electromechanical behavior of the crystal. We assume, for simplicity, that the free energy of the unstrained, unpolarized crystal is equal to zero; we can refer the strains to the equilibrium configuration of the crystal in the absence of stresses. Insofar as the strains are small, as is usually the case, only quadratic terms need be retained, which, in elastic treatments, amounts to the acceptance of Hook's law. However, in view of the application of the thermo-dynamic theory to ferroelectric crystals, in which the polarization assumes unusually large values, we are going to include terms in higher orders of the polarization components. The effect of temperature is taken into account by assuming that the coefficients of the power series are functions of temperature.

We thus postulate the following expression for the free energy of the crystal in terms of strains and polarization:

$$A(x, P) = \tfrac{1}{2} \sum_{1}^{3} \chi_{ij} P_i P_j + \tfrac{1}{3} \sum_{1}^{3} \omega^x_{ijk} P_i P_j P_k +$$

$$+ \tfrac{1}{4} \sum_{1}^{3} \xi^x_{ijkl} P_i P_j P_k P_l + \tfrac{1}{5} \sum_{1}^{3} \psi^x_{ijklm} P_i P_j P_k P_l P_m +$$

$$+ \tfrac{1}{6} \sum_{1}^{3} \zeta^x_{ijklmn} P_i P_j P_k P_l P_m P_n + \qquad\qquad \text{(I-1)}$$

$$+ \tfrac{1}{2} \sum_{1}^{3} c^P_{ijkl} x_{ij} x_{kl} +$$

$$+ \sum_{1}^{3} a_{ijk} x_{ij} P_k + \tfrac{1}{2} \sum_{1}^{3} q_{ijkl} x_{ij} P_k P_l + \cdots\cdots$$

where the summations are carried out for all possible combinations of the sub-scripts running from 1 to 3 (according to the usual convention that, for example, $P_x = P_1$, $P_y = P_2$, $P_z = P_3$). The superscripts of the coefficients are necessary to specify the conditions under which such coefficients are observed: the super-script $x$ means that the coefficient is observed at constant strain, the superscript $P$ that it is observed at constant polarization. Some of the coefficients of the free energy expansion (I-1) have a familiar physical meaning. $\chi_{ij}$ represents the re-ciprocal dielectric susceptibility of the unpolarized crystal, $c_{ijkl}$ the tensor of the elastic constants, $a_{ijk}$ that of the piezoelectric constants, $q_{ijkl}$ that of the electro-strictive constants.

It may be helpful to recall that the piezoelectric constants $a_{ijk}$ form a third-rank tensor, while the elastic constants $c_{ijkl}$ and the electrostrictive constants $q_{ijkl}$ are fourth-rank tensors. In this notation, there are twenty-seven components of the piezoelectric coefficients and eighty-one components of the elastic constants (see, e.g., the book of Nye (N 1) and the article of Smith (S 3)). The full tensor notation is useful when one wants to display the true character, and particularly the transformation properties, of the coefficient involved. But for calculations in particular problems it is advantageous to reduce the number of suffixes as much as possible, i.e. to use the matrix notation. In the case of a third-rank tensor, the first suffix is maintained, but the second and third suffixes in the full tensor notation are replaced in the new notation by a single suffix running from 1 to 6 as follows:

| tensor notation | 11 | 22 | 33 | 23,32 | 31,13 | 12,21, |
|---|---|---|---|---|---|---|
| matrix notation | 1 | 2 | 3 | 4 | 5 | 6 |

In the case of a fourth-rank tensor, the first two suffixes are abbreviated into a single one running from 1 to 6, the last two are abbreviated in the same way, according to the same scheme given above. The same formal change is made for the stress and strain components in order to reduce them from the two-suffixes to the one-suffix notation.* It must be remembered, however, that in spite of their appearance, with two suffixes, the piezoelectric and elastic constants do not transform like the components of a second-rank tensor. With the matrix notation, in the most general case (triclinic symmetry), there are six dielectric, twenty-one elastic, eighteen piezoelectric and thirty-six electrostrictive terms, but consideration of the symmetry elements represented in the point group of the crystal reduces the number of the non-zero coefficients.

For a crystal belonging to a centrosymmetrical class, the terms involving odd powers of the polarization vanish, as well as the piezoelectric term. In the case of a piezoelectric crystal, we may, in the first approximation, neglect the electrostrictive term.

In Eq. (I-1) we have not included, for simplicity, the terms which describe the thermal, thermoelastic and pyroelectric properties of the crystal. Moreover, it should be noticed that, in general, the derivatives of the above free-energy function with respect to the strain $x$ and the polarization $P$ are *not* the externally applied stress $X$ and field $E$, respectively, as in the usual thermodynamic formalism. The derivative with respect to $x$ is rather the *total* stress, consisting of two parts: (1) the externally applied stress that would produce the prescribed strain if $P = 0$; (2) the stress caused piezoelectrically by the polarization $P$ (C 1). Similarly, the derivative of $A(x, P)$ with respect to $P$ is not the externally applied field, but rather the field that would produce, in a clamped crystal, the same total polarization which is given by the prescribed values of $P$ and $x$. If the strain $x = 0$, and there is no spontaneous polarization, then $[\partial A(x, P)/\partial P]_{x=0}$ is equal to the externally applied field; if the polarization $P = 0$, then $[\partial A(x, P)/\partial P]_{P=0}$ is the field that would cause, in a clamped crystal, the polarization due to $x$ (C 1).

We first want to consider the case of a piezoelectric, but not ferroelectric, crystal. We therefore neglect the terms of order higher than the second in $P$, and the electrostrictive term. Also, we eliminate, for simplicity, the suffixes and the summation signs and write the free energy formally in the following way:

$$A(x, P) = \tfrac{1}{2} c^P x^2 + \tfrac{1}{2} \chi^x P^2 + aPx. \tag{I-2a}$$

Similarly, we can write the free energy in terms of stress $X$ and polarization $P$ thus:

$$A(X, P) = \tfrac{1}{2} s^P X^2 + \tfrac{1}{2} \chi^X P^2 - bPX, \tag{I-2b}$$

---

* The same convention applies of course to the piezoelectric and elastic coefficients to be introduced later, in particular the piezoelectric moduli $d_{ijk}$ and the elastic compliances $s_{ijkl}$ It should be kept in mind, however, that in certain cases it is necessary to introduce factors of 2 (or 4) when changing from the tensor to the matrix notation (see e.g. the book of Nye (N 1)). As an example, for a third-rank tensor:

$$d_{ijk} = d_{in} \text{ when } n = 1, 2 \text{ or } 3, \text{ but}$$
$$2d_{ijk} = d_{in} \text{ when } n = 4, 5 \text{ or } 6.$$

where $s^P$ is the elastic compliance at constant polarization and $\chi^X$ is the reciprocal dielectric susceptibility of the crystal under a constant stress. (For $X = 0$, this is the reciprocal susceptibility of the "free" crystal, as distinguished from $\chi^{x=0}$ which is the reciprocal susceptibility of the "clamped" crystal.)

On the other hand, choosing $x$ and $E$, or $X$ and $E$ as independent variables, respectively, we obtain

$$A(x, E) = \tfrac{1}{2} c^E x^2 + \tfrac{1}{2} k^x E^2 + e E x, \qquad (\text{I-3a})$$

and

$$A(X, E) = \tfrac{1}{2} s^E X^2 + \tfrac{1}{2} k^X E^2 - d E X, \qquad (\text{I-3b})$$

respectively.

From these free-energy functions, we obtain the following fundamental equations of the piezoelectric crystal. From (I-2a):

$$\left.\begin{aligned} X &= -c^P x + a P, \\ E &= -a x + \chi^x P. \end{aligned}\right\} \qquad (\text{I-4a})$$

From (2b):

$$\left.\begin{aligned} x &= -s^P X + b P, \\ E &= b X + \chi^X P. \end{aligned}\right\} \qquad (\text{I-4b})$$

From (3a):

$$\left.\begin{aligned} X &= -c^E x + e E, \\ P &= e x + k^x E, \end{aligned}\right\} \qquad (\text{I-5a})$$

and from (3b):

$$\left.\begin{aligned} x &= -s^E X + d E, \\ P &= -d X + k^X E. \end{aligned}\right\} \qquad (\text{I-5b})$$

The relations between the various coefficients of these equations can be easily obtained by suitable substitutions. For example, if we substitute formally $P$ from (I-5b) into the expression for $x$ in (I-4b), and compare the result with the expression for $x$ in (I-5b) we obtain:

$$\frac{d}{k^X} = b, \qquad s^E - s^P = b d. \qquad (\text{I-6})$$

Other useful relations are given in Table III-2.

These relations are particularly interesting in the case where one of the coefficients is found experimentally to exhibit an anomaly at a given temperature. From (I-6), for example, we deduce that if the dielectric susceptibility $k^X$ shows an anomaly (as at the transition temperature of a piezoelectric crystal to a ferroelectric state), either one of the two piezoelectric coefficients $d$ or $b$ should also be anomalous. The experiment shows, as we will see in the following, that it is the piezoelectric modulus $d$ which has an anomaly at the transition temperature. Accordingly, the coefficient $b$ is practically temperature independent, as is the elastic compliance at constant polarization, $s^P$. Thus, the anomalies of the elastic and piezoelectric behavior can be predicted on the basis of the above thermodynamic treatment.

Let us now make a second step and examine the dielectric behavior of the crystal when it undergoes a ferroelectric transition. To simplify the treatment, we are going to deal only with the free energy as an explicit function of stresses and polarization, $A(X, P)$, and assume that:

(i) all stresses are equal to zero, $X = 0$;

(ii) the polarization vector $P$ is directed along one of the crystallographic axes only; and

(iii) the non-polar phase is centrosymmetrical.

Taking into account the higher-order terms in $P$, we obtain:

$$A(P) = \tfrac{1}{2}\,\chi^X P^2 + \tfrac{1}{4}\,\xi^X P^4 + \tfrac{1}{6}\,\zeta^X P^6, \qquad (\text{I-7})$$

where the coefficients are functions of temperature.

Suppose first that all the coefficients are positive in the non-polar phase. For $P = 0$, $A(0) = 0$, and this value corresponds to a minimum of the free energy.

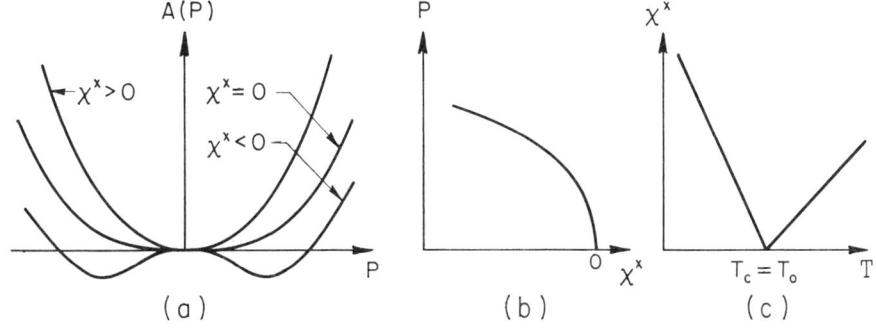

Fig. I-5. Second-order transition (schematic).

(a) Free energy as a function of polarization at temperatures larger than, equal to and lower than the Curie temperature.

(b) Spontaneous polarization as a function of temperature.

(c) Reciprocal susceptibility as a function of temperature.

If we assume at first that only $\chi^X$, the reciprocal free susceptibility, depends on the temperature, while $\xi^X$ and $\zeta^X$ are temperature independent, we see that as soon as $\chi^X$ becomes negative, the function $A(P)$ will have a maximum for $P = 0$ and two minima for a non-zero value of $P$. This situation is explained schematically in Fig. I-5(a). The stable state of the crystal will correspond to $P = 0$ for $\chi^X > 0$ and to $P \neq 0$ for $\chi^X < 0$; i.e. the crystal will undergo a ferroelectric transition. The order of the transition is determined, among other things, by the way in which the polarization goes from the value zero to a finite value, i.e. whether the onset of $P$ is continuous or discontinuous. In the present case, $P$ is continuous at the transition: this is a transition of the *second order*. The behavior of $P$ as a function of $\chi^X$, and thus of temperature, is depicted schematically in

Fig. I-5(b). Under these conditions, we can eventually do without the sixth-order term in (I-7) and write for the field $E$:

$$E = \frac{\partial A}{\partial P} = \chi^X P + \xi^X P^3 \tag{I-8}$$

whence, putting $E = 0$ (*spontaneous* polarization):

$$P^2 = -\frac{\chi^X}{\xi^X}. \tag{I-9}$$

It follows that the reciprocal suspectibility is, *above* the transition:

$$\left(\frac{\partial^2 A}{\partial P^2}\right)_{P=0} = \chi^X,$$

*below* the transition:

$$\left(\frac{\partial^2 A}{\partial P^2}\right) = \chi^X + 3\,\xi^X P^2$$

i.e. using (I-9),

$$\left(\frac{\partial^2 A}{\partial P^2}\right) = -2\,\chi^X. \tag{I-10}$$

The temperature dependence of $\chi^X$ is depicted schematically in Fig. I-5(c); the dielectric susceptibility of the free crystal becomes infinite at the transition temperature.

Equation (I-8) gives us information about the $P$–$E$ relationship below the transition temperature $T_c$. Owing to the fact that $\chi^X$ becomes negative below $T_c$, the $P$–$E$ curve has the character shown in Fig. I-6 by the curve $ABCD$. The portion $BC$

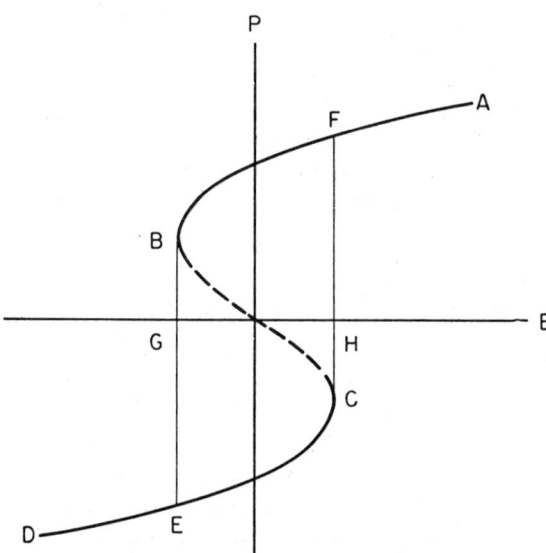

FIG. I-6. Second-order transition: polarization as a function of field below the Curie point.

of the curve corresponds to an unstable state; the crystal jumps directly from the state $B$ to $E$ and from $C$ to $F$. A coercive field $E_c$ is thus predicted with the magnitude given by $GH/2$. The theoretical value of the coercive field, however, always turns out to be one or two orders of magnitude larger than the observed value. This is because the present theoretical treatment assumes that the reversal of the polarization of the whole crystal takes place in one single step. In reality, the reversal mechanism consists of nucleation and growth of new domains (see, e.g. Section II-7), processes which proceed by steps and require time. Thus, the experimental coercive field $E_c$ is not only dependent on the amplitude but also on the frequency of the applied a.c. field.

Abandoning now the assumption that all coefficients in (I-7) are positive above the transition, we assume, for example, that $\xi^X$ is negative and $\zeta^X$ is positive. This means that it is possible for the free-energy function to have two equal minima, one for $P = 0$ and the other for $P \neq 0$, at the same value of $\chi^X > 0$, i.e. at the same temperature. The stable state of the crystal will jump from one with $P = 0$ to one with $P = P_0$ *discontinuously* at this (transition) temperature: this transition is of the *first order* (Fig. I-7).*

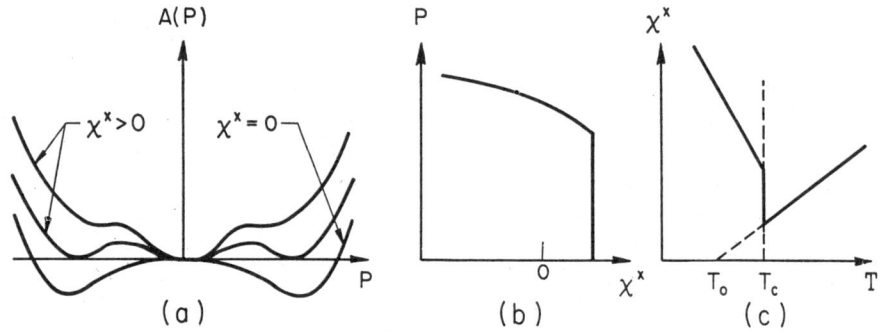

FIG. I-7. First-order transition (schematic).

(a) Free energy as a function of polarization at temperatures larger than, equal to and lower than the Curie temperature.

(b) Spontaneous polarization as a function of temperature.

(c) Reciprocal susceptibility as a function of temperature.

We compute the values of $P = P_0$ and $\chi^X = \chi_0^X$ at the transition by considering that the values of the free energy for $P = 0$ and $P = P_0$ must be equal at this temperature:

$$A(0) = \tfrac{1}{2}\,\chi^X\,P^2 + \tfrac{1}{4}\,\xi^X\,P^4 + \tfrac{1}{6}\,\zeta^X\,P^6 = 0. \tag{I-11}$$

* Experimentally, it is often difficult to establish whether or not the polarization (or other quantities such as volume, entropy, etc.) is discontinuous at the transition temperature. For this reason, it has been suggested that a better definition of a first order transition is coexistence of the two phases in equilibrium at the transition temperature, so that a definite phase boundary is formed (see, e.g. Smoluchowski (S 4)). This definition, however, is also subject to experimental limitations.

On the other hand, the field $E$ must be zero for the polarization to be spontaneous:

$$\frac{\partial A}{\partial P} = E = \chi^X P + \xi^X P^3 + \zeta^X P^5 = 0. \tag{I-12}$$

From (I-11) and (I-12) we obtain:

$$P_0^2 = \frac{3}{4} \left( -\frac{\xi^X}{\zeta^X} \right), \tag{I-13}$$

$$\chi_0^X = \frac{3}{16} \left( \frac{\xi^X}{\zeta^X} \right)^2, \tag{I-14}$$

showing that a discontinuous change in $P$ occurs at the transition (Fig. I-7b) and the value of $\chi^X$ at the transition, $\chi_0^X$, is positive. The temperature dependence of $\chi^X$ is shown schematically in Fig. I-7(c): the dielectric suspectibility of the free crystal is finite but discontinuous at the transition temperature.

The simple thermodynamic treatment reported above is thus able to describe the essential features of a ferroelectric transition of the first or second order by assuming that only $\chi^X$, the reciprocal dielectric susceptibility, is a function of the temperature. Actually, this is not strictly correct, as the other coefficients of the free-energy expansion could also be temperature dependent, but this fact does not much change the overall picture. We are going to see in the following chapters that the above thermodynamic theory can be extended to predict, for example, the heat of transition, the effect of mechanical stresses or electric fields on the transition, the optical anomalies, etc. Also, the treatment will be extended to those types of transition which involve a change in direction of polarization, as in the ferroelectrics of the perovskite class.

Two further remarks are necessary to this introductory treatment of the thermodynamics of crystals. The first is that the phenomenological theory of ferroelectrics outlined above by its nature is only an approximation. This is so not only because we make use of a finite number of terms of the free-energy expansion, but also because we assume that the same function can be used both for the non-polar and the polar phase. The latter assumption can be justified, at this stage, by the fact that the polar phases are obtained through very slight distortions of the non-polar phases.

The second remark is concerned with the implication contained in the above treatment, that the ferroelectric phenomenon has a dielectric origin. Thus, the coefficient of $P^2$ in the free-energy expansion bears the significant temperature dependence, and the anomalies of the piezoelectric, elastic and other properties are thought to be a mere consequence of the anomalous dielectric behavior. This is certainly true for a number of ferroelectric crystals, such as Rochelle salt, potassium di-hydrogen phosphate, barium titanate and tri-glycine sulfate, but, as pointed out in Section 1, may be incorrect for other, less spectacular, ferroelectrics.

The thermodynamic treatment outlined above will be extended and applied to specific cases in the following chapters, in particular, to the most typical ferroelectrics just mentioned. For a quick survey of where, in this book, the thermodynamic and other theories are discussed, the reader is referred to Appendix II.

## 6. Antiferroelectricity

In considering the alignment of electric dipoles within a ferroelectric crystal, we must keep in mind that this co-operative phenomenon is governed by two essentially different types of interaction forces. Forces due to chemical bonds, van der Waals attraction, repulsion forces and others have all such short ranges that, usually, interaction between nearest neighbors only need be considered. In contrast to these forces, those due to dipolar interaction have a very long range. Consequently, an accurate calculation of the interaction between a particular dipole and all the other dipoles of a given sample is generally quite complicated.

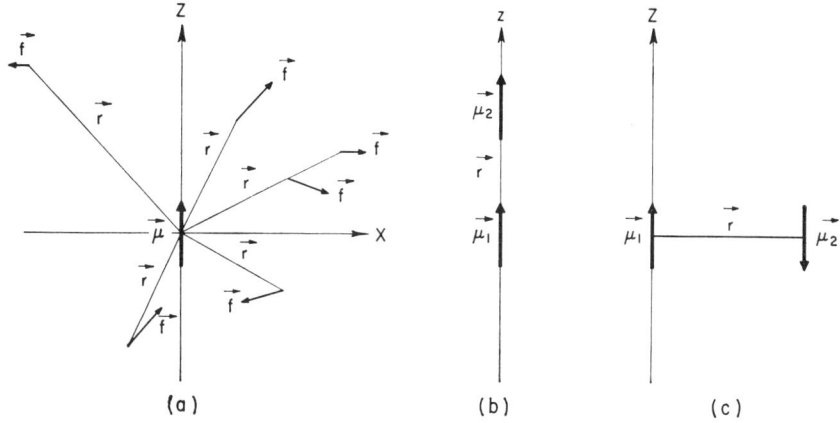

FIG. I-8. (a) The intensity of the dipole field $f$ (according to Böttcher (B 1)). (b) Parallel dipoles (see text). (c) Antiparallel dipoles (see text).

It may be useful to recall the characteristics of the dipole field. Given a point dipole at the origin of a rectangular co-ordinate system $x, y, z$, the field strength $f$ caused by this dipole at a point located at the distance $r$ from the origin is given by the formula

$$f = \frac{3\mu_1 \cdot r}{r^5} \cdot r - \frac{\mu_1}{r^3}, \tag{I-15}$$

where $\mu_1$ is the dipole moment. If the $z$-axis is along the dipole vector, the relative values of $f$ at some points in the $xz$ plane are shown in Fig. I-8(a). If we now bring another point dipole $\mu_2$ at the point with co-ordinates $r(x, y, z)$, the interaction energy between the two dipoles $\mu_1$ and $\mu_2$ is given by the work required to bring the second dipole from infinity at the location $r$ (see, e.g., Böttcher (B 1)):

$$W = -f \cdot \mu_2 = -\frac{3}{r^5}(\mu_1 \cdot r)(\mu_2 \cdot r) + \frac{\mu_1\mu_2}{r^3}. \tag{I-16}$$

It is easily seen that the minimum value of the interaction energy $W$ is obtained when $\mu_1$, $\mu_2$ and $r$ all have the same direction (Fig. I-8b), in which case:

$$W = -\frac{2\mu_1\mu_2}{r^3}. \tag{I-17}$$

The maximum value of $W$ is reached when one of the dipoles has the same direction as $r$ and the other the opposite direction:

$$W = \frac{2\mu_1 \mu_2}{r^3}.$$

When both dipoles are perpendicular to $r$ and antiparallel to each other (Fig. I-8 c), the interaction energy is

$$W = -\frac{\mu_1 \mu_2}{r^3}, \tag{I-18}$$

but if they are parallel to one another, $W = \mu_1\mu_2/r^3$.

Thus, it is seen that when both dipoles are parallel to $r$, then the parallel arrangement is stable; when both are perpendicular to $r$, then the antiparallel arrangement is more stable. This is true, of course, only for two isolated point dipoles.

If we next bring a large number of point dipoles (which, for simplicity, we will assume to be all equal) at the lattice points of a given lattice, then the total

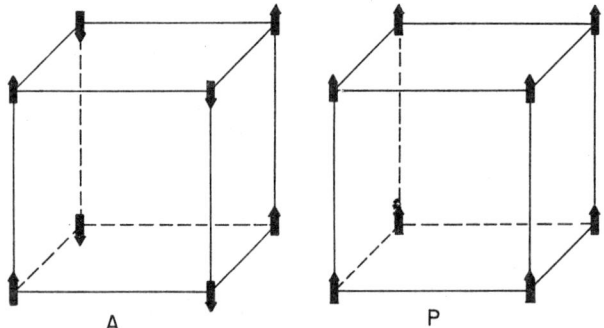

A                    P

Fig. I-9. Antiparallel (A) and parallel (P) dipole arrays in a simple cubic lattice.

dipolar interaction energy of the crystal thus obtained will obviously be a function of the orientations chosen for each dipole. If we allow either parallel or anti-parallel orientation only, we are theoretically in the position to compute whether, for the given lattice, the arrangement of all dipoles parallel to each other corresponds to a lower value of the interaction energy than any other arrangement, involving antiparallel dipole orientations. These calculations are obviously quite complicated, but in a few simple cases they can be performed. In the case of a simple cubic lattice, for example, Luttinger and Tisza (L 1) were able to show that the antiparallel dipole array indicated with the letter $A$ in Fig. I-9 has the lowest energy of all possible arrays in this lattice, including that involving parallel orientation of the dipoles ($P$ in Fig. I-9).

The unit cell of the antiparallel dipole lattice $A$, however, is no longer primitive. The periodicity of this lattice must be multiplied by two in the directions perpendicular to the dipole axis. The lattice can be looked upon as being built by two interpenetrating sublattices, each of which is occupied by dipoles orientated parallel to each other, but the polarities of the two sublattices are antiparallel to each other. Just as we call *polar* the crystal which has the lattice denoted by the letter $P$ in Fig. I-9, we are going to call *antipolar* the crystal characterized by the lattice $A$ in the same figure. Thus we call antipolar the crystals which are characterized by the existence of two oppositely polarized sublattices.

Antiparallel dipole arrangements actually do occur, in lattices which are more complicated than the simple cubic. Moreover, it is possible to encounter substances which, at a certain temperature, undergo a transition from a non-polar to an antipolar state. Dielectrically, such a transition may or may not be detectable, but calorimetric measurements and, above all, accurate structural analyses ought to be able to detect this type of transition. A typical example of the case in which a transition from a non-polar into an antipolar phase can be detected by dielectric measurements is that of ammonium di-hydrogen phosphate, $NH_4H_2PO_4$, isomorphous with $KH_2PO_4$, whose dielectric constant drops anomalously, upon cooling, at the transition temperature of about $-126$ °C.

It should be noticed that, according to the definition, an antipolar crystal should always have a superlattice below the transition from a non-polar phase. However, it is not always necessary for such a superlattice to be characterized by a multiple unit cell, as was the case, above, for a simple cubic lattice. If the unit cell of the non-polar phase happens to contain an even number of equivalent molecular groups, a multiple unit cell may not be necessary in the antipolar state.

So far, our definition of an antipolar crystal was given according to a logical parallelism with that of a polar crystal. The latter case occurs, obviously, when the energy corresponding to a parallel dipole orientation is a minimum. We know that, in most such cases, the dipole interaction is such as to give a rigid dipole array, i.e. a pyroelectric crystal. From the dielectric point of view, however, such cases are not very interesting. They become interesting only when the contributions to the interaction energy are so delicately balanced as to give a rather loosely bound array of parallel dipoles. In this case, the direction of the spontaneous polarization can be reversed with an external field and we have a ferroelectric crystal. In the thermodynamic language, we may say that, in a ferroelectric crystal, the free energy of the polar state is comparable to the free energy of the non-polar state.

Similarly, in an antipolar crystal the dipole interaction may be such as to give a rigid array of antiparallel dipoles. Again, such a case is rarely interesting from the dielectric point of view. It becomes interesting only if the coupling between the dipoles is such that the interaction energy is comparable to that of a polar crystal. In terms of thermodynamics, we may say that the free energy of the antipolar state is comparable to that of the polar state. When this is the case, then we say that the crystal considered is *antiferroelectric*. Accordingly, we define an antiferroelectric crystal as an antipolar crystal whose free energy is comparable to that of a polar crystal. This definition is in line with that of a ferroelectric crystal. Just as x-ray methods, for example, may be able to establish the polarity of a given crystal, but only dielectric methods can establish its ferroelectric activity, so can the antipolar character of a crystal be established, eventually, by x-ray methods, but only dielectric measurements can definitely establish its antiferroelectric character.

How can this be done? There are two consequences of our definition of an antiferroelectric crystal which affect its dielectric behavior. In the first place, when a crystal undergoes a transition from a completely unpolarized to such an antiferroelectric state, it will show a pronounced dielectric anomaly at the antiferroelectric transition point. This large anomaly is not directly associated with the antipolar phase, but is a consequence of the existence of a ferroelectric state with comparable free energy. In this case, we are experimentally able to observe the dielectric anomaly at the transition, but we will not observe dielectric hysteresis in the low-temperature phase. It is actually *because* of the observability of these particular antiferroelectric transitions that the concept of antipolarity and antiferroelectricity has become important in the physics of dielectric materials and is usually closely associated to the phenomenon of ferroelectricity.

The second consequence of only a small difference between the free energies of ferro- and antiferro-electric states is that it is possible, in antipolar phases of this kind, to decrease the free energy of the polarized state more than that of the antipolarized state by means of external mechanical or electrical stresses. At sufficiently high field strengths, antiferroelectric crystals become ferroelectric, although not necessarily along the same axis. Generally, the application of large electric fields induces a phase transition; the crystal lattice undergoes a rearrangement and the polar axis of the induced phase is not directed along the former antipolar direction. In antiferroelectric, *orthorhombic* $PbZrO_3$, for example, application of strong electric fields can induce a ferroelectric *rhombohedral* phase.

It should be emphasized, at this point, that the mere fact that a crystal exhibits a dielectric anomaly but no dielectric hysteresis on either side of the transition temperature is not sufficient to label the crystal as antipolar or antiferroelectric. Generally, a detailed structure analysis is required to establish whether the atomic shifts along a certain crystallographic direction are antiparallel. Alone, this analysis is again not sufficient to prove the existence of an antiparallel "polarization" but this does not invalidate the fact that the phenomenological behavior of the crystal can be understood in terms of sublattices with opposite polarizations.

From the above description it should be clear that the phenomena of ferroelectricity and antiferroelectricity are related to a given crystallographic *direction* within a crystal. In other words, a crystallographic direction may be polar, antipolar or non-polar. Thus, it is not difficult to conceive spatial arrays which may cause a crystallographic direction to be polar and another, perpendicular to it, antipolar. In this case, the corresponding crystal may exhibit ferroelectric properties along the one axis and antiferroelectric properties along the other axis. A new name has been introduced for such crystals, namely *ferrielectric*, but what we have said above should make this increase in nomenclature unnecessary.

With the exception of a few special cases, this book is not going to be concerned further with the description and the theories of antiferroelectric crystals. Only the cases of the perovskite antiferroelectric crystals lead zirconate, $PbZrO_3$, and sodium niobate, $NaNbO_3$ are going to be discussed (see Chapter V). The general problem of antiferroelectricity has been treated by Kittel (K 3) and Takagi (T 1) and reviewed by Känzig (K 1) and Forsbergh (F 1).

## BIBLIOGRAPHY

(B 1)  BÖTTCHER, C. J. F., *Theory of Electric Polarization*, Elsevier, Amsterdam (1952).

(B 2)  BROWN, W. F., *Dielectrics, Handbuch der Physik*, Vol. 17, pp. 1—154, Springer-Verlag, Berlin (1956).

(C 1)  CADY, W. G., *Piezoelectricity*, McGraw-Hill, New York (1946).

(D 1)  DEKKER, A. J., *Solid State Physics*, Prentice-Hall, Englewood Cliffs, N. J. (1958).

(D 2)  DEVONSHIRE, A. F., Theory of ferroelectrics, *Phil. Mag.* Suppl. 3, 85 (1954).

(F 1)  FORSBERGH, P. W., JR., *Piezoelectricity, Electrostriction and Ferroelectricity, Handbuch der Physik*, Vol. 17, pp. 264—392, Springer-Verlag, Berlin (1956).

(F 2)  FRÖHLICH, H., *Theory of Dielectrics*, Clarendon Press, Oxford (1949).

(J 1)  JAYNES, E. T., *Ferroelectricity*, Princeton University Press, Princeton (1953).

(K 1)  KÄNZIG, W., *Ferroelectrics and Antiferroelectrics, Solid State Physics*, Vol. 4, pp. 1—197, Academic Press, New York (1957).

(K 2)  KITTEL, C., *Introduction to Solid State Physics*, John Wiley, New York (1956).

(K 3)  KITTEL, C., *Phys. Rev.* **82**, 729 (1951).

(L 1)  LUTTINGER, J. M. and TISZA, L., *Phys. Rev.* **70**, 954 (1946).

(M 1)  MEGAW, H. D., *Ferroelectricity in Crystals*, Methuen, London (1957).

(M 2)  MUELLER, H., *Ann. N. Y. Acad. Sci.* **40**, 321 (1940).

(N 1)  NYE, J. F., *Physical Properties of Crystals*, Clarendon Press. Oxford (1957).

(S 1)  SAWYER, C. B. and TOWER, C. H., *Phys. Rev.* **35**, 269 (1930).

(S 2)  SMYTH, C. H., *Dielectric Behavior and Structure*, McGraw-Hill, New York (1955).

(S 3)  SMITH, C. S., *Macroscopic Symmetry and Properties of Crystals, Solid State Phyiscs*, Vol. 6, pp. 175—249, Academic Press, New York (1958).

(S 4)  SMOLUCHOWSKI, R., *Phase Transformations in Solids, Handbook of Physics*, pp. 8—110, McGraw-Hill, New York (1958).

(T 1)  TAKAGI, Y., *Phys. Rev.* **85**, 315 (1952).

# CHAPTER II

# TRI-GLYCINE SULFATE AND ISOMORPHOUS CRYSTALS

## 1. Introduction

Tri-glycine sulfate is a relatively young member of the fast-growing family of ferroelectric crystals. It has immediately attracted the attention of many researchers because it exhibits ferroelectric properties at room temperature, and it can be grown easily in large samples. Chemically and crystallographically, it is by far not the simplest ferroelectric known, but its phenomenological behavior is very simple. It represents, in fact, one of the most typical examples of a ferroelectric whose behavior fits quite perfectly the description that we have given in Chapter I. This is the reason why we treat it first.

The ferroelectric activity of tri-glycine sulfate, $(NH_2CH_2COOH)_3 \cdot H_2SO_4$, (abbreviated TGS), was discovered by Matthias *et al.* (M 1) in 1956. The Curie temperature lies at 49 °C. The phase above the transition has monoclinic symmetry and belongs to the centrosymmetrical class $2/m$. Below the transition temperature, the mirror plane disappears and the crystal belongs to the polar point group 2 of the monoclinic system. Ferroelectricity is found along the direction of the two-fold polar axis (monoclinic *b* axis). The transition is of the second order.

Ferroelectricity occurs also in two crystals isomorphous with TGS. They are: tri-glycine selenate, $(CH_2NH_2COOH)_3 \cdot H_2SeO_4$, (abbreviated TGSe), with transition temperature 22 °C (M 1), and tri-glycine fluoberyllate, $(CH_2NH_2COOH)_3 \cdot H_2BeF_4$ (abbreviated TGFB), with Curie temperature 70 °C (P 1). The characteristics of these transitions are similar to those found in TGS.

Large crystals of these compounds can be grown easily from water solutions prepared by reacting an aqueous glycine solution with the proper amount of the corresponding (sulfuric, selenic or fluoberyllic) acid. The growth can be achieved by slow evaporation of the solvent at constant temperature or by slowly lowering the temperature at constant supersaturation. A more sophisticated method has been described by Nitsche (N 1), who reported also the temperature dependence of the solubility curves.

The habit of the crystals obtained depends somewhat on the growing procedure and is generally quite complex owing to the presence of many faces. The *b* faces, perpendicular to the monoclinic *b* axis, are mostly absent or very small, but the crystals have a pronounced cleavage plane parallel to (010). Thus, crystal plates oriented perpendicularly to the polar axis can be easily obtained by cleavage of larger crystals with no need of knowing the directions of the monoclinic *a* and *c* axes.

However, the investigation of the dielectric, piezoelectric and elastic tensors requires the choice and the identification of a reference system of co-ordinates. Since the most convenient monoclinic angle $\beta$ is approximately equal to 105° and the predominant face is often the $c$ face, the following orientation of a reference systems of orthogonal co-ordinates is selected in relation to the most common crystal habit (K 1). The $Y$ axis is parallel to the polar axis, the $Z$ axis is parallel to the natural edge which forms an angle 105° with the predominant ($c$) face of the crystal, and the $X$ axis is perpendicular to Y and Z to form a right-handed system, as indicated in Fig. II-1.

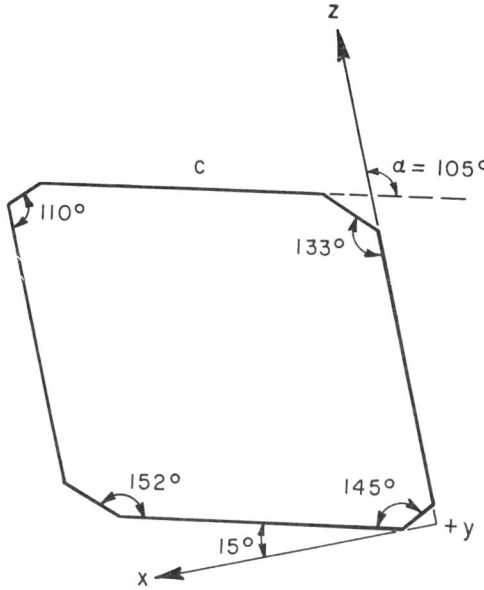

FIG. II-1. Reference system of co-ordinates in a crystal of tri-glycine sulfate (according to Konstantinova *et al.* (K 1)).

## 2. Dielectric Properties

The components of the dielectric constant tensor of TGS have the following values at 23 °C (measuring field 1 V/cm, frequency 500 kc/s) (K 1):

$$\varepsilon_a = 8.6, \quad \varepsilon_b = 43, \quad \varepsilon_c = 5.7*$$

The temperature dependence of these quantities, measured at 1 kc/s and 1 V/cm is shown graphically in Fig. II-2 (H 1). The component $\varepsilon_{22} = \varepsilon_b$, measured along the polar axis, is the most important quantity and will therefore be referred to, in the following, as the dielectric constant $\varepsilon$. It shows a very pronounced anomaly in the vicinity of the transition temperature $T_c$. The actual peak value is determined by unavoidable inhomogeneities of temperature and stress distri-

* It should be noted that $\varepsilon_c$ is not the dielectric constant measured perpendicularly to the $c$ face but that measured along the $Z$ axis of Fig. II-1.

butions, and also by the quality of the electrodes, as these may cause partial clamping of the crystal. Peak values of $\varepsilon$ greater than $10^5$ have been measured (T 1).

The constants $\varepsilon_{11} = \varepsilon_a$ and $\varepsilon_{33} = \varepsilon_c$ are practically independent of temperature.

The dielectric constant $\varepsilon$ follows the Curie–Weiss law $\varepsilon \cong C/(T - T_0)$ in the vicinity of the transition temperature, as shown by the plot of $1/\varepsilon$ vs. temperature (T 1) (Fig. II-3). The Curie constant $C$ is equal to 3200 °K, but variations of

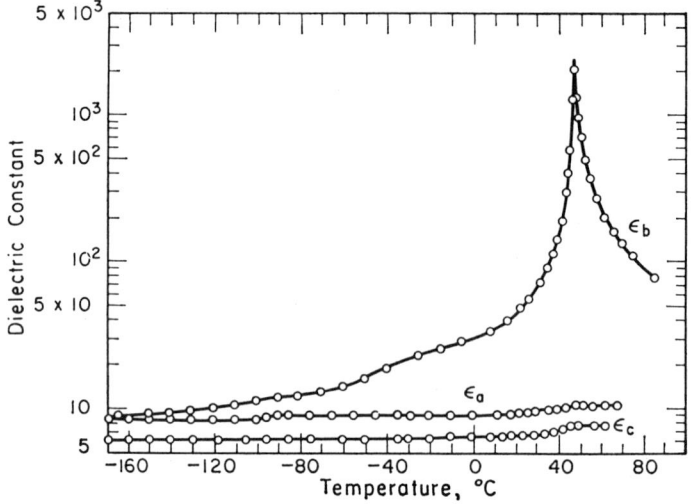

FIG. II-2. Dielectric constant of tri-glycine sulfate as a function of temperature (according to Hoshino *et al.* (H 1)).

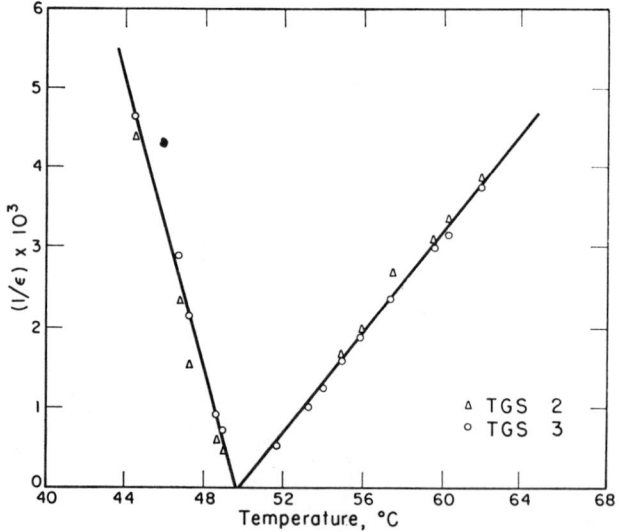

FIG. II-3. Curie–Weiss law for tri-glycine sulfate (according to Triebwasser (T 1)).

about 10% can be found in different specimens (H 1). The Curie–Weiss temperature $T_0$ agrees practically with the Curie temperature $T_c = 49$ °C. High-frequency measurements of the dielectric constant were carried out by Nishioka and Takeuchi (N 2). At 9600 Mc/s, the room-temperature values of the dielectric constant and the loss factor were reported to be $\varepsilon = 20$ and $\tan \delta = 0.043$, respectively, indicating that dielectric relaxation is already effective in the microwave region. The dielectric anomaly at the Curie temperature is, however, still present at the frequency employed (see also Lurio and Stern (L 1) and Nakamura and Furuichi (N 3)).

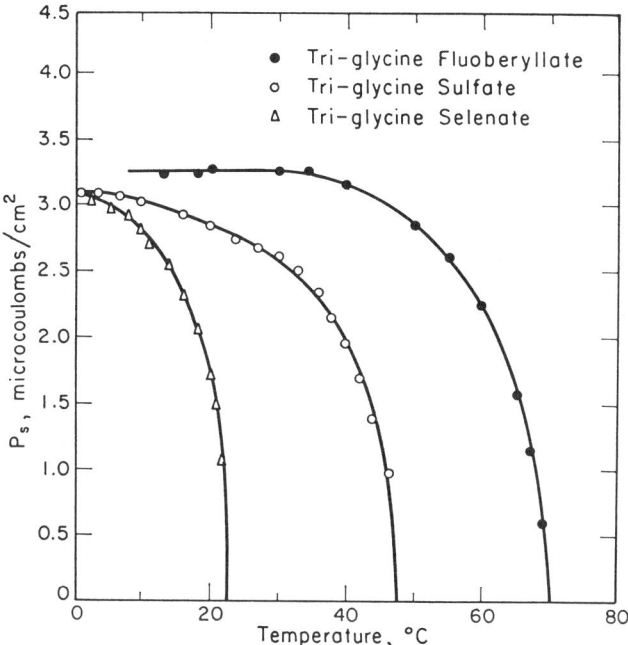

FIG. II-4. Spontaneous polarization $P_s$ of tri-glycine sulfate (H 1), tri-glycine fluoberyllate (H 1) and tri-glycine selenate as a function of temperature

The temperature dependence of the spontaneous polarization $P_s$ is depicted in Fig. II-4, and is typical of a transition of the second order. The value of $P_s$ at room temperature is about $2.8 \times 10^{-6}$ C/cm² (H 1), but the data reported by different authors may fluctuate within $\pm 20\%$ of this value (T 2), (T 3), (D 1), (Z 1). Below 0 °C, the spontaneous polarization increases only slowly with decreasing temperature, reaching a value of approximately $4.3 \times 10^{-6}$ C/cm² at $-140$ °C (C 4).

The pyroelectric effect was investigated at room temperature by Savage and Miller (S 2), and down to $-140$ °C by Chynoweth (C 4), with the dynamic technique first introduced by Chynoweth and described in Section IV-2. The pyroelectric hysteresis loops may sometimes appear to be biased in the direction of the polarization axis, when precautions are not taken to eliminate electrode-edge effects. The technique allowed Chynoweth to establish that the polarization can still be reversed at temperatures as low as $-140\ °C$, but reversal may take several seconds for applied fields of the order of $10^3$ V/cm. Above the Curie point, where the structure of TGS is centrosymmetric, no pyroelectric effect should be expected, but Chynoweth (C 4) has found that small pyroelectric signals can be generated, temporarily, at temperatures appreciably higher than $T_c$. Upon heating a previously polarized crystal, the pyroelectric signal first passes through a sharp peak at the Curie point, then drops rapidly to zero but finally builds up again in the opposite direction. As the temperature continues to rise, this residual signal passes through a maximum and then decays to zero in a few minutes. If the temperature is then lowered, the pyroelectric signal does not reappear until the Curie point is approached. The cause of the residual signal has thus been "annealed out". It has also been found that the sign of the residual signal cannot be affected by applying polarizing fields with opposite direction at room temperature. These unexpected effects have been explained by Chynoweth in terms of compensation charges accumulated around residual domains. It has namely been established that a number of narrow domains may persist in a crystal at room temperature even when strong d.c. fields are applied (see Section 7). Free compensation charges are expected to accumulate on the domain boundaries of these residual domains, and when the spontaneous polarization vanishes, as it does when the crystal is heated through the Curie point, these charges are left behind and take some time to disappear. While they last, they *induce* a polarization that is the cause of the residual pyroelectric signals. It was found that fields of the order of $10^2$–$10^3$ V/cm had to be applied to the crystal above the Curie point to induce a pyroelectric signal comparable to the residual signal. The annealing out of the residual signal would thus be ascribed to the gradual annihilation of the compensation charges around the old domain locations.

It has also been established that a TGS crystal cooled through the Curie temperature shows a marked tendency to repolarize in the same direction in which it was previously poled. This effect has been assumed to be due to the existence of a thin layer with no (or zero net) polarization at the surface of the crystal. Bearing no spontaneous polarization, such a surface layer would not take part in the process of polarization reversal, and hence would give rise to compensation charges at the interface between bulk and surface layer. The field produced by these interface charges, while they last, throughout the bulk of the crystal may be small but quite sufficient to cause the whole crystal to repolarize in the same sense as it cools through the Curie point, since the coercive field vanishes at the Curie point. The nature of the postulated surface layer is not well understood, at present, but may tentatively be conceived of as a chemical disturbance or a mechanical distortion of the crystal structure at the surface. The thickness of this layer was tentatively estimated to be within $10^{-7}$ cm and $10^{-5}$ cm (C 4).

Measurements of the coercive field $E_c$ of TGS have been carried out by a number of authors (T 3), (D 1), (T 2). As was pointed out in Section I-5, this quantity is strongly dependent upon the amplitude and frequency of the applied

field, at any given temperature below the Curie point. At room temperature, the value of $E_c$ measured with a field of 1500 V/cm and a frequency of 50 c/s is equal to 430 V/cm (D 1). The dependence of the 50 c/s coercive field upon the amplitude of the applied field has been tentatively written by Toyoda *et al.* (T 3) in the form
$$E_c = \text{constant} \times E_{\max}^b,$$

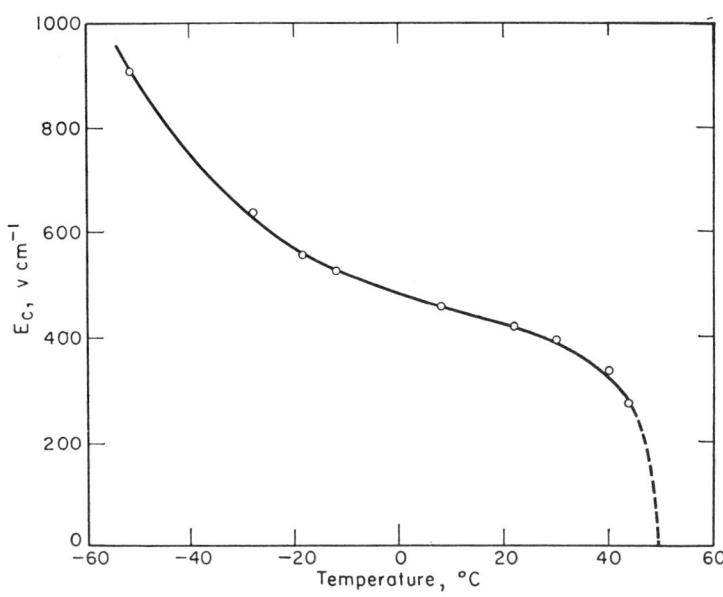

FIG. II-5. Coercive field $E_c$ of tri-glycine sulfate, measured with a field of 1500 V/cm at 50 c/s as a function of temperature (according to Domanski (D 1)).

where the exponent $b$ equals 0.5 at 20 °C and 0.4 at $-73$ °C. The temperature dependence of $E_c$ measured with a 50 c/s field of 1500 V/cm is depicted in Fig. II-5. The sharp increase occurring below $-10$ °C is not indicative of phase transitions, but of the fact that the relaxation time associated with the reversal process increases rapidly with decreasing temperature.

### 3. Thermodynamic Theory and Dielectric Non-linearity

As we mentioned in Section 1, the phenomenological theory of TGS is very simple. Qualitatively, it is already contained in the general thermodynamic treatment outlined in Section I-5. We will presently consider the quantitative aspects of this treatment and we will see, on the basis of the agreement between theory and experiment, that the essential features of a ferroelectric transition such as that of TGS can be described satisfactorily on the basis of the expansion of the free-energy function *in terms of polarization only*.

The experimental results described in the preceding section are strongly indicative of the fact that the ferroelectric transition in TGS is of the second order. We refer, in particular, to the fact that the spontaneous polarization $P_s$

is a continuous function of temperature and the dielectric constant becomes extremely large at the Curie temperature. In addition, thermal expansion measurements have revealed no abrupt change in volume at the Curie point (E 2).

Since we are only interested in the dielectric behavior along the monoclinic $b$ axis, we are going to consider the polarization and field components in this direction only, i.e. we put

$$P_1 = P_3 = 0, \quad E_1 = E_3 = 0 \quad \text{and} \quad P_2 = P, \quad E_2 = E.$$

We then can write the free energy of the stress-free crystal in the non-polar phase (Eq. I-7) as follows:

$$A(P, T) = \frac{2\pi}{C}(T - T_0)P^2 + \tfrac{1}{4}\xi P^4 + \tfrac{1}{6}\xi P^6. \tag{II-1}$$

The coefficient of $P^2$ is written in such a way as to take care of the experimentally proven existence of the Curie–Weiss law. The superscript $X$ of the coefficients $\xi^X$ and $\zeta^X$ (Eq. I-7), indicating that these quantities are measured at constant (zero) stress, is dropped for convenience. Moreover, we assume, in the first approximation, that the coefficients $\xi$ and $\zeta$ are independent of temperature, at least in the vicinity of the transition.

Under these assumptions, we can derive the temperature dependence of the spontaneous polarization $P_s$ from (II-1) and see how it fits the experimental data, as indicated by Triebwasser (T 1). The partial derivative $(\partial A/\partial P)_T = E$ gives us an equation for $P_s$ when we put $E = 0$. A simple solution of this equation can be written if we assume that we are sufficiently close to the transition temperature for the polarization to be small. We obtain (T 1):

$$P_s^2 = \frac{4\pi}{\xi C}(T_0 - T) - \frac{16\pi^2}{C^2}\frac{\zeta}{\xi}(T_0 - T)^2. \tag{II-2}$$

By the least square fitting of the experimental $P_s^2$ vs. $(T_0 - T)$ curve in the vicinity of the transition temperature, the coefficients $\xi$ and $\zeta$ can be determined. The results are (T 1):

$$\xi = 8.0 \times 10^{-10} \text{ (e.s.u./cm}^2\text{)}^{-2}; \quad \zeta = 5.04 \times 10^{-18} \text{ (e.s.u./cm}^2\text{)}^{-4}.$$

We are now in the position to check quantitatively how well the power series expansion (II-1) converges in the temperature range of interest to us. Evaluating the terms in $P^2$, $P^4$ and $P^6$ at about 40 °C, where $P \cong 2 \times 10^{-6}$ C/cm$^2$, we find that they contribute $7.3 \times 10^5$ ergs/cm$^3$, $2.6 \times 10^5$ ergs/cm$^3$, and $0.4 \times 10^5$ ergs/cm$^3$, respectively. Thus, we are justified in dropping the term in $P^6$ from expression (II-1), for convenience. In this approximation, Eq. (II-2) predicts simply a linear relationship between $P_s^2$ and $T$, a fact which is confirmed by the experimental results in a temperature range of about 10° below the transition point (H 1). The coefficient $\xi$ can then be determined simply from the slope of the $P^2$ vs. $T$ curve in the vicinity of the transition temperature. The result is: $\xi = 9.3 \times 10^{-10}$ (e.s.u./cm$^2$)$^{-2}$. The main part of the difference between this value and that reported above is due to the rather wide range of experimental accuracy (10–15%).

There is another method to determine the coefficients of the power expansion (II-1). This method makes use of the nonlinear properties of the dielectric constant near the transition point *in the non-polar phase*. From (II-1), dropping the sixth-order term, we obtain immediately

$$\frac{\partial E}{\partial P} \cong \frac{4\pi}{\varepsilon} = \chi + 3\,\xi\,P^2. \tag{II-3}$$

Thus, if one can measure the dielectric constant as a function of polarization in the non-polar state, one has a way to determine $\xi$. Such measurements were carried out by Triebwasser (T 1) using a method described by Drougard *et al.* (D 2) for the investigation of the polarization dependence of the capacitance of a sample. The experimental arrangement is depicted schematically in Fig. II-6. The capa-

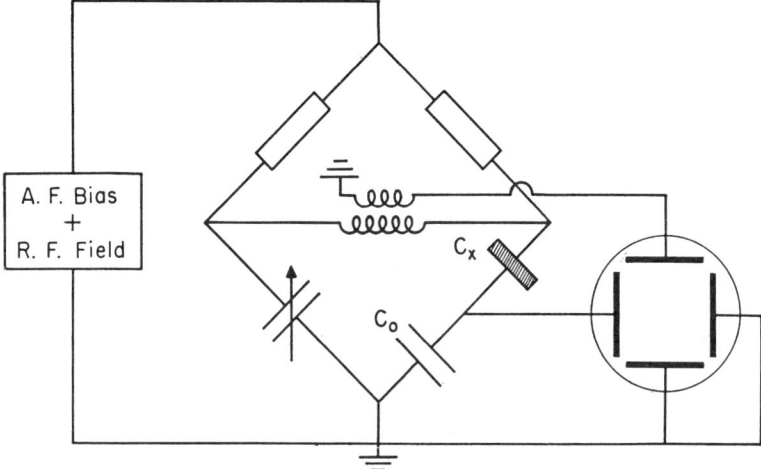

FIG. II-6. Schematic diagram for the investigation of the polarization dependence of the dielectric constant.

citance bridge is fed with a very small-amplitude radiofrequency field (e.g. 50 kc/s) superimposed on a large-amplitude low-frequency field (e.g. 60 c/s). The voltage across the linear capacitor $C_0$, in series with the crystal $C_x$, is proportional to the polarization developed across the sample by the biasing low-frequency field, and is applied to the $X$ axis of an oscilloscope. The bridge output is applied to the $Y$ axis of the scope, so that the unbalance of the bridge is displayed against the polarization and the bridge can thus be balanced for any value of the polarization.

The results of Triebwasser's measurements are shown in Fig. II-7. According to Eq. (II-3), the slope of the straight line in this figure yields the value of $\xi$. The result is $\xi = 9.2 \times 10^{-10}$ (e.s.u./cm²)⁻². This value necessitates a correction due to the adiabatic conditions of the measurement (see Section 5), after which the value of $\xi$ becomes equal to $7.7 \times 10^{-10}$ (e.s.u./cm²)⁻², in excellent agreement with the value obtained from the measurements in the polar phase. The con-

sistency of the values of $\xi$ determined above and below the transition temperature is a quantitative confirmation of the applicability of the thermodynamic
theory.

The low-frequency biasing field used in the above experiment can be replaced
by a d.c. biasing field. The non-linearity of the dielectric constant can thus
be determined by measuring the capacitance of the sample with a small a.c.
field superimposed on a large d.c. biasing field. The result can be predicted from
the free-energy function: given a value of $P$, we can compute the dielectric constant
from (II-3) and the electric field from the partial derivative of the free energy

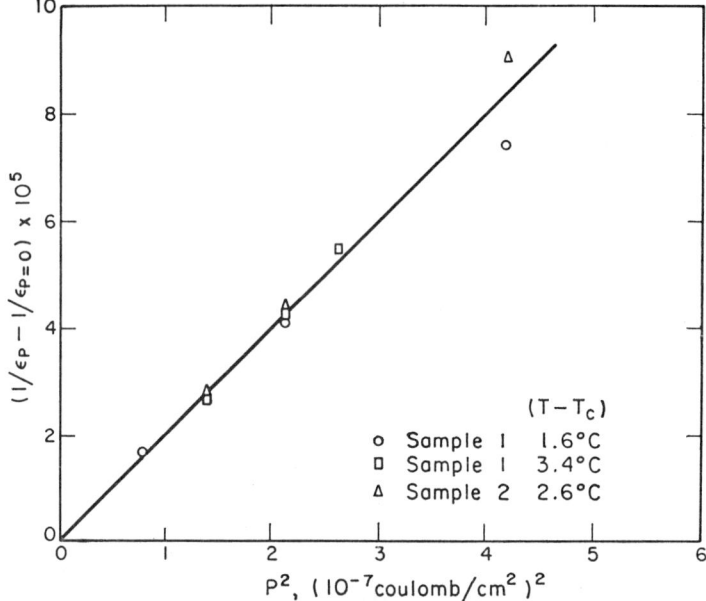

FIG. II-7. Polarization dependence of the dielectric constant of tri-glycine sulfate
(biasing field 200 c/s sine wave, measuring field 50 kc/s) (according to Triebwasser (T 1)).

(II-1) with respect to polarization. The result is depicted in Fig. II-8. Measurements of the small-signal dielectric constant under the action of d.c. biasing
fields were also carried out by Chapelle and Taurel (C 5). These measurements
indicate a small variation of the $\xi$ coefficient with temperature.

### 4. Specific Heat

The specific heat $c_p$ of TGS was measured by Hoshino *et al.* (H 1) in the temperature range from 10 to 65 °C, i.e. through the transition point. The experimental results are shown graphically in Fig. II-9, and clearly indicate an anomaly
of the specific heat over a relatively wide temperature range in the vicinity of
the transition. The shape of the curve is particularly interesting not only because

it is indicative of a second order transition, a fact already established by the dielectric data, but especially because it does not show the characteristics of a $\lambda$-type transition.

The heat of transition $\Delta Q$ and the corresponding entropy change $\Delta S$ can be determined by integrating the experimental curve of the excess specific heat. The results are (H 1)

$$\Delta Q = 150 \text{ cal/mole},$$

$$\Delta S = 0.48 \text{ cal/mole } °C.$$

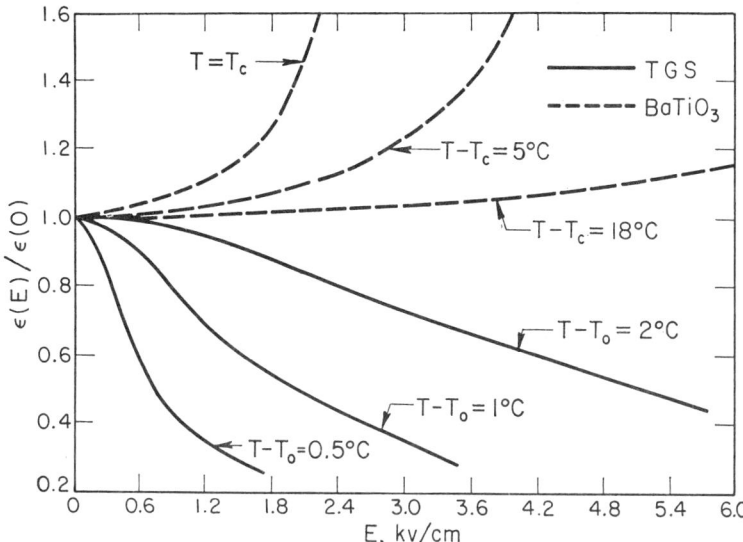

FIG. II-8. Field dependence of the dielectric constant of tri-glycine sulfate. The case of BaTiO$_3$ is also shown as an example of the behavior expected theoretically for a first-order phase change (according to Triebwasser (T 1)).

Let us now see how well these results compare with those expected on the basis of the thermodynamic theory. Since the entropy $S$ is equal to $-(\partial A/\partial T)_P$ and we have assumed that the coefficients $\xi$ and $\zeta$ are temperature independent, we obtain for the entropy change:

$$\Delta S = S_0 - S = \frac{2\pi}{C} \Delta P^2, \qquad (II-4)$$

where $S$ and $S_0$ represent the entropies of the polar and non-polar states, respectively. Using the experimental data for the Curie constant and the spontaneous polarization at 23 °C, we obtain

$$\Delta S = 0.64 \text{ cal/mole } °C,$$

in fair agreement with the above value determined from the specific heat curve. The accuracy involved in this comparison is poor, owing to the fairly small transition energy of TGS, which is strongly affected by the choice of the base line in Fig. II-9.

We can further compute the excess specific heat from Eq. (II-4):

$$\Delta c_p = c_p - c_p^0 = \frac{2\pi}{C} T \frac{\partial P^2}{\partial T}, \tag{II-5}$$

where $c_p$ and $c_p^0$ represent the specific heats of the polar and non-polar states respectively. Equation (II-5) allows us to compute the expected temperature dependence of $c_p$ from the dielectric data and compare it with the direct measurement (Fig. II-9). It is quite clear that the flat top of the theoretical curve is a consequence of the linear temperature dependence of $P^2$ in the immediate vicinity of the transition (Eq. II-2). The agreement between the shape of the experimental curve and that calculated from the dielectric data is, however, not quite as satisfactory as one would wish. It has been pointed out (H 1) that the thermodynamic treatment based on the expansion of the free energy in terms of polarization only may be incomplete, as it does not take into account short-range order effects.

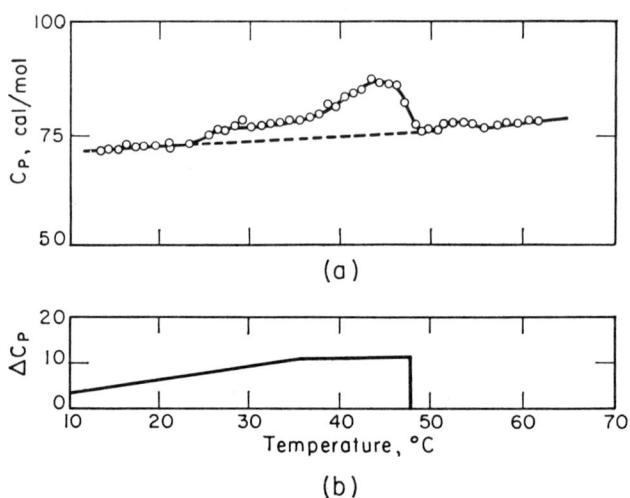

Fig. II-9. Specific heat of tri-glycine sulfate as a function of temperature.
(a) Experimental. (b) Theoretical (according to Hoshino *et al.* (H 1)).

## 5. Adiabatic and Isothermal Dielectric Constant

The fact that the polarization is a rapidly varying function of temperature and field in the vicinity of the transition point has other important consequences, e.g. the differentiation between the values of the dielectric constant measured isothermally and adiabatically.

We have seen in Chapter I that a certain relationship is expected between the dielectric constant below and the dielectric constant above the transition. This relationship, Eq. (I-10), yields the value of the reciprocal dielectric susceptibility $(\partial^2 A/\partial P^2)$ below the Curie temperature and is rewritten here for convenience:

$$\frac{\partial^2 A}{\partial P^2} = 2\chi^x. \tag{II-6}$$

Since the reciprocal dielectric constant is a linear function of temperature above (Curie–Weiss law) as well as below the transition point (Fig. II-3), Eq. (II-6) requires that the downward slope be just *twice* as steep as the upward slope. However, the ratio between the slopes of the experimental lines is found to be 2.7, with an accuracy of about $\pm 10\%$. The reason for this discrepancy is to be sought in the fact that the dielectric constant measured with an a.c. field of, say, 10 kc/s in the ferroelectric range represents an adiabatic value, while the theory assumes isothermal conditions.

In the following, we want to derive an expression relating the adiabatic dielectric constant $\varepsilon_S$ to the isothermal dielectric constant $\varepsilon_T$ (D 3), (T 1), (W 3). Considering the electric field $E$ as an explicit function of temperature $T$, entropy $S$ and polarization $P$, i.e. $E(T, S, P)$; writing the total differential $dE$ in terms of partial derivatives, and putting $dS = 0$ for an adiabatic process, we obtain:

$$\left(\frac{\partial E}{\partial P}\right)_S = \left(\frac{\partial E}{\partial T}\right)_P \left(\frac{\partial T}{\partial P}\right)_S + \left(\frac{\partial E}{\partial P}\right)_T. \tag{II-7}$$

The partial derivative $(\partial T/\partial P)_S$ can be written in another way as follows. From the expression for the free energy $A$ of a stress-free crystal:

$$dA = TdS + EdP$$

we have $T = (\partial A/\partial S)_P$ and $E = (\partial A/\partial P)_S$, and hence the Maxwell's relation

$$\left(\frac{\partial T}{\partial P}\right)_S = \left(\frac{\partial E}{\partial S}\right)_P,$$

which latter can be written

$$\left(\frac{\partial E}{\partial S}\right)_P = \left(\frac{\partial E}{\partial T}\right)_P \Big/ \left(\frac{\partial S}{\partial T}\right)_P,$$

so that Eq. (II-7) becomes:

$$\left(\frac{\partial E}{\partial P}\right)_S = \left(\frac{\partial E}{\partial P}\right)_T + \left(\frac{\partial E}{\partial T}\right)_P^2 \Big/ \left(\frac{\partial S}{\partial T}\right)_P. \tag{II-8}$$

Now, $(\partial E/\partial P)_S = \chi_S$, or the reciprocal adiabatic dielectric susceptibility; $(\partial E/\partial P)_T = \chi_T$, or the reciprocal isothermal dielectric susceptibility; and, finally, by definition, $(\partial S/\partial T)_P = c_P/T$, where $c_P$ represents the specific heat per unit volume at constant polarization $P$. We can then write Eq. (II-8) thus:

$$\chi_S = \chi_T + \frac{T}{c_P}\left(\frac{\partial E}{\partial T}\right)_P^2. \tag{II-8a}$$

The experiment yields the adiabatic value $\chi_S$: how can we compute $\chi_T$ by means of Eq. (II-8a)?

It is easy to obtain an expression for $(\partial E/\partial T)_P$ from our free-energy function (II-1) and the previously assumed temperature independence of the coefficients $\xi$ and $\zeta$. We obtain $(\partial E/\partial T)_P = (4\pi/C) P$, so that Eq. (II-8) can be written

$$\chi_S = \chi_T + \frac{16\pi^2}{C^2} \cdot \frac{T}{c_P} P^2. \tag{II-9}$$

We can correlate this expression with the free-energy expansion (II-1) by noting that $\chi_T = (\partial^2 A / \partial P^2)_T$ and thus, neglecting again the $P^6$ term of $A(P, T)$:

$$\chi_S = \frac{4\pi}{C}(T - T_0) + 3\xi P^2 + \frac{16\pi^2}{C^2}\frac{T}{c_P}P^2, \qquad \text{(II-9a)}$$

or since $\chi_S \cong 4\pi / \varepsilon_S$:

$$\frac{1}{\varepsilon_S} = \frac{T - T_0}{C} + \left(\frac{3}{4\pi}\xi + \frac{4\pi T}{C^2 c_P}\right)P^2. \qquad \text{(II-10)}$$

It may be worthwhile to discuss this expression in more detail:

(i) *Above the transition temperature*, the dielectric constant measured with a *very small* a.c. field is adiabatic in principle, however, in practice we can neglect the correction term as the polarization is also very small. Thus the dielectric constant is characterized by the Curie–Weiss law

$$\left(\frac{1}{\varepsilon}\right)_{T > T_c} = \frac{T - T_0}{C}$$

(first term in Eq. (II-10)). If we apply a biasing field, causing a polarization $P$, we measure an adiabatic value of the dielectric constant. Thus, the slope of the straight line in Fig. II-7 is equal to the whole expression in the parenthesis on the right-hand side of Eq. (II-10) and the value of $\xi$ can be computed only if the adiabatic correction term is known. The direct measurement of the specific heat at constant pressure reported above (Section 4) yields the value of $c_{E=0}$, i.e. the specific heat at zero field. Above the transition temperature, where $P = 0$ (and also far below this temperature, where the polarization is almost constant), the values of $c_{E=0}$ and $c_P$ are equal and we can thus evaluate the adiabatic correction term, with the results reported in Section 3.

(ii) *Below the transition temperature*, the existence of a spontaneous polarization $P_s$ makes a distinction between adiabatic and isothermal dielectric constant always necessary. In the immediate vicinity of the transition temperature, $P_s^2$ is given by Eq. (II-2), where we neglect the second-order term. Introducing this expression into (II-10) and arranging the terms we obtain

$$\left(\frac{1}{\varepsilon}\right)_{T < T_c} = -2\frac{T - T_0}{C}\left(1 + \frac{8\pi^2}{C^2 c_P \xi}T\right). \qquad \text{(II-11)}$$

Differentiating with respect to $T$, we obtain for the slope of the curve below and near the transition (where $T \approx T_0$):

$$\frac{d}{dT}\left(\frac{1}{\varepsilon}\right)_{T < T_c} = -\frac{2}{C}\left(1 + \frac{8\pi^2}{C^2 c_P \xi}T_0\right),$$

while, above the transition temperature:

$$\frac{d}{dT}\left(\frac{1}{\varepsilon}\right)_{T > T_c} = \frac{1}{C}.$$

Hence the ratio of the slopes is, in the vicinity of $T_0$:

$$\frac{\mathrm{d}}{\mathrm{d}T}\left(\frac{1}{\varepsilon}\right)_{T<T_c} \bigg/ \frac{\mathrm{d}}{\mathrm{d}T}\left(\frac{1}{\varepsilon}\right)_{T>T_c} = -2\left(1+\frac{8\pi^2}{C^2 c_P \xi}T_0\right). \qquad \text{(II-12)}$$

If we neglect the adiabatic correction term, Eq. (II-12) agrees with Eq. (II-6). Consideration of this term implies a correction of about 20% thus predicting a slope ratio of 2.4, to be compared with the experimental value 2.7 (T 1). The residual discrepancy, which may be outside the range of the experimental error, may be explained by the fact that, below that transition temperature, the crystal consists of many domains polarized antiparallel to each other which may cause partial clamping (H 1). The phenomenon of domain clamping will be discussed in Chapter IV.

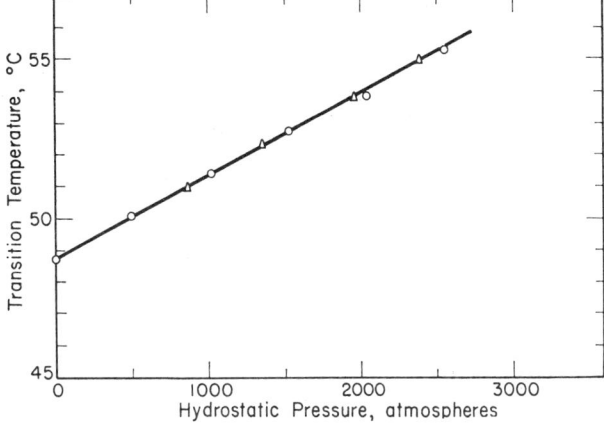

FIG. II-10. Transition temperature of tri-glycine sulfate as a function of hydrostatic pressure (according to Jona and Shirane (J 1)).

## 6. Effects of Stresses and Radiation

The dielectric properties of ferroelectric crystals can be strongly affected by a number of agents, such as external stresses, fields and radiations. It is interesting to study these effects because they will help to shed light on the microscopic mechanism leading to ferroelectricity. In the case of tri-glycine sulfate, two such effects have been studied in detail, viz., the effect of hydrostatic pressures and that of radiation damage.

### Pressure Effects

Application of hydrostatic pressures up to 2700 atm shows that the Curie temperature of TGS is displaced linearly upwards by the pressure $p$, according to the simple formula(J 1):

$$T_c = T_c^0 + Kp, \qquad \text{(II-13)}$$

where $T_c^0$ is the transition temperature at atmospheric pressure and $K = 2.6 \times 10^{-3}$ °C/atm (Fig. II-10). All the characteristics of the ferroelectric transition (Curie–Weiss law, onset and value of the spontaneous polarization) remain unaltered with respect to the new transition temperature.

The displacement of the critical temperature $T_c$ with pressure $p$ in the second-order transition of TGS can be predicted according to the well-known Ehrenfest relation (E 1):

$$\frac{dT_c}{dp} = \frac{\alpha - \alpha'}{c_p - c_p'} \frac{T_c}{\varrho}, \qquad (II-14)$$

where $\alpha$ and $\alpha'$ designate the volume expansion coefficients above and below $T_c$, respectively; $c_p$ and $c_p'$ the corresponding specific heats at constant pressure and $\varrho$ the density of the crystal.

The same displacement can be predicted, on the other hand, on the basis of the expansion of the free-energy function of stress and polarization, $A(X, P)$, formally similar to expression (I-1). This expansion can be simplified in the present centrosymmetrical case, as the only non-zero components of the stress tensor are $X_{11} = X_{22} = X_{33} = p$, and, of course, the only non-zero component of the polarization vector is $P_2 = P$. Thus we obtain the following expression for the free energy in terms of hydrostatic pressure $p$ and polarization $P$:

$$A(p, P, T) = Sp^2 + KpP^2 + A(P, T),$$

where the quantities $S$ and $K$ are functions of the elastic and electrostrictive constants, and $A(P, T)$ is the free-energy function of polarization $P$ for zero stress [Eq. (II-1)]. Collecting terms, we obtain:

$$A(p, P, T) = Sp^2 + \frac{2\pi}{C}[T - (T_c^0 + Kp)]P^2 + \frac{1}{4}\xi P^4. \qquad (II-15)$$

Thus, we can predict that:

(i) The transition temperature depends linearly on the applied hydrostatic pressure $p$.

(ii) The reciprocal dielectric constant of the non-polar phase follows a Curie–Weiss law even under hydrostatic pressure and the Curie constant is unaffected by pressure.

(iii) An expression for $P_s^2$ can be obtained from (II-15) which is similar to (II-2), namely:

$$P_s^2 = -\frac{4\pi}{C\xi}[T - (T_c^0 + Kp)].$$

This shows that, in the vicinity of the transition temperature and with the assumption that the coefficient $\xi$ is independent of pressure, the square of the spontaneous polarization is a linear function of both temperature and pressure.

The experiment confirms the above predictions (J 1) within the experimental accuracy, thus justifying the assumptions made for their derivation.


### Radiation Effects

Radiation damage effects on TGS have been reported by Chynoweth (C 1) who used 30 keV X-rays as damaging radiation and studied its effect on the hysteresis loops. Small X-ray dosages are reportedly sufficient to cause very

remarkable deformations of the hysteresis loop. Figure II-11 is a typical form
assumed by an originally rectangular hysteresis loop; the curve splits irto two
separate loops, each fairly rectangular, and which are biased by equal and opposite
amounts along the field axis. There appears to be no appreciable threshold energy
required of the X-rays in order to produce the effects, if the bombardment times

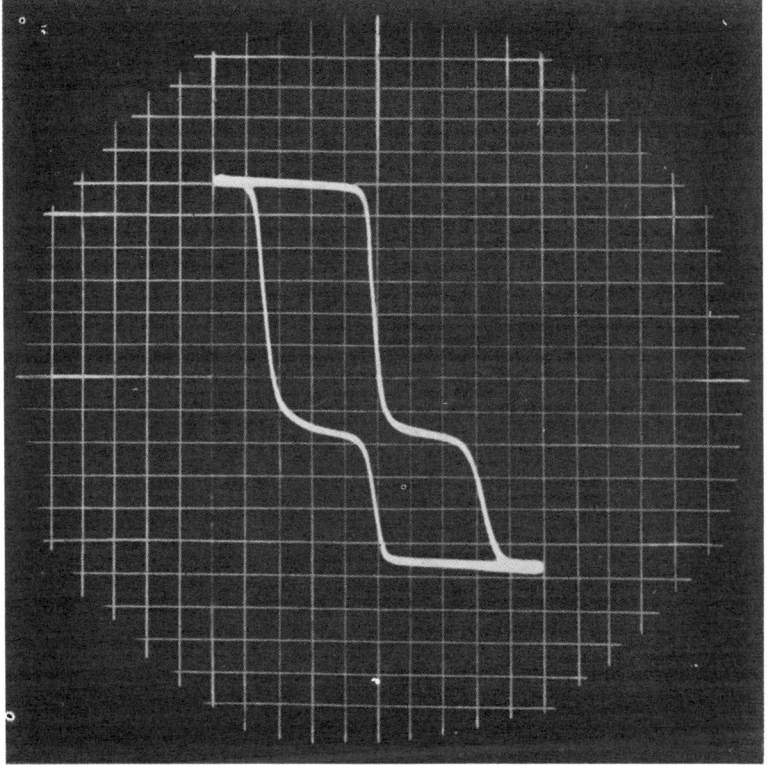

Fig. II-11. Typical distortion of hysteresis loop of tri-glycine sulfate as a consequence of X-ray bombardment (according to Chynoweth (C 1)).

are adjusted accordingly. This suggests that the damage results from effects
of ionization rather than from ionic displacements. The effect is also the same
irrespectively of whether the crystal is bombarded above or below the transition
point, but application of an electric a.c. field during or immediately after the
bombardment seems to relieve the distortion temporarily. A diagram showing
the pattern of the radiation damage and annealing effects is depicted in Fig. II-12.
There is evidence for very severe strain in the crystals exposed to radiation,
while the experimental tests seem to deny any explanation of the effects on the
basis of space-charge fields.

The detailed origin of the radiation damage effects is not known to date
and will have to await more detailed investigations of the effect of radiation

damage on other characteristics of the ferroelectric transition in TGS, such as the dielectric constant, the transition temperature, the elastic constants etc. A tentative explanation has recently been proposed by Abe (A 1) in terms of the strains caused by lattice imperfections due to the radiation. Similar radiation damage effects have been observed in Rochelle salt crystals, using both X-rays

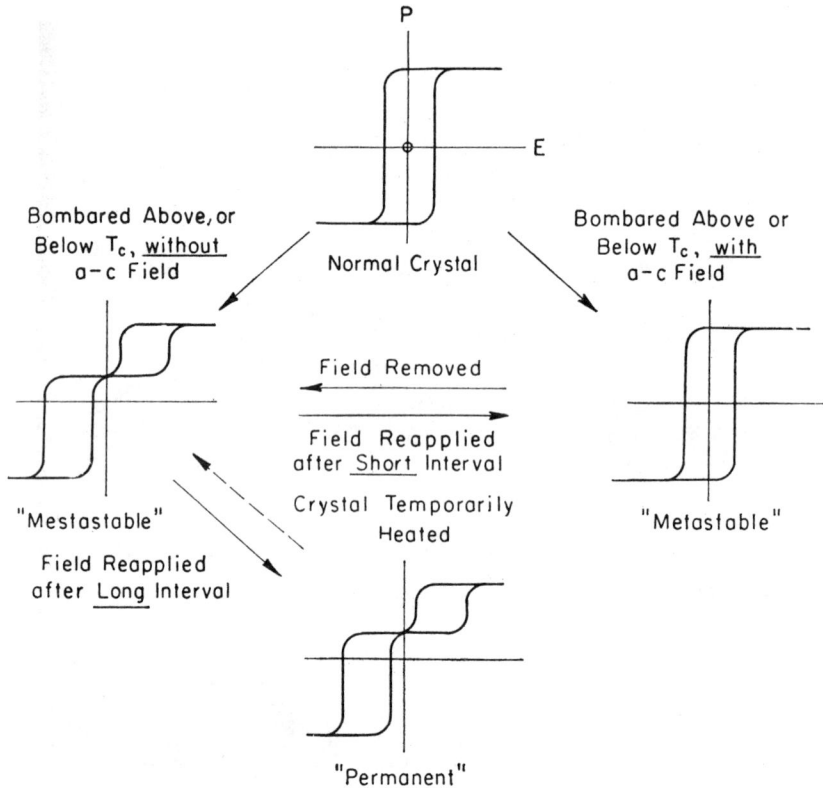

FIG. II-12. Pattern of radiation damage and annealing effects on the hysteresis loop of tri-glycine sulfate (according to Chynoweth (C 1)).

and γ-radiation (see Chapter VII). Unfortunately, we do not know, to date, how much this damage affects the structure of the crystals investigated. If this is the case, an interesting point arising from these investigations is that (C 1), as the X-ray dosages required to produce large changes in the ferroelectric properties are small compared with those normally used in crystallographic structure determinations, there is some doubt whether the structure determined by X-ray crystallography (see Section 9) represents the structure of an undamaged tri-glycine sulfate crystal.

The effect of neutron bombardment upon the dielectric properties of TGS has been recently investigated by Fatuzzo (F 5).

## 7. Domains

### General Considerations about Ferroelectric Domains

In an ideally perfect ferroelectric crystal, the existence of domains can be understood on the basis of qualitative energetic considerations. Suppose that we start from a perfect, non-conducting, insulated ferroelectric crystal in vacuum, which is required to have a uniform spontaneous polarization throughout. For the sake of simplicity, we assume that the shape of this hypothetical "single-domain" crystal is spherical. Then the charges induced on the external surfaces will create a depolarizing electric field whose energy is proportional to the volume of the crystal and the square of the polarization. Such a depolarizing field makes it impossible for the crystal to remain in the uniformly polarized state hypothesized initially. The result is a destruction of the uniform polarization: the crystal is consequently subdivided into regions, or domains, having antiparallel directions of spontaneous polarization, a state which is energetically more stable because it reduces the depolarizing field. This process, however, will not proceed indefinitely, because a certain amount of energy is going to be stored in the boundary layers, or "walls" between the domains. When the overall wall energy has increased to the point of balancing the decrease in energy of the depolarizing field, then an equilibrium configuration is reached and this is going to be the stable domain configuration at the temperature considered.

This, however, is only true if the crystal is ideally *perfect*. In a *real* crystal, the domain configuration resulting from the energetic balance mentioned above is hardly stable at any temperature. A real crystal is never an ideal insulator: the charges induced by the spontaneous polarization are partially compensated by bulk and surface conductivity. Furthermore, a real crystal contains a relatively large number of vacancies, dislocations, impurities, in one word, imperfections, which disturb the uniformity of the polarization and, therefore, of the depolarizing field. Finally, most crystals are in a state of non-uniform strain as a consequence of the conditions of the growth process. Therefore, the domain configuration actually observed in a given crystal is the result of a compromise between the energetic requirements of an ideally perfect crystal and the perturbing effects of conductivity, strain and imperfections in a real crystal. The latter effects may proceed very slowly with time, and therefore the observed domain configuration of a real crystal may be only metastable (aging effects).

In addition, it should be emphasized that one of the most important factors affecting a given domain pattern in a real crystal is the way in which the domains are created when the crystal is cooled through the Curie temperature. The process of domain formation at this temperature involves nucleation. This is most important for crystals which undergo a transition of the first order but may also play an important role in those which undergo a second-order transition, if the temperature distribution throughout the crystal is non-uniform. As the nucleation process is also dependent on strains and imperfections, the domain pattern which is formed at first does not always correspond to an absolute minimum of the free energy. This state is only metastable.

The theoretical calculation of the domain wall energy is a very complicated problem that can be solved only in special simple cases, and connected with it is the problem of the thickness of the domain wall. Again, rather simple energetic considerations are sufficient to convince one that the wall is going to be thin. This is in striking contrast with the situation occurring in ferromagnetic crystals, where the domain wall consists of a transition region in which the magnetization vector turns over gradually from a given direction to the opposite one (see Fig. II-13a). This is so because the thickness of a ferromagnetic domain wall is a result of the compromise between the exchange energy and the anisotropy energy. The former would require a domain boundary as wide as possible, the latter attempts to do just the contrary, in order to avoid the directions of "hard magnetization", but the exchange term is orders of magnitude larger than the

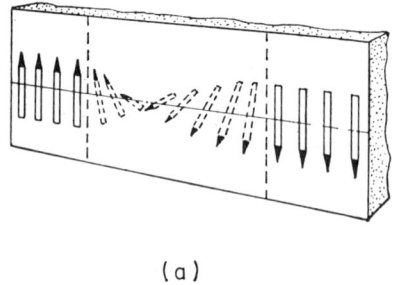 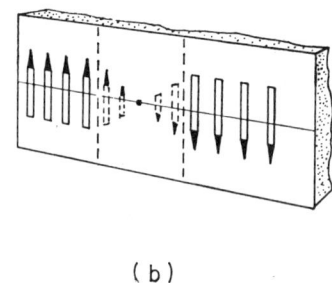

(a)                                                        (b)

FIG. II-13. Difference between ferromagnetic and ferroelectric 180° domain walls (schematic).
(a) Ferromagnetic wall. In iron, the thickness of the transition layer (comprised between dashed lines) is approximately 300 lattice constants.
(b) Ferroelectric wall. In most ferroelectrics, the thickness of the wall is of the order of one lattice constant.

anisotropy term and the result is a rather thick domain wall. In ferroelectric crystals, there is no real counterpart of the magnetic exchange energy. The polarization may be ionic or electronic, and the main role is played by the electrostatic interaction between polarized ions. We have already mentioned in Chapter I that in a ferroelectric crystal the interaction energies of parallel and antiparallel dipole arrays do not differ much from each other, thus allowing a narrow domain wall. Moreover, the anisotropy of most ferroelectrics, especially those with low symmetry like TGS, is so high that the direction of the spontaneous polarization can hardly deviate from that of the ferroelectric axis. Hence, the polarization vector will not rotate within the wall, but rather will decrease in magnitude, passing through zero, and increase on the other side with opposite sign (see Fig. II-13 b). The elastic energy plays also an important role, owing to the strain caused by the polarization through the piezoelectric effect, and this contribution will also favor, in general, a rather thin domain wall.

We can only hope to learn more about the factors causing the domain configuration actually observed in a ferroelectric crystal from a direct study of the ferroelectric domains. Such a study involves the observation of the geometry of the domains, i.e. their static properties, and the direct or indirect observation of the motion of the domains, i.e. their kinematic characteristics. The former can be carried out with optical or X-ray methods, the latter both with optical and electrical means.

### The Geometry of Domains in TGS

The ferroelectric domains of TGS cannot be observed directly under the polarizing microscope, as long as the investigated sample is unstrained. This is due to the fact that the symmetry of the non-polar phase is centrosymmetric (non-piezoelectric), so that the (electrostrictive) strain caused by an electric field along the $b$ axis remains the same, in magnitude and sign, when the field is reversed. Polarization reversal changes the crystal from a right-handed to a left-handed one, or vice versa, but does not displace the optic plane (S 1), (T 6).

The domains can, however, be suitably identified by means of etching techniques. This method, which was first applied to the study of ferroelectric domains by Hooton and Merz (H 2) on barium titanate (see Chapter IV) makes use of the well-known fact that the two ends of a polar axis have different properties. Thus, the dissolving (etching) rate of the positive end ("head") of the polar axis is different from that of the negative end ("tail") of the same axis. Clearly, this method enables us only to study the *surface* of a TGS plate cleaved perpendicularly to the polar axis, but we can, in this way, see how large the cross section of the domains is.

The effects of various etchants, such as water, methanol, acetone etc., on TGS plates have been reported by Toyoda and co-workers (T 4), (T 6), (T 7), who established that the positive end of the polar axis etches faster, and thus appears uneven, than the negative end which, therefore, has a smooth appearance. Figure II-14 shows a photograph of the etch patterns obtained by Konstantinova *et al.* (K 2) on the top and bottom surfaces of a thin plate of TGS cut perpendicular to the polar axis. One pattern is the mirror image of the other, showing that most of the domains penetrate throughout the crystal. The cross-sectional dimensions vary from a few-tenths of a millimeter to several millimeters. The polarity of adjacent domains can be checked with an electrometer, thus establishing that the sign is reversed in going from one domain to the adjacent one. The cross-section is mostly oval shaped and irregular.

A detailed investigation of domain patterns in TGS has been recently carried out by Chynoweth and Feldmann (C 6) using water as an etchant. The technique consists in immersing the polished sample in water for periods of time varying from a few to several seconds, depending on the temperature of the water bath. Despite its obvious limitations due to the high solubility of TGS in water, this technique has the advantage of a markedly high resolution in revealing fine details of the domain structure. As already indicated by the photograph in Fig. II-14, the small domains appear to have most frequently elliptical or lenticular shape, the major axes of the ellipses lying approximately in the direction of the optic plane (Fig. II-15a). Only the larger domains appear to run through the crystal from one surface to the other, but the domain walls are not always exactly parallel to the ferroelectric axis. The smaller domains terminate within the crystal plate in the form of spikes. This has been verified by etching the side of the crystal parallel to the ferroelectric axis and is shown in the photomicrograph of Fig. II-15(b). It is, incidentally, rather surprising, at first, to see that domains can be revealed in this way, as the polarization for either sign

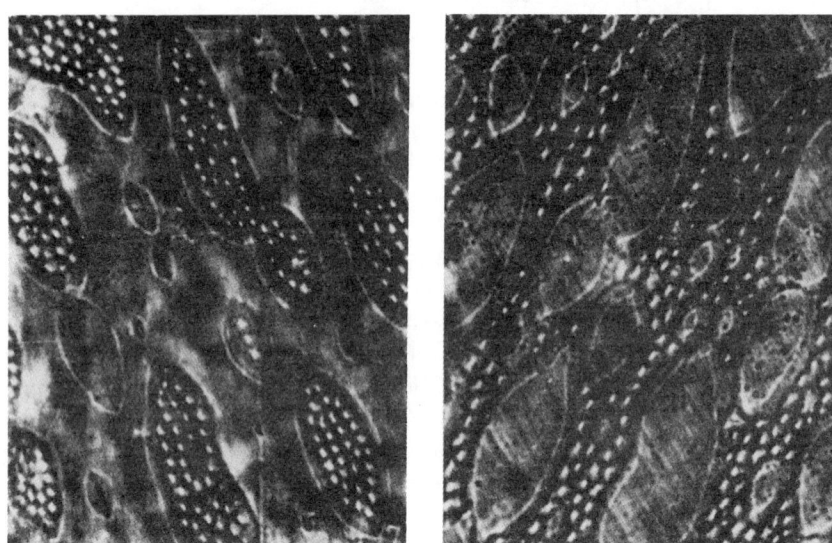

FIG. II-14. Photograph of the etched top and bottom surfaces of a plate of tri-glycine sulfate perpendicular to the polar axis (according to Konstantinova *et al.* (K 2)).

of domain is supposedly parallel to the surface. However, this fact may possibly be understood on the basis of the structural mechanism involved in the reversal of polarity (see Section 9). This technique of etching the side of the crystal confirms the fact that the larger domains run right through the crystal and, in addition, reveals often domains which are completely internal. As shown in Fig. II-16(a), these internal domains appear as long, thin, cigar-shaped regions pointed at both ends, lying along the ferroelectric axis, and are most probably lenticular in cross-section.

Chynoweth and Feldmann (C·6) were able to show that very regular arrays of domains are introduced into the crystals when they are subjected to thermal shocks. A cooling shock gives rise to many small spike-shaped nuclei extending into the crystal from the surfaces. A warming shock results in a regular array of considerably larger domains. The following model was proposed to account for the observed phenomena. Consider a non-electroded crystal plate initially in thermal equilibrium at the temperature $T_1$ and then subjected to a sudden cooling shock to a temperature $T_2(T_2 < T_1)$. Obviously, the surface of the plate will be the first to attain the new temperature $T_2$, and there will be a more or less sharp temperature boundary moving rapidly into the interior of the crystal plate. Owing to the temperature difference across the boundary there will be a

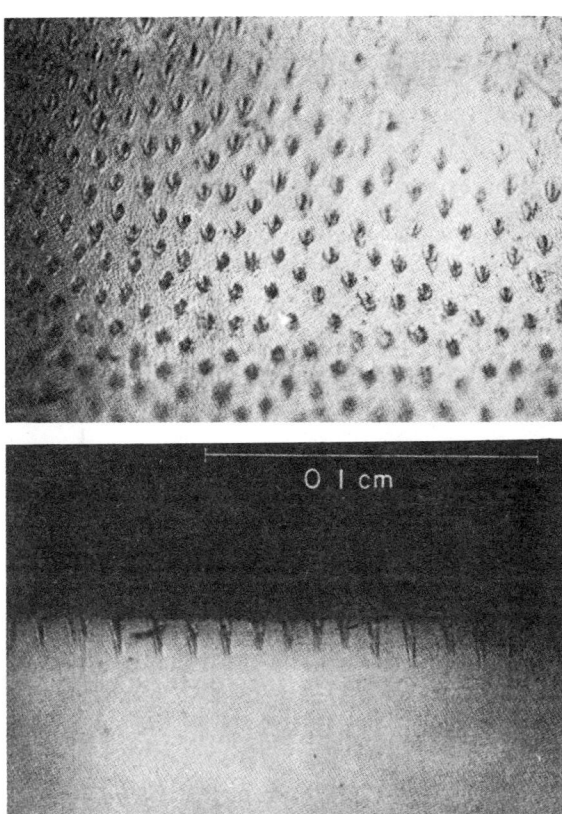

Fig. II-15. Fine detail of domain pattern of tri-glycine sulfate as revealed by the water-etching technique.
(a) Face perpendicular to the ferroelectric axis,
(b) Face parallel to the ferroelectric axis.
The photographs show that the fine dots in (a) represent the basis of small spike-shaped domains (according to Chynoweth and Feldman (C 6)).

discontinuity in the spontaneous polarization. The polarization discontinuities give rise to bound charges which will be compensated to some extent by free charges external to the crystal. These charges, in turn, give rise to fields both in the regions between the crystal surface and the temperature boundary, and in the interior of the plate between the two temperature boundaries. It is easy to see that the former fields oppose the polarization and so will cause domains to nucleate at the crystal surfaces; the latter fields (in the interior) are in the same direction as the polarization and so will not give rise to reverse polarized domains. In the warming shock, however ($T_2 < T_1$), by a similar argument the fields in the surface regions will be in the same direction as the polarization, but in the interior the field will·oppose the polarization and so can cause the nucleation of internal domains. Under the influence of the interior field these domains will tend to grow often as far as the surface of the crystal plate (C 6). This model

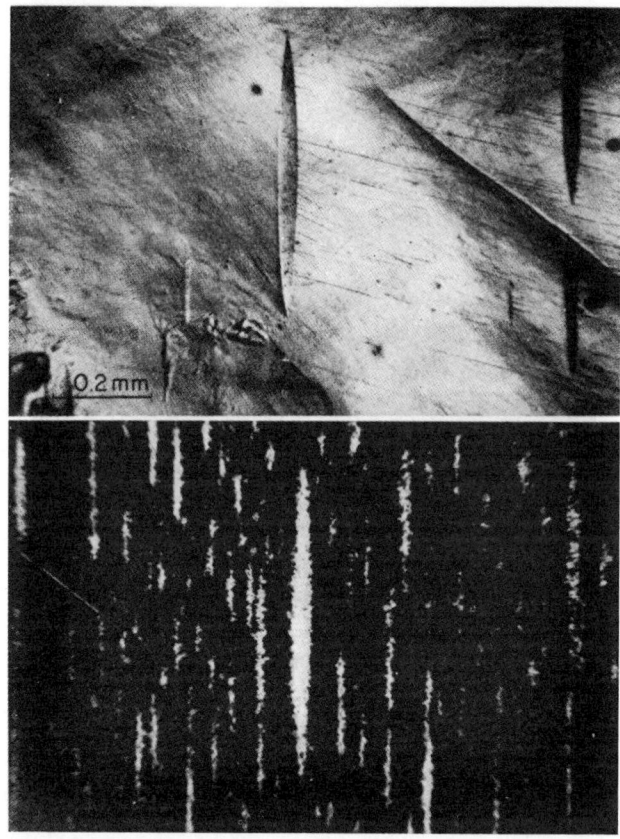

FIG. II-16. Ferroelectric domains in tri-glycine sulfate as revealed on a side of the crystal parallel to the ferroelectric axis, (a) by the water-etching technique, (b) by the powder deposition method. Note that each of the domains in (a) correlates with a domain in (b) but there are also many others in (b) which do not show in (a). This demonstrates the ability of the powder technique to reveal domains submerged below the crystal surface (according to Chynoweth and Feldman (C 6)).

seems to provide a satisfactory qualitative explanation for the domain patterns observed experimentally after thermal shocks.

Another technique for delineating ferroelectric domain patterns was reported by Pearson and Feldmann (P 2) and applied, among other crystals, to TGS. The technique consists in using a colloidal suspension of electrostatically charged powders in insulating organic liquids. It is found that yellow sulfur suspended in hexane deposits best on negative dipole ends, while red lead oxide deposits satisfactorily on positive dipole ends. The "powder patterns" of domains are obtained by placing a few drops of colloid solution on the crystal surface. After the hexane has evaporated, the powders are fixed in place indefinitely by the electrostatic charges. Figure II-17 shows a comparison between the results obtained with the powder technique (Fig. II-17 a) and those obtained with the etching technique (Fig. II-17 b). The correlation between the two patterns is complete. Figure II-16(b), on the other hand, shows that the powder deposition technique can also be successfully applied to the study of internal domains that do not cross the surface of (010) plates.

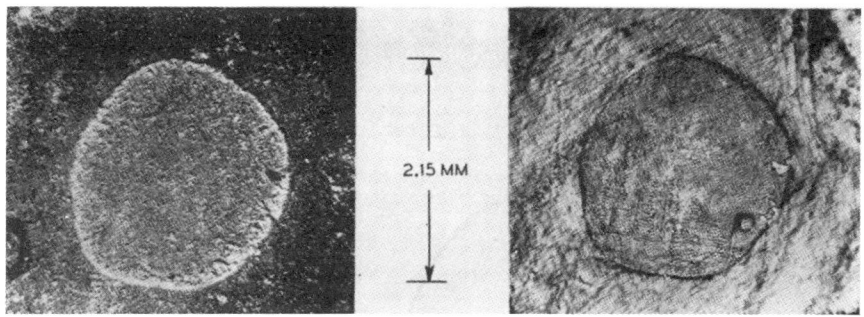

Fig. II-17. Comparison of powder and etch patterns on tri-glycine sulfate crystals. (a) Sulfur powder, (b) Same crystal etched in water. Circular areas are negative domains (according to Pearson and Feldman (P 2)).

### The Process of Polarization Reversal

It is quite evident that when the polarization of a ferroelectric crystal is reversed by means of an external electric field, the domain configuration must undergo radical changes. It has therefore become customary to speak of either polarization reversal, dynamic properties of the domains, or switching characteristics, when referring to the same physical process, namely the way in which polarity is reversed by an external field. A large amount of work has been done on various ferroelectrics, including TGS, in order to understand this process. Most of the methods of investigations have been developed for the work on $BaTiO_3$ (see Section IV-8), and subsequently applied to other ferroelectrics. These methods may be classified roughly into two main groups, viz. (i) direct observation of the motion of domain walls during polarization reversal, and (ii) measurement of the switching transient under pulsing conditions.

Investigations of the latter type have been carried out on TGS by Wieder (W 1), Fatuzzo (F 1), Pulvari and Kuebler (P 3), Toyoda et al. (T 3), Kiyasu Zen'iti et al. (Z 1) and Fatuzzo and Merz (F 2). The technique mostly employed is that first developed by Merz (M 2) for similar investigations on barium titanate. This technique consists in applying to the crystal a train of rectangular voltage pulses of successive opposite polarities in such a way that each pulse reverses the sign of the previous remanent polarization. The rise time of the pulses must be short (of the order of $10 \, m\mu/sec$) as compared to the time required for the field to reverse the polarization of the sample. The amplitude and the duration of the pulses must be sufficiently large to cause the amount of reversed polarization to be substantially $2P_s$. Finally, the intervals between pulses must be long enough for the crystal to reach an equilibrium state between successive pulses. The pulses are applied to the crystal in series with an ohmic resistance, and the switching current transient developed across this resistance is displayed on an oscilloscope. Figure II-18 shows the switching transient observed under these conditions and defines the switching time $t_s$ as the time required for the switching current to drop to zero. The faster the dipoles switch, the larger the peak current $i_{max}$.

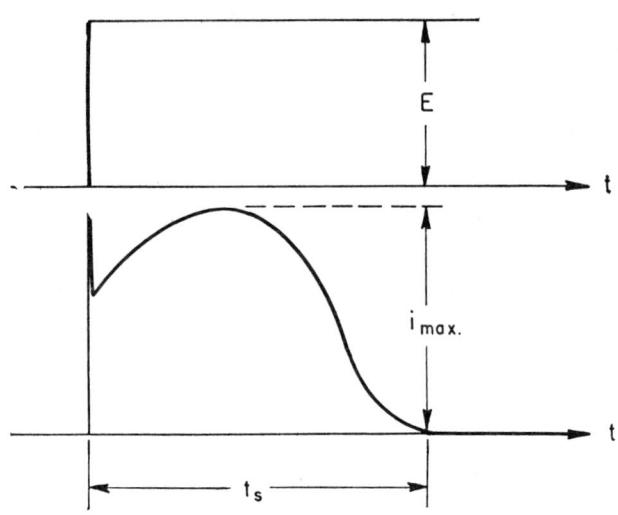

Fig. II-18. Pulsing field and pulsing current as functions of time (schematic).

Figure II-19 shows a plot of the reciprocal of the switching time $t_s$ in TGS as a function of field (F 2). A similar figure can be obtained by plotting the peak current $i_{max}$ as a function of field (W 1). The figure shows that there are three regions of different behavior of the crystal, depending on the applied field. At low field strengths (smaller than about 7 kV/cm) the switching time $t_s$ follows an exponential law of the form

$$\frac{1}{t_s} = \frac{1}{t_\infty}\,e^{-\alpha/E}, \tag{II-16}$$

where the activation field $\alpha$ is a function of temperature: at 30 °C, $\alpha = 3.8$ kV/cm (W 1). At higher fields the switching mechanism seems to change: we find a linear behavior of $1/t_s$ (and $i_{max}$) with $E$:

$$\frac{1}{t_s} = \text{constant} \times E. \tag{II-17}$$

These results can be interpreted in terms of two mechanisms responsible for the switching process: nucleation of domains, and domain wall motion. Following Fatuzzo and Merz (F 2), we define the nucleation time $t_n$ as the time necessary to form all nuclei, and the domain wall motion time $t_d$ as the time necessary for one domain to grow through the sample. Then, we can write that the total switching time is approximately given by

$$t_s \cong t_n + t_d. \tag{II-18}$$

Thus, if the two quantities $t_n$ and $t_d$ are not of the same order of magnitude, the switching time $t_s$ is determined principally by the slower of the two mechanisms.

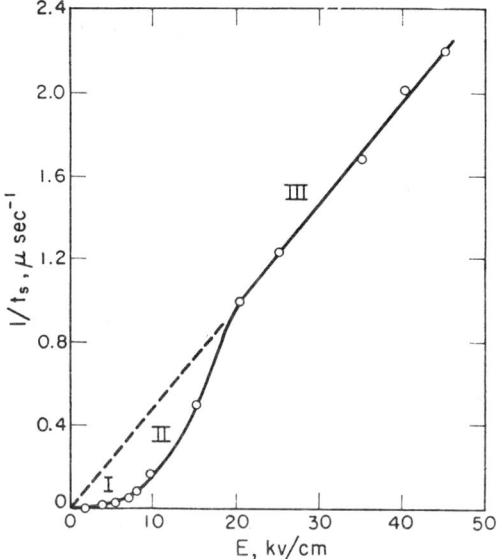

Fig. II-19. Reciprocal switching time $t_s$ of tri-glycine sulfate as a function of field (according to Fatuzzo and Merz (F 2)).

At low fields (region I in Fig. II-19), the rate of nucleation is so low that the switching is primarily governed by nucleation of new domains $(t_n \gg t_d)$. Assuming that, at low fields, the probability of forming new domains depends exponentially upon the applied field, thus:

$$p_n = p_0\, e^{-\alpha/E}, \tag{II-19}$$

it follows that

$$\frac{1}{t_s} \cong \frac{1}{t_n} = \frac{1}{t_\infty}\, e^{-\alpha/E}. \tag{II-20}$$

At high field strengths, on the other hand (region III of Fig. II-19), the nucleation rate is so large that the switching time is primarily determined by the velocity of the domain walls $(t_d \gg t_n)$. Assuming that the domain wall motion can be described by (F 2):

$$v = \frac{d}{t_d} = \mu E \tag{II-21}$$

where $d$ is the distance covered by the wall in motion (thickness of the sample), and $\mu$ the mobility, we obtain

$$\frac{1}{t_s} \simeq \frac{1}{t_d} = \frac{\mu}{d}\, E = \frac{\mu}{d^2}\, V. \qquad (\text{II-22})$$

This implies that, in this region of high fields, the switching time should depend quadratically upon the thickness of the sample, a dependence that was confirmed by the experimental results of Fatuzzo and Merz for applied fields larger than about $10^4$ V/cm. The mobility $\mu\,(T)$ can be determined as a function of temperature from graphs such as Fig. II-19 plotted for different temperatures. At 30 °C, $\mu = 2.2$ cm²/V sec (W 1). The temperature dependence of the reciprocal mobility $\beta = 1/\mu$ is depicted in Fig. II-20. The rapid increase of $\beta$ below about $-10$ °C

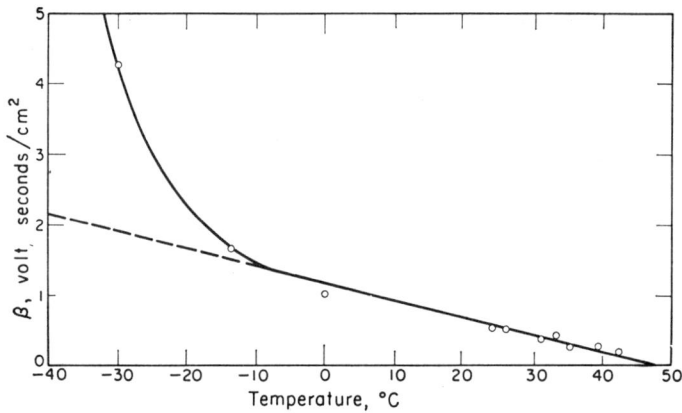

FIG. II-20. Reciprocal mobility $1/\mu = \beta$ in tri-glycine sulfate as a function of temperature (according to Wieder (W 1)).

may be compared with the analogous behavior of the coercive field at these temperatures (Fig. II-5). If the domain wall mobility $\mu$ is independent of the applied field, then the observed quadratic dependence of the switching time upon crystal thickness indicates that, at high fields, the forward motion of the domain walls is very much faster than any sidewise motion.

At low fields, on the other hand, sidewise expansion of the nucleated domains may bring a considerable contribution to the switching transient. This has been established by Chynoweth and Abel (C 2) using fields of the order of 30–35 V/cm on specimens provided with liquid electrodes (a saturated solution of TGS and sodium sulfate in normal octyl alcohol), rather than noble metal electrodes. The domains were observed, after any suitable electrical treatment of the sample, by means of the powder deposition technique described above. The experiments proved that, in the field range indicated above, polarization reversal in TGS is usually accomplished by the formation and sideways expansion of a few domains; in some extreme cases, switching was found to occur by the sideways growth of a single domain. The mechanism by which such a sidewise motion of the wall can occur is not completely understood, at this stage, but there are indications that it may be the result of a continuous formation of new nuclei along the edges of a switched region (C 2), (F 2). This, in turn, may indicate that the probability of formation of new nuclei is affected by the number of growing and already grown domains, an effect that has been called the "nucleus domain interaction" by Fatuzzo and Merz (F 2), and seems to play a role in all ferroelectric reversals, moderate in TGS, very important in $BaTiO_3$ (see Section IV-8), and $LiH_3(SeO_3)_2$ (see Section VIII-8).

It would seem reasonable to expect that the sites at which new domains are nucleated could be predetermined by locally increasing the applied field or decreasing the coercive strength. However, it was proved by Chynoweth and Abel (C 3) that this is not the case. Local increase of the applied field could be achieved by sandblasting small dimples on one surface of the sample; local decrease of the coercive force could be attained by locally damaging a poled crystal with X-rays (C 1). In neither case was it possible to predetermine the nucleation sites at low applied fields. It appears, therefore, that these sites are primarily determined by singularities that are somehow built into the crystal. They seem to have a stronger influence on domain nucleation than the local singularities produced artificially, and are to be considered the cause for the observed reproducibility (C 2) of the domain patterns of crystals that have been alternatively switched partially and totally. It is not possible to say, at this stage, just what these built-in singularities are, but there are reasons to believe that they may be more concentrated on the surface of the sample. In fact, Fatuzzo and Merz (F 2) were able to conclude, from studies of the effect of inhomogeneous temperature changes upon the shape of the switching transients, that heating of the surface alone accelerates the nucleation of new domains, whereas heating of the bulk of the crystal enhances the domain wall motion. It has been suggested (W 1) that the nucleation sites may be crystal imperfections and, as such, may vary in number from sample to sample. Wieder (W 1) has in fact found variations of the value of the activation field $\alpha$ up to 50% in different crystals.

The model of a switching mechanism based upon nucleation and growth of antiparallel domains was investigated in detail by Fatuzzo and Merz (F 2). The nucleated domains grow through the thickness of the crystal in the form of cones until the apices reach the opposite electrode. At this point the walls become perpendicular to the electrode surfaces and the domains assume a more cylindrical shape. Sideway expansion is thereafter aided, most probably, by the nucleus domain interaction which enhances nucleation adjacent to an already existing wall. An analytical treatment of this model let Fatuzzo and Merz (F 2) to satisfactory agreement between calculated and observed switching transients in TGS.

## 8. Piezoelectric and Elastic Properties

Elastic and piezoelectric measurements of TGS crystals were carried out at room temperature by Konstantinova et al. (K 1), who reported complete sets of the elastic constants, elastic compliances and piezoelectric moduli. A partial list of values of these coefficients was also published by Husimi and Kataoka (H 3). There appears to be no agreement among the values reported by different authors. Husimi and Kataoka (H 3) do not specify the reference system of co-ordinates used in their work, and it is possible that it may differ from that assumed by Konstantinova *et al.* (K 1). However, the discrepancies between the published values of the piezoelectric modulus $d_{22}$, and those of the elastic compliances $s_{22}$ seem to indicate that the disagreement is likely to be due to causes other than axial choice. In no case is it specified, for example, whether the elastic quantities reported represent constant-field or constant-polarization values. The temperature dependence of some of the piezoelectric moduli $d_{ik}$ was found to be strongly anomalous at the Curie point (H 3).

## 9. Structural Characteristics

Crystallographic data for tri-glycine sulfate, $(NH_2CH_2COOH)_3 \cdot H_2SO_4$, were reported by Wood and Holden (W 2) as follows:

$$a' = 9.15 \text{ Å}; \quad b' = 12.69 \text{ Å}; \quad c' = 5.73 \pm 0.03 \text{ Å}; \quad \beta = 105° 40' \pm 20'.$$

The space group, at room temperature, is $C_2^2 - P2_1$ and the experimental density $\varrho = 1.69$ g cm$^{-3}$. There are two formula units in the unit cell. The crystal is optically negative, with $2V = 40 \pm 5°$. The optic plane is almost parallel to the $(\bar{1}02)$ plane. The refractive indices are 1.59, 1.49, 1.56, respectively (T 5).

A detailed structural investigation of the polar room-temperature phase was carried out by Hoshino *et al.* (H 4) using three-dimensional X-ray diffraction data. The axial settings selected for this work,

$$a = 9.42 \text{ Å}; \quad b = 12.64 \text{ Å}; \quad c = 5.73 \text{ Å}; \quad \beta = 110° 23',$$

are related to those reported above by the transformation formulas:

$$\boldsymbol{a} = \boldsymbol{a}' + \boldsymbol{c}', \quad \boldsymbol{b} = \boldsymbol{b}', \quad \boldsymbol{c} = -\boldsymbol{c}'.$$

A projection of the structure along the $c$ axis is given in Fig. II-21.

The understanding of this structure may be facilitated by a preliminary discussion of the crystal–chemical characteristics of the glycine molecule, $NH_2CH_2COOH$. When pure glycine crystallizes in its most common modification, the molecular groups assume the so-called zwitter-ion structure, which is commonly written in the form $NH_3^+ CH_2COO^-$ to indicate the simultaneous occurrence of positive and negative charges at different points of the molecular group. Furthermore, it is found that the two carbon and the two oxygen atoms lie all in the same plane, while the nitrogen atom is displaced out of the plane (M 3). This is, however, not the only configuration assumed by the glycine group in crystals. It is possible to encounter structures in which the carbon, oxygen and nitrogen atoms are coplanar. For example, the structure of di-glycine chloride shows that both configurations of the glycine group are present in this crystal: the completely planar configuration and the partially planar configuration (with the nitrogen atom displaced from the plane containing the remaining four atoms) (H 5).

Now, the structure of TGS shows that, of the three glycine groups contained in the assymmetric unit (designated with the symbol $A$ in Fig. II-21), two assume the completely planar configuration and one assumes the partially planar configuration. In Fig. II-21, the glycine groups designated with I and III are completely planar, the glycine group designated with II is only partially planar: the nitrogen atom is displaced from the plane of the two C's and the two O's of the same group by 0.27 Å. Furthermore, the completely planar groups are

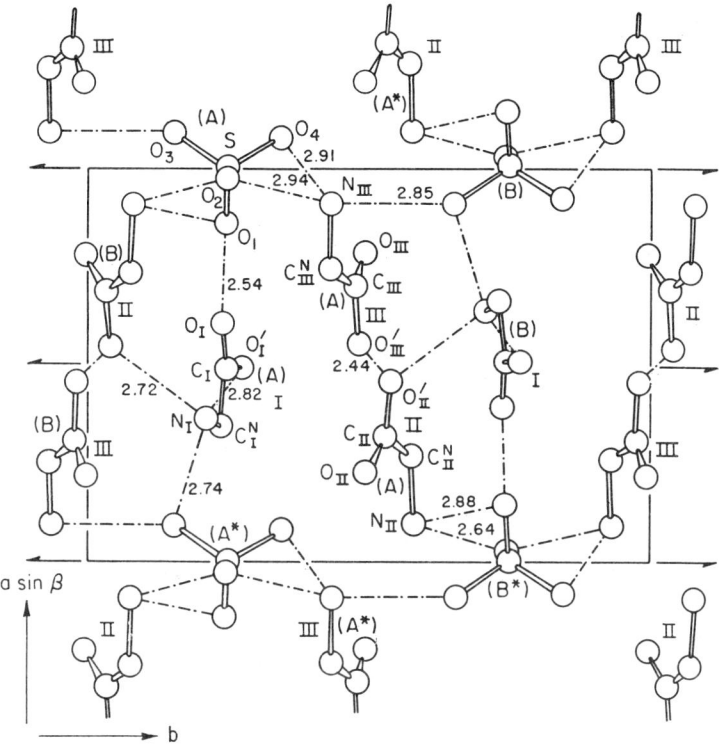

Fig. II-21. Projection of the structure of tri-glycine sulfate along the $c$ axis (according to Hoshino *et al.* (H 4)).

monoprotonated, having taken the protons from the $H_2SO_4$ group, and are thus best designated as glycinium ions, whereas the partially planar glycine group assumes the zwitter-ion configuration. For this reason, Hoshino *et al.* (H 4) propose to call the compound glycine di-glycinium sulfate and describe it with the chemical formula $(NH_3^+ CH_2COO^-)(NH_3^+ CH_2COOH)_2 \cdot SO_4^{2-}$.

The molecular plane of the completely planar glycinium ions I makes an angle of 12.5° with the plane at $b = 1/4$. This plane (and that at $b = 3/4$) becomes a crystallographic mirror plane above the Curie temperature, but only statistically, due to random orientation of the glycinium ions I.

As far as the rest of the structure is concerned, the X-ray data indicate a distortion of the sulfate ion from tetrahedral symmetry, at room temperature, and lead to the *assumption* of a rather complicated hydrogen bond system which is tentatively suggested by the chain lines inserted between atoms in Fig. II-21. Confirmation of this proposed hydrogen-bond system will have to await the results of neutron diffraction analyses. One short hydrogen bond, however, appears to be already established with certainty (H 4), and this is the bond between $O'_{III}$ ($A$) and $O'_{II}(A)$ in Fig. II-21 (bond length 2.44 Å).

In the configuration depicted in this figure, according to the formulation of Hoshino *et al.*, the proton is closer to the oxygen $O'_{III}(A)$ of the completely planar glycinium ion III. This is actually the reason *why* the group III is considered the glycinium ion, while the glycine group II is considered the zwitter-ion.

The molecular mechanism of polarization reversal in TGS is then described as follows (H 4). A reversal of the H association from group III to group II along the short $O'_{III}(A) - H - O'_{II}(A)$ bond involves a reversal of the former glycinium–glycine roles of these two groups: This means that group III becomes the zwitter-ion and group II becomes the glycinium ion. The $N_{II}(A)$ atom moves back into the plane of the remaining atoms, while the $N_{III}(A)$ atom moves away from its corresponding plane. These changes affect the position of the glycinium ion I via the complex hydrogen-bond system relating the nitrogens of the glycine groups with the oxygens of the sulfate groups. The coplanar glycinium ion I, inclined at 12.5° with respect to the plane at $b = 1/4$, flips over into the symmetrical position with respect to this same plane. As the nitrogen atom of this glycinium group lies about 0.5 Å away from the plane at $b = 1/4$, this would imply a remarkably large shift of the nitrogen (about 1 Å) along the direction of the monoclinic $b$ axis. However, since the reported temperature factor of this atom is quite large (H 4) it is possible that an eventual interference between temperature factor and positional parameter would make the reported value of the latter rather inaccurate.

The importance of the role played by the glycine groups in the process of polarization onset and reversal is indirectly confirmed by the Raman investigations of TGS. The interpretation of the Raman spectra (T 5), (K 3) confirms the X-ray result that the $SO_4$ groups do not have tetrahedral symmetry at room temperature. The observed Raman lines also allow the conclusion that these groups maintain a mirror plane at room temperature. Thus, it follows that the $SO_4$ groups cannot bear an electric moment parallel to the twofold axis and the phenomenon of ferroelectricity must be attributed to the glycine groups (T 5).

The essential features of the molecular mechanisms proposed on the basis of the structural investigations appear, therefore, quite reasonable. No model theory has been developed, to date, on this ground.

## Character of the Transition

A theoretical and experimental study of the characteristics of the ferroelectric transition in TGS has been carried out by Shibuya and Mitsui (S 3) through investigation of the diffuse X-ray scattering that is caused by disorder in the crystal. Both the dielectric data and the result of the structural analysis indicate that the ferroelectric transition in TGS (similarly to that in $KH_2PO_4$ and in Rochelle salt), is probably of the order–disorder type, in contrast to the so-called displacive transitions exhibited by the ferroelectric perovskites. The basic features of an order–disorder transition is the existence of rotatable dipoles, or ions with two equilibrium positions, which become ordered below a critical temperature. Such a co-operative phenomenon has some aspects in common with the order–disorder transitions in alloys and the ferromagnetic phase changes, but several features of the ferroelectric phenomenon differ drastically from the other co-operative processes. One of these features is the strong anisotropy of the dipole interaction forces. In TGS, the structural studies have revealed that dipole interaction is strong along the $b$ and $c$ axes, but very weak along the $a$ axis. For the case of isotropic dielectrics, Fröhlich (F 3) has shown that the dielectric constant can be related to thermal fluctuations of the polarization. More recently, the theory of thermal fluctuations of electrical polarization in ferroelectrics has been considered by Burgess (B 1) with special attention to temperatures in the vicinity of the Curie point. It seems thus reasonable to think that one can get some information about the character of the transition from a study of diffuse scattering of X-rays.

The procedure followed by Shibuya and Mitsui (S 3) is to *assume* that the structure of TGS consists of a mixture of two basic structures, which differ from one another in the positive or negative polarity of certain statistical units. These statistical units are assumed to be the crystallographic unit cells. Above the Curie point, the two basic structures are present in the same proportions, thus producing, in a macroscopic crystal, a center of symmetry. Below the Curie point, the proportions change in favor of one or the other of the two basic structures, thus giving rise to a spontaneous polarization. This implies the formation of dipole clusters, and the degree of order can be characterized by a long-range order parameter $S$ defined as

$$S = \frac{N_+ - N_-}{N_+ + N_-},$$

where $N_+$ and $N_-$ are the number of cells with positive and negative polarity, respectively. $S$ can be determined experimentally by noting its relation to the spontaneous polarization:

$$S = \frac{(N_+ - N_-)\mu}{(N_+ + N_-)\mu} = \frac{P_s}{P_0}$$

where $\mu$ is the dipole moment and $P_0$ is the saturation value of the spontaneous polarization at 0 °K, or at very low temperatures. It turns out that, at 22 °C, $S = 0.77$, i.e. at this temperature the actual structure of a TGS crystal consists of 11.5% of one and 88.5% of the other basic structure.

Assuming, however, as a first approximation, that the structure revealed by the X-ray analysis is that of a basic structure, one can derive theoretical expressions for the intensities of both Bragg reflections and diffuse scattering. Taking the origin of a reference system at the center of a symmetry of the superposition of the two basic structures, the X-ray structure factors of the structures with positive and negative polarities can be written as

$$F_+ = A - iB \text{ and } F_- = A + iB,$$

and the intensity of the diffuse scattering by

$$I_c \propto B^2 \times \varepsilon.$$

It should be noted that, owing to the choice of the origin, $B$ is not the imaginary part of the structure factor of the actual crystal but that of the basic structure and thus $B$ has a finite value also above the Curie point. The direct proportionality of $I_c$ with the dielectric constant predicts an anomaly of the diffuse scattering at the Curie point.

The experimental observations are in satisfactory agreement with these theoretical predictions (S 3). The observed diffuse scattering exhibits a pronounced peak at the Curie point. On the basis of these results, it can thus be confirmed that the ferroelectric phenomenon in TGS is due to a transition from a disordered to an ordered state.

### 10. Isomorphous Compounds

As mentioned in Section 1, two compounds isomorphous with TGS exhibit the same type of ferroelectric transition as TGS, viz. tri-glycine selenate (M 1) and tri-glycine fluoberyllate (P 1) Crystallographic data for these compounds were reported by Pepinsky *et al.* (P 1). The phenomenological behavior, so far as known, is similar in every respect to that of TGS (H 1) and the same holds, most probably, for the molecular mechanism involved. Table II-1 lists some of the pertinent data concerning TGS and its ferroelectric isomorphs. Pulsing experiments on the fluoberyllate have been carried out by Pulvari and Kuebler (*P*3) and the effect of hydrostatic pressure on the selenate has been studied by Jona and Shirane (J 1). Mixed crystals of tri-glycine sulfate and selenate have been grown and investigated by Fatuzzo and Nitsche (F 4). The Curie temperature was found to vary linearly with composition, so that it is possible to prepare crystals having any desired Curie temperature in the range between 22° C and 49 °C. The dielectric properties of partially deuterated tri-glycine sulfate have been studied by Chapelle and Taurel (C 7).

TABLE II-1. DATA ON FERROELECTRIC TRI-GLYCINE SULFATE AND ISOMORPHS

| Name of compound (Abbreviation) Chemical formula | Transition temperature (°C) | Spontaneous polarization $P_s$ at $T$ °C (C/cm²) | $\xi$ coefficient of free energy expansion (e.s.u./cm²)⁻² | Curie constant (°K) | Reference |
|---|---|---|---|---|---|
| Tri-glycine sulfate (TGS) $(NH_2CH_2COOH)_3 \cdot H_2SO_4$ | 49 | $2.8 \times 10^{-6}$ at 20 °C | $8.5 \times 10^{-10}$ | 3200 | M 1, H 1, T 1 |
| Tri-glycine selenate (TGSe) $(NH_2CH_2COOH)_3 \cdot H_2SeO_4$ | 22 | $3.2 \times 10^{-6}$ at 0 °C | $4.7 \times 10^{-10}$ | 4000 | M 1, J 1 |
| Tri-glycine fluoberyllate (TGFB) $(NH_2CH_2COOH)_3 \cdot H_2BeF_4$ | 70 | $3.2 \times 10^{-6}$ at 20 °C | $13.5 \times 10^{-10}$ | 2350 | H 1 |

## BIBLIOGRAPHY

(A 1)  ABE, R., *J. Phys. Soc. Japan* **15**, 795 (1960).

(B 1)  BURGESS, R. E., *Can. J. Phys.* **36**, 1569 (1958).

(C 1)  CHYNOWETH, A. G., *Phys. Rev.* **113**, 159 (1959).

(C 2)  CHYNOWETH, A. G. and ABEL, J. L., *J. Appl. Phys.* **30**, 1073 (1959).

(C 3)  CHYNOWETH, A. G. and ABEL, J. L., *J. Appl. Phys.* **30**, 1615 (1959).

(C 4)  CHYNOWETH, A. G., *Phys. Rev.* **117**, 1235 (1960).

(C 5)  CHAPELLE, J. and TAUREL, L., *Compt. rend.* **249**, 378 (1959).

(C 6)  CHYNOWETH, A. G. and FELDMAN, W. L., *J. Phys. Chem. Solids* **15**, 225 (1960).

(C 7)  CHAPELLE, J. and TAUREL, L., *Compt. rend.* **249**, 1332 (1959).

(D 1)  DOMANSKI, S., *Proc. Phys. Soc. (London)* **B 72**, 306 (1958).

(D 2)  DROUGARD, M. E., LANDAUER, R. and YOUNG, D. R., *Phys. Rev.* **98**, 1010 (1955).

(D 3)  DEVONSHIRE, A. F., Theory of Ferroelectrics, *Phil. Mag.* Suppl. **3**, 85, (1954).

(E 1)  EHRENFEST, P., *Proc. Acad. Sci. Amsterdam* **36**, 153 (1933).

(E 2)  EZHKOVA, Z. I., ZHDANOV, G. S. and UMANSKIJ, M. M., *Kristallografiya* **4**, 249 (1959).

(F 1)  FATUZZO, E., *Helv. Phys. Acta* **31**, 309 (1958).

(F 2)  FATUZZO, E. and MERZ, W. J., *Phys. Rev.* **116**, 61 (1959).

(F 3)  FRÖHLICH, H., *Theory of Dielectrics*, ch. II, sec. 7, Oxford University Press, London (1958).

(F 4)  FATUZZO, E. and NITSCHE, R., *Z. Elektrochem.* **63**, 970 (1959).

(F 5)  FATUZZO, E., *Helv. Phys. Acta* **33**, 501 (1960).

(H 1)  HOSHINO, S., MITSUI, T., JONA, F. and PEPINSKY, R., *Phys. Rev.* **107**, 1255 (1957).

(H 2)  HOOTON, J. A. and MERZ, W. J., *Phys. Rev.* **98**, 409 (1955).

(H 3)  HUSIMI, K. and KATAOKA, K., *J. Phys. Soc. Japan* **14**, 105 (1959).

(H 4)  HOSHINO, S., OKAYA, Y. and PEPINSKY, R., *Phys. Rev.* **115**, 323 (1959).

(H 5)  HALM, T. and BUERGER, M. J., *Z. Krist.* **108**, 419 (1957).

(J 1)  JONA, F. and SHIRANE, G., *Phys. Rev.* **117**, 139 (1960).

(K 1)  KONSTANTINOVA, V. P., SIL'VESTROVA, I. M. and ALEKSANDROV, K. S., *Kristallografiya* **4**, 69 (1959).

(K 2)  KONSTANTINOVA, V. P., SIL'VESTROVA, I. M. and YURIN, V. A., *Kristallografiya* **4**, 125 (1959).

(K 3)  KRISHNAN, R. S. and BALASUBRAMIAN, K., *Proc. Indian Acad. Sci.* **A 48**, 138 (1958).

(L 1)  LURIO, A. and STERN, E., *J. Appl. Phys.* **31**, 1125 (1960).

(M 1)  MATTHIAS, B. T., MILLER, C. E. and REMEIKA, J. P., *Phys. Rev.* **104**, 849 (1956).

(M 2)  MERZ, W. J., *Phys. Rev.* **95**, 690 (1954).

(M 3)  MARSH, R. E., *Acta Cryst.* **11**, 654 (1958).

(M 4)  MILLER, R. C. and HEIDENREICH, R. D., *J. Appl. Phys.* **29**, 957 (1958).

(N 1)  NITSCHE, R., *Helv. Phys. Acta* **31**, 306 (1958).

(N 2)  NISHIOKA, A. and TAKEUCHI, M. *J., Phys. Soc. Japan* **14**, 971 (1959).

(N 3)  NAKAMURA, E. and FURUICHI, J., *J. Phys. Soc. Japan* **15**, 2101 (1960).

(P 1)  PEPINSKY, R., OKAYA, Y. and JONA, F., *Bull. Am. Phys. Soc.* **2**, No. 4, 220 (1957).

(P 2)  PEARSON, G. L. and FELDMAN, W. L., *J. Phys. Chem. Solids* **9**, 28 (1959).

(P 3)  PULVARI, C. F. and KUEBLER, W., *J. Appl. Phys.* **29**, 1742 (1958).

(S 1)  SHUVALOV, L. A., ALEKSANDROV, K. S., and ZHELUDEV, I.S., *Kristallografiya* **4**, 130 (1959).

(S 2)  SAVAGE, A. and MILLER, R. C., *J. Appl. Phys.* **30**, 1646 (1959).

(S 3)  SHIBUYA, I. and MITSUI, T., *J. Phys. Soc. Japan* **16**, 479 (1961).

(T 1)  TRIEBWASSER, S., *I.B.M.J. Research Developm.* **2**, 212 (1958).

(T 2)  TAUREL, L., POUREL, E. and THOMASSIN, F., *Compt. rend.* **246**, 70 (1958).

(T 3)  TOYODA, H., WAKU, S., SHIBATA, H. and TANAKA, Y., *J. Phys. Soc. Japan* **14**, 109 (1959).

(T 4)  Toyoda, H., *J. Phys. Soc. Japan* **14**, 376 (1959).

(T 5)  Taurel, L., Delain, C. and Guérin, C., *Compt. rend.* **246**, 3042 (1958).

(T 6)  Toyoda, H., Waku, S. and Hirabayashi, H., *J. Phys. Soc. Japan* **14**, 1003 (1959).

(T 7)  Toyoda, H., *J. Phys. Soc. Japan* **15**, 1539 (1960).

(W 1)  Wieder, H. H., U. S. Naval... Ordnance Lab., Corona, Calif., Tech. Memo No. 42–6, September (1957).

(W 2)  Wood, E. A. and Holden, A. N. *Acta Cryst.* **10**, 145 (1957).

(W 3)  Wieder, H. H., *J. Appl. Phys.* **30**, 1010 (1959).

(Z 1)  Kiyasu Zen'iti, Husimi, K. and Kataoka, K. *J. Phys. Soc. Japan* **13**, 661 (1958).

# CHAPTER III

# POTASSIUM DIHYDROGEN PHOSPHATE AND ISOMORPHOUS CRYSTALS

## 1. Introduction

Potassium di-hydrogen phosphate, $KH_2PO_4$, often referred to in abbreviated form as KDP, crystallizes in the tetragonal system at room temperature. This phase belongs to the non-centrosymmetrical point group $\bar{4}2m$; therefore, it is piezoelectric. A transition occurs at 123 °K into a ferroelectric phase with ortho-rhombic symmetry and point group $mm$. The polar axis lies along the direction of the $c$ axis of the tetragonal phase. The ferroelectric activity was first reported by Busch and Scherrer (B 1) in 1935, and since then a very large number of papers have been published on the subject of the physical properties of this crystal.

A number of isomorphous compounds, obtained by substitution of potassium by rubidium or cesium, and phosphorous by arsenic, have also been found to undergo ferroelectric transitions at low temperatures. These compounds are dis-cussed in Section 6 of this Chapter. Substitution of potassium by ammonium leads to the isomorphous ammonium di-hydrogen phosphate, $NH_4H_2PO_4$ (abbreviated ADP) which undergoes a transition to an antipolar state at $-125$ °C, the anti-polar direction being parallel to one of the $a$ axis of the room-temperature tetra-gonal phase.

Large, colorless single crystals of KDP can be grown easily from water solu-tions, mainly by slow evaporation of a saturated solution at constant temperature. A detailed study of the conditions under which very large, clear crystals can be grown was carried out by Walker (W 1) for the case of ADP and the results are useful in part for the case of KDP. The crystals are prismatic, elongated along the [001] axis, with large (100) and (101) faces.

## 2. Dielectric Properties and Non-linearity

The dielectric constant measured along the $a$ axis ($\varepsilon_a$), and that along the $c$ axis ($\varepsilon_c$), are both of the order of 50 at room temperature. Upon cooling, $\varepsilon_c$ in-creases hyperbolically and reaches a very high value, of the order of $10^5$, at the transition temperature $T_c$ (see Fig. III-1). Below this temperature, $\varepsilon_c$ drops quite sharply, but not discontinuously. This drop can be explained on the basis of the dielectric saturation effects observed above the Curie point (see Section II-3), since the dielectric constant decreases when the crystal is polarized with an external d.c. field. In the ferroelectric region, the polarization is spontaneous and causes a decrease of $\varepsilon_c$ with increasing polarization.

The component $\varepsilon_a$ also shows an anomaly at the transition temperature, although much less pronounced than that exhibited by $\varepsilon_c$. This is an indication that the dielectric co-operative phenomenon also affects the component along the tetragonal $a$ axis.

In a temperature range of about $50\,^\circ$ C above the transition temperature, the dielectric constant $\varepsilon_c$ follows the Curie–Weiss law: $\varepsilon_c \cong C/(T - T_0)$ with $C = 3250\,^\circ$ and $T_0 = -150\,^\circ$C (B 3). Thus, the Curie–Weiss temperature $T_0$ coincides with the transition temperature $T_c$.

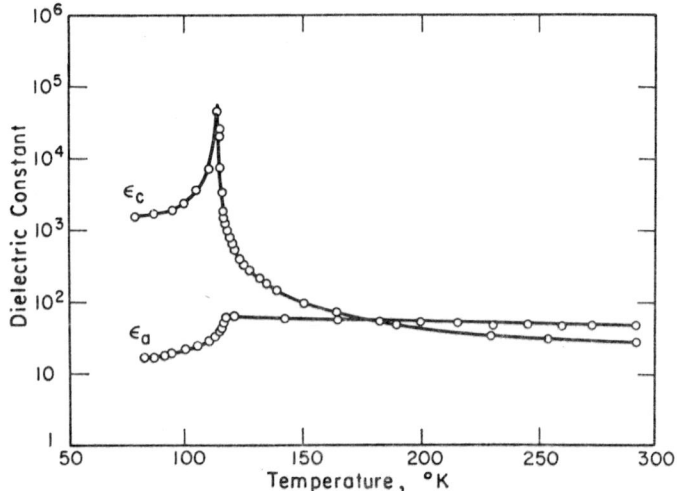

Fig. III-1. Dielectric constant of $KH_2PO_4$ as a function of temperature (according to Busch (B 2)).

The Curie–Weiss law is obeyed for measuring field frequencies up to the order of $10^7$ c/s (B 3). No relaxation of the dielectric constant $\varepsilon_c$ is found, except for the dispersion due to piezoelectric resonance, up to frequencies of the order of $10^{10}$ c/s (M 1). The measurement of the dielectric constant with a low-amplitude a.c. field of very high-frequency yields information about the dielectric behavior of the clamped crystal. At frequencies that are above the piezoelectric dispersion, mechanical resonance is suppressed and the crystal is clamped by inertia. It should be noted, however, that below the Curie temperature, the crystal is only partially clamped because the spontaneous polarization cannot be suppressed with this technique. Therefore, the application of high-frequency fields does not affect the Curie temperature. In order to "clamp" the spontaneous polarization we should suppress the spontaneous strain. This is impossible to achieve in practice because, as we already know, the crystals split into domains, each of which keeps its spontaneous strain although the crystal as a whole has zero strain. Nevertheless, Baumgartner's measurements of the clamped crystal (B 3), using frequencies of $10^7$ c/s, yield two important results, viz:

(i) The Curie–Weiss temperature of the clamped crystal lies about 4 °C lower than that of the free crystal, thus indicating that the effect of total clamping would reduce the Curie temperature by this amount.

(ii) The fact that the Curie–Weiss law is unaffected by the use of high frequencies, i. e. the Curie constants of the clamped and the free crystal are the same, indicates that the difference between the reciprocal susceptibilities of the clamped and the free crystal is essentially temperature independent. We can express this difference in terms of the constants introduced in the fundamental equations of a piezoelectric crystal as defined in Section I-5. In fact, from Eqs. (I-4a), (I-4b), (I-5a), (I-5b) we obtain

$$\chi^x = \chi^X + ab = \chi^X (1 + ad), \tag{III-1}$$

and we conclude, therefore, from the experimental data that the product of the pertinent constants $ab$, or $\chi^X ad$, is temperature independent.

Baumgartner also performed very extensive measurements of the dependence of the dielectric constant $\varepsilon_c$ upon field and polarization in a narrow temperature region above the transition temperature (B 4). These were the first measurements of this kind ever carried out on ferroelectric crystals and the experimental techniques used were the following. The field dependence of the dielectric constant was investigated by applying a large d.c. biasing field upon the crystal; as we have seen in Section II-5, this technique yields the value of the adiabatic dielectric constant for zero stress as a function of the field (and, of course, the temperature). The polarization dependence of the dielectric constant was investigated by first measuring the polarization as a function of the field with a ballistic galvanometer, then differentiating the results to obtain the isothermal dielectric constant as a function of polarization (and temperature). Comparison of the two sets of data yields the difference between isothermal and adiabatic values of the dielectric constant, which is discussed in detail in Section II-5. This difference is quite large in the vicinity of the transition point, because KDP shows a very marked electrocaloric effect; about 1 °C above the transition temperature, a field of $10^4$ V/cm causes an adiabatic temperature change of more than 1 °C. The experimental results can be written in terms of the adiabatic dielectric susceptibility in the following way:

$$\frac{1}{k_S} = \frac{4\pi}{C}(T - T_0) + A_H''(P,O) + \frac{16\pi^2}{C^2}\frac{T}{c_P}P^2, \tag{III-2}$$

and the similarity between this expression and Eq. (II-9a) is quite obvious. Thus, the correction terms consist of two contributions, viz. a saturation term and an adiabatic term. The *saturation* term $A_H''(P,O)$ is the second derivative of the higher order terms of the free energy function with respect to polarization. Unfortunately, the coefficients of the terms with higher orders of $P$ in the free energy expansion are not known for KDP, but they could be computed from Baumgartner's data. It appears that the contribution of the sixth and higher power term in the free energy expansion cannot be neglected. The *adiabatic* term,

$$\frac{16\pi^2}{C^2}\frac{T}{c_P}P^2,$$

can be computed as soon as the specific heat at constant polarization, $c_P$, is known. Baumgartner's data yield $c_P = 0.977$ J/cm³ °C.

The two correction terms of the dielectric constant are plotted as functions of polarization in Fig. III-2. It is seen that, for polarization values lower than about $2 \times 10^{-6}$ C/cm², the adiabatic term is larger than the saturation correction, but the situation is reversed at higher polarizations.

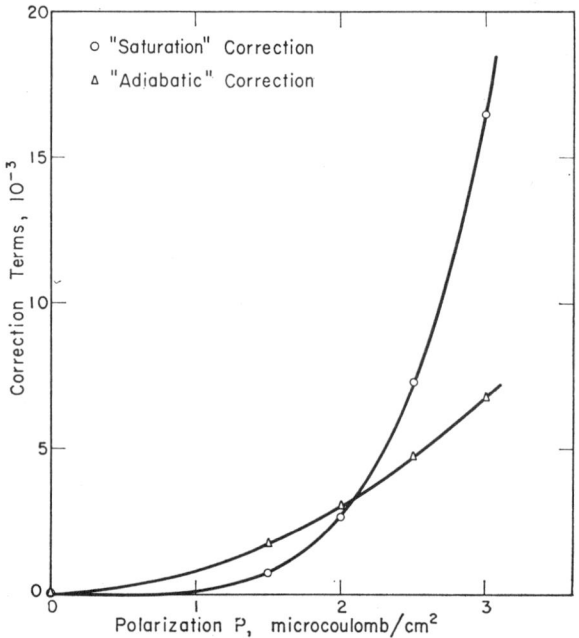

Fig. III-2. Saturation and adiabatic correction terms of the dielectric constant of KDP, as computed from Baumgartner's measurements (B 4) by Devonshire (D 1).

The field dependence of the dielectric constant $\varepsilon_c$ is depicted in Fig. III-3, which is a logarithmic plot of $\varepsilon_c$ vs. the reduced temperature $(T - T_c)$ for different values of the d.c. biasing field. This is a different representation of the behavior depicted in Fig. II-8 for tri-glycine sulfate. It should be noted that the effect of an electric field is to blur the transition, so that we can no longer speak of a transition in the usual sense (D 1). However, we see that the peak of the dielectric constant is displaced toward higher temperatures by the biasing field.

The temperature dependence of the spontaneous polarization $P_s$ is shown in Fig. III-4. At 100 °K, i.e. 23° below the transition point, $P_s = 4.7 \times 10^{-6}$ C/cm². Two qualitative observations can be made on the behavior of the spontaneous polarization of KDP, viz:

(i) the onset of $P_s$ at the transition point occurs continuously, the transition is of the second order; yet,

(ii) the rise of $P_s$ at the transition temperature is quite rapid, as compared for example to that shown by tri-glycine sulfate. Qualitative comparison with

Eq. (II-2) would seem to indicate that, for KDP, the coefficient $\xi$ is small and the coefficient $\zeta$ is more relevant (K 4).

The low-frequency coercive field $E_c$ is of the order of 2 kV/cm at approximately 100 °K and apparently shows marked fluctuations from sample to sample. It is interesting that at approximately 60 °K a sharp break is observed in the tempera-

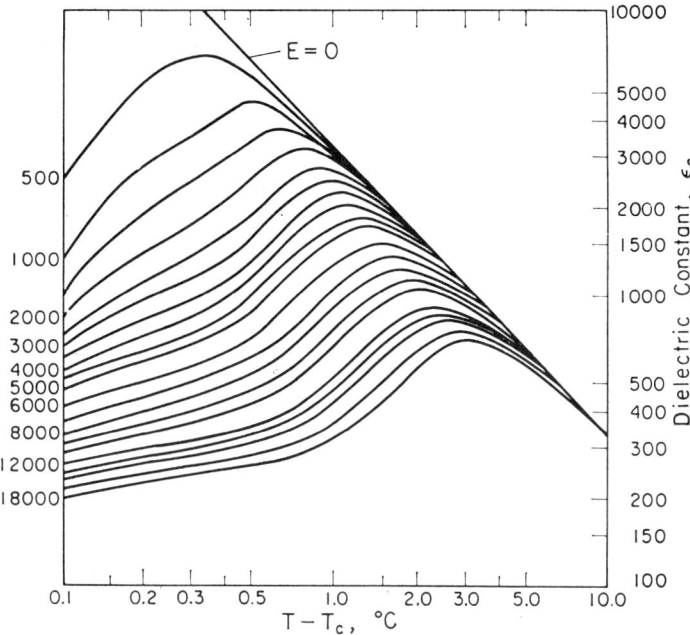

FIG. III-3. Dielectric constant $\varepsilon_c$ of KDP as a function of reduced temperature $(T - T_c)$ for different values of the d.c. biasing field $E$. The $E$ values are given for each curve on the left hand side (according to Baumgartner (B 4)).

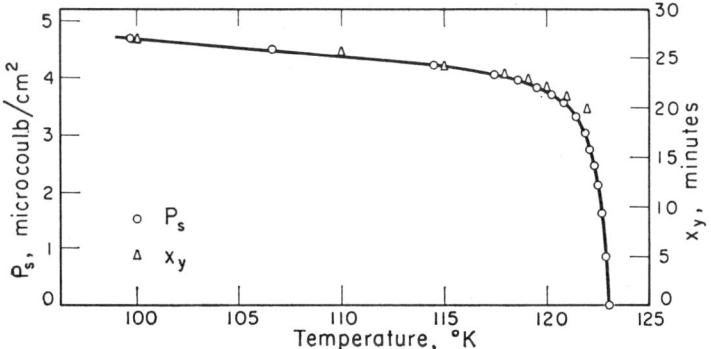

FIG. III-4. Spontaneous polarization of KDP as a function of temperature (according to von Arx and Bantle (V 1)). The triangles represent measurements of the spontaneous shear and are referred to the ordinate axis on the right (according to de Quervain (D 2)).

ture dependence of $E_c$, and this quantity increases rapidly at lower temperatures (Fig. III-5). This is probably due to hindered domain–wall motions at these temperatures.

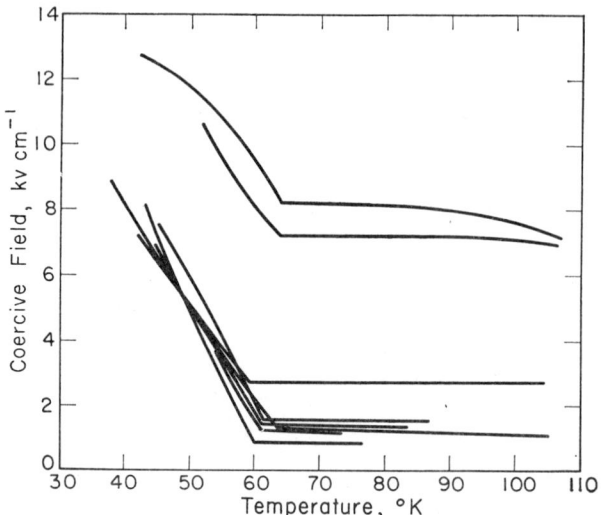

FIG. III-5. Coercive field $E_c$ of different samples of KDP as a function of temperature (according to Barkla and Finlayson (B 5)).

## 3. Specific Heat and Lattice Distortion

### The Specific Heat

The specific heat of KDP was measured by a number of authors. Considerable discrepancies are found both in the details of the shape of the specific heat curve and in the value of the transition heat obtained by integration. Figure III-6 shows the temperature dependence of the specific heat at constant pressure as measured by Stephenson and Hooley (S 1). At the transition point, an anomaly with the typical $\lambda$-shape is observed. The transition energies of $KH_2PO_4$ and the iso-morphous compounds $KD_2PO_4$ and $KH_2ASO_4$ are listed in Table III-1.

It should be pointed out, in connection with the typical $\lambda$-shape of the specific heat anomaly, that the thermodynamic relation (II-5)

$$\Delta c_p = \frac{2\pi}{C} T \frac{\partial P^2}{\partial T},$$

is verified satisfactorily by the experimental data (V 2) for KDP.

### The Lattice Distortion

The linear expansion coefficients of KDP were measured by de Quervain (D 2), as functions of temperature, with X-ray methods. The volume expansion coefficient $(1/V)(dV/dT)$ was computed to be equal to $8.2 \times 10^{-5}$. The relative variation of the lattice spacings with temperature is depicted in Fig. III-7. In order to understand the structural distortion that occurs spontaneously at the transition temperature, we first need to clarify the coordinate system to which we refer. Crystal morphology and other crystallographic considerations favor the choice

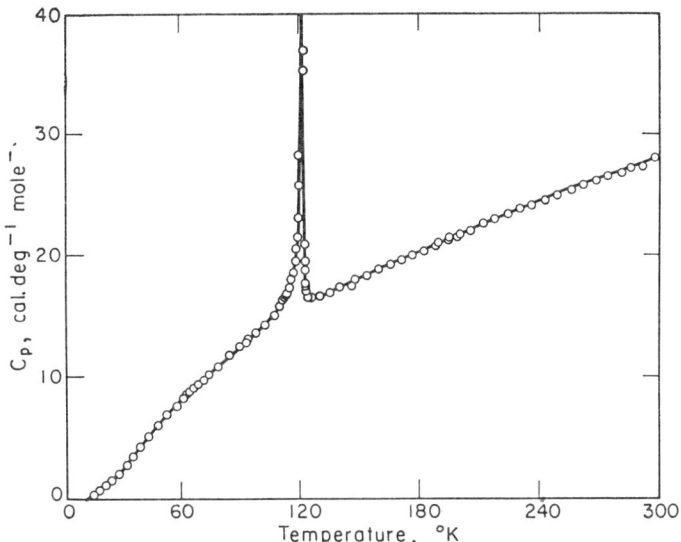

FIG. III-6. Specific heat of KDP as a function of temperature (according to Stephenson and Hooley (S 1)).

TABLE III-1. TRANSITION ENERGIES OF $KH_2PO_4$, $KD_2PO_4$ AND $KH_2AsO_4$

| Substance | Curie temperature (°K) | Heat of transition $\Delta Q$ (cal/mole) | Transition entropy $\Delta S$ (cal/mole °C) | References |
|---|---|---|---|---|
| $KH_2PO_4$ | 123 | 57 | 0.47 | BANTLE (B 6) |
| | | 87 | 0.74 | STEPHENSON and HOOLEY (S 1) |
| | | 87 | 0.74 | DANNER and PEPINSKY (D 3) |
| $KD_2PO_4$ | 213 | 100 | 0.47 | BANTLE (B 6) |
| $KH_2AsO_4$ | 97 | 84 | 0.87 | BANTLE (B 6) |
| | | 84 | 0.90 | STEPHENSON and ZETTLEMOYER (S 6) |

of a reference system in which the $a_1$ and $a_2$ axes are perpendicular to the major faces of the crystal. The corresponding unit cell has a square cross-section with $a_1 = a_2 = 7.453$ Å. Figure III-8(a) shows schematically this cross-section and the lattice distortion which occurs at the transition temperature, the $c$ axis being perpendicular to the plane of the figure. The spontaneous strain consists of a slight change in length of $a_1$ and a shear $x_{12} = x_6 = x_y$, which amounts to 27' at approximately 20° below the transition temperature (D 2). It would appear logical to describe the distorted cell in terms of pseudo-monoclinic axes $a_1$ forming an angle of $90° + x_y$ with each other. However, the true symmetry of the low-temperature phase is orthorhombic, and the unit cell has a rectangular cross-

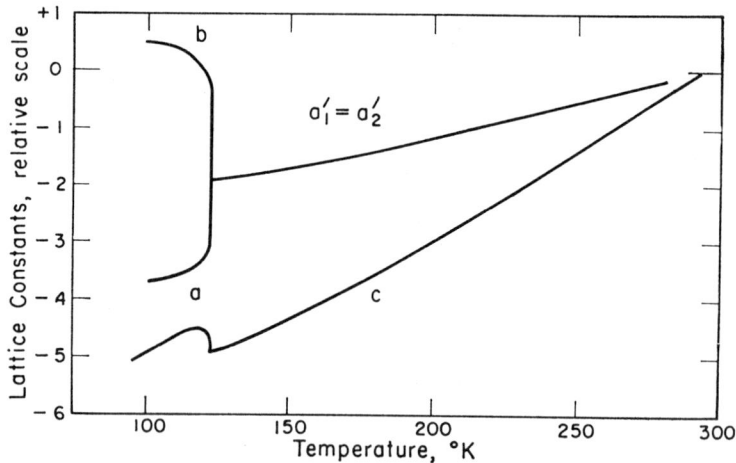

FIG. III-7. Temperature dependence of the lattice constants of KDP. The ordinate scale represents the deviation from the room temperature values in units of $10^{-2}$ Å (according to de Quervain (D 2)).

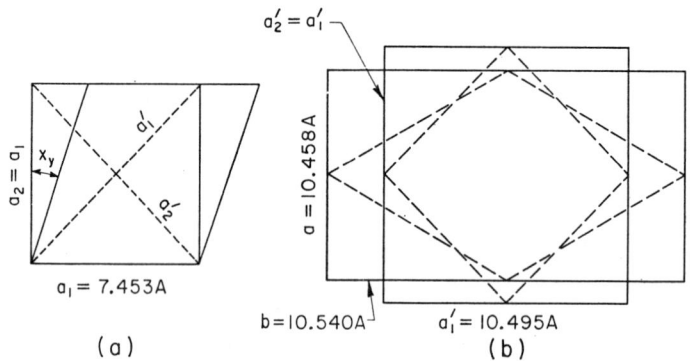

FIG. III-8. Unit cells and lattice distortion in KDP (schematic). (a) The lattice distortion consists of a slight change in $a_1$ and a shear $x_y$, (b) The lattice distortion is an elongation of $\mathbf{a}_1'$ to $a$ and a compression of $a_2'$ to $\mathbf{b}$.

section approximately along the directions $a_1'$, $a_2'$ of the original [110] axes. With this in mind, it may be more convenient, for certain purposes, to adopt the larger tetragonal cell (with edges $a_1' = \sqrt{(2)}a_1$) from the beginning. Then the spontaneous strain consists in a slight elongation of the $a_1'$ axis (which becomes the ortho-rhombic $b$ axis) and a simultaneous contraction of the $a_2'$ axis (orthorhombic $a$ axis) (Fig. III-8b). It is the relative behavior of the $a_1' = a_2'$ and the $c$ axes, which is shown in Fig. III-7. The relation between the orthorhombic parameters $a$, $b$ and the monoclinic parameters $a_1$, $(90° + x_y)$ is the following:

$$a = 2\,a_1 \cos \frac{90° + x_y}{2},$$

$$b = 2\,a_1 \sin \frac{90° + x_y}{2}.$$

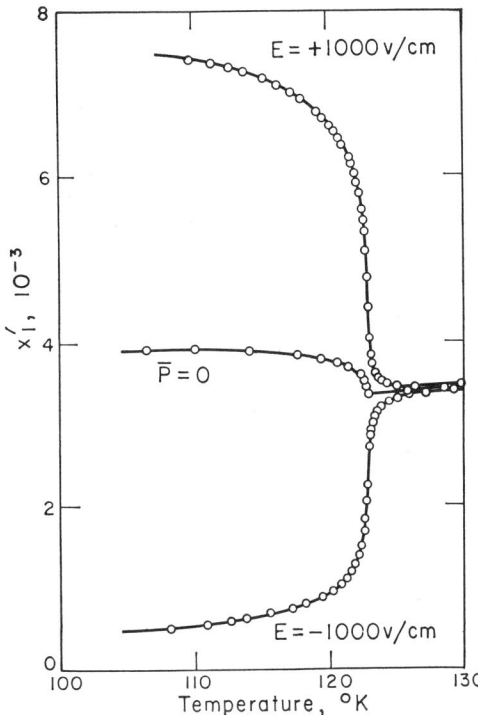

FIG. III-9. Elongation of $a_1'$ and contraction of $a_2'$ in KDP measured interferometrically in the vicinity of the transition temperature (according to von Arx and Bantle (V 1)).

The lattice distortion was also measured by von Arx and Bantle (V 1) using Fizeau's interference method to determine the elongation of $a_1'$ and the contraction of $a_2'$ at the transition temperature. Figure III-9 shows the result of these measurements. A d.c. field of 1000 V/cm was applied in order to polarize the crystal uniformly along the $c$ direction. The *mean value* of the two curves also shows an anomaly at 123 °K, indicating that the expansion of $a_1'$ and the contraction of $a_2'$ are not equal and opposite, as would be required by a pure shear $x_y$. The crystal shows in addition to the shear $x_y$, a distortion which is about an order of magnitude smaller than $x_y$ and which turns out to be proportional to the square of the spontaneous polarization (see Section 4). The deformation of the insulated crystal coincides perfectly with the mean curve (curve $\bar{P} = 0$ in Fig. III-9). The experimental results can be expressed in the following way: the expansion of a single domain crystal along $a_1'$ is

$$x_1' = \frac{b}{2} P + \alpha P^2, \tag{III-3}$$

with $b \cong 2.0 \times 10^6$ c.g.s units, $\alpha = 3.84 \times 10^{-12}$ c.g.s units; and the expansion along $c$ (see Fig. III-7) is

$$x_3 = \beta P^2, \tag{III-4}$$

with $\beta \cong 2.9 \times 10^{-12}$ c.g.s. units.

The dependence of the spontaneous shear $x_y$ upon temperature was determined by de Quervain's X-ray investigation from the data shown in Fig. III-7., The thermal behavior of $x_y$ is strikingly similar to that of the spontaneous polarization, as shown in Fig. III-4, where the triangles refer to the values of the spontaneous shear. Thus, it is evident that the spontaneous *deformation is directly proportional to the spontaneous polarization.* The question that immediately arises is whether the constant of proportionality between these spontaneous quantities in the ferroelectric region is the same as that occurring in the piezoelectric effect above the transition temperature. The answer can be given only after studying the piezoelectric behavior of the crystal above $T_c$, which we are going to do in the next section, and is significantly affirmative. For the time being, we limit ourselves to stating that de Quervain's data yield the following value for the proportionality constant between spontaneous deformation and polarization (at 110 °K):

$$\frac{P_s}{x_y} = 6.15 \times 10^{-4} \text{ C/cm}^2 = 1.85 \times 10^6 \text{ c.g.s units.} \tag{III-5}$$

## 4. Piezoelectric and Elastic Properties

The definition of the coefficients which characterize the piezoelectric and elastic properties of a crystal is given by the fundamental equations established in Section I-5, namely Eqs. (I-4a), (I-4b), (I-5a), (I-5b). As mentioned previously, the most convenient set of equations to use from the experimental viewpoint is that involving the stress tensor $X$ and the field vector $E$ as *independent* variables, i.e. Eqs. (I-5b):

$$\left. \begin{array}{l} P = -dX + k^X E, \\ x = -s^E X + dE. \end{array} \right\} \tag{III-6}$$

In the case of KDP, which belongs to the point group $42m$ at room temperature, the scheme of the piezoelectric moduli $d_{ik}$, referred to the co-ordinate system $a_1$, $a_2$, $c$ (see Fig. III-8) has the following form: (C 1)

$$\begin{array}{cccccc} 0 & 0 & 0 & d_{14} & 0 & 0 \\ 0 & 0 & 0 & 0 & d_{14} & 0 \\ 0 & 0 & 0 & 0 & 0 & d_{36} \end{array}$$

Thus, there are only two non-zero piezoelectric moduli, $d_{14}$ and $d_{36}$.

The non-zero dielectric susceptibilities are of course $k_{11}^X = k_1^X = k_a^X$ and $k_{33}^X = k_3^X = k_c^X$, and the scheme of the elastic compliances $s_{ik}^E$ is (neglecting the superscript $E$):

$$\begin{matrix} s_{11} & s_{12} & s_{13} & 0 & 0 & 0 \\ s_{12} & s_{11} & s_{13} & 0 & 0 & 0 \\ s_{13} & s_{13} & s_{33} & 0 & 0 & 0 \\ 0 & 0 & 0 & s_{44} & 0 & 0 \\ 0 & 0 & 0 & 0 & s_{44} & 0 \\ 0 & 0 & 0 & 0 & 0 & s_{66} \end{matrix}$$

thus involving only six non-zero coefficients.

The experimental evidence (to be discussed below) shows that some of the piezoelectric moduli, $d_{ik}$, and elastic compliances, $s_{ik}^E$, behave anomalously at the Curie temperature. This is in accordance with Eq. (I-6). We may now make a step further and ask: which one of the coefficients $d_{ik}$ and $s_{ik}^E$ is going to behave anomalously as the transition temperature is approached? The answer is that anomalies may occur for those piezoelectric and elastic coefficients which are related to the polarization along the ferroelectric direction. These are, in the present case of KDP: the piezoelectric modulus $d_{36}$, which relates the shear stress $X_y = X_6$ to the polarization $P_c = P_3$; and the elastic compliance $s_{66}^E$, which relates the shear stress $X_6$ to the shear deformation $x_y = x_6$. Eq. (III-6) can be specified in this case, thus:

$$\left. \begin{aligned} P_3 &= -d_{36} X_6 + k_3^X E_3, \\ x_6 &= -s_{66}^E X_6 + d_{36} E_3. \end{aligned} \right\} \tag{III-7}$$

### Piezoelectric Properties

From the first of Eqs. (III-7), we see that if $E = 0$, then $P_3 = d_{36} X_6$, and we can therefore measure the modulus $d_{36}$ by applying a known stress $X_y$ to the crystal and measuring the polarization $P_3$ induced piezoelectrically on the $c$ faces of the crystal. Such measurements were carried out by Bantle and Caflish (B 7).

On the other hand, from the second of Eqs. (III-7), we see that $x_6 = d_{36} E_3$, if $X_6 = 0$. Thus, if we apply a field $E_3$ on the stress-free crystal and measure the piezoelectric strain, $x_y$, we again have a way to determine $d_{36}$, in this case from the converse piezoelectric effect. Such measurements were carried interferometrically by von Arx and Bantle (V 1).

The results of both the measurements of the direct and the converse piezoelectric effect are in excellent agreement with each other, as required by thermodynamics, and yield the temperature dependence of the modulus $d_{36}$ which is depicted in Fig. III-10. It follows from the results that $d_{36}$ obeys a Curie–Weiss law of the type:

$$d_{36} = d_{36}^0 + \frac{B}{T - T_0} \tag{III-8}$$

where $B = 1.26 \times 10^{-4}$ c.g.s. units, $d_{36}^0 \cong -8 \times 10^{-8}$ c.g.s. units and $T_0 = T_c$. The temperature-independent constant $d_{36}^0$ can, however, be neglected in the vicinity of the Curie temperature. It is evident from Fig. III-10 that the piezoelectric modulus $d_{36}$ shows a marked anomaly at the transition temperature.

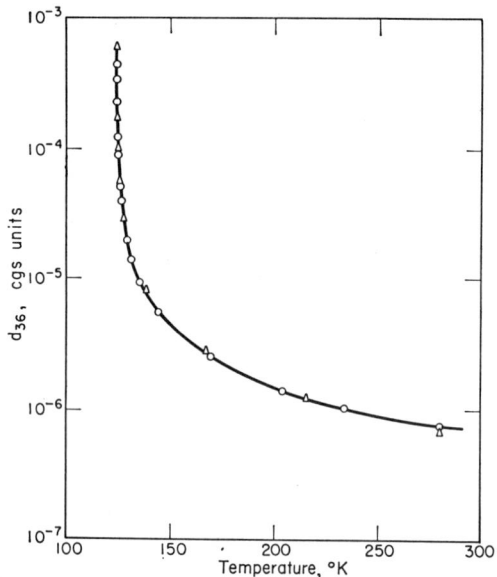

FIG. III-10. Piezoelectric modulus $d_{36}$ of KDP as a function of temperature according to the measurements of the direct effect (triangles) and the converse éffect (open circles) (according to Bantle and Caflish (B 7) and von Arx and Bantle (V 1)).

The first important consequence of this result is the following. We see from Eqs. (III-7) that, for the stress-free crystal ($X = 0$), the polarization $P_3$ and the deformation $x_y$ are related in the following way:

$$P_3 = \frac{k_c^X}{d_{36}} x_6 .\qquad\qquad\text{(III-9)}$$

Recalling that the dielectric susceptibility also follows a Curie–Weiss law:

$$k_c^X \cong \frac{C}{4\pi(T - T_0)},$$

we can see that the ratio

$$\frac{k_c^X}{d_{36}} \cong \frac{C}{4\pi B} = 2.05 \times 10^6.$$

This value has the same order of magnitude as in normal piezoelectric crystals, e.g. quartz. Moreover, this value is very close to the value determined by de Quervain (D 2) for the ratio between the *spontaneous* polarization $P_s$ and the *spontaneous* strain $x_y$ below the transition temperature (Eq. III-5). The value of $k_c^X/d_{36}$ is very slightly temperature dependent, as shown in Fig. III-11, whence it appears that de Quervain's result fits perfectly on the extrapolation of the line below the transition point. Thus, we conclude that the proportionality between polarization and deformation, a fundamental property of all piezoelectric crystals, is not disturbed by the onset of ferroelectricity. *The piezoelectric anomalies are only a consequence of the dielectric anomalies.*

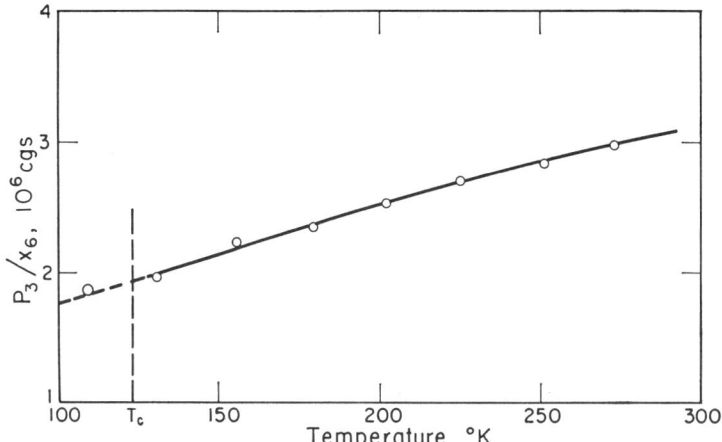

FIG. III-11. Temperature dependence of the ratio between polarization $P_3 = P_c$ and shear strain $x_6 = x_y$ in KDP at zero stress (according to de Quervain (D 2)). The experimental points above the Curie point $T_c$ were calculated from the ratio $(k_3^X/d_{36})$ between the dielectric susceptibility of the free crystal along $c\,(k_3^X)$ and the piezoelectric modulus $d_{36}$. The point below $T_c$ was determined from the ratio between *spontaneous* polarization and *spontaneous* strain, as measured with X-rays.

Very closely above the transition temperature, therefore, non-linearity of the piezoelectric modulus and saturation effects are observed (V 1), which are fully analogous to those exhibited by the dielectric constant (B 4) (see Section 3). In the ferroelectric region, on the other hand, the relation between piezoelectric polarization and stress is no longer unique. If we apply large enough stresses to displace the domain walls, we can observe a *piezoelectric hysteresis* curve (V 1). The direct measurement of $d_{36}$ below the transition point can, therefore, only be carried out on a single-domain crystal. If the measurement is to be carried out via the determination of the strain caused by an applied field $E_3$, care should be exercised to correct for the so-called *quadratic piezoelectric effect*. This effect arises because the symmetry of the crystal is lowered in the polar phase, so that new piezoelectric moduli come into being. As this effect is inherent to all ferroelectric crystals in their polar phases, we discuss it in some detail in the following.

The actual measurement of the piezoelectric strain $x_y$ is carried out indirectly by determining the elongation along the [110] direction, or, in other words, the elongation along the $a_1'$ axis as discussed in connection with Fig. III-8. Calling this elongation $x_1'$, it can be shown that, above the Curie point, $x_1' = x_y/2$. In this respect, it is easier to refer the properties to the primed co-ordinate system $a_1'$ and $c$ rather than to $a_1$ and $c$. In the new system, the scheme of the piezoelectric moduli is written in the following way:

$$
\begin{array}{cccccc}
0 & 0 & 0 & 0 & d'_{15} & 0 \\
0 & 0 & 0 & -d'_{15} & 0 & 0 \\
d'_{31} & -d'_{31} & 0 & 0 & 0 & 0
\end{array}
$$

and it is the modulus $d'_{31}$ which is going to be anomalous at the transition temperature. It is: $d'_{31} = d_{36}/2$ and $x'_1 + x'_2 = 2x'_1 = x_6$.

Below the transition point, the symmetry is orthorhombic, point group $mm$, and the corresponding scheme of the piezoelectric moduli is

$$
\begin{array}{cccccc}
0 & 0 & 0 & 0 & d''_{15} & 0 \\
0 & 0 & 0 & d''_{25} & 0 & 0 \\
d''_{31} & d''_{32} & d''_{33} & 0 & 0 & 0
\end{array}
$$

Thus, we see that $d''_{31} \neq d''_{32}$, and then $x''_1 \neq x''_2$, and $x''_1$ is no longer equal to $x_y/2$. Moreover, a new non-zero modulus appears, $d''_{33}$; the polarization $P_3$ can be affected by a pressure along the $c$ axis. The occurrence of these moduli represents simply the normal (linear) piezoelectric effect of the low-symmetry phase, but it is called the quadratic piezoelectric effect of the higher-symmetry phase because these new moduli are proportional to the polarization. Written in terms of polarization $P$ (Eq. I-4b), the elongation $x'_1$ of a single domain crystal takes the form

$$
x'_1 = \frac{b_{36}}{2} P_3 + \alpha P_3^2 .
$$

This explains the results obtained for the expansion of an *insulated* crystal in the ferroelectric region (Fig. III-9 and Eq. (III-3)) as being due to the quadratic piezoelectric effect. The insulated crystal splits into equal volumes of positive and negative domains and only the quadratic effect can be detected.

Before closing this discussion of the piezoelectric properties of KDP, it should be pointed out, in connection with Eq. (III-9), that several interesting conclusions can be drawn from the fact that the ratio $k_c^X/d_{36}$ is almost temperature independent. Since according to Eq. (I-6), $k^X/d = b$, the piezoelectric coefficient $b_{36}$ must be practically temperature independent. Moreover, as the difference between clamped and free dielectric susceptibility is also independent of temperature, Eq. (III-1) $(\chi^x - \chi^x = ab)$ predicts that the corresponding coefficient $a_{36}$ is also practically independent of temperature. For this reason, the coefficients $a$ and $b$ are often called the "true" constants of the piezoelectric crystal, expressing the fact that the "true" piezoelectric behavior of the crystal is normal throughout the ferroelectric transition.

The behavior of the piezoelectric modulus $d_{14}$ is also interesting. This modulus relates the shear stress $Y_z = Z_y = X_4$ to the polarization in the $a$ direction $P_1$. Since the dielectric constant $\varepsilon_a$ has a small anomaly at the transition temperature (Fig. III-1), the modulus $d_{14}$ is also expected to behave anomalously. Figure III-12 shows that this is in fact the case.

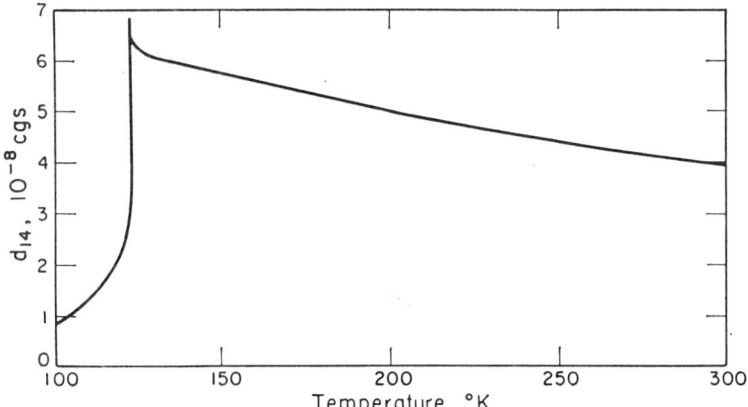

FIG. III-12. Temperature dependence of the piezoelectric modulus $d_{14}$ of KDP (according to Ess (E 1)).

### Elastic Properties

Turning now to the elastic properties of KDP, we are in the position to predict what anomalies will occur. We know that, if $E = 0$, a relatively small shear stress $X_y$ causes a very large polarization $P_3$ along the $c$ axis shortly above the Curie point ($d_{36}$ is very large). We know also that this polarization $P_3$ is proportional to the shear strain $x_y$. Thus, a relatively small shear stress $X_6$ causes a very large shear deformation $x_6$ in the vicinity of the transition point. Since, from (III-7), $x_6 = s_{66}^E X_6$, it follows that the elastic compliance at constant (zero) field must show a large anomaly at the transition, similarly as do $\varepsilon_3$ and $d_{36}$. This anomaly is therefore a direct consequence of the anomalous piezoelectric properties and, in the final instance, of the anomalous dielectric properties.

The condition $E = 0$ is very important. It is experimentally fulfilled by coating the $c$ faces of a KDP plate with conducting material and shorting the two electrodes. Accordingly, the behavior of the crystal at $E = 0$ is often referred to as that of the *shorted* crystal. With respect to the shear $X_6$, the shorted crystal becomes very *soft* in the vicinity of the transition point.

The experimental method consists in measuring the resonance frequency of suitable plates and bars excited piezoelectrically. It can be shown (M 1) that the resonance frequency $\nu_R$ is proportional to $1/l \sqrt{(\varrho s_{66})}$, where $l$ is the length and $\varrho$ the density of the crystal bar.

If, on the other hand, the condition $E = 0$ is not fulfilled, i.e. the crystal is *insulated*, then the polarization $P_3$ caused by a shear $X_6$ induces charges on the $c$ faces of the crystal. These charges, in turn, cause a strong depolarizing field $E_3 = -4\pi P_3$, which reduces the polarization practically to zero. The *insulated* crystal corresponds to the condition $D = 0$, where $D$ is the dielectric displacement. In ferroelectric crystals, owing to the large values of the dielectric constant, the condition $D = 0$ is practically the same as the condition $P = 0$.

In this case, according to Eq. (I-4b) ($x_6 = -s_{66}^P X_6$), we measure the elastic compliance at constant (zero) polarization. Since the polarization is suppressed, the strain $x_6$ caused by the stress $X_6$ will remain normally small and the compliance $s_{66}^P$ will have no anomaly at the transition temperature. The *insulated* crystal remains *hard* through the ferroelectric transition. The compliance $s_{66}^P$ is a "true" constant in the sense introduced above for the piezoelectric coefficients $a$ and $b$.

Experimentally, the compliance $s_{66}^P$ can be measured by exciting vibrations of the unplated crystal in the wide gap of a condenser (B 4). Let $t_g$ be the total thickness of the gap between the condenser plates and the surface of the crystal plate, and $t_c$ the thickness of the crystal plate. As the condenser plates are short-circuited, we can write:

$$t_g E_g + t_c E_c = 0,$$

where $E_g$ and $E_c$ are the field strengths across the gap and the crystal, respectively. The continuity of the dielectric displacement at the crystal surfaces requires that:

$$E_g = E_c' + 4\pi P.$$

Eliminating $E_g$ from the two above equations and introducing into (III-7), we obtain

$$x_6 = \left[ s_{66}^{E=0} - \frac{4\pi d_{36}^2}{\varepsilon_3 + \dfrac{t_c}{t_g}} \right] X_6,$$

and the expression in the brackets can be looked upon as being the elastic compliance of the crystal in a gap. Letting the gap thickness $t_g$ go to infinity ($t_g \to \infty$), we obtain the elastic compliance of the insulated crystal:

$$s_{66}^{\text{insulated}} = s_{66}^{E=0} - \frac{4\pi d_{36}^2}{\varepsilon_3} = s_{66}^P. \qquad \text{(III-10)}$$

The latter equality can be derived directly from Eqs. (I-4a)–(I-5b). In the vicinity of the transition temperature, where $\varepsilon_3 > 10^3$, by choosing a gap such that $t_g = t_c$ we actually measure $s^P$, since the error involved becomes smaller than $0.1\%$ (B 4).

However, if the frequency of the mechanical vibrations of the crystal becomes very high, the pattern of the surface charges generated piezoelectrically may be such as to reduce the depolarizing field even if the crystal is electrically insulated. Another method used for the determination of the elastic constants of ferro-electric crystals consists, namely, in exciting the crystal to very high-frequency vibrations ($\sim 20$ Mc/s) by means of an external transducer. The ultrasonic waves travelling within the crystal give rise to local variations of the refractive index and represent, therefore, a sort of three-dimensional grating which causes diffraction of a monochromatic light beam sent through the crystal. This is the method of Schäfer–Bergmann (B 8), (Z 1), (J 1). In this case, the sample is unplated, but it can be shown that the elastic constants determined with this method approach the values of the shorted crystal as the measuring frequency is increased (J 1).

The experimental results of the elastic measurements show very nicely the difference between $s_{66}^P$ and $s_{66}^E$. Figure III-13 depicts the result of Mason's resonance investigation (M 2). The coefficients reported in the figure are the elastic constants $c_{66}^E$ and $c_{66}^D$ ($\cong c_{66}^P$). It is easy to show from Eqs. (I-4a) to (I-5b) that $s_{66}^E = 1/c_{66}^E$ and, of course, $s_{66}^P = 1/c_{66}^E$. The analysis of the data reported graphically in Fig. III-13 shows that a Curie–Weiss law holds for the quantity $s_{66}^E - s_{66}^P$:

$$s_{66}^E - s_{66}^P = \frac{D}{T - T_0},\qquad\text{(III-11)}$$

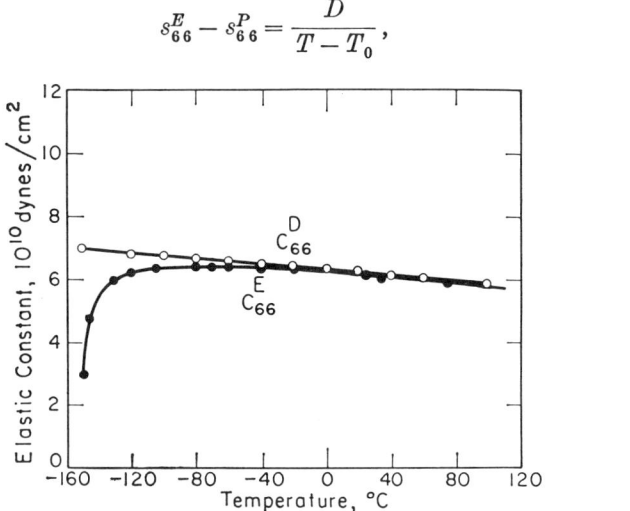

FIG. III-13. Elastic constants $c_{66}^E$ and $c_{66}^D$ ($= c_{66}^P$) of KDP as a function of temperature (according to Mason (M 2)).

where $D = 4.6 \times 10^{-11}$ cm² °C/dyn and $T_0$ is again the transition temperature $T_c$. This elastic Curie–Weiss law is to be expected from Eq. (III-10), whence taking into account the piezoelectric Curie–Weiss law (III-8) and the dielectric Curie–Weiss law, it follows that

$$D = \frac{4\pi B^2}{C}.\qquad\text{(III-11a)}$$

From this, the expected value of D is $6.1 \times 10^{-11}$ cm² °C/dyn. This value is in satisfactory agreement with the value of the elastic Curie constant measured directly, if we take into account the experimental accuracy.

It should also be pointed out that the dynamic method often used for the determination of the elastic (and piezoelectric) coefficients makes it necessary to distinguish between adiabatic and isothermal conditions, just as in the case of the dielectric constant. This problem was also investigated by Baumgartner (B 4). The ratio between isothermal and adiabatic values of the piezoelectric or elastic coefficients is equal to that between isothermal and adiabatic dielectric constants (see Section II-5).

The elastic properties of KDP which are not related to the anomalous shear in the (001) plane are quite normal. Figure III-14 shows the thermal behavior of the $c_{ik}$ constants (except $c_{66}$) measured by Zwicker (Z 1) with the Schäfer–Bergmann method. The $c_{ik}$ coefficients (or "elastic constants") are defined in Eqs. (I-4a) and (I-5a), and their matrix is equal to that of the $s_{ik}$. The indeterminacy of the constants $c_{11}$ and $c_{33}$, below the transition temperature, is due to the lowering of the symmetry in the ferroelectric phase. The small anomaly of the $c_{12}$ constant is not reliable, owing to the large experimental error involved in the measurement of this coefficient.

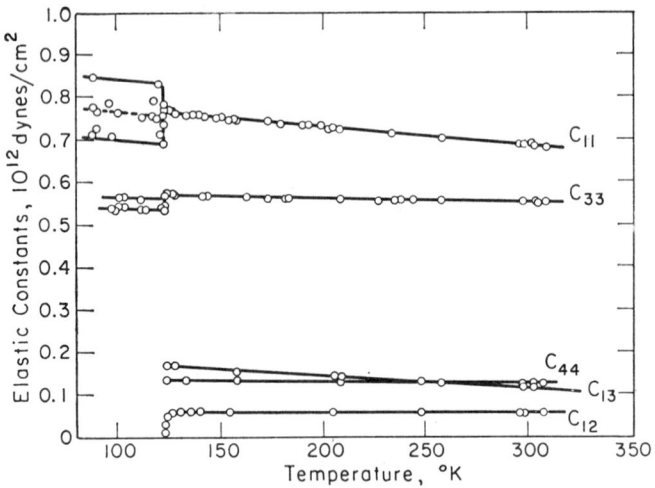

FIG. III-14. Elastic constants $c_{ik}$ (except $c_{66}$) of KDP as a function of temperature (according to Zwicker (Z 1)).

### The "True" Constants

What are now the conclusions that can be drawn from the study of both the piezoelectric and elastic properties of KDP? We have learned that some of the coefficients $d_{ik}$ and $s_{ik}^E$ (and $c_{ik}^E$) behave anomalously at the transition temperature (piezoelectric and elastic Curie–Weiss laws) and that these anomalies are consequences of the dielectric anomaly (dielectric Curie–Weiss law). We have also learned that both the piezoelectric coefficients $a_{ik}$ and $b_{ik}$, as well as the elastic quantities at constant polarization, $s_{ik}^P$ and $c_{ik}^P$, do *not* exhibit anomalies at the Curie temperature. It is, therefore, more logical, for theoretical considerations, to describe the properties of KDP in terms of the system of Eqs. (I-4a) and (I-4b), involving "true" constants of the crystal; namely

$$\left. \begin{aligned} X_6 &= - c_{66}^P x_6 + a_{36} P_3, \\ E_3 &= - a_{36} x_6 + (1/k_3^x) P_3, \end{aligned} \right\} \quad \text{(III-12)}$$

and:

$$\left. \begin{aligned} x_6 &= - s_{66}^P X_6 + b_{36} P_3, \\ E_3 &= b_{36} X_6 + (1/k_3^X) P_3. \end{aligned} \right\} \quad \text{(III-13)}$$

In the latter system, it is only the susceptibility $k_3^X$ of the free crystal which behaves anomalously; in the former, it is the susceptibility $k_3^x$ of the clamped crystal. For convenience, a list of the anomalous and non-anomalous coefficients and their interelations is presented in Table III-2.

The relation between the reciprocal free- and clamped-susceptibilities was already given in Eq. (III-1). We see from this equation that since $a = d/k^X s^P$ and $b = a s^P$, we can write:

$$\frac{1}{k_3^x} = \frac{1}{k_3^X} + \frac{(d_{36})^2}{(k_3^X)^2 s_{66}^P}.$$

Hence, if we assume that a Curie–Weiss law holds for the free susceptibility, namely that $k_3^X = (C/4\pi)/(T - T_0)$, we are led, after several steps, to a Curie–Weiss law for the clamped crystal (B 13):

$$k_3^x = \frac{(C/4\pi)}{T - [T_0 - \{(d_{36})^2\, C/4\pi\, (k_3^X)^2\, s_{66}^P\}]}. \tag{III-14}$$

Thus, we see that the Curie constant remains the same, and the transition of the clamped crystal is predicted to be lower by the amount

$$(d_{36})^2\, C/4\pi (k_3^X)^2 s_{66}^P.$$

This factor is independent of temperature and is computed to be approximately 4°, in agreement with the experimental results reported in Section 2. Thus, the effect of clamping is to reduce the range of the ferroelectric state.

The transition of the free crystal occurs when $\chi_3^X = 1/k_3^X$ becomes equal to zero, i.e. from Eq. (III-1), when the reciprocal susceptibility of the clamped crystal becomes equal to

$$\chi_3^x = \frac{1}{k_3^x} = a_{36} b_{36} = \frac{(d_{36})^2}{(k_3^X)^2 s_{66}^P},$$

TABLE III-2. ANOMALOUS AND NON-ANOMALOUS ("TRUE") COEFFICIENTS OF FERROELECTRIC CRYSTALS THAT ARE PIEZOELECTRIC IN THE NON-POLAR PHASE

| Piezoelectric–Elastic equations | |
|---|---|
| $P = -dX + k^X E$ <br> $x = -s^E X + dE$ <br> $P = ex + k^x E$ <br> $X = -c^E x + eE$ | $E = bX + \chi^X P$ <br> $x = -s^P X + bP$ <br> $E = -ax + \chi^x P$ <br> $X = -c^P x + aP$ |

| | Anomalous coefficients | | | Non-anomalous coefficients |
|---|---|---|---|---|
| | General symbol | $KH_2PO_4$ | Rochelle salt | |
| Dielectric | $k^X,\ k^x$ | $k_3^X,\ k_3^x$ | $k_1^X,\ k_1^x$ | |
| Piezoelectric | $d,\ e$ | $d_{36},\ e_{36}$ | $d_{14},\ e_{14}$ | $b,\ a$ |
| Elastic | $c^E,\ s^E$ | $c_{66}^E,\ s_{66}^E$ | $c_{44}^E,\ s_{44}^E$ | $c^P,\ s^P$ |

Relations between the coefficients

$$\chi^X = 1/k^X,\ \chi^x = 1/k^x$$
$$d = e s^E$$
$$c^E = 1/s^E$$
$$k^X - k^x = ed$$

$$b = a s^P$$
$$c^P = 1/s^P$$
$$\chi^x - \chi^X = ab$$

$$d/k^X = b$$
$$e/k^x = a$$
$$s^E - bd = s^P$$

which corresponds to a dielectric constant of approximately 700 (K 4).

In conclusion, since the piezoelectric and elastic anomalies of KDP are consequences of the dielectric anomaly of the clamped crystal, it is clear that a successful theory of the ferroelectric effect in KDP should have as its first aim an explanation of the latter anomaly only.

## 5. Optical Properties and Electro-optic Effect

KDP crystals are negative uniaxial. The refractive indices for the sodium $D$ lines are, at 15 °C, $n_1 = n_2 = n_a = 1.5095$ and $n_3 = n_c = 1.4684$. The birefringence ($n_3 - n_1 = 0.0411$) is larger than that of quartz (0.0091) but smaller than that of calcite (0.1721). The birefringence increases with decreasing temperature and shows an anomaly at the transition point (Z 2) (see Fig. III-16). This anomaly is due to the so-called "spontaneous Kerr effect" resulting from spontaneous strain and polarization $via$ piezo-optic and electro-optic effects. These effects are usually very small in normal crystals but become large enough to be observed in ferroelectric crystals. We therefore give in the following a concise treatment of these effects with special reference to the case of KDP. The theory originates from Pockels (P 1) and is reported briefly in Cady's book (C 1).

It is well known that the optical properties of a crystal are characterized by the index-ellipsoid, or indicatrix, defined as

$$\frac{x^2}{n_1^2} + \frac{y^2}{n_2^2} + \frac{z^2}{n_3^2} = 1, \tag{III-15}$$

where $n_1$, $n_2$, $n_3$ are the principal refractive indices. The optical properties of any substance can be affected by external mechanical stresses: this is the effect of $photoelasticity$, which, in crystals, is usually called the $elasto\text{-}optic$ or $piezo\text{-}optic\ effect$ and occurs in $all$ crystals. In addition, in crystals which belong to a non-centrosymmetrical point group, the optical properties, viz. the refractive indices, can be affected by external electric fields. This is the $electro\text{-}optic$ effect. Both the piezo-optic and the electro-optic effect consist in a distortion of the indicatrix (III-15) under the influence of mechanical or electrical stresses. The equation of the deformed indicatrix is, generally:

$$a_{11}x^2 + a_{22}y^2 + a_{33}z^2 + 2a_{23}yz + 2a_{31}zx + 2a_{12}xy = 1 \tag{III-16}$$

where the $a_{ik}$'s (not to be confused with the piezoelectric constants of Eq. (I-4 a)) are called the $polarization\ constants$. The theoretical treatment of the piezo-optic and electro-optic effect, as given by Pockels, consists of expressing the $change$ in the polarization constants in terms of the strains $x_{jl}$ and the polarization $P$, thus:

$$a_{ik} - \delta_{ik}\frac{1}{n_1^2} = \sum_1^3 p_{ikjl}^P x_{jl} + \sum_1^3 r_{ikj}^x P_j \tag{III-17}$$

where $\delta_{ik}$ is the Kronecker symbol, the $p_{ikjl}^P$ are the piezo-optic coefficients at constant polarization and $r_{ikj}^x$ are the electro-optic coefficients at constant strain. The three equations (III-17) for which the Kronecker symbol is equal to 1 ($\delta_{ii}$)

represent a change in length of the principal axes of the indicatrix, the remaining three equations (for which $\delta_{ik} = 0$) represent a rotation of the indicatrix around its axes.

For experimental purposes, it is better to express the change of the polarization constants in terms of stress $X$ and field $E$. We do this in two steps. First we write the change of the polarization constants in terms of strain and field:

$$a_{ik} - \delta_{ik} \frac{1}{n_i^2} = \sum_1^3 p_{ikjl}^E x_{jl} + \sum_1^3 e_{ikj}^x E_j. \tag{III-18}$$

The first term on the right-hand side represents the piezo-optic effect; the second term represents the electro-optic effect, occurring only in non-centrosymmetrical classes. The electro-optic coefficients $e_{ikj}^x$ are, however, not accessible to direct measurements, because an external electric field $E$ causes elastic deformations as a consequence of the piezoelectric effect and the electrostriction, and these deformations, in turn, affect the refractive indices through the piezo-optic effect.

The second step consists of replacing the strains with the stresses and the fields. An expression for the strains in terms of $X$ and $E$ can be obtained from the fundamental expansion of the free energy $A(X,E)$ (Eq. I-3b). Dropping subscripts and summation signs for the sake of simplicity, we obtain for the strains:

$$x = -sX + dE + ME^2. \tag{III-19}$$

Substituting (III-19) into (III-18), we obtain

$$a - \delta_{ik} \frac{1}{n^2} = -psX + (e + pd)E + pME^2 = -gX + fE + hE^2, \tag{III-20}$$

In the most general case (triclinic symmetry), we will have thirty-six non-zero piezo-optic constants $g = ps$; eighteen electro-optic moduli of the *first order* $f = e + pd$; and thirty-six electro-optic moduli of the *second order* $h = pM$. In the special case of KDP, there are only two non-zero electro-optic moduli of the first order, namely $f_{41}$ and $f_{63}$, and seven non-zero electro-optic moduli of the second order, namely $h_{11}$, $h_{12}$, $h_{13}$, $h_{31}$, $h_{33}$, $h_{44}$ and $h_{66}$. The equation of the indicatrix can now be written by replacing the polarization constants from (III-20) into (III-16).

In practice, this equation can be considerably simplified, as it was done in the case of KDP by Zwicker and Scherrer (Z 2). The measurements were carried out on the stress-free crystal ($X = 0$) and the second-order piezoelectric effect could be neglected in all temperature ranges except very close to the transition point. Under these conditions, the equation of the indicatrix can be written as:

$$\frac{x^2 + y^2}{n_1^2} + \frac{z^2}{n_3^2} + 2f_{41}(yE_1 + xE_2)z + 2f_{63}xyE_3 = 1, \tag{III-21}$$

where the variables $x, y, z$ are referred to the principal axes $a_1, a_2, c$ of the crystal according to the choice discussed in connection with Fig. III-8. With Eq. (III-21), we are in the position to discuss all electro-optical effects of the first order in KDP.

The most important effect is that obtained by the application of a field $E_3$ along the tetragonal $c$ axis: $E_1 = E_2 = 0$. In this case, it is clear from (III-21) that the length of the principal axes of the indicatrix along the crystallographic axes $a_1$, $a_2$, and $c$ is unchanged by the field. This means that if we send light in the direction of $a_1$ or $a_2$ we will not be able to observe a first-order electro-optic effect. However, if we send light in the direction of the tetragonal axis $c$, we will be able to notice that the field along $c$ has caused an elongation of the $a_1'$ axis and a contraction of the $a_2'$ axis (Fig. III-8). It is therefore more convenient to rewrite the equation of the indicatrix as referred to the coordinate system $a_1'$, $a_2'$, $c$ introduced in Fig. III-8, thus:

$$\frac{x'^2}{n_1^2}\left(1 + f_{63}\, E_3\, n_1^2\right) + \frac{y'^2}{n_1^2}\left(1 - f_{63}\, E_3\, n_1^2\right) + \frac{z^2}{n_3^2} = 1. \qquad \text{(III-22)}$$

We can easily derive from this equation the birefringence induced by the field in the $a_1'\, a_2'$ plane:

$$n_{a_1'} - n_{a_2'} \cong n_1^3\, f_{63}\, E_3. \qquad \text{(III-23)}$$

Since, in this experiment, the light beam is parallel to the electric field, one speaks of a *longitudinal* electro-optic effect.

If, on the other hand, the light beam is sent in the direction of the $a_1'$ or $a_2'$ axis, i.e. perpendicular to the field (*transversal* electro-optic effect), Eq. (III-22) yields:

$$\left. \begin{aligned} \dot{n}_c - n_{a_1'} &\cong (n_3 - n_1) + \tfrac{1}{2}\, n_1^3\, f_{63}\, E_3\,, \\ n_c - n_{a_2'} &\cong (n_3 - n_1) - \tfrac{1}{2}\, n_1^3\, f_{63}\, E_3\,. \end{aligned} \right\} \qquad \text{(III-24)}$$

Thus, we see that the linear transversal electro-optic effect is superimposed on the natural birefringence of the crystal $(n_3 - n_1)$, and that the effect is half as large as in the case of the longitudinal effect.

If we measure both the longitudinal and the transversal effects as functions of temperature, we have two ways of determining the electro-optic modulus $f_{63}$, i.e. from (III-23) and (III-24). The experimental results of Zwicker and Scherrer (Z 2) are shown in Fig. III-15. It is evident that $f_{63}$ follows a law of the Curie–Weiss type and is therefore anomalous at the transition. Thus, the ratio $f_{63}/\varepsilon_3$ (where $\varepsilon_3$ is the anomalous dielectric constant along $c$) is essentially temperature independent. This indicates that, if we had maintained the treatment in terms of polarization $P$ rather than field $E$ (following Eq. III-17), we would have obtained equations involving non-anomalous, "true" electro-optic coefficients. Such equations would also be more significant from the microscopic point of view, since the primary effect on light is certainly given by the polarization of the lattice rather than the externally applied electric field. We write, in other words, that the change in birefringence $\Delta n$ is proportional to the polarization, thus:

$$\Delta n = \alpha P_3 = \alpha\, \frac{\varepsilon_3 - 1}{4\pi}\, E_3\,,$$

where $\alpha$ is a constant which can be determined experimentally. Comparison of this expression for $\Delta n$ with (III-23) yields a relation between the dielectric constant $\varepsilon_c$ and the electric-optic modulus $f_{63}$ and enables one to "compute" the dielectric constant from the electro-optic effect. The agreement between computed and observed dielectric constant is quite satisfactory (Z 2) and the small discrepancies are probably due to inaccurate values of the observed dielectric constant.

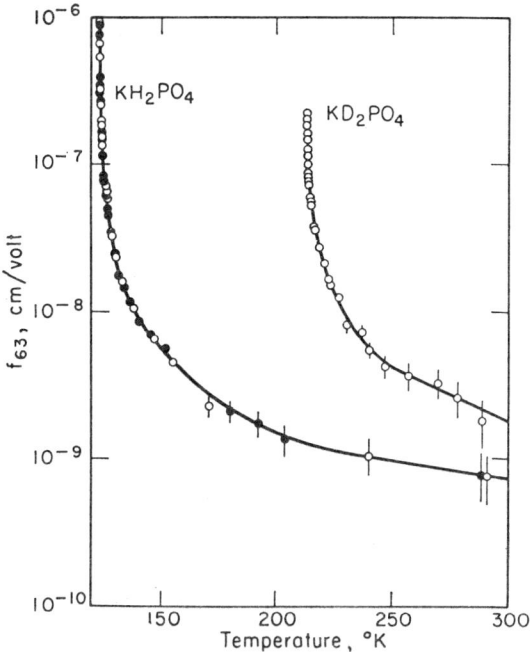

FIG. III-15. Electro-optic modulus $f_{63}$ of KDP as a function of temperature. Open circles: transversal effect; full circles: longitudinal effect (according to Zwicker and Scherrer (Z 2)).

It is interesting to point out, in this connection, that the ratio of $f_{63}/\varepsilon_3$ for KDP is about $1.7 \times 10^{-8}$ c.g.s., while the same quantity for other piezoelectric crystals, e.g. quartz ($f_{41}/\varepsilon_1 = 1.6 \times 10^{-10}$ c.g.s.), is 100 times smaller. The same order of magnitude is found for the difference between the Kerr effect of polar liquids and that of normal, non-polar liquids. This may indicate that the mechanism of dielectric polarization in KDP differs considerably, even above the transition point, from that of normal crystals (K 4).

In the vicinity of the transition temperature, Zwicker and Scherrer observed pronounced deviations from the expected linear effect described by Eqs. (III-23) and (III-24). These deviations are due to effects which arise in part from the saturation effects of the dielectric constant (see Section 2) and in part from the second-order electro-optic effect superimposed upon the first-order effect discussed above.

Below the transition, the onset of spontaneous polarization along the $c$ axis causes the occurrence of a spontaneous electro-optic effect. The spontaneous change in birefringence in the $a_1'c$ or $a_2'c$ plane can be observed, if the crystal is made a single domain by means of an external field along $c$. Figure III-16 shows the experimental results (Z 2). It is significant that, when the crystal is insulated, it splits into equal volumes of positive and negative domains and the experimental curve coincides with the mean value of the curves for a single positive and a single negative domain. The linear effect is compensated and only the quadratic

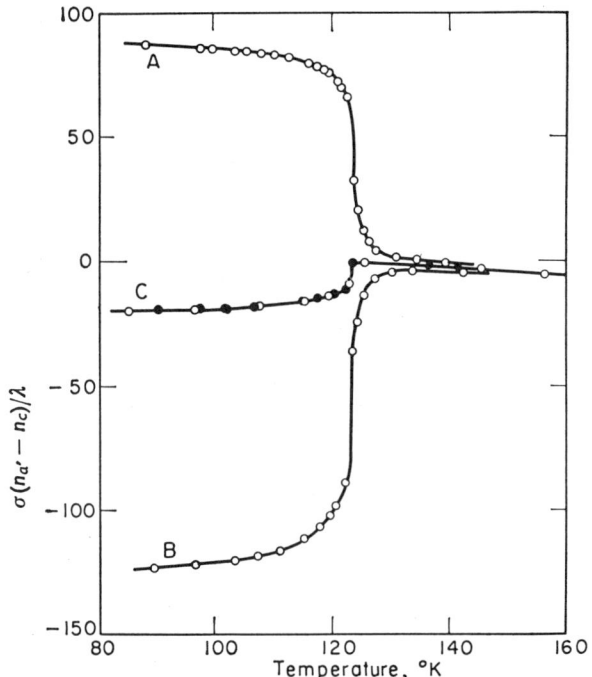

FIG. III-16. Spontaneous change of birefringence $(n_{a_1'} - n_e)/\lambda$ in KDP as a function of temperature. (A) single domain achieved with a bias of $+3000$ V/cm; (B) single domain achieved with a bias of $-3000$ V/cm; (C) Mean value of curves (A) and (B) (full circles) and experimental points obtained on insulated crystal (open circles) (according to Zwicker and Scherrer (Z 2)).

effect can be observed. It should be noticed that the same arguments hold for this quadratic effect which are valid for the spontaneous strain (see Section 4); the lowering of the symmetry involves the occurrence of new electro-optic moduli which are proportional to the polarization.

The experiments show that the ratio between the first-order *spontaneous* change in birefringence and the *spontaneous* polarization has the same value as the ratio between the first-order *induced* change in birefringence and the *induced* polarization above the transition temperature. Thus, the electro-optic effect provides a tool for the measurement of the spontaneous polarization. The birefringence shows hysteresis effects in the ferroelectric region (electro-optic hysteresis). Electro-optic measurements also allow one to determine the temperature dependence of the coercive field, the anomaly of the specific heat, etc.

The behavior of the electro-optic modulus $f_{41}$ would, therefore, provide information about the polarization along the tetragonal $a$ axis. However, this modulus is substantially smaller than $f_{63}$ and its measurement would require the experimental determination of a change in birefringence smaller than $10^{-9}$. For this reason, $f_{41}$ could not be measured by Zwicker and Scherrer.

## 6. Isomorphous Crystals and Isotope Effects

As mentioned in Section 1, a number of crystals isomorphous with $KH_2PO_4$ also undergo transitions to a phase with lower symmetry at low temperatures. With a very few exceptions, the ferroelectric properties of the low-temperature

TABLE III-3. TRANSITION TEMPERATURES $T_c$ OF
CRYSTALS ISOMORPHOUS WITH $KH_2PO_4$

| Substance | $T_c$ (°K) | References |
|-----------|-----------|-----------|
| $KH_2PO_4$ | 123 | V 1 |
| $KD_2PO_4$ | 213 | B 6 |
| $KH_2AsO_4$ | 97 | B 6 |
| $KD_2AsO_4$ | 162 | S 2 |
| $RbH_2PO_4$ | 147 | M 3, B 16 |
| $RbD_2PO_4$ | 218 | M 3 |
| $RbH_2AsO_4$ | 110 | S 2 |
| $RbD_2AsO_4$ | 178 | S 2 |
| $CsH_2PO_4$ | 159 | S 3 |
| $CsH_2AsO_4$ | 143 | S 2 |
| $CsD_2AsO_4$ | 212 | S 2 |

phase were confirmed by direct dielectric investigations. In these cases, the behavior was found to be similar to that of KDP.

A very large effect of isotopic substitution (replacement of hydrogen by deuterium) was found in all crystals of this group of ferroelectrics. Table III-3 lists all the ferroelectrics so far investigated in the KDP family. It is seen that the isotope effect consists of a very large change in the transition temperature; the deuterated salts undergo the ferroelectric transition at higher temperatures than the corresponding hydrogen salts, the amount of the displacement of transition temperature varying from 90 °C in $KH_2PO_4$ to 67 °C in $KH_2AsO_4$. The magnitude of this isotope effect is an indication of the vital role played by the hydrogen bonds in the mechanism leading to ferroelectricity.

A number of solid solutions between the crystals listed in Table III-3 were studied by Matthias and Merz (M 7). Starting with $KH_2PO_4$, the Curie temperature is increased by substituting K partially with Tl, while it is decreased when K is partially replaced by Rb, Cs and $NH_4$. For example, a content of 7.1% Tl in $KH_2PO_4$ raises the Curie point to 126 °K, a much larger (underterminined) content of Cs in $KH_2PO_4$ lowers the Curie temperature to 108 °K.

It may be pointed out again here, as mentioned in Section 1, that the ammonium salts of this series of phosphates and arsenates undergo transitions to antipolar states, which we do not discuss in the present book. Large isotope effects are observed for these transition temperatures as well, namely from $-125\,°C$ for $NH_4H_2PO_4$ to approximately $-28\,°C$ for $ND_4D_2PO_4$, and from $-57\,°C$ for $NH_4H_2AsO_4$ to $+31\,°C$ for $ND_4D_2AsO_4$. For a general discussion of the properties of these crystals, the reader is referred to the review articles by Shirane et al. (S 7), by Känzig (K 4) and to the book of Megaw (M 4).

## 7. Structural Characteristics

The most important contribution to an understanding of the atomistic mechanism occurring at the ferroelectric transition of $KH_2PO_4$ came from structural investigations by means of X-ray and neutron diffraction. The first X-ray structural study was done by West (W 2), continued by de Quervain (D 2) and later refined by Frazer and Pepinsky (F 1), who concentrated their attention to the structural changes occurring in a narrow temperature range immediately above and below the transition point. Although the X-ray results gave strong indications as to the location of the hydrogen atoms in the lattice from the interatomic distances between oxygen atoms, it was not until application of neutron diffraction techniques that the hydrogen positions could be accurately located in the structure. The essential role played by neutron diffraction analysis in problems of this sort is due to the difference between the scattering processes of X-rays and neutrons. X-rays are scattered by electrons, thus the scattering factor of the various elements increases with increasing atomic number and is smallest for hydrogen. Neutron scattering, on the other hand, occurs at the nucleus and thus the scattering factors of the chemical elements do not differ much from each other. For details of the neutron diffraction methods, the reader is referred to the book of Bacon (B 9).

Detailed neutron diffraction analyses of $KH_2PO_4$ were carried out by Bacon and Pease (B 10), (B 11), Peterson et al. (P 2) and Levy et al. (L 1). The overall picture of the transition occurring in $KH_2PO_4$ at 123 °K, as obtained both from X-ray and neutron studies, reveals the role played by the hydrogen atoms in the co-operative phenomenon which leads to ferroelectricity. It should be mentioned here that the most successful theory of $KH_2PO_4$ was developed by Slater (S 4) on a model hypothesized on the basis of the early West's structure (see Section 8). This model assumed the very ordering of the hydrogen atoms which was later beautifully confirmed by the neutron analysis.

The symmetry of the room-temperature phase of $KH_2PO_4$ is tetragonal. The space group and the dimensions of the unit cell depend on whether we refer it to the coordinate system $a_1$, $a_2$, $c$ or $a_1'$, $a_2'$, $c$ of Fig. III-8. In the former case, the space group is $I\bar{4}2d$ and the lattice dimensions are

$$a_1 = a_2 = 7.453 \text{ Å}, \quad c = 6.959 \text{ Å},$$

the unit cell containing four formula units. In the second case, the space group is $F\bar{4}d2$ and the new cell dimensions are:

$$a_1' = a_2' = 10.534 \text{ Å}, \quad c = 6.959 \text{ Å},$$

the cell containing now eight formula units. The latter cell, as we know, remains rectangular below the transition and the space group of the polar orthorhombic phase is $Fdd$. The changes in cell dimensions with decreasing temperature are summarized in Table III-4.

TABLE III-4. TEMPERATURE DEPENDENCE OF THE LATTICE CONSTANTS
OF $KH_2PO_4$

| Temperature (°K) | Space group | Lattice constants (Å) | | | References |
|---|---|---|---|---|---|
| | | $a$ | $b$ | $c$ | |
| 295 | $F\bar{4}d2$ | 10.534 | — | 6.959 | B 10 |
| 132 | $F\bar{4}d2$ | 10.495 | — | 6.919 | B 11 |
| 77 | $Fdd2$ | 10.458 | 10.540 | 6.918 | B 11 |
| 126 | $F\bar{4}d2$ | 10.48 | — | 6.90 | F 1 |
| 116 | $Fdd2$ | 10.44 | 10.53 | 6.90 | F 1 |

The framework of the room-temperature structure is depicted in Fig. III-17 as obtained from West's data. Each phosphorous atom is surrounded by four oxygens at the corners of a tetrahedron which is almost regular (being com-

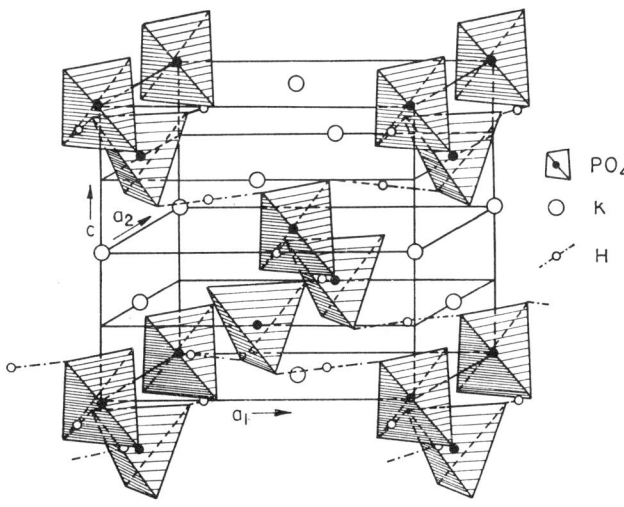

FIG. III-17. Structure of $KH_2PO_4$. Unit cell in space group $I\bar{4}2d$ (according to West (W 1)).

pressed by approximately 2% along the $c$ axis). These $PO_4$ groups, together with the potassium atoms, build up the structure in such a way that K and P atoms alternate each other at a distance $c/2$ in the direction of the $c$ axis. Every $PO_4$ group is linked to four other $PO_4$ groups, spaced $c/4$ apart along $c$, by hydrogen bonds. Thus, the linkage is such that there is a hydrogen bond between one "upper" oxygen of one $PO_4$ group and one "lower" oxygen of the neighboring $PO_4$ group, and each hydrogen bond lies very nearly perpendicular to the $c$ axis.

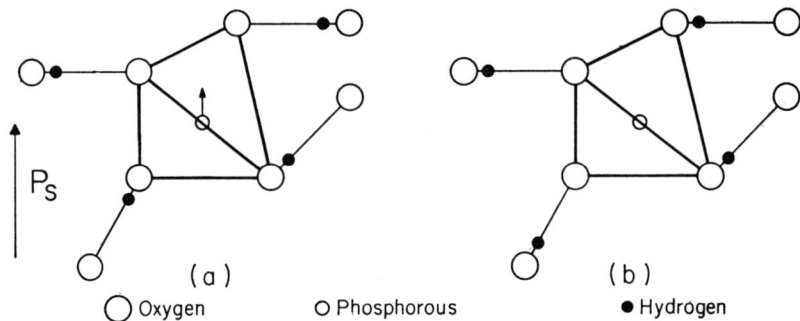

FIG. III-18. Configurations of the $(H_2PO_4)^-$-groups. (a) Dipole moment oriented along the positive $c$ axis. Arrow at the P atom indicates the direction of the shift of this atom below the Curie temperature; (b) No dipole along the $c$ axis.

As revealed by the neutron diffraction studies, only two hydrogens are located nearest any one $PO_4$ group; therefore, as a group they form $(H_2PO_4)^-$ ions. There are six ways in which two hydrogens can do so, and this gives rise to six different configurations of the $H_2PO_4$ groups, two when the hydrogens are nearest to both "upper" or both "lower" oxygens and four when the hydrogens are nearest to an "upper" and a "lower" oxygen (Fig. III-18). The $PO_4$ tetrahedra are also linked by the potassium atoms. Each K is surrounded by.eight oxygens and is somewhat closer to four of them than to the other four. The interatomic distances of the room temperature structure are given in the first column of Table III-5 and a schematic projection of the structure on the (001) plane is depicted in Fig. III-19.

Information about the location of the hydrogens in the room temperature structure is given by the Fourier projection on the (001) plane from neutron diffraction data, Fig. III-20(a). The hydrogen nuclei appear as kidney-shaped dotted contours between oxygens of adjacent $PO_4$ tetrahedra. These contours are elongated along the axis of the bond, a fact that can be explained by either one of the following hypotheses:

(i) the hydrogens perform strong anisotropic vibrations along the bond axis; or

(ii) the hydrogens are distributed statistically off center, the distance between the two positions being about 0.35 Å. The diffraction technique is not able to distinguish between these two alternatives, and a decision must be made on other grounds (see below).

In going through the transition the length of the hydrogen bonds do not change much (Table III-5), but below the transition temperature the hydrogens are ordered in such a way that in a single-domain crystal they are, for example, all near "upper oxygens" or near "lower oxygens", depending on the polarity (see Fig. III-18a). Reversal of the polarity, for example by means of an electric field, produces a shift of the hydrogen atoms along the $O-H-O$

bonds from one set of ordered positions to the other. Figure III-20(b) shows the Fourier projection on (001) at 77 °K. The hydrogen peaks are displaced about 0.20 Å from the center of the O−H−O line. The structure can be looked upon as being built of K atoms and $H_2PO_4$ groups.

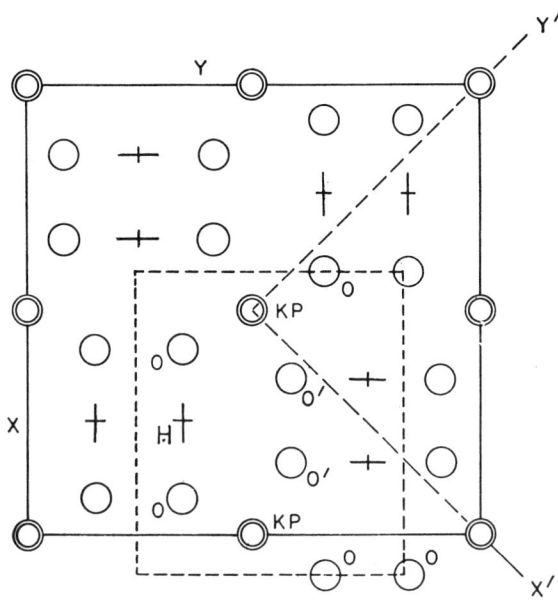

FIG. III-19. Schematic projection of the $KH_2PO_4$ structure on (001) (according to Bacon and Pease (B 10)).

TABLE III-5. BOND LENGTHS IN $KH_2PO_4$ [ACCORDING TO BACON AND PEASE (B 11)]

(The lengths have been recomputed in ångström units from the original data (M 4); $O_H$ = oxygen with a close hydrogen.)

| Bond | Room temperature | 132 °K | 77 °K |
|---|---|---|---|
| $P-O_H$<br>$P-O$ | $1.541 \pm 0.05$ | $1.541 \pm 0.004$ | $1.586 \pm 0.02$<br>$1.511 \pm 0.02$ |
| $O_H-O_H$<br>$O-O$ | $2.533 \pm 0.007$ | $2.533 \pm 0.009$ | $2.526 \pm 0.006$<br>$2.554 \pm 0.006$ |
| $O-O'_H$<br>$O'_H-O$ | $2.508 \pm 0.008$ | $2.508 \pm 0.005$ | $2.524 \pm 0.02$<br>$2.517 \pm 0.02$ |
| $O-H-O$ | $2.492 \pm 0.005$ | $2.480 \pm 0.004$ | $2.491 \pm 0.004$ |
| $O_H-H$ | $1.07 \pm 0.01$ | $1.07 \pm 0.01$ | $1.05 \pm 0.014$ |
| $O-H$ | $1.42 \pm 0.01$ | $1.41 \pm 0.01$ | $1.43 \pm 0.014$ |
| $K-O_H$<br>$K-O$ | $2.894 \pm 0.006$ | $2.876 \pm 0.003$ | $2.902 \pm 0.03$<br>$2.839 \pm 0.03$ |
| $K-O'_H$<br>$K-O'$ | $2.825 \pm 0.004$ | $2.808 \pm 0.005$ | $2.818 \pm 0.01$<br>$2.783 \pm 0.01$ |

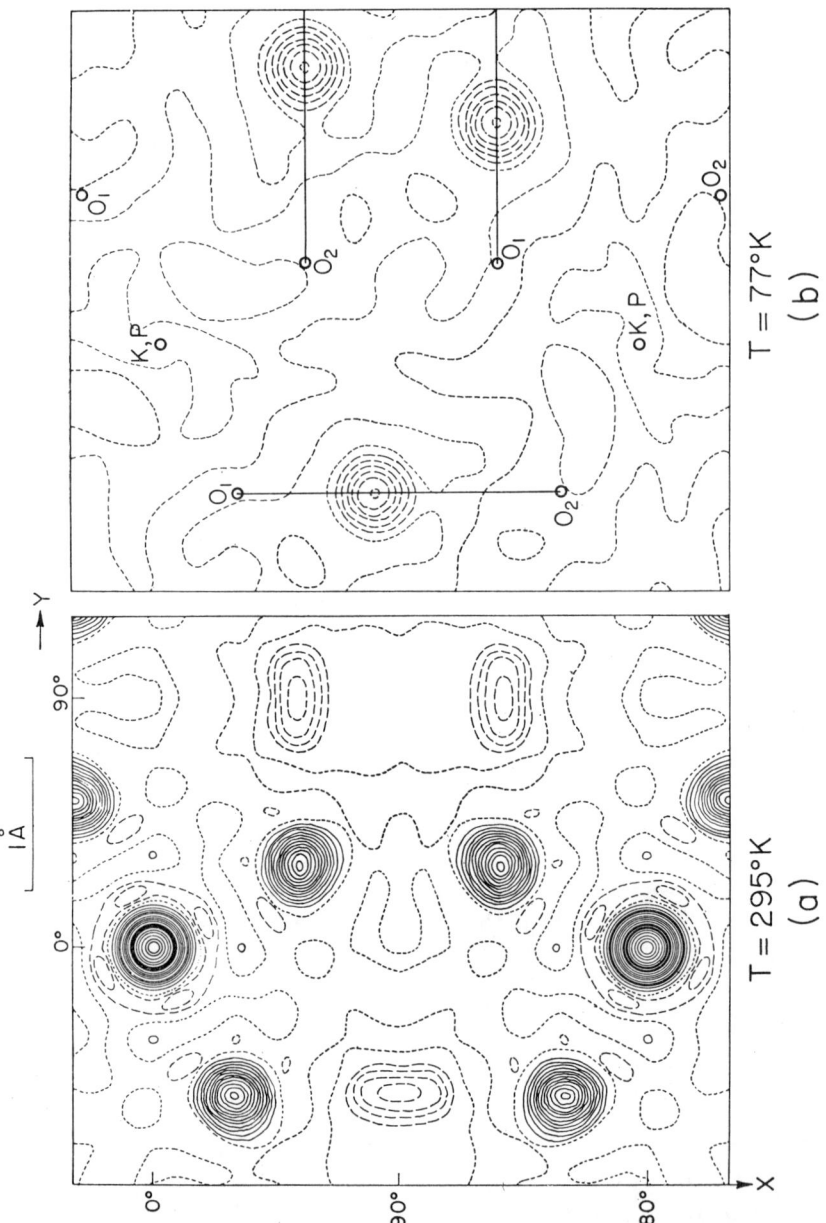

FIG. III-20. Fourier projection on (001) of $KH_2PO_4$. Solid lines are positive, broken lines negative and dotted lines zero contours. This figure corresponds to the region outlined by the dotted lines in Fig. III-19 (according to Bacon and Pease (B 10)): (a) at room temperature; (b) at 77 °K, hydrogen atoms only.

This ordering of the hydrogen atoms is connected with a significant, although small, shift of the other atoms from their equilibrium position in the tetragonal phase. The overall effect of the transition on the other atoms is the following. The framework of the oxygen atoms is very little affected, except for the small shear of the lattice as a whole. Referred to the oxygen framework, the potassium and phosphorous atoms move along the $c$ axis in opposite directions by about 0.04 Å and 0.08 Å, respectively, so that both are farther from those oxygens which have hydrogen atoms close to them (B 11). The X-ray investigation of Frazer and Pepinsky (F 1) yields somewhat different results for the displacement of the potassium and phosphorous atoms, namely 0.05 Å and 0.03 Å in opposite directions, respectively.

Since the hydrogen bonds are only about 0.5° out of parallelism with the (001) plane, the $c$ component of the hydrogen shift is far too small to account for the total polarization. Thus, it seems more likely that the polarization arises from the displacements of the potassium and phosphorous atoms in the direction of the $c$ axis. We can actually calculate the value of the (ionic) polarization, if we make certain assumptions about the effective charge of each ion in the structure. Assuming for example, such electronic states as $P^{5+}$, $K^{1+}$ and $O^{2-}$, we can calculate from the atomic positions reported by Bacon and Pease that the value of the (ionic) spontaneous polarization ought to be $5.0 \times 10^{-6}$ C/cm². This result is in fairly good agreement with the experimental value $P_s = 4.7 \times 10^{-6}$ C/cm². In the case assumed, the direction of the spontaneous polarization is the same as that of the shift of the $P^{5+}$ ion, as indicated in Fig. III-18(a). More precisely, the polarization is pointing from the oxygen ions that have a hydrogen close to them toward the phosphorous ion. Thus, when the polarization is "up", the hydrogens are nearest to the "lower" oxygens of each $PO_4$ group.

Bacon and Pease made a different assumption about the electronic states of potassium and phosphorous, namely $K^{1+}$ and $P^{3-}$. In this way, they calculated a magnitude of $4.7 \times 10^{-6}$ C/cm² for the spontaneous polarization, which is in better agreement with the experimental data but results in a direction of the spontaneous polarization opposite to that described immediately above. There is admittedly no *a priori* reason for rejecting the model advanced by Bacon and Pease, but their charge assignment is quite difficult to justify.

The ordinary X-ray and neutron analyses are not in a position to give an answer to this problem, as they cannot distinguish between positive and negative directions. The technique based upon the anomalous dispersion of X-rays, however, can do so and it allows the determination of the absolute configuration of polar structures (see Chapter X). The absolute configuration of $KH_2PO_4$ has recently been determined in this way by Unterleitner et al. (U 4). The results of this investigation are in accordance with the first assumption made above, namely that of electronic states $K^{1+}$ and $P^{5+}$.

Thus, the mechanism of the transition in $KH_2PO_4$ appears to be rather clearly elucidated by the X-ray and neutron diffraction analyses. Discrepancies between the two methods are found only with regard to the thermal vibrations of the atoms, in that X-ray evidence seems to point toward strong anisotropic vibrations of some atoms (namely K and P), while the neutron data reveal no significant difference, either above or below the transition point, for the amplitude of vibration of any atom in different directions. The latter is true also for the hydrogen atoms, if the elongated peak observed in the (001) projection at room temperature (Fig. III-20a) is assumed to be due to a double minimum potential curve, with half-hydrogen, on the average, in each minimum position.

As stated above, a distinction between the alternatives of anisotropic thermal oscillations and statistical disorder of the hydrogens in the tetragonal phase cannot be made on the basis of the diffraction data alone. The problem has been attacked by the study of infrared and Raman spectra (K 1), (L 2), (L 3), (M 5), (O 1), (R 1). Unfortunately, however, the $O-H$ stretching frequency in crystals containing short hydrogen bonds, such as KDP, is often difficult to identify because of the large frequency displacement and breadth of the bands. In addition, the bands are characterized by low peak intensity, rather flat peaks and often the appearance of several maxima of comparable intensity. These factors make the assignment of a single frequency to the band difficult and somewhat arbitrary. In spite of these difficulties, comparison between the infrared spectra of $KH_2PO_4$ and $KD_2PO_4$ led various authors to assign wavenumbers of the order of 2800 cm$^{-1}$ to 2200 cm$^{-1}$ to the hydrogen vibration. Blinc and Hadzi (B 14) gave an interpretation of the particularly interesting spectral region around 2800 cm$^{-1}$ in terms of tunneling of the protons between two minima of potential energy. The minima were taken to be of equal depth in the non-polar phase and unsymmetrical in the ferroelectric phase. A quantum-mechanical treatment of this vibrational problem yielded results which are in agreement with the experimental data.

It should be mentioned, on the other hand, that recent observations, by Pelah et al. (P 3), of hydrogen vibration frequencies in $KH_2PO_4$, by means of inelastic scattering of cold neutrons, indicate that the assignment of the infrared bands around 2800 cm$^{-1}$ to hydrogen vibrations may be questionable. According to the scattering data, it is the infrared reflectance peak at 540 cm$^{-1}$ which should be assigned to the hydrogen vibration mode. However, the information provided by the neutron inelastic scattering data is too scarce, at the present time, to allow a clarification of the contradiction with the infrared results.

The important point is that, irrespective of which of the infrared peak should be identified as due to hydrogen bond vibrations, all infrared and Raman investigation are in accordance in reporting *no change* of the pertinent bonds through the transition point. This means that the $O-H \cdots O$ bond does not change and the $O-H \cdots O$ distances remain essentially the same through the transition. The infrared peaks simply become more intense in the ordered ferroelectric phase.

The infrared measurements, on the other hand, are in agreement in assigning two reflection peaks at about 1100 cm$^{-1}$ and 900 cm$^{-1}$ to the $PO_4$ vibration (M 5). This is confirmed by the neutron inelastic scattering experiments carried out, for comparison, on $K_3PO_4$, $K_2HPO_4$ and $KH_2PO_4$ (P 3).

Nuclear magnetic resonance experiments also show no significant changes in the line width and relaxation time of the proton magnetic resonance at the transition. Furthermore, recent experiments concerned with the deuteron magnetic resonance spectrum and relaxation on $KD_2PO_4$ establish that the electric field gradient tensor at the site of the deuteron does not change appreciably with temperature over the transition point (B 12). This indicates again that the nature of the hydrogen bond is not significantly affected by the transition.

## 8. Theoretical Treatments

The recognition that the anomalies of the piezoelectric, elastic and electro-optic properties of $KH_2PO_4$ are due, in the final instance, to the anomaly of the dielectric properties of the clamped crystal led the first theoretical efforts toward an explanation of the phenomenon on the basis of *permanent* electric dipoles. The theory of dielectric polarization due to permanent dipoles was developed earlier by Debye from an adaptation of Langevin's theory of paramagnetism to the study of dielectrics. Accordingly, the early theoretical approach to the ferroelectricity of $KH_2PO_4$ was based on the Langevin–Weiss theory of ferromagnetism. A review of such a treatment can be found in an article by Baumgartner et al. (B 13). The interaction between the dipoles was taken into account by way of the Lorentz formula for the internal field: $F = E + \gamma P$, where $E$ is the external field, $P$ the polarization and $\gamma$ the Lorentz factor. We already know that, if the medium is isotropic or has cubic symmetry, $\gamma$ has the value $4\pi/3$. In crystals of lower symmetry, $\gamma$ has values differing from $4\pi/3$, although of the same order of magnitude, and, moreover, it has tensor character. Straightforward application of Boltzmann statistics to an assembly of dipoles in the Lorentz field $F$ leads to the Langevin function and to the prediction of a Curie temperature. However, the quantitative agreement between theory and experiment is so poor that it casts considerable doubt over the validity of Lorentz's internal field concept as a means to express the interaction between the dipoles. The formula of Onsager is a better approximation of this interaction*, but it has

---

* The explicit assumption made in the Lorentz treatment is that the dipole moments of all atoms are parallel, an assumption which is obviously valid for the *induced* moments in structures with sufficiently high symmetry, but is not valid for *permanent* dipole moments oriented at random. Onsager's model is that of a spherical molecule with radius $a$ and a point dipole $\mu$ at its center. The other molecules of the dielectric are treated as a continuum with dielectric costant $\varepsilon$, equal to the actual (unknown) dielectric constant of the substance considered. Onsager's formula is then:

$$\frac{(\varepsilon - 1)(2\varepsilon + 1)}{12\pi\varepsilon} = \frac{N}{1 - f\alpha}\left[\alpha + \frac{\mu^2}{(1 - f\alpha)\,3kT}\right],$$

where $\alpha$ is the polarizability and $\quad f = \dfrac{1}{a^3} \cdot \dfrac{2(\varepsilon - 1)}{2(\varepsilon + 1)}.$

This formula can be simplified markedly for the case of negligible polarizability ($\alpha = 0$), thus

$$\varepsilon = \frac{1}{4}\left[\left(1 + \frac{4\pi N\mu^2}{kT}\right) + \sqrt{\left\{\left(1 + \frac{4\pi N\mu^2}{kT}\right)^2 + 8\right\}}\right],$$

whence one can see, incidentally, that the "$4\pi/3$ catastrophe" is avoided, since $\varepsilon = \infty$ only at $T = 0$.

been shown (P 4) that this formula can only predict a first-order transition from a polar into a non-polar state.

Thus, it appears that the consideration of long-range forces, electrostatic in character, arising from the fields of the dipoles themselves, is insufficient for a quantitatively satisfactory explanation of the ferroelectric phenomenon in $KH_2PO_4$. A considerably better picture is obtained with Slater's theory (S 4), in which short-range forces provide the entire interaction force. As pointed out previously, Slater's approach is based on the statististical treatment of a model which was derived from West's unrefined structure of $KH_2PO_4$ long *before* neutron diffraction could prove it correct. Slater's theory represents the first successful attempt to explain the ferroelectricity of $KH_2PO_4$ from a molecular point of view.

### The Theory of Slater

The structural studies (Section 7) have revealed the role played by the hydrogen atoms constituting the bonds which link neighboring $PO_4$ groups. These $PO_4$ groups are distributed on a tetragonal diamond-type lattice and each group is surrounded tetrahedrally by four neighboring groups. On every hydrogen bond, there are two equilibrium positions for the hydrogens located symmetrically with respect to the center of the bond. Slater assumes that the distribution of the hydrogens between these positions is subject to two restricting conditions:

(i) There is one and only one hydrogen on each bond.

(ii) There are only two hydrogens near any one $PO_4$ group.

These two conditions are very similar to those postulated by Pauling for the structure of ice in order to explain its zero-point entropy (P 5). Accordingly, they mean that the structure is built up of $(H_2PO_4)^-$ groups (and K) and that molecular groups such as $(H_3PO_4)$ or $(HPO_4)^{2-}$ are so unlikely to occur that they can be neglected completely. We have seen in the preceding section that there are six possibilities to arrange two hydrogens nearest to a given $PO_4$ group. Where the two hydrogens are closest to the two "upper" oxygens of the $PO_4$ group the resulting $(H_2PO_4)^-$ group is considered to be a dipole pointing in the negative direction of the tetragonal $c$ axis; when the two hydrogens are nearest to the "lower" oxygens, the $(H_2PO_4)^-$ dipole points toward the positive end of the $c$ axis; and when one hydrogen is nearest to one "upper" and the other to one "lower" oxygen of the $PO_4$ group (there being four different ways in which this case can be realized), the $(H_2PO_4)^-$ dipole is oriented perpendicularly to the $c$ axis.

These six configurations are not equivalent. Slater proposes that the two configurations for which either both "upper" or both "lower" oxygens have hydrogens nearest to them (corresponding to dipoles pointing down or up) are energetically equivalent to each other and associated to an energy which is normalized to zero. The remaining four configurations, corresponding to dipoles oriented perpendicular to the $c$ axis, are also assumed to be energetically equivalent to one another and associated with an energy parameter $\varepsilon^S$. It is assumed, in other words, that the energy of the crystal in the absence of external electric fields is given by the number of those dipoles which are perpendicular to the $c$ axis multiplied by the constant parameter $\varepsilon^S$. If however, an electric field $E$ is applied along the positive $c$ axis, then the crystal energy is decreased by an amount which is equal to the product of $E$ and the electric moment of the crystal along $+c$. Calling $\mu$ the dipole moment, it follows that the internal energy of the crystal in the presence of a field $E$ is:

$$U = N_0 \varepsilon^S - (N_+ - N_-) \mu E, \qquad \text{(III-25)}$$

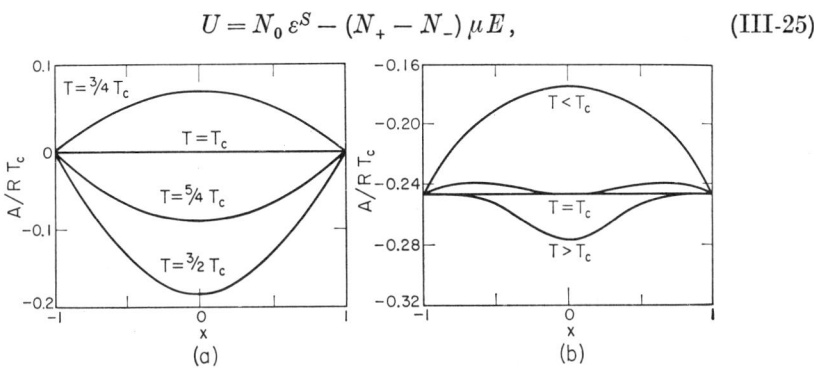

FIG. III-21. Free energy as a function of relative polarization: (a) according to Slater's theory (S 4); (b) according to Yomosa and Nagamiya (Y 1).

where $N_0$, $N_+$ and $N_-$ are the numbers of $(H_2PO_4)^-$ dipoles pointing across, up, and down the tetragonal $c$ axis, respectively. ($N_0 + N_+ + N_- = N$, where $N$ is the total number of dipoles.) There are many ways in which a state $N_+, N_-, N_0$ can be realized, and the problem is to find the number $W(N_+, N_-, N_0)$ of arrangements of the hydrogens which lead to such a state. If we can solve this problem, we can compute the entropy of the system as $S = k \cdot \ln W(N_+, N_-, N_0)$, where $k$ is the Boltzmann constant, and apply the usual thermodynamic formalism to compute the dielectric and thermal properties of the crystal.

Slater finds an expression for the function $W(N_+, N_-, N_0)$ by counting the number of the pertinent hydrogen arrangements with a method that consists of building up the crystal molecule by molecule. He assumes that, at a given step in the process of building up, the crystal already contains a certain number of molecules and that he adds another molecule to the surface. Considering the various ways in which the hydrogens of this new molecule can be arranged, he then finds how the number of arrangements increases on account of the addition of this molecule. He finally arrives at a formula for $W(N_+, N_-, N_0)$. Takagi (T 3) obtained exactly the same formula using a different counting procedure.

When the function $W$ is known the free energy $A(N_+, N_-, N_0) = U - TS$ can be computed. The condition for a minimum of the free energy allows the elimination of $N_0$ and the expression of the free energy in terms of the variable $x = (N_+ - N_-)/N$. Figure III-21 (a) is a graphical representation of $A(x)$ for the case $E = 0$: $x = \pm 1$ corresponds to the fully polarized state and $x = 0$ to the unpolarized state. The latter is stable for temperatures $T > T_c$, the former occurs for $T < T_c$. At the transition temperature $T_c$, the crystal goes from the unpolarized to the *fully* polarized state.

Thus, the function $W$ leads directly to the *existence* of a phase change. The Curie temperature $T_c$ is given by the formula:

$$kT_c = \frac{\varepsilon^S}{\ln 2}, \qquad (\text{III-26})$$

and is, therefore, directly proportional to the energy parameter $\varepsilon^S$. The theory accounts for the temperature dependence of the dielectric suspectibility in a satisfactory way, leading to the Curie–Weiss law:

$$\frac{\partial P}{\partial E} = \frac{N\mu^2}{k\ln 2}\,\frac{1}{T-T_c}. \qquad (\text{III-27})$$

The entropy difference between the polarized and the unpolarized state is computed to be 0.69 cal/mole °C at the Curie temperature. Above the transition, the theory predicts an additional entropy increase, spread over a wide temperature range, such as to give a total entropy difference of 0.805 cal/mole °C. These entropy changes are in satisfactory agreement with the experimental results (see Table III-1).

The character of the transition predicted by Slater's theory is very remarkable. It is a very unusual type of transition, as every intermediate state between the fully unpolarized and the completely polarized state is just as stable as the initial and the final state. It is similar to the usual transition of the first order insofar as polarization and energy undergo discontinuous changes and there is latent heat. On the other hand, it is similar to a transition of the second order, insofar as the phase change takes place in a homogeneous way and thus every quantity is continuous in time; it could be considered as the limiting case of a second order transition.

Experimentally, as we know, the transition is not sudden, but takes place through a finite temperature range. Slater attributes the observed gradual increase of the spontaneous polarization and the $\lambda$-type anomaly of the specific heat to irregular distribution of stresses within the crystal, caused by the spontaneous strain, which give rise to different transition temperatures for different domains.

In spite of this explanation, some objections have been raised against Slater's theory because of the discrepancy with the experimental data concerning the character of the transition. Since the theory is treating the tetragonally clamped crystal, it is reasonable to ask whether the free crystal could have a second-order transition. Forsbergh (F 2) has shown that this is in fact the case, but only if the relation between strain and polarization is complemented with saturation terms, as in Eq. (III-3).

One may also ask whether Slater's result is due to the counting procedure followed for the derivation of the function $W(N_+, N_-, N_0)$. This question was answered by Takahashi (T 1), who approached the same model of Slater in a different way. He formally replaced the hydrogen bonds with "arrows" and studied how many configurations of arrow "chains" are possible, within the crystal, for a certain number of positive and negative chains to exist. In this terminology, Slater's second assumption (ii) is equivalent to saying that, at each

$PO_4$ group, there are two in-going and two out-going "arrows". Takahashi computed the partition function rigorously for all possible chain configurations and arrived at exactly the same result as Slater, i.e. a discontinuous transition for which, at $T = T_c$, the states involving any arbitrary ratio between the numbers of positive and negative chains are equally stable.

One then could think of changing the character of Slater's transition by modifying the theory in such a way as to include the deformation of the crystal, due to piezoelectric interaction, thus accounting also for the anomalous piezoelectric effect. This approach is due to Yomosa and Nagamiya (Y 1). If the crystal is strained with the shear $x_y$, the $PO_4$ groups will change their orientation and the configurational energy of the crystal will also change. This fact is taken into account by assuming that the energy difference between a $(H_2PO_4)^-$ group polarized along $+c$ or $-c$ and a group polarized perpendicular to $c$ is a linear function of the strain. Calling $\varepsilon_+$ the energy of a positive group, $\varepsilon_-$ that of a negative group, it is postulated that:

$$\varepsilon_+ = -\varepsilon^S - \beta x_y$$

$$\varepsilon_- = -\varepsilon^S + \beta x_y$$

The fact that, in the strained crystal, the four orientations of a $(H_2PO_4)^-$ group polarized perpendicular to the $c$ axis are no longer equivalent is neglected. Slater's formula (III-25) becomes, under these assumptions:

$$U = N_+ \varepsilon_+ + N_- \varepsilon_- - (N_+ - N_-)\mu E + \tfrac{1}{2} a N x_y^2 - \tfrac{1}{2} k E^2 - f x_y E \quad \text{(III-28)}$$

where the fourth term represents that part of the elastic energy which is independent of the hydrogen-bond configurations, the fifth term is the energy due to the polarization induced by an external field (independently of the hydrogen bond configuration), and the last term represents the interaction energy between the polarization induced piezoelectrically by the strain $x_y$ (not involving the hydrogen bonds), and the field $E$.

Combining this $U$ function with Slater's $W$ function, a free energy function is found which allows the discussion of the dielectric, piezoelectric and elastic properties of the crystal. This theory yields very satisfactory results in that it predicts a difference of 3.7 °C between the transition temperature of the free crystal and that of the clamped crystal, in agreement with the experimental results (see Section 2), and it predicts a spontaneous deformation of the free crystal, below the transition, in the form of a shear $x_y = 26'$, also in excellent agreement with the experimental data (see Section 3). The theory predicts, in addition, an anomaly of the elastic compliance $s_{66}^E$ at the transition temperature; the thermal behavior of $s_{66}^E$ is quantitatively in good agreement with the measurements. The character of the transition is changed, with respect to that given by Slater's theory, insofar as there is a range in temperature in which the states with $x = 0$ and $x = \pm 1$ coexist. Figure III-21 (b) shows the free energy as a function of $x$ as obtained from this theory. The transition is now typically of the first order. Thus, although the modifications suggested by Yomosa and Nagamiya lead to considerable improvements of the original Slater's theory, the discontinuous character of the transition is still in disagreement with the experimental results.

This discrepancy, however, can be avoided by only slightly relieving the infinitely tight interrelations of hydrogen positions expressed by Slater's assumptions (i) and (ii) (see p. 96). Relief of the first assumption was undertaken by Shirane and Oguchi (S 5), who allowed a small probability that two or no hydrogens can temporarily reside on any one of the bonds. Relief of the second assumption was studied by Takagi (T 2), who allowed a small probability for the existence of $(H_3PO_4)$ or $(HPO_4)^{2-}$ groups. In both cases, an energy parameter $\varepsilon_i$ is assigned to what, in each case, can be considered a lattice imperfection, and the effect of a change in the ratio $\varepsilon_i/\varepsilon^S$ from zero to infinity on the character of the transition is studied. It is found that Slater's step function is rounded off with decreasing values of $\varepsilon_i/\varepsilon^S$, and satisfactory agreement with the experimental data is reached when $\varepsilon_i/\varepsilon^S$ ranges from 5 to 10. This result is in keeping with the expectation that the probability of occurrence of these lattice imperfections must be rather low. However, there is no direct proof that the value of this ratio is appropriate.

A different approach to Slater's model has recently been made by Grindlay and ter Haar (G 1). A positive or negative dipole moment is assigned to a hydrogen bond according to whether the proton occupies one or the other of the two possible equilibrium positions along the bond. Each bond interacts with three other bonds on each of the two $PO_4$ groups it connects. If only nearest-neighbor interactions are considered, the problem is essentially one of the Ising type. Treatment of this problem with the Bethe method (B 15) under consideration of Slater's assumption (ii) yields results which are fully equivalent to those of Slater's calculations. The treatment is thereafter extended along the path indicated by Takagi (T 2), with the difference that not only configurations such as $H_3PO_4$ and $(HPO_4)^{2-}$ are allowed, but also configurations such as $(H_4PO_4)^+$ and $(PO_4)^{3-}$. The energy parameters associated with these four configurations are chosen to be $\varepsilon_i$, $\varepsilon_i$, $2\varepsilon_i$ and $2\varepsilon_i$, respectively, while the energy associated with the configuration $(H_2PO_4)^-$ is normalized to zero, as in Slater's theory. The result of the theory of Grindlay and ter Haar shows that the order of the transition depends on the ratio $\varepsilon_i/\varepsilon^S$ where $\varepsilon^S$ is Slater's energy parameter. If $\varepsilon_i/\varepsilon^S > \dfrac{3}{2}$, the transition is of the first order; if $\varepsilon_i/\varepsilon^S < \dfrac{3}{2}$, the transition is of the second order. The authors do not report the value of the ratio $\varepsilon_i/\varepsilon^S$ which yields quantitative agreement with the experimental data, but this value must necessarily be smaller than the value of 10 obtained by Takagi. The discrepancy may arise from the fact that the latter assumed the probability for the configurations $(H_4PO_4)$ and $(PO_4)^{3-}$ to be zero. This approximation, however, may be closer to the truth than the rather high probability assumed for these configurations by Grindlay and ter Haar.

In conclusion, the objections raised against Slater's theory because of the character of the transition that it predicts should not be considered very serious. The theory can be made to predict a second-order transition, as we have seen, upon consideration of very reasonable points, such as the existence of a small concentration of lattice imperfections.

A more serious objection to Slater's theory is rather the fact that it cannot explain the large isotope effect, i.e. the almost doubling of the transition temperature in going from $KH_2PO_4$ to $KD_2PO_4$. Since in Slater's theory the transition temperature is given by Eq. (III-26) this means that, at the transition temperature, $\varepsilon^S = KT_c \ln 2$, and there is no apparent reason why the energy difference $\varepsilon^S$ between dipoles pointing along and perpendicular to $c$ should nearly be doubled when H is replaced by D. It is true that the deuterated crystal shows an increase in the lattice constant along the $a$ axis by 0.25% with respect to the hydrogenated compound (U 2), but it cannot be said whether or not this is sufficient to explain the doubling of the value of $\varepsilon^S$.

An attempt at explaining the isotope effect within the framework of Slater's theory has recently been made by Senko (S 8) by introducing into the model the contributions due to dipole–dipole interactions. The new free energy function relates with one another the changes in Curie temperature and in spontaneous polarization caused by deuterium substitution. This improved model accounts rather satisfactorily for the shift in Curie point and for the temperature and field dependence of the dielectric constant, but is rather inaccurate in its predictions of the transition entropy and the temperature dependence of the spontaneous polarization in the vicinity of the Curie temperature.

### Theory of Pirenne

Another possibility, due to Pirenne, is to consider the protons as anharmonic oscillators moving in a certain potential well $V$ (interaction between protons and neighboring atoms). The spontaneous polarization is assumed to be entirely due to dipole–dipole interaction. The actual crystal polarization is divided into two parts:

(i) the polarization $P_p$ due to displacements of the protons, which carry an effective charge $ne$ ($n < 1$); and

(ii) the polarization $P_b$ of the "background" due to all other causes (electronic polarization, displacements of ions other than H). $P_b$ is assumed proportional to $P_p$.

The dipole–dipole interaction is taken care of by means of a molecular field $\gamma P_p$. Then, for a polarized state to be in equilibrium, $V$ must be anharmonic and the simplest case is to treat the oscillators quantum-mechanically as particles in a square well potential. Now, when H is replaced by D, every vibrational energy level is lowered by a factor 2, since the mass is doubled, and thus becomes twice more polarizable. The energy $E_0$ of the first vibration level is found to be $E_0 = 0.683 \times 10^{-2}$ eV. This predicts an absorption band of energy $3E_0$, i.e. in the region 166 cm$^{-1}$ for $KH_2PO_4$ and 83 cm$^{-1}$ for $KD_2PO_4$, which is in fair agreement with the infrared data, as pointed out by Pirenne (P 4).

The theory yields a value of $\gamma = 5.97$, which is not too far from the Lorentz value $4\pi/3 = 4.18$, and a value of $n = 0.27$, which does not differ much from the value in water, $n = 0.40$. The agreement between theory and experiment is good as far as the saturation polarizations, the Curie constants and the heat of transitions of the hydrogenated and the deuterated crystals are concerned. The ratio of proton polarization to total polarization, $P_p/P$, is found to be rather small, 0.206, suggesting that the background polarization is playing an important role.

Concluding, Pirenne's theory is certainly successful in its attempt to explain the isotope effect. It is probable, however, as mentioned by Forsbergh (F 2), that all other possibilities suggested to explain the isotope effect also contribute and co-operate to give the large change observed.

While Pirenne used a single-minimum potential field, Blinc (B 17), more recently, assumes a double-minimum potential in which tunneling of the protons in the hydrogen bonds is allowed. Blinc's theory is essentially a quantum-mechanical treatment of the simple local-field theory developed by Mason and Devonshire (see below). In addition to explaining the infrared and the nuclear magnetic resonance data (see also Section III-7), the theory accounts fairly well for the isotope effects. Its weakness lies mainly in the fact that it does not take into account short-range interactions, which have proven so effective in Slater's model.

### Long-range Interactions

Insofar as Pirenne's theory introduces a molecular field proportional to the polarization, to account for the dipolar interaction, this theory can be considered one involving long-range forces, in contrast to Slater's treatment. Two other theories have been proposed for KDP using the long-range interactions, namely the theory of Mason (M 1) and that of Grindlay and ter Haar (G 1).

Mason's theory associates a dipole moment $\mu$ to each hydrogen bond. The bond potential is assumed to be symmetrical. The interaction between the dipoles is taken care of by means of an internal field

$$F = E + \frac{4\pi}{3} P_b + \gamma P_d,$$

where $P_d$ is the dipole polarization and $P_b$ the background polarization. The latter is put equal to $\alpha_b F$, where $\alpha_b$ is the polarizability per unit volume due to the background only (without the hydrogens). The quantity $\alpha_b$ can be determined from the temperature-independent part of the dielectric constant:

$$\frac{4\pi}{3} \alpha_b = \frac{\varepsilon_0 - 1}{\varepsilon_0 + 2}.$$

An expression for the *clamped* dielectric constant is obtained:

$$\varepsilon^x = \varepsilon_0 + \frac{C}{T - T_0},$$

where $C$ and $T_0$ are functions of the number of dipoles per unit volume, the dipole moment $\mu$ and the Lorentz factor $\gamma$. Comparison with the experimental data leads to the following values of $\mu$ and $\gamma$:

$$\mu = 0.81 \times 10^{-18} \text{ c.g.s.}; \quad \gamma = 0.567,$$

where $\mu$ is independent of temperature. The theory predicts a polarization much larger than that observed. Mason justifies this discrepancy by judging that the experimental values of the spontaneous polarization are too small, owing to the fact that at low temperatures the largest fields applied are not sufficient to reverse all the domains.

Devonshire (D 1) has pointed out that the model involving dipoles with two positions of equilibrium does not fit the case of KDP very well. The potential well responsible for the dipoles is probably one intermediate between the two extreme cases depicted schematically in Fig. I-2.

The theory of Grindlay and ter Haar (G 1) assumes a dipolar hydrogen lattice interpenetrating *ionic* lattices of potassium, phosphorous and oxygen. The long-range interactions are taken into account by the Lorentz internal-field approximation and the ionic parameters of the K, P and O lattices are computed with the method of Born and Huang. The theory leads to a qualitative understanding of the ferroelectric behavior of KDP but includes two serious quantitative inconsistencies. Since the driving fields from the hydrogen lattice on to the potassium and phosphorous sites have the same sign, the theory requires that the displacements of K and P occur in the same direction, in contrast to the results of the structural analyses. In order for these displacements to be in opposite directions, the K and P ions must have effective charges of opposite sign, a fact which is in disagreement with the expected behavior of electropositive ions. Furthermore, the theory cannot account for both the temperature dependent and the temperature independent part of the dielectric constant. The authors (G 1) point out that the failure of the ionic model is in line with the expectation that the ionic content of KDP is rather low.

## 9. Domains and Effects of Particle Size

Direct evidence for the existence of domains in the ferroelectric phase of $KH_2PO_4$ was first given by the X-ray investigations of de Quervain (D 2), and of Ubbelohde and Woodward (U 3) and by the optical study of Zwicker and Scherrer (Z 2). Domain patterns were later observed under the polarizing microscope by Mitsui and Furuichi (M 6), but no systematic investigations were carried out, mainly because the low temperatures at which the polar phase appear make such an experimental study quite difficult. It is not difficult, on the other hand, to predict what the domain structure should look like from our knowledge of the low-temperature symmetry and the lattice distortion. Figure III-22 shows a schematic drawing of two equivalent domain configurations which are compatible with the spontaneous strain $x_y$. Direct observation of these domain patterns should be possible, under favorable conditions, with the polarizing microscope, because the domains with opposite polarization have, in the position depicted in Fig. III-22, slightly different extinction positions. However, no experimental information is available, to date, concerning the details of the domain structure and the dynamic properties of domains in KDP.

Some information about the energy and the thickness of domain walls can be gained indirectly from experiments concerned with the effect of particle size. Such experiments were carried out by Känzig and co-workers (K 2), (K 3), (J 2) and were stimulated by the following argument. Since a rigorous mathematical treatment of a practical case is almost hopeless, the case of an insulated non-conducting spherical crystal, which is forced to be uniformly polarized, was considered. The depolarizing energy of such a single-domain crystal would be:

$$U_E = \tfrac{1}{2} \frac{4\pi}{3} P^2 V$$

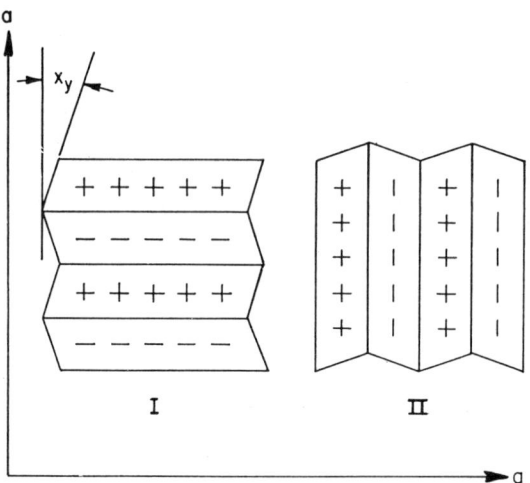

FIG. III-22. Expected domain patterns in $KH_2PO_4$ (schematic) (according to Känzig (K 4)).

where $V$ is the volume, while the energy of transition,

$$U_Q = \frac{2\pi}{C} T_c P^2 V$$

(see Eq. II-4) is so much smaller than $U_E$ that the hypothesized polarized state cannot occur unless the electrostatic energy is reduced either by conduction or formation of domains. Since conduction is negligible at low-temperatures, the only mechanism left to decrease the depolarizing energy in a *perfect* $KH_2PO_4$ crystal is that of domain formation. The domain wall energy is a surface energy; if the dimensions of the crystal are decreased, it is reasonable to expect that this surface energy will become important, with respect to the volume energies, below a certain critical size of the crystal. It will then be energetically more favorable to do without the domain walls and the whole crystal will be forced to become a single domain. Then the polarization will be annihilated by the depolarizing field and the crystal will no longer be ferroelectric. It should be noted (K 4) that this effect has no counterpart in ferromagnetic crystals, as the magnetostatic self-energy is always about $10^3$ times smaller than the transition energy and thus cannot suppress the spontaneous magnetization.

Experiments were carried out on colloidal particles of $KH_2PO_4$ imbedded in an insulating medium with an effective dielectric constant of about 5. X-ray and electron diffraction techniques, as well as dielectric measurements were used to detect the ferroelectric transition. The results show that no spontaneous deformation occurs in particles with diameter smaller than 1500 Å. It is concluded from this that particles of this size no longer polarize spontaneously. On the other hand, particles larger than 4000 Å still show the normal ferroelectric behavior of large crystals. As expected, the critical particle size is found to decrease with increasing dielectric constant of the imbedding medium until, when the latter becomes conducting, no critical size exists (particles with diameter of about 500 Å are still found to be ferroelectric in a conducting medium).

The evaluation of these results in terms of wall energy and wall thickness requires the development of a molecular theory of the wall on the basis of a model. Since the treatment of a realistic model is exceedingly difficult, Känzig and Sommerhalder (K 3) considered the simple model of a body-centered cubic lattice occupied by point dipoles pointing along [001] at the absolute zero of temperature. The domain wall was assumed to be parallel to the (100) plane and its energy calculated from the difference between the interaction energies of the state with the wall and the state without the wall, respectively. The dipolar wall energy was found to be $\sigma_{dip} = 0.88\,P^2 a$, where $P$ is the value of the polarization inside the domains and $a$ is the lattice constant. This formula could be used for a crude estimate of the order of magnitude of the wall energy in $KH_2PO_4$ using the critical particle size determined experimentally. It was concluded that the wall energy is of the order of 40 ergs/cm$^2$ and the wall thickness from 2 to 3 unit cells.

Other molecular models for the domain wall in $KH_2PO_4$ were proposed (B 5), involving both "polarized" walls (where the dipole vectors are rotated by 90° at the time) and "neutral" walls (where the dipole moments are reduced to zero within the wall), but their rigorous mathematical treatment is very difficult.

## BIBLIOGRAPHY

(B 1)  BUSCH, G. and SCHERRER, P., *Naturwiss.* **23**, 737 (1935).
(B 2)  BUSCH, G., *Helv. Phys. Acta* **11**, 269 (1938).
(B 3)  BAUMGARTNER, H., *Helv. Phys. Acta* **24**, 326 (1951).
(B 4)  BAUMGARTNER, H., *Helv. Phys. Acta* **23**, 651 (1950).
(B 5)  BARKLA, H. M. and FINLAYSON, D. M., *Phil. Mag.* **44**, 109 (1953).
(B 6)  BANTLE, W., *Helv. Phys. Acta* **15**, 373 (1942).
(B 7)  BANTLE, W. and CAFLISH, C., *Helv. Phys. Acta* **16**, 235 (1943).
(B 8)  BERGMANN, L., *Der Ultraschall*, Hirzel Verlag, Zürich (1954).
(B 9)  BACON, G. E., *Neutron Diffraction*, Clarendon Press, Oxford (1955).
(B 10) BACON, G. E. and PEASE, R. S., *Proc. Roy. Soc. (London)* **A 220**, 397 (1953).
(B 11) BACON, G. E. and PEASE, R. S., *Proc. Roy. Soc. (London)* **A 230**, 359 (1955).
(B 12) BJORKSTAM, J. L. and UEHLING, E. A., *Phys. Rev.* **114**, 961 (1959).
(B 13) BAUMGARTNER, H., JONA, F. and KÄNZIG, W., *Ergeb. exakt. Naturwiss.* **23**, 235 (1950).

(B 14) BLINC, R. and HADZI, D., *Molec. Phys.* **1**, 391 (1958).

(B 15) BETHE, H., *Proc. Roy. Soc. (London)* **A 150**, 552 (1935).

(B 16) BECK, M. and GRÄNICHER, H., *Helv. Phys. Acta* **23**, 522 (1950).

(B 17) BLINC, R., *J. Phys. Chem. Solids* **13**, 204 (1960).

(C 1) CADY, W. G., *Piezoelectricity*, McGraw-Hill, New York (1946).

(D 1) DEVONSHIRE, A. F., *Theory of Ferroelectrics, Phil. Mag.* Suppl. **3**, 85 (1954).

(D 2) DE QUERVAIN, M., *Helv. Phys. Acta* **17**, 509 (1944).

(D 3) DANNER, H. R. and PEPINSKY, R., *Phys. Rev.* **99**, 1215 (1955).

(E 1) ESS, H., Thesis, Eidg. Techn. Hochschule, Zürich (1946). Unpublished.

(F 1) FRAZER, B. C. and PEPINSKY, R., *Acta Cryst.* **6**, 273 (1953).

(F 2) FORSBERGH, P. W., JR., *Piezoelectricity, Electrostriction and Ferroelectricity, Handbuch der Physik*, vol. 17, pp. 264–392, Springer-Verlag, Berlin (1956).

(G 1) GRINDLAY, J. and TER HAAR, D., *Proc. Roy. Soc. (London)* **A 250**, 266 (1959).

(J 1) JONA, F., *Helv. Phys. Acta* **23**, 795 (1950).

(J 2) JACCARD, C., KÄNZIG W. and PETER, M., *Helv. Phys. Acta* **26**, 521 (1953).

(K 1) KETELAAR, J. A. A., *Acta Cryst.* **7**, 691 (1954).

(K 2) KÄNZIG, W., *Phys. Rev.* **87**, 385 (1952).

(K 3) KÄNZIG, W. and SOMMERHALDER, R., *Helv. Phys. Acta* **26**, 603 (1953).

(K 4) KÄNZIG, W., *Ferroelectrics and Antiferroelectrics, Solid State Physics*, vol. 4, pp. 1–197, Academic Press, New York (1957).

(L 1) LEVY, H. A., PETERSON, S. W. and SIMONSEN, S. H., *Phys. Rev.* **93**, 1120 (1954).

(L 2) LANDSBERG, G. S. and BARYSHANSKAYA, F. S., *Doklady Akad. Nauk S.S.S.R.* **61**, 1027 (1948).

(L 3) LORD, R. C. and MERRIFIELD, R. E., *J. Chem. Phys.* **21**, 166 (1953).

(M 1) MASON, W. P., *Piezoelectric Crystals and their Application to Ultrasonics*, Van Nostrand, New York (1950).

(M 2) MASON, W. P., *Phys. Rev.* **69**, 173 (1946).

(M 3) MATTHIAS, B. T., *Phys. Rev.* **85**, 723 (1952).

(M 4) MEGAW, H. D., *Ferroelectricity in Crystals*, Methuen, London (1957).

(M 5) MURPHY, G. M., WEINER, G. and OBERLY, J. J., *J. Chem. Phys.* **22**, 1322 (1954).

(M 6) MITSUI, T. and FURUICHI, J., *Phys. Rev.* **90**, 193 (1953).

(M 7) MATTHIAS, B. and MERZ, W., *Helv. Phys. Acta* **19**, 227 (1946).

(N 1) NEWMAN, R., *J. Chem. Phys.* **18**, 669 (1950).

(O 1) OBERLY, J. J. and WEINER, G., *J. Chem. Phys.* **20**, 740 (1952).

(P 1) POCKELS, F., *Lehrbuch der Kristalloptik*, Teubner, Leipzig (1906).

(P 2) PETERSON, S. W., LEVY, A. H. and SIMONSEN, S. H., *J. Chem. Phys.* **21**, 2084 (1953).

(P 3) PELAH, I., LEFKOWITZ, I., KLEY, W. and TUNKELO, E., *Phys. Rev. Letters* **2**, 94 (1959).

(P 4) PIRENNE, J., *Helv. Phys. Acta* **22**, 479 (1949).

(P 5) PAULING, L., *J. Am. Chem. Soc.* **57**, 2680 (1935).

(P 6) PIRENNE, J., *Physica* **15**, 1019 (1949).

(R 1) RUNDLE, R. E. and PARASOL, M., *J. Chem. Phys.* **20**, 1487 (1952).

(S 1) STEPHENSON, C. C. and HOOLEY, J. G., *J. Am. Chem. Soc.* **66**, 1397 (1944).

(S 2) STEPHENSON, C. C., CORBELLA, J. M. and RUSSELL, L. A., *J. Chem. Phys.* **21**, 1110 (1953).

(S 3) SEIDL, F., *Tschermak's mineral. u. petrog. Mitt.* **1**, 432 (1950).

(S 4) SLATER, J. C., *J. Chem. Phys.* **9**, 16 (1941).

(S 5) SHIRANE, G. and OGUCHI, T., *J. Phys. Soc. Japan* **4**, 172 (1949).

(S 6) STEPHENSON, C. C. and ZETTLEMOYER, A. C., *J. Am. Chem. Soc.* **66**, 1402 (1944).

(S 7) SHIRANE, G., JONA, F. and R. PEPINSKY, *Proc. I.R.E.* **43**, 1738 (1955).

(S 8) SENKO, M., *Phys. Rev.* **121**, 1599 (1961).

(T 1)  TAKAHASHI, H., *Proc. Phys. Math. Soc. Japan* **23**, 1069 (1941).

(T 2)  TAKAGI, Y., *J. Phys. Soc. Japan* **3**, 271 (1948).

(T 3)  TAKAGI, Y., *Proc. Phys. Math. Soc. Japan* **23**, 44 (1941).

(U 1)  UBBELOHDE, A. R. and WOODWARD, I., *Proc. Roy. Soc. (London)* **A 188**, 358 (1947).

(U 2)  UBBELOHDE, A. R. and WOODWARD, I., *Proc. Roy. Soc. (London)* **179**, 399 (1942).

(U 3)  UBBELOHDE, A. R. and WOODWARD, I., *Nature* **156**, 20 (1945).

(U 4)  UNTERLEITNER, F., OKAYA, Y., VEDAM, K. and PEPINSKY, R., *Abstracts Annual Meeting Am. Cryst. Assoc.*, Ithaca, New York, July (1959).

(V 1)  VON ARX, A. and BANTLE, W., *Helv. Phys. Acta* **17**, 298 (1944).

(V 2)  VON ARX, A. and BANTLE, W., *Helv. Phys. Acta* **16**, 211 (1943).

(Y 1)  YOMOSA, S. and NAGAMIYA, T., *Progr. Theoret. Phys.* **4**, 263 (1949).

(W 1)  WALKER, A. C., *J. Franklin Inst.* **250**, 481 (1950).

(W 2)  WEST, J., *Z. Krist.* **74**, 306 (1930).

(Z 1)  ZWICKER, B., *Helv. Phys. Acta* **19**, 523 (1946).

(Z 2)  ZWICKER, B., and SCHERRER, P., *Helv. Phys. Acta* **17**, 346 (1944).

# CHAPTER IV

# BARIUM TITANATE

## 1. Introduction

Barium titanate ($BaTiO_3$) is, to date, the most extensively investigated ferro-electric material. It is extremely interesting from the viewpoint of the solid state scientist because its structure is far simpler than that of any other ferroelectric known, thus offering great promise for a better understanding of the ferroelectric phenomenon. It is also interesting from the viewpoint of practical applications, firstly, because it is chemically and mechanically very stable, secondly, because it exhibits ferroelectric properties at and above room temperature, and finally because it can be easily prepared and used in the form of ceramic polycrystalline samples.

The anomalous dielectric properties of barium titanate were, in fact, discovered on ceramic specimens through studies carried out independently from each other around 1943 by Wainer and Salomon in the United States, Ogawa in Japan, and Wul and Goldman in Russia. The ferroelectric activity of $BaTiO_3$ was reported in 1945–46 by von Hippel and co-workers (V 1), and independently by Wul and Goldman (W 1). Since then, a very considerable portion of the extensive literature concerning the ferroelectricity of barium titanate has involved investigations of polycrystalline specimens. The first single crystals of barium titanate were produced in Switzerland in 1947, but it was not until the publication of Remeika's method (R 9) in 1954 that large single crystals became easily available, facilitating sub-stantial progress in this field. Owing to the importance of this subject, Section 12 of the present chapter is devoted to the problem and the techniques of growing single crystals of barium titanate. Accordingly, the properties reported and dis-cussed in this chapter are mainly those pertinent to single crystals. A brief survey of the characteristics of ceramic barium titanate is given in Section 13.

The Curie temperature of $BaTiO_3$ is at 120 °C. The symmetry of the non-polar phase is cubic (point group $m3m$), thus centrosymmetrical and non-piezoelectric. This phase has a *perovskite-type* structure. This structure is common to a large family of compounds with the general formula $ABO_3$, whose representative in nature is the mineral $CaTiO_3$, called perovskite. Figure IV-1 shows the unit cell of the cubic perovskite-type lattice.

The symmetry of the polar phase, i.e. the phase below 120 °C, is tetragonal, with point group $4mm$. The fourfold rotation axis is thus the polar direction, identified with the tetragonal $c$ axis. This axis lies parallel to the direction of one of the original cubic $\langle 100 \rangle$ directions. The tetragonal unit cell results from the following distortion of the original cubic cell: one of the cube edges is

elongated to become the tetragonal $c$ axis, the two other cube edges are compressed to become the tetragonal $a$ axes. This distortion is depicted schematically
in Fig. IV-2(b). Since there are six equivalent $\langle 100 \rangle$ axes in the cubic phase, the
polar axis can be parallel to any one of these six equivalent directions. This gives
rise to more complicated domain patterns than those encountered in tri-glycine
sulfate and in potassium dihydrogen phosphate (Chapters II and III, respectively).

The tetragonal phase of barium titanate has been the object of most of the
investigations pertinent to this crystal, since this phase is structurally simple
and also stable at room temperature.

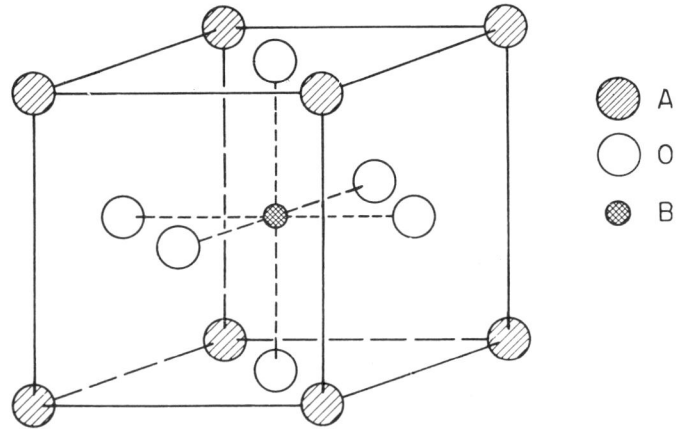

FIG. IV-1. Cubic perovskite-type structure $ABO_3$.

Barium titanate differs from the ferroelectrics that we have considered up
to this point in that other phase transitions occur at lower temperatures. The
tetragonal phase is stable from 120 °C to about 5 °C. Below 5 °C, a new phase
appears, which has orthorhombic symmetry belonging to point group $mm$. This
phase is still ferroelectric, but the direction of the spontaneous polarization is
now parallel to one of the original cubic $\langle 110 \rangle$ directions. It is useful to think of
the orthorhombic unit cell stable below 5 °C as a distortion of the original cubic
cell which is stable above 120 °C. This distortion consists in an elongation of one
of the face diagonals of the original cube, which becomes the orthorhombic $a$
direction (polar), and a compression of the other face diagonal to become the
orthorhombic $b$ direction. Again, since there are twelve equivalent $\langle 110 \rangle$ directions
in the cubic phase there are twelve possible directions of the spontaneous polarization in the orthorhombic phase. For some purposes, it is convenient to retain
axes almost parallel to the original cube edges. In this case, the unit cell has
monoclinic geometry (Fig. IV-2 c).

The orthorhombic phase of barium titanate is stable from 5 °C to about
− 90 °C. At − 90 °C, a third phase transition occurs and the symmetry changes,
upon cooling, from orthorhombic to rhombohedral. The point group of this phase
is $3m$ and the polar axis lies along one of the original cubic $\langle 111 \rangle$ directions. The

rhombohedral distortion consists in a stretch of the original cubic cell along one of the body diagonals (Fig. IV-2d). As there are eight equivalent ⟨111⟩ axes in the cubic lattice, there are as many allowed directions of the spontaneous polarization in the rhombohedral phase.

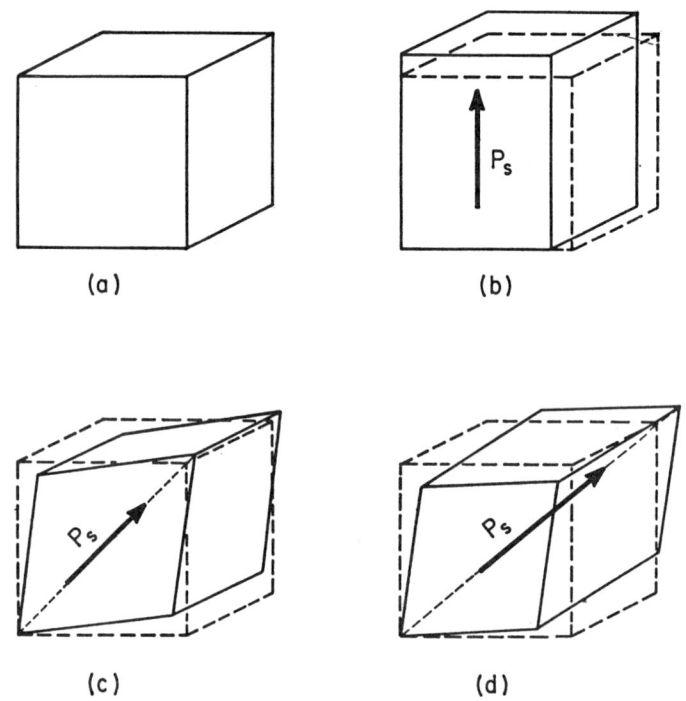

(a)                                           (b)

(c)                                           (d)

FIG. IV-2. Unit cells of the four phases of BaTiO$_3$.

        (a) Cubic, stable above 120 °C.
        (b) Tetragonal, stable between 120 °C and 5 °C.
        (c) Orthorhombic, stable between 5 °C and −90 °C.
        (d) Rhombohedral, stable below −90 °C.

The dotted lines in (b), (c) and (d) delineate the original cubic cell. The heavy arrows indicate the direction of the spontaneous polarization $P_s$ in each phase.

## 2. Dielectric and Optical Properties

### The Dielectric Constant

In the cubic phase there is only one principal dielectric constant, while in the tetragonal phase there are two, denoted by $\varepsilon_a$ and $\varepsilon_c$. Both these constants can be measured on a *single-domain* crystal. In the orthorhombic and rhombohedral phases it is very difficult to obtain single-domain crystals, so that the measurement of the principal dielectric constants is practically impossible without the help of biasing fields.

The dielectric properties of $BaTiO_3$ single crystals were first investigated extensively by Merz (M 1). Subsequently, Cross (C 3) and a number of other authors (see below) extended and improved the results obtained by Merz. Figure IV-3 shows the temperature dependence of the dielectric constant measured with a small a.c. field along the pseudo-cubic edge (M 1). In this figure, only the values of the dielectric constant in the tetragonal phase have a clear meaning, as they were measured on carefully selected single-domain crystals with the proper orientations. In the orthorhombic phase, where the crystal generally consists of a

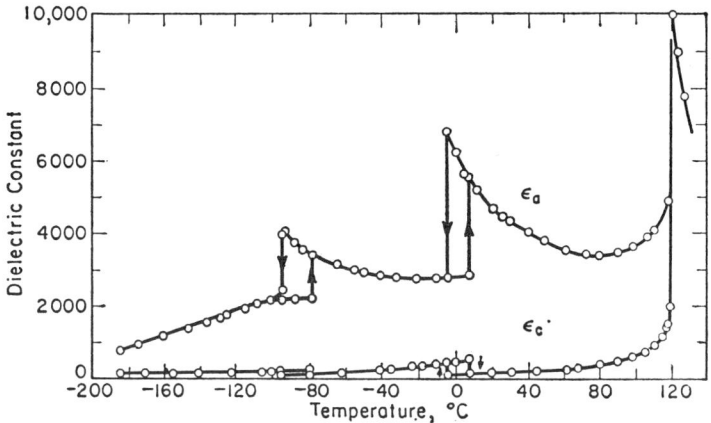

FIG. IV-3. Dielectric constants of $BaTiO_3$ as functions of temperature. The values of $\varepsilon_a$ and $\varepsilon_e$ in the tetragonal phase refer to single-domain crystals (according to Merz (M 1)).

mixture of domains with different orientations of the polar axis, one might at first expect that the measured values of the dielectric constant fluctuate around a certain mean value. However, we observe in Fig. IV-3 that there are two distinct branches of the dielectric constant curve between 5 °C and − 90 °C. The existence of these two branches adjoining the curves of the tetragonal modification can be understood with the help of Fig. IV-4. The crystal $B$ which was used for the measurement of $\varepsilon_c$ in the tetragonal phase has a larger probability to assume the configuration $D$ rather than $E$. This is because the former implies a jump of the polar axis of 45°, while, for the latter, this jump is of 90°. The crystal $C$ which was used for the measurement of $\varepsilon_a$ in the tetragonal phase, can, on the other hand, assume the configurations $D$ or $E$ with equal probability. The dielectric constant measured along the pseudo-cubic [001] direction will thus essentially be $\varepsilon_{45° a,b}$, for the lower branch of Fig. IV-3 in the orthorhombic range. The upper branch, continuation of tetragonal $\varepsilon_a$, represents an average of orthorhombic $\varepsilon_c$ and $\varepsilon_{45° a,b}$, depending on the relative number of domains having orientations $E$ and $D$, respectively. In the rhombohedral phase, the dielectric constant measured along the pseudo-cubic edge should be single-valued no matter how many domains exist, as the only possible configuration is now that denoted by $F$ in Fig. IV-4. The existence of two branches of the dielectric constant below $-90$ °C is thus not understandable at this stage. We will discuss this point in some more detail in Section 6.

A few interesting observations should be made in Fig. IV-3. In the first place, we notice that the dielectric constant $\varepsilon_c$ along the polar axis is much smaller than that perpendicular to it, in contrast to the situation encountered in TGS and KDP. From Merz's curves in Fig. IV-3, we read values of approximately 200 and 4000

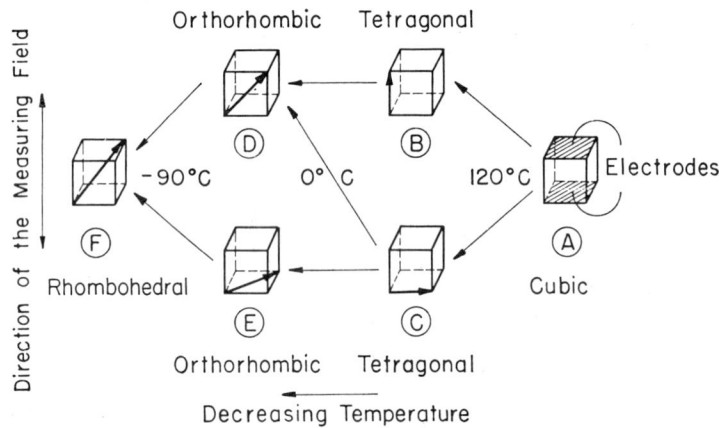

FIG. IV-4. Schematic drawing of the possible orientations of the polar axis with respect to a small measuring field applied along the pseudo-cubic [001] direction of BaTiO$_3$. The heavy arrow within each cube indicates the direction of the polar axis. The light arrows between the cubes point toward the most probable configuration at each transition.

for $\varepsilon_c$ and $\varepsilon_a$, respectively, at room temperature. The most reliable values available to date for these quantities are $\varepsilon_c = 160$ (L 1) and $\varepsilon_a = 4100$ (B 2). We also note that the dielectric constant exhibits pronounced anomalies at the transitions from tetragonal to orthorhombic and from orthorhombic to rhombohedral states. Since these are transitions between ordered states (first-order transitions), one observes thermal hystereses in both cases over a temperature spread of about 10°. Finally, we see that the Curie transition is of the first order. A thermal hysteresis of about 2 degrees is connected with this transition (K 1), (D 1), although it does not appear in Fig. IV-3.

In the cubic phase, the dielectric constant follows the Curie–Weiss law $\varepsilon = C/(T - T_0)$ over a temperature range of approximately 100 °C. The values of the constants appearing in this law can best be determined on ceramic specimens with high purity and low porosity, as the crystals available to date for dielectric measurements are never very pure (see Section IV-12). The most reliable results are those obtained by Roberts (R 1), from ceramic materials, as follows:

$C = 1.54 \times 10^5$ degrees; $T_0 = 118\,^{\circ}\text{C}$; $T_c = 130\,^{\circ}\text{C}$. In single crystals, the value of the Curie constant $C$ is of the order of $10^5$ degrees, but there is disagreement among different investigators about the exact value of $C$, for which figures ranging from 1.5 to 6.5 $\times 10^5$ degrees have been reported (V 1), (R 1), (G 1), (K 1), (M 2), (D 1). These discrepancies are most probably due to various amounts of different impurities contained in the crystals investigated. The more recent results of Merz (M 2), $C = 1.56 \times 10^5$ degrees, and Drougard and Young (D 1), $C = 1.73 \times 10^5$ degrees, ought to apply best to the barium titanate crystals usually grown at the present time.

In contrast to the case of TGS and KDP, the Curie–Weiss temperature $T_0$ of BaTiO$_3$ is always lower than the Curie temperature $T_c$. This is in keeping with the behavior of the dielectric constants at a transition of the first order (see Section I-5). The most reliable value for the difference $T_c - T_0$ is approximately $10\,^{\circ}\text{C}$ (M 2), (D 1).

Dielectric constant and loss measurements while traversing the hysteresis loop were carried out by Drougard et al. (D 14).

### Dielectric Properties of BaTiO$_3$ at High Frequencies

In discussing the frequency dependence of the dielectric constant of BaTiO$_3$, one should keep in mind that the behavior is expected to be different in the various phases. In the tetragonal phase, the dielectric constant is expected to exhibit a drop at the frequency corresponding to piezoelectric resonance of the sample. Below this frequency, one measures the dielectric constant of the free crystal, above this frequency, one measures the dielectric constant of the clamped crystal. At still higher frequencies, an additional drop in dielectric constant may occur, which is to be ascribed to dielectric relaxation. In the cubic phase, on the other hand, no piezoelectric resonance can occur, as this phase is centrosymmetric: the values of clamped and free dielectric constants are equal to one another. Dielectric relaxation, however, will still cause a drop in dielectric constant at a given high frequency.

Measurements by Bond et al. (B 1) on multidomain crystals indicate that piezoelectric resonance occurs in the vicinity of $10^7$ c/s in plates having an area of approximately 1 cm$^2$ and a thickness of approximately 1 mm. The dielectric constant drops by a factor of 2 at the piezoelectric resonance frequency. Similar measurements were also carried out by Danielson et al. (D 2). The results of Fousek (F 6) indicate that the ratio between the dielectric constant measured at $10^9$ c/s and that measured at $10^3$ c/s is of the order of $0.7 - 0.8$.

The most reliable investigation of the high-frequency dielectric properties of BaTiO$_3$ single crystals was made by Benedict and Durand (B 2) (see also Lurio and Stern (L 12) and Nakamura and Furuichi (N 7)). Single domain crystals were investigated in the temperature range from 25 $^{\circ}$C to 170 $^{\circ}$C at a frequency of $2.4 \times 10^{10}$ c/s. Figure IV-5 shows the results of this investigation. At room temperature, the dielectric constant $\varepsilon_a$ has a value approximately 1/2 smaller than that measured at low frequencies, while the high frequency value of tan $\delta$ is only 0.1. This means that the value of $\varepsilon_a$ measured at $2.4 \times 10^{10}$ c/s is still equal to that measured immediately above the piezoelectric resonance. Consequently, the resonance frequency of the potential well characteristic of BaTiO$_3$ must be larger than $2.4 \times 10^{10}$ c/s, and no relaxation is observed yet at this frequency. In fact, no relaxation seems to occur below a frequency of $5 \times 10^{10}$ c/s (T 1). The Curie–Weiss law is still obeyed at the frequency used by Benedict and Durand (B 2).

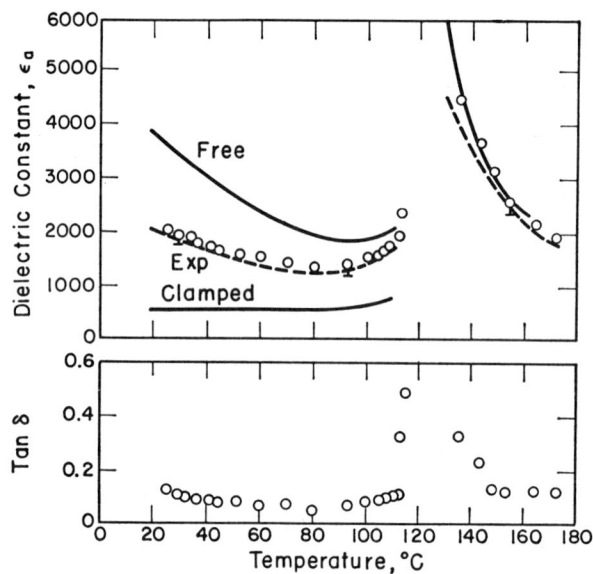

Fig. IV-5. Temperature dependence of the dielectric constant $\varepsilon_a$ and of the loss tan $\delta$ of BaTiO$_3$. The solid lines give the theoretical values of the free and clamped dielectric constant from Devonshire's theory (D 4). The dashed line gives the average value of $\varepsilon_a$ measured at 24 kMc/s on five crystals. The circles represent experimental points for one of the crystals studied (according to Benedict and Durand (B 2)).

The results obtained on ceramic specimens of BaTiO$_3$ differ from the data pertinent to single crystals in the following way (P 1), (V 3). The dielectric constant drops from a value of 1500 to approximately 300 in the frequency range from $10^9$ to $10^{10}$ c/s, while tan $\delta$ increases from 0.01 to approximately 1. The reasons for this drop in the dielectric constant of ceramic BaTiO$_3$ are not completely understood. Explanations have been given in terms of inertia of the domain boundaries (K 2) as well as piezoelectric resonance of the crystallites (V 9), (V 10), (D 4). In the cubic phase, on the other hand, no drop in the dielectric constant of ceramic BaTiO$_3$ was found by Schmitt (S 3) up to $9.4 \times 10^9$ c/s, and a small drop was detected by Powles and Jackson (P 1).

The low-temperature behavior of the high-frequency characteristics of ceramic (Ba, Pb)TiO$_3$ was investigated by Rabenhorst and Melichercik (R 2) in the temperature range from $+20\,°C$ to $-180\,°C$ at a frequency of $9.1 \times 10^9$ c/s. At room temperature, the high-frequency value of the dielectric constant was found to be about one-quarter of the low frequency value, in agreement with the results discussed above. Anomalies in the high-frequency dielectric constant were observed at the transitions from tetragonal to orthorhombic and from ortho-rhombic to rhombohedral states.

It follows from the present description of the dielectric properties of BaTiO$_3$ at high frequencies that a large amount of work remains to be done in this area. The fact that no relaxation is observed (in single crystals) below $5 \times 10^{10}$ c/s does, however, indicate that the non-linear dielectric properties of BaTiO$_3$ remain unaffected up to that frequency range.

### The Spontaneous Polarization

The temperature dependence of the spontaneous polarization $P_s$ of BaTiO$_3$ single crystals is depicted in Fig. IV-6 as measured by Merz (M 1) with the circuit

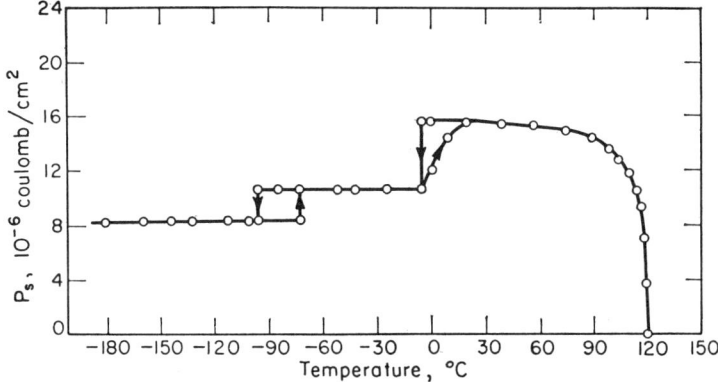

FIG. IV-6. Spontaneous polarization of BaTiO$_3$, measured along the pseudo-cubic edge as a function of temperature (according to Merz (M 1)).

of Sawyer and Tower (see Section I-1). The experimental points were obtained from measurements of the spontaneous polarization along the pseudo-cubic [100] direction. In the orthorhombic phase, as the $P_s$ vector is oriented along the pseudo-cubic [110] direction, the magnitude of this vector is obtained by multiplying the corresponding values of Fig. IV-6 with $\sqrt{2}$. Similarly, in the rhombohedral phase, the experimental results reported in Fig. IV-6 must be multiplied with $\sqrt{3}$ in order to obtain the magnitude of $P_s$. The onset of the spontaneous polarization at the Curie point appears to be rather continuous, but this was later found to be due to imperfections and inhomogeneities in the crystals used. The investigations of Känzig and Meier (K 3) and those of Merz (M 2), which were carried out on very good crystals, indicate that the onset of spontaneous polarization is discontinuous (Fig. IV-7). The value of the spontaneous polarization is $26 \times 10^{-6}$ C/cm$^2$ at room temperature. The smaller values of the spontaneous polarization reported earlier (Fig. IV-6) must be ascribed to the inferior quality of the crystals used, in which a number of domains remained unreversed by the applied electric field.

A method for the measurement of the spontaneous polarization of ferroelectric crystals has been described, and applied to BaTiO$_3$, by Chynoweth (C 1).
The method consists in subjecting the crystal to periodic temperature changes
by means of (infrared) light flashes, and in measuring the (amplified) pyroelectric
current obtained. The basic idea is the following: calling $P_s$ the spontaneous
polarization of the crystal at the temperature $T$, a change $dP_s$ can be caused

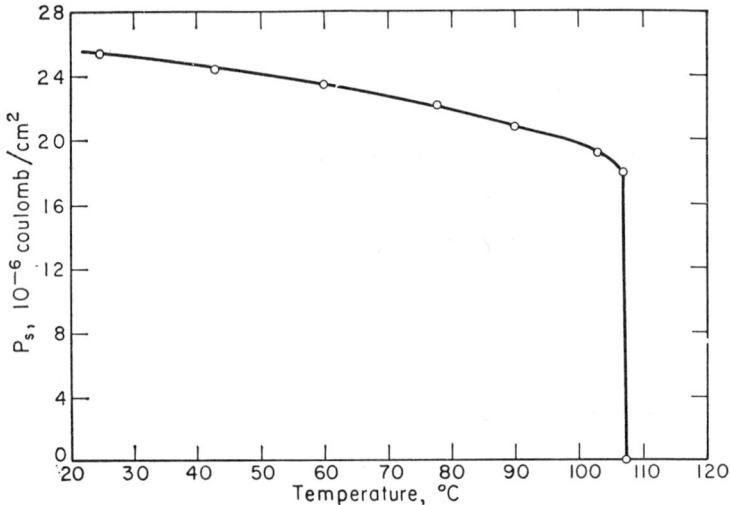

FIG. IV-7. Spontaneous polarization of tetragonal BaTiO$_3$ as a function of temperature (according to Merz (M 2)).

by a change $dT$. If this temperature change is effected in a time $dt$, the polarization changes of the rate $dP_s/dt$, which is equivalent to a current $i$, flowing in
an external circuit. Thus, the current density is:

$$i = \frac{dP_s}{dt} = \left[ \left( \frac{dP_s}{dT} \right) \left( \frac{dT}{dt} \right) \right]_T . \tag{IV-1}$$

If the rate of change in temperature ($dT/dt$) is maintained constant throughout
the measurements at different values of $T$, then the current density $i$ is proportional to $dP_s/dT$. Thus, upon integrating the curve of $i$ vs. $T$ one can obtain $P_s$ as
a function of temperature. This, however, does not yield the absolute value
of $P_s$ and the curve obtained must be normalized.

It should be noted that Chynoweth's method differs from the classical pyroelectric measurements (see, e.g., Cady's book (C 2)) in that it is dynamic, rather
than static. The response depends on the *rate of change* in temperature ($dT/dt$),
rather than the actual change in temperature. In the case of BaTiO$_3$, it is computed that the voltage $\Delta V$ generated across a crystal capacitor by a temperature
change $\Delta T$ is of the order of 15 V/°C at 30 °C. Using suitable amplifiers, temperature
changes of the order of $10^{-5}-10^{-6}$ °C can easily be detected (C 1).

Measurements of the pyroelectric current $i$ as a function of temperature in the tetragonal and cubic phase show that $i$ varies in the manner to be expected from the variation of $P_s$ with temperature (C 1). This investigation of the pyroelectric effect has the advantage that it allows the study of the polarization in a non-destructive way: it is not necessary to upset the pattern of domain orientation in a given crystal as in displaying hysteresis loops. It should be noted, however, that the pyroelectric signal is determined primarily by $(\mathrm{d}P/\mathrm{d}T)$, the pyroelectric coefficient, which does not necessarily depend on the amount of *reversible* polarization.

The study of the pyroelectric effect confirms the result that the Curie transition of $BaTiO_3$ is of the first order. A pyroelectric effect can be induced, in the cubic phase, by means of an electric field (C 1). This field-induced effect at temperatures above the Curie point is in keeping with the predictions of the thermodynamic theory of ferroelectrics (D 5) (see Section 3).

Another method for the measurement of the spontaneous polarization has been applied to $BaTiO_3$ by Husimi and co-workers (K 14), (H 9), (H 10). The technique used by these authors consists in the determination of the piezoelectric modulus of the crystal with an ultrasonic method. The principle is as follows: when a stress is applied to a single-domain crystal of $BaTiO_3$ in a direction parallel to the polar axis, charges are liberated on the (001) surfaces owing to the piezoelectric effect (see Section 5). If the crystal contains equal numbers of positive and negative domains, the charges will obviously cancel. If, however, the crystal is only partially polarized, the piezoelectric modulus $d$ measured under these conditions is related to the actual polarization $P$ of the sample in the same way as the moduls $d_{31}$ of a single-domain crystal to the spontaneous polarization $P_s$ i.e., $d/P = d_{31}/P_s$.

Thus, measurement of the piezoelectric modulus can give information about the polarization. The values reported by Husimi for the spontaneous polarization of $BaTiO_3$ are very much larger than those obtained from hysteresis loops. This is probably due to charges of non-piezoelectric nature caused by the prolonged application of biasing fields.

### The Coercive Field

The hysteresis loops of good single crystals of $BaTiO_3$ exhibit rather sharp corners and a marked rectangular appearance. The value of the coercive field, measured at room temperature, varies from a minimum of 500 V/cm to a maximum of about 2000 V/cm (M 2) (W 2) (L 1). These experimental values are much smaller than the value expected on the basis of the thermodynamic theory (see Section I-5). The theoretical value is of the order of $10^5$ V/cm (L 1), and represents the critical field at which the polarization vector goes from a metastable state antiparallel to the field to a stable state parallel to the field. The relatively small experimental values of the coercive field tell us that the crystal actually leaves the metastable state long before this state becomes unstable. In order to understand how this comes about we must study the mechanism by which the reversal of the polarization takes place. We are going to discuss this problem in detail in Section 8, but we may, at this stage, anticipate some of the results which have a bearing on the value of the coercive field.

Merz (M 11) found experimentally that the rate at which the polarization reverses itself is proportional, at low fields, to $\exp(-\alpha/E)$, where $E$ is the applied field and $\alpha$ a quantity dependent upon temperature. This means that we cannot speak of a well-defined coercive field. The field at which the polarization reverses its direction depends on the time that it is permitted to take for the reversal. Thus, the shape of the hysteresis loop depends on the rate at which the loop is traversed. This point has been demonstrated particularly well by Landauer *et al.* (L 1). On the basis of the exponential law found experimentally by Merz, these authors have assumed that the rate at which polarization is reversed may be given by the formula:

$$\frac{\mathrm{d}P}{\mathrm{d}t} = F(P)\,\mathrm{e}^{-\alpha/E(t)},$$

where the function $F(P)$ takes into account the fact that the switching rate depends on the extent to which the crystal has already reversed its polarization. In order to be able to integrate the above equation, certain assumptions must be made about the functions $F(P)$ and $E(t)$. Since, in most hysteresis loop observations, the voltage varies sinusoidally with time, i.e. $E(t) = E_0 \sin \omega t$, the function $E(t)$ can be approximated by

$$E = E_0 \omega t = E't,$$

provided that the peak voltage is sufficiently larger than the coercive field. Thus, $E'$ represents the rate at which the field increases with time. Integration of the equation for $\mathrm{d}P/\mathrm{d}t$ leads, under these conditions, to the results depicted in Fig. IV-8. This figure illustrates how the shape of the hysteresis loop, and therefore the coercive field depend on the rise rate of the applied field. At slow rates, the crystal is given enough time to reverse its polarization, and the reversal is practically completed long before the peak voltage is reached. At fast rates, the crystal is not given sufficient time for the process, and the coercive field becomes larger. This is another way of saying that the coercive field is not a material constant but rather depends on the rise rate $E'$ of the fields, that is, in the final instance (since $E' = E_0 \omega$) on the amplitude $E_0$ and the frequency $\omega$ of the sinusoidal driving field. The frequency dependence of the coercive field of $BaTiO_3$ was studied directly by Wieder (W 3) and by Campbell (C 18).

The temperature dependence of the coercive field is depicted in Fig. IV-9 as measured by Wieder (W 2) from 60 c/s hysteresis loops. It turns out that the 60 c/s coercive field is strongly dependent upon the thickness of the sample measured (M 3). The experimental results can be described by an equation of the type:

$$E_c = E_\infty + \frac{A}{d}. \tag{IV-2}$$

Here, $E_\infty$ is the coercive field strength which can be obtained on very thick samples, $d$ is the thickness of the sample with the coercive field $E_c$, and $A$ is a constant. From Fig. IV-10, $A = 2.2$ V (M 3). This thickness dependence of the coercive field is attributed to the existence of a surface layer in BaTiO$_3$ crystals, as originally suggested by Känzig (K 4). The thinner the crystal, the more important this surface layer, provided, of course, that the thickness of the layer is indepen-

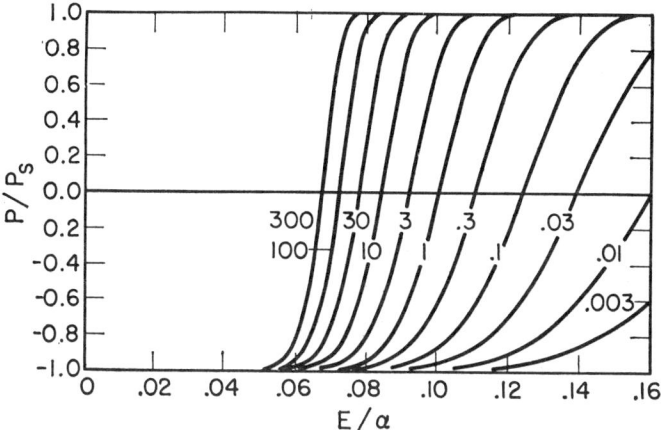

FIG. IV-8. Dependence of the coercive field of BaTiO$_3$ on frequency and amplitude of the applied field. On the ordinate axis, the polarization varies from a value $-P_s$ to a value $+P_s$. On the abscissa axis, the reduced field $E/\alpha$ varies linearly with time. The parameters of the curves represent *approximately* the rise rates of the applied field, given in terms of the time taken for the field to reach 1000 V/cm (according to Landauer *et al.* (L 1)).

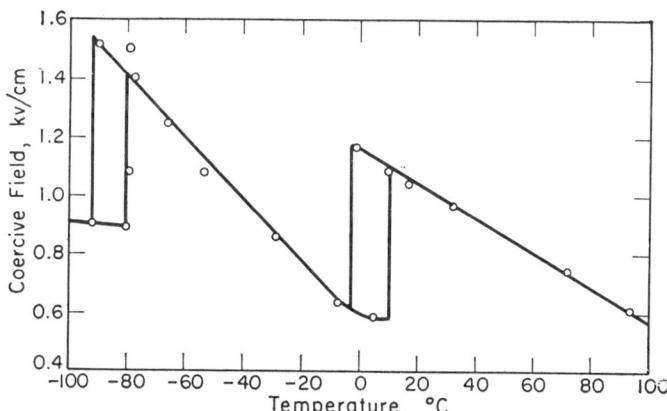

FIG. IV-9. Coercive field of BaTiO$_3$ as a function of temperature (according to Wieder (W 2)).

dent of the crystal thickness. This problem will be discussed in detail in Section 9.

A theory of the coercive field of BaTiO$_3$ was given by Janovec (J 2) on the basis of the picture, originally put forward by Merz (M 11), describing polarization reversal as a process of nucleation and growth of antiparallel domains. This theory defines the coercive field as the field for which the probability of nucleating antiparallel domains starts increasing very rapidly. The critical dimensions of antiparallel nuclei depend on the thickness of the crystal. The theory leads to a satisfactory order-of-magnitude agreement with the experimental values of the coercive field and its dependence upon the sample thickness.

A more recent theoretical treatment by Abe (A 5) is rather based on the assumption that the coercive field is determined by the velocity of forward growth

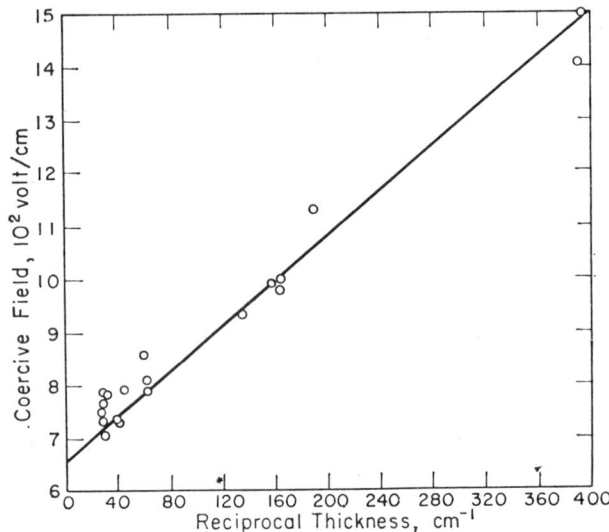

FIG. IV-10. 60 c/s coercive field strength of BaTiO$_3$ as function of reciprocal thickness $d$ of the sample (according to Merz (M 3)).

of the domains. The calculated dependences of the coercive force upon field, crystal thickness and temperature are also in fair agreement with the experimental results.

### Optical Properties

The refractive index of the cubic phase is unusually large, if compared with most other solids: $n \cong 2.4$. Its temperature dependence is depicted in Fig. IV-11, as measured interferrometrically by Hofmann, Gaydou and Gränicher (G 2). The refractive index $n_a$ of the tetragonal phase is practically independent of temperature and shows no measurable anomaly at the Curie point. In the cubic phase, the refractive index increases with decreasing temperature, indicating that the electronic polarizability increases as the Curie point is approached from above. The interferometric measurements indicate also that, from the transition temperature up to about 400 °C, the "optical" thickness of a BaTiO$_3$ plate does not change, as the decrease in refractive index is compensated exactly by the linear thermal expansion (G 2). In Fig. IV-11, the behavior of the refractive index $n_c$ was not measured directly but rather calculated from the measured values of $n_a$ and the birefringence data reported by Meyerhofer (M 4) (see below). Thus, it appears that the broad anomaly of the refractive index of BaTiO$_3$ reported earlier by Busch *et al.* (B 3) does not represent the true behavior of a single crystal, but was probably caused by domain rearrangements and anisotropic strains in the crystals used for that investigation.

The anisotropy of the optical absorption coefficients was measured by Casella and Keller (C 19). At the absorption edge (approximately 3.2 eV), the absorption coefficient for light polarized parallel to the *a* axis was found to be slightly larger than that for light polarized parallel to the *c* axis. It was pointed out that these

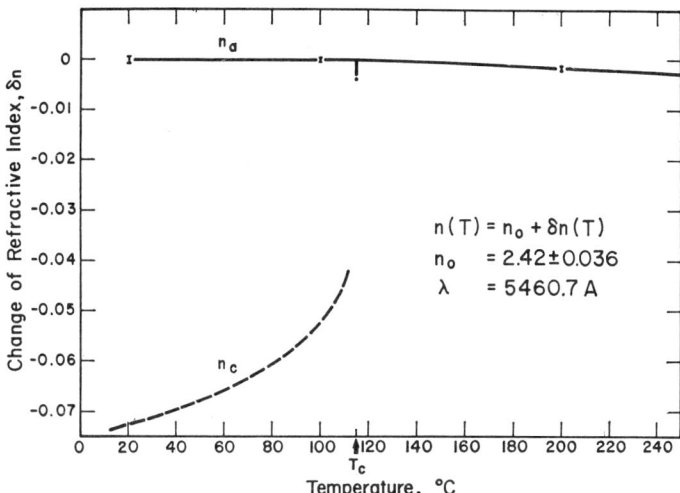

FIG. IV-11. Refractive index of BaTiO$_3$ as a function of temperature. Solid curve: Refractive index along the *a* axis, measured interferometrically and corrected for thermal expansion with Megaw's data (M 5). Dashed curve: Refractive index along the *c* axis, as computed from the solid curve and the birefringence data of Meyerhofer (M 4) (according to Hofmann, Gaydou and Gränicher (G 2)).

results can be qualitatively understood by making certain reasonable assumptions about the energy band structure of polar crystals such as BaTiO$_3$.

The temperature dependence of the birefringence $\Delta n = n_c - n_a$ of the tetragonal phase is depicted in Fig. IV-12, as measured on four different crystals (M 4). At 20 °C the average value of $\Delta n$ is $-0.073$. The crystal is, therefore, optically negative. The discontinuity and the thermal hysteresis at the Curie temperature appear clear from this measurement. Also, a very close similarity between the thermal behavior of $\Delta n$ and that of $P_s$ is observed. It was pointed out by Merz (M 1) that the birefringence of the tetragonal phase is due to the spontaneous Kerr effect arising from the spontaneous polarization. Following the treatment given in Section III-5 for the piezo-optic and electro-optic effects, we see that the change in the polarization constants is:

$$\Delta a = px, \qquad (IV\text{-}3)$$

in the centrosymmetrical non-polar phase, and

$$\Delta a = px + rP, \qquad (IV\text{-}4)$$

in the tetragonal polar phase. Substituting the strain $x$ from Eq. (I-4b) into (IV-4) we find that:

$$a = -psX + (bp + r)P + qP^2. \qquad \text{(IV-5)}$$

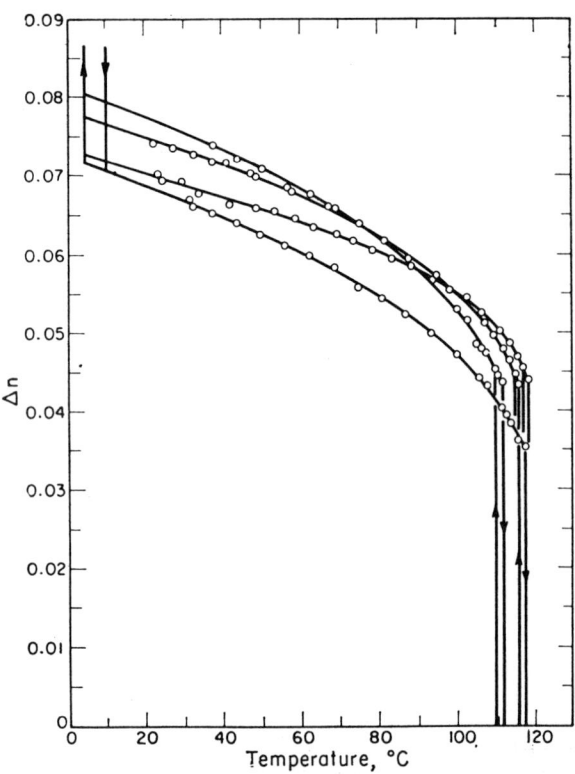

Fig. IV-12. Birefringence of four different crystals of tetragonal BaTiO$_3$ as a function of temperature (according to Meyerhofer (M 4)).

Observing that the occurrence of the electro-optic and piezoelectric coefficients $r$ and $b$, respectively, is a consequence of the lowering of the symmetry, so that both $r$ and $b$ are proportional to $P$, we can write for the birefringence $\varDelta n$ of the stress-free crystal ($X = 0$):

$$\varDelta n = \text{constant} \times P^2, \qquad \text{(IV-6)}$$

expressing the fact that the birefringence can be regarded as the consequence of the quadratic electro-optic effect. Eq. (IV-6) is satisfied by the optical and electric data (M 1). The measured values of $P^2/\varDelta n$ lie between $0.9 \times 10^{-8}$ and $1.25 \times 10^{-8}$ C$^2$/cm$^4$, according to Meyerhofer (M 4).

The birefringence behaves similarly in the orthorhombic phase (M 4). At 0°C, $P = 31 \times 10^{-6}$ C/cm$^2$, $\varDelta n (= n_a - n_b) = 0.115$ and $P^2/\varDelta n = 0.83 \times 10^{-8}$ C$^2$/cm$^4$. The general behavior of the birefringence of BaTiO$_3$ over a wide range of temperature is depicted in Fig. IV-13. The lower values reported in this figure (K 5) with respect to Fig. IV-12 may be ascribed to difference in quality and impurity content of the crystals used for the two investigations. In the orthorhombic phase, the two values of birefringence reported in Fig. IV-13 are those which are possible when the light travels along one of the pseudo-cubic axes perpendicular to a single-domain crystal plate. If the polar axis (ortho-rhombic $a$) lies within the plane of the plate, extinction will occur at 45° to the pseudo-cubic edges (symmetrical extinction) and the birefringence measured in this case is $n_a - n_b$. Otherwise, the polar axis is inclined at 45° to the plane of the plate and extinction will then occur parallel to the pseudo-cubic edges

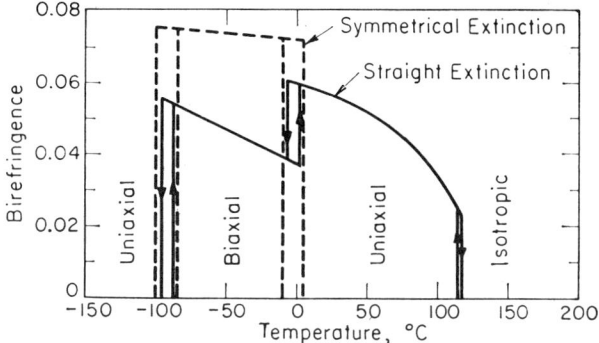

Fig. IV-13. Birefringence of BaTiO$_3$ as a function of temperature (according to Kay and Vousden (K 5)).

(straight extinction): the observed birefringence is $n_c - n_{ab}$, where $n_{ab}$ is the refractive index of a light wave polarized along the direction at 45° to $a$ and $b$.

A theoretical calculation of the birefringence of tetragonal BaTiO$_3$ was recently carried out by Kinase et al. (K 23). These authors took into account the actual distortion of the lattice but assumed spherical symmetry for the electronic configuration of the atoms. The total electronic dipole moment of the unit cell was computed from the local Lorentz fields acting on each atom and the known values of the polarizabilities of Ba, Ti and O; hence the optical dielectric constants were derived. The theoretical value of the birefringence was found to be $n_c - n_a = 0.0191$, implying that under the simplifying assumptions made for the calculations, tetragonal BaTiO$_3$ should be optically positive and weakly birefringent, in contrast to the experimental results. This discrepancy was attri-buted by the authors to the fact that their theory did not take into account the additional birefringence due to the electro-optic effect.

## 3. Specific Heat and Lattice Distortion

### The Specific Heat

Since the Curie transition in BaTiO$_3$ is of the first order, it should be charac-terized by a latent heat. Direct evidence for the existence of a latent heat has been given by Roberts (R 3), who measured the temperature change in polycrystalline specimens caused by an electric-field step function applied in the vicinity of the Curie temperature. A few degrees above this temperature, the tetragonal phase can be induced by a sufficiently large electric field (see Section 4) and a temperature rise of 0.5 °C is measured during this process. When the field is turned off, a temperature drop of the same amount occurs. These tem-

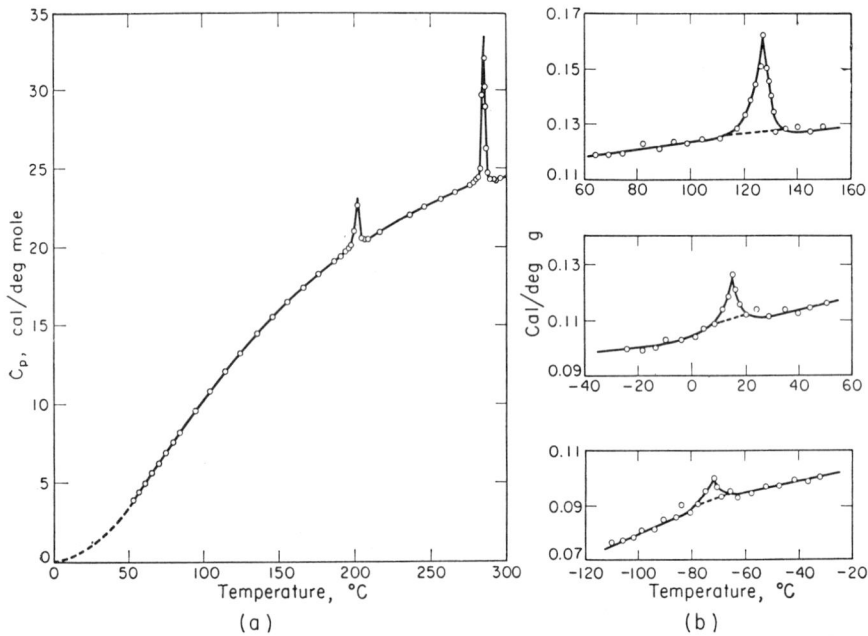

FIG. IV-14. Specific heat of BaTiO₃ as function of temperature (a) according to Todd and Lorenson (T 2), (b) according to Shirane and Takeda (S 4).

perature changes have been interpreted as those accompanying the adiabatic transition from the non-polar to the polar phase. If the material is cooled from higher temperatures into the region of the thermal hysteresis, where the non-polar phase is only metastable, and *then* an electric field is applied, the transition into the polar phase takes place irreversibly and this process is also accompanied by a temperature rise of 0.5 °C. This indicates that a latent heat can be associated with the transition, and Roberts (R 3) estimated it to be larger than 15 cal/mole.

However, the specific heat measurements do not indicate abrupt changes. This is due to the fact that internal strains, surface effects, temperature inhomogeneities and other causes co-operate in spreading the anomalous heat changes over a finite temperature range. One then measures, at best, $\lambda$-shaped peaks. Fig. IV-14 shows the behavior of the specific heat of polycrystalline BaTiO₃ in the temperature range of interest. Table IV-1 summarizes the results of a number of investigations concerning the values of the transition heats and entropy changes.

We are now in the position to check the thermodynamic relation (II-4) between the entropy change $\Delta S$ and the dielectric quantities:

$$\Delta S = (2\pi/C)\,P^2, \qquad\qquad\qquad \text{(IV-7)}$$

where $C$ is the Curie constant and $P^2$ the square of the spontaneous polarization in the vicinity of the Curie temperature. In evaluating (IV-7) one should keep in mind that the experimental value of $\Delta S$ is generally obtained by integrating the specific heat curve between two temperatures $T_1$ and $T_2$, and the value of $P^2$ to be used in (IV-7) is actually equal to $[P(T_1)^2 - P(T_2)^2]$. Since generally $T_2 > T_c$ and hence $P(T_2) = 0$, the bracketed expression reduces to the square of the polarization measured at the temperature $T_1$. Table IV-2 shows the comparison between the values of $(2\pi/C)\,P^2$ and the experimental values of $\Delta S$ obtained from specific heat measurements. The table includes also the pertinent data of a few other ferroelectric materials.

TABLE IV-1.  TRANSITION HEATS $\Delta Q$ AND ENTROPY CHANGES $\Delta S$ OF BaTiO$_3$

| Transition | $\Delta Q$ (cal/mole) | $\Delta S$ (cal/mole °C) | Reference |
|---|---|---|---|
| Cubic to tetragonal | $50 \pm 5$ | 0.125 | SHIRANE and TAKEDA (S 4) |
| | 47 | 0.12 | VOLGER (V 4) |
| | 47 | 0.12 | BLATTNER et al. (B 4) |
| Tetragonal | $22 \pm 4$ | 0.076 | SHIRANE and TAKEDA (S 4) |
| to | 15.5 | 0.054 | VOLGER (V 4) |
| orthorhombic | 16 | 0.058 | BLATTNER et al. (B 4) |
| | 26 | 0.091 | TODD and LORENSON (T 2) |
| Orthorhombic | $8 \pm 2$ | 0.04 | SHIRANE and TAKEDA (S 4) |
| to | 14.3 | 0.07 | VOLGER (V 4) |
| rhombohedral | 12 | 0.06 | TODD and LORENSON (T 2) |

TABLE IV-2.  TEST OF THE RELATION $\Delta S = \dfrac{2\pi}{C}\,P^2$ FOR A FEW FERROELECTRIC CRYSTALS

($C$ is the Curie constant appearing in the Curie—Weiss law of the dielectric constant: $\varepsilon = C/(T - T_0)$)

| Substance | $C$ (°K) | $P$ near Curie point ($10^{-6}$ C/cm$^2$) | $\dfrac{2\pi}{C}\,P^2$ (cal/mole °C) | $\Delta S$ (cal/mole °C) |
|---|---|---|---|---|
| BaTiO$_3$ | $1.7 \times 10^5$ | 20.0 | 0.12 | 0.12 |
| KNbO$_3$ | $2.4 \times 10^5$ | 26.0 | 0.15 | $0.19 \sim 0.28$ |
| Cd$_2$Nb$_2$O$_7$ | $1.0 \times 10^5$ | 5.0 | 0.02 | 0.01 |
| KH$_2$PO$_4$ | $3.3 \times 10^3$ | 5.0 | 0.60 | $0.47 \sim 0.72$ |
| Tri-glycine sulfate | $3.3 \times 10^3$ | 2.0 | 0.64 | 0.48 |
| Tri-glycine fluoberyllate | $2.3 \times 10^3$ | 2.0 | 1.11 | 1.17 |
| Methylammonium aluminum alum | $5 \times 10^2$ | 1.0 | 0.79 | $\sim 1.0$ |

### The Lattice Distortion

As mentioned in Section 1, the three phase changes occurring in BaTiO$_3$ are characterized by an expansion of the original cubic lattice in the direction of the spontaneous polarization and a contraction in the direction perpendicular to it. The temperature dependence of the lattice constants of BaTiO$_3$ in the four phases is depicted in Fig. IV-15 as determined by Kay and Vousden (K 5) on the basis of X-ray diffraction data. At room temperature, the axial ratio $c/a$ is 1.01, showing that the spontaneous strain amounts to 1% (M 5). In the orthorhombic phase, the lattice distortion can be described in terms of a slight

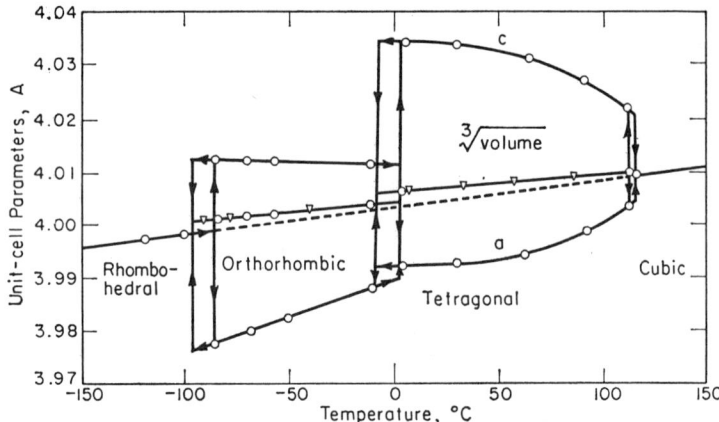

FIG. IV-15. Lattice constants of BaTiO$_3$ as functions of temperature (according to Kay and Vousden (K 5)).

change in length of the original cube edges and a shear in one of the cubic (001) planes (see Fig. IV-2c). The resulting unit cell has thus monoclinic geometry. We choose to call $c_m$ the unique axis of this pseudo-monoclinic cell, owing to a formal analogy with the unit cell of the tetragonal phase. The monoclinic angle will therefore be called $\gamma$, and the two remaining monoclinic axes will be equal to one another: $a_m = b_m$. Other choices of axes may be found in the literature for describing the pseudo-monoclinic cell of the orthorhombic modification of perovskite-type crystals. We are going to adhere to the choice described above throughout this book in all other modifications of double-oxide compounds. The monoclinic parameters $a_m = b_m$, $c_m$ and $\gamma$ are related to the orthorhombic parameters $a_0$, $b_0$ and $c_0$ by the formulas:

$$a_0 = 2 a_m \sin \frac{\gamma}{2}$$

$$b_0 = 2 a_m \cos \frac{\gamma}{2}$$

$$c_0 = c_m.$$

In discussing the crystallographic properties of a pseudo-symmetric phase the subscripts $m$ for monoclinic and $o$ for orthorhombic may often be omitted,

when no confusion can arise. In Fig. IV-15, the parameters shown in the orthorhombic phase are those referred to the pseudo-monoclinic cell. The monoclinic
angle $\gamma$ is larger than 90° by 12'.

In the rhombohedral phase, the distortion of the original cubic cell consists
in an elongation along the [111] direction. The angle between the rhombohedral
axes differs from 90° by 8'.

The curve showing the behavior of the cube root of the cell volume,
in Fig. IV-15, indicates a volume expansion at the Curie transition and a volume

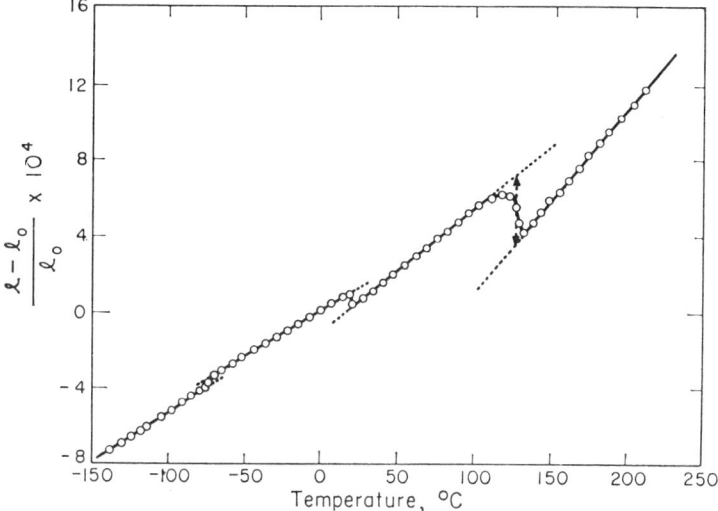

FIG. IV-16. Thermal expansion of polycrystalline BaTiO$_3$ as a function of temperature (according to Shirane and Takeda (S 4)).

contraction at the two lower transitions, for decreasing temperature. These
results are contradicted, in part, by the results of dilatometric measurements
carried out on polycrystalline specimens of BaTiO$_3$ by Shirane and Takeda (S 4).
Fig. IV-16 shows, according to these investigations, that the transitions from
cubic to tetragonal and from tetragonal to orthorhombic states are both connected
with a volume expansion, while the lowest transition from the orthorhombic
to the rhombohedral phase is connected with a small volume contraction. The
discrepancy involves, therefore, only the volume anomaly occurring at the transition between tetragonal and orthorhombic states. Although the macroscopic
volume change measured on polycrystalline samples is more easily affected by
domain reorientations and other causes than the volume change measured by
X-rays, it should be noted that the pressure dependence of the transition temperature speaks in favor of the results of the dilatometric tests. In fact, the shift
of the transition temperature due to hydrostatic pressure can be predicted from
the equation of Clausius–Clapeyron:

$$\frac{\Delta T_c}{\Delta p} = T_c \frac{\Delta V}{L}, \qquad\qquad \text{(IV-8)}$$

where $p$ is the hydrostatic pressure, $T_c$ the transition temperature, $\Delta V$ the volume anomaly and $L$ the latent heat. Table IV-3 summarizes the results of the dilatometric tests and the values of $\Delta T_c/\Delta p$ as computed from these data by using Eq. (IV-8). The experimental results of Merz (M 6), to be described in Section 4, show very good agreement with the calculated data.

Regarding the calculated values of $\Delta T_c/\Delta p$ listed in Table IV-3, it should be pointed out that the Clausius–Clapeyron equation (IV-8) requires the use of the latent heat $L$ and the discontinuous change of the unit cell volume $\Delta V$ just at the Curie point. It is very difficult to compute the contribution of the latent heat to the continuous anomaly of the specific heat measured experimentally. An estimate of this contribution may be done by observing that the transition energy is proportional to $P^2$. As the experimental value is obtained by integration over a certain temperature range, this value should be decreased in proportion to the ratio between the total change in $P^2$ occurring in the same range and the discontinuous change in $P^2$ occurring at the Curie point. From the spontaneous polarization curve (Fig. IV-7), we can compute this ratio and thus estimate that the latent heat at the Curie point is of the order of 35 cal/mole. The same argument could be made in connection with the estimate of the discontinuous volume change from Fig. IV-16. However, since the Clausius–Clapeyron equation involves the ratio $\Delta V/L$ and both $L$ and $\Delta V$ are proportional to $P^2$, it is reasonable to expect that the ratio between the integrated values of volume anomaly and transition energy is approximately equal to the ratio of volume discontinuity and latent heat (S 4).

TABLE IV-3.　VOLUME ANOMALIES AND PRESSURE DEPENDENCE OF THE TRANSITION TEMPERATURES IN $BaTiO_3$ (ACCORDING TO SHIRANE AND TAKEDA (S 4))

| Transition | Transition heat (cal/mole) | Volume anomaly ($Å^3$) | $\Delta T_c/\Delta p$ ($10^{-9}$ °C cm²/dyn) | |
|---|---|---|---|---|
| | | | Calculated | Observed |
| Cubic to tetragonal | 50 | − 0.062 | − 6.7 | − 5.7 |
| Tetragonal to orthorhombic | 22 | − 0.014 | − 2.6 | − 2.9 |
| Orthorhombic to rhombohedral | 8 | − 0.006 | + 2.0 | not. meas. |

Turning back to Fig. IV-15, it will again appear evident that there is a striking similarity between the thermal behavior of the spontaneous strain and that of the spontaneous polarization. This we would expect on the basis of what we have learned in the case of $KH_2PO_4$, with the difference that in $BaTiO_3$ the piezoelectric effect is a consequence of the electrostrictive distortion of the cubic structure. Thus the spontaneous strain is proportional to the square of the spontaneous polarization (M 1), (B 5). As we will see in Section 5, the ratio between the spontaneous strain and the square of the spontaneous polarization is nothing else but the electrostrictive coefficient. For the moment, we limit ourselves to noticing that:

$$\frac{x_3}{P_3^2} = 1.23 \times 10^{-12} \text{ c.g.s.}; \quad \frac{x_1}{P_3^2} = -0.56 \times 10^{-12} \text{ c.g.s.} \tag{IV-9}$$

The signs of these ratios express the fact that $BaTiO_3$ has a positive electrostriction, i.e. it elongates in the direction of its polarization and it contracts in the direction perpendicular to it. It was pointed out by Kinase and Takahashi (K 6) that this is not the result that one would expect if the polarized crystal were composite of rigid electric dipoles parallel to each other. In fact, in such a case the attraction of the dipoles would rather cause a contraction in the direction of polarization and an elongation perpendicular to it. The opposite effect observed in $BaTiO_3$ (and in all perovskite ferroelectrics) is due to the fact that the polarization is caused by the shifts of ions from their symmetrical positions and these shifts depend on the lattice spacing.

This can be understood quantitatively in the following way: if we write the total energy $U$ of a perovskite-type crystal at the absolute zero of temperature as (K 6):

$$U = U_1 + U_2, \tag{IV-10}$$

where $U_1$ represents the energy of a fictitious crystal which has the same lattice parameters as the real crystal but *no* spontaneous polarization; and $U_2$ is the energy due to the dipole interaction, we see that $U_2$ must be equal to zero as long as the lattice parameter is smaller than a critical value $a_{crit}$. Below such a value, in fact, no ionic shift would occur. Figure IV-17 depicts schematically

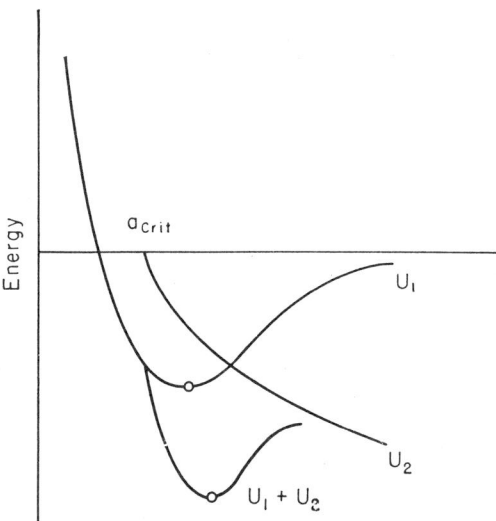

FIG. IV-17. Lattice energy $U_1$, dipole interaction energy $U_2$ and total energy $U = U_1 + U_2$ as functions of the lattice parameter in arbitrary units. The minimum of $U$ corresponds to larger lattice parameter than that of $U_1$ (according to Kinase and Takahashi (K 6)).

the relation between lattice parameter and energy. When the lattice parameter exceeds the value $a_{crit}$, spontaneous shifts of ions occur and the additional energy $U_2$ is added. The minimum of $(U_1 + U_2)$ is thus shifted to the right, i.e. toward larger values of the lattice parameter, with respect to the minimum of $U_1$.

Kinase and Takahashi (K 6) have developed a molecular quantitative theory of the spontaneous deformation of BaTiO$_3$ which is worth considering in some detail. Two basic simplifications were made:

(i) the effect of temperature was neglected, i.e., the crystal was treated at 0 °K; and

(ii) only one kind of ions was allowed to shift from its equilibrium position, namely the Ti ions.

Thus, the lattice energy $U_1$ [Eq. (IV-10)] is written as:

$$U_1 = U_{overlap} + U_{van\ der\ Waals} + U_{Coulomb}. \qquad (IV-11)$$

The overlap energy of two atoms $A$ and $B$ at distance $r$ from each other is taken as $(\lambda_{AB} r^{-9})$ and the van der Waals energy as $(-\mu_{AB} r^{-6})$, where the inter-atomic distances $r$ can be expressed as functions of the lattice parameters of the tetragonal crystal: $(a + \Delta c)$ and $(a + \Delta a)$. The Coulomb interaction energy between two ions with charges $e_A$ and $e_B$ is taken to be (D 3) $(e_A e_B r^{-1})$. Thus, $U_1$ can be written as a function of the lattice deformations $\Delta c$ and $\Delta a$.

With the assumption that the Ti ion is displaced by a distance $x$ from the center of the tetragonal unit cell, the potential energy of this ion is written as:

$$U_2 = ax^2 + bx^4 + a'x^2. \qquad (IV-12)$$

The first two terms represent the potential of the Ti ion owing to the overlap and van der Waals interaction, the term $a'x^2$ represents the Coulomb inter-action. Again, the coefficients $a$ and $b$ can be expressed as functions of the defor-mations $\Delta c$ and $\Delta a$, while the dipole interaction coefficient $a'$ is, in addition, a function of the "effective charge" $ne$ of the Ti ion ($n = 4$ would correspond to the ionized state Ti$^{4+}$).

The equilibrium state is then found from the conditions:

$$\frac{\partial U}{\partial \Delta c} = 0 \quad \text{and} \quad \frac{\partial U}{\partial \Delta a} = 0.$$

The resulting lattice deformations are expressed in terms of the dipole inter-action coefficient $a'$, i.e. in the final instance, in terms of the "effective charge" $ne$ of the Ti ions. Figures IV-18 (a) and IV-18 (b) depict the results of the theoretical calculations of $\Delta c$ and $\Delta a$ and of the volume excess $(\Delta c + 2 \Delta a)$, respectively, where the values of $n$ are indicated on the abscissa axis. Comparison of these figures with Figs. IV-15 and IV-16, respectively, seems to indicate that the effect of temperature is something equivalent to the variation of the dipole-interaction coefficient.

Quantitative evaluation of the theory of Kinase and Takahashi yields results which are in fair agreement with the experimental data if $n$ is chosen to be approximately 2, i.e. the Ti ions are only doubly charged. Assuming $n = 2.12$ the theoretical results are the following (K 11): the tetragonality of $BaTiO_3$ is computed as approximately 1.6%, to be compared with the experimental value of about $1 - 1.5\%$ (M 5), (K 5), (A 1). The volume excess is computed as 0.27%, to be compared with the experimental value of about 0.2% obtained as the difference between the volume at room temperature and that extrapolated from the values of the cubic phase above 120 °C (see Fig. IV-16). Finally, the spontaneous polari-

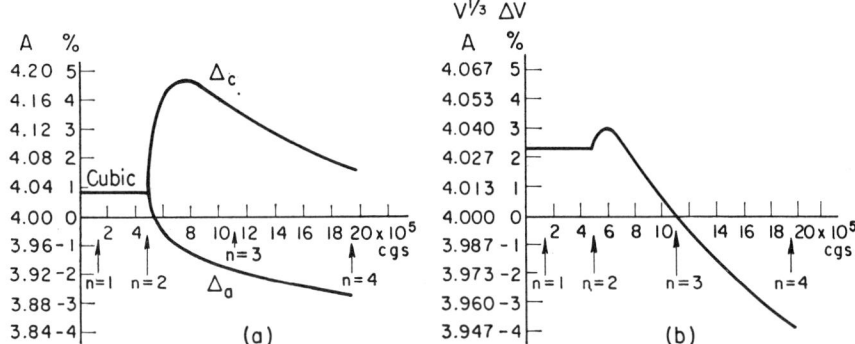

FIG. IV-18. (a) Theoretical value of the lattice deformation parallel ($\Delta c$) and perpendicular ($\Delta a$) to the polarization of $BaTiO_3$ as functions of dipole interaction. (b) Theoretical value of volume excess ($\Delta c + 2\Delta a$) of $BaTiO_3$ as a function of dipole interaction (according to Kinase and Takahashi (K 6)).

zation is computed as being approximately $24 \times 10^{-6}$ C/cm$^2$, to be compared with the experimental value of $26 \times 10^{-6}$ C/cm$^2$ (see also Kinase (K 24)).

In order to obtain agreement between theory and experiment, an effective charge of about 2 has to be assumed for the Ti ions. This is not inconsistent with the structure of $BaTiO_3$: a value of $n = 4$ would, in fact, correspond to a strictly ionic compound, which $BaTiO_3$ is not. Slater's theory of $BaTiO_3$ leads also to a value of $n$ which is of the order of 1 or 2 (see Section 11). Finally, it should be kept in mind (K 6) that the theory of Kinase and Takahashi is valid only at the absolute zero of temperature, where on the other hand, the real crystal is polarized along the pseudocubic [111] direction and the spontaneous deformation has rhombohedral geometry.

## 4. Non-linear Properties and the Effects of Fields and Stresses

### Non-linear Properties and the Effects of Biasing Fields

As we have seen in the case of tri-glycine sulfate (Chapter II) and $KH_2PO_4$ (Chapter III), the non-linear characteristics of ferroelectric crystals are studied experimentally by applying large electric fields upon the samples to affect the dielectric constant and the transition temperature. The case of $BaTiO_3$ differs from those treated in the previous chapters in that the Curie transition of $BaTiO_3$ is of the first order. This difference is best understood by studying the effect of an electric field along the polar axis on the curve of polarization vs. temperature, as shown by Devonshire (D 5) (Fig. IV-19). In the case of a second-order transition, the effect of an electric field is to blur the phase change so that the derivative of the curve of $P$ vs. temperature no longer has discontinuities. The polarization induced by the field in the non-polar phase gradually goes over into the spontaneous polarization upon decreasing the temperature, and, as mentioned in Chapter III, we can no longer speak of a Curie point in this case. A first order transition, on the other hand, will remain a first-order transition in the presence of a field, if the latter is not too large. Here again, strictly speaking, the concept of a Curie

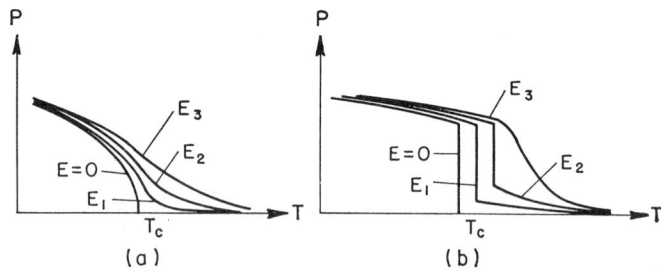

Fig. IV-19. Polarization as a function of temperature with and without a biasing field $E$: (a) near a second-order transition; (b) near a first-order transition (according to Devonshire (D 5)).

point as representing a transition from a non-polar to a polar state cannot be applied, because the crystal is polarized by the field in the otherwise non-polar phase. But we do observe a discontinuous change in polarization when we decrease the temperature of the sample, and this discontinuity occurs at a temperature $T_c + \Delta T_c$ which is generally higher than the Curie temperature. When the field strength exceeds a certain critical value (see below) the change in polarization occurs gradually rather than discontinuously (Fig. IV-19b).

These effects, as all the others that we are going to discuss in the present section, can be explained and understood on the basis of the thermodynamic theory outlined in Section I-5. In the light of this theory, the experimental results can then be used for computing the values of the coefficients appearing in the expansion of the free energy. It may, therefore, be convenient to recall here the fundamental equations which help in understanding the physical effects to be described in this section.

We write the expansion of the free energy of a stress-free crystal ($X = 0$) of $BaTiO_3$, from Eq. (I-1), in the following way (D 5):

$$A(P) = \tfrac{1}{2} \chi^X (P_1^2 + P_2^2 + P_3^2) + \tfrac{1}{4} \xi_{11}^X (P_1^4 + P_2^4 + P_3^4) +$$
$$\tfrac{1}{2} \xi_{12}^X (P_2^2 P_3^2 + P_3^2 P_1^2 + P_1^2 P_2^2) +$$
$$\tfrac{1}{6} \zeta_{111}^X (P_1^6 + P_2^6 + P_3^6) + \hspace{3cm} \text{(IV-13)}$$
$$\tfrac{1}{2} \zeta_{112}^X [P_1^2 (P_2^4 + P_3^4) + P_2^2 (P_3^4 + P_1^4) + P_3^2 (P_1^4 + P_2^4)] +$$
$$\tfrac{1}{2} \zeta_{123}^X P_1^2 P_2^2 P_3^2 + \cdots\cdots\cdots$$

where $P_1$, $P_2$, $P_3$ are the polarization components along the crystallographic axes $x$, $y$, $z$ respectively.

In the present section, we are going to be concerned mostly, although not exclusively, with the properties of $BaTiO_3$ in the cubic and tetragonal phase with fields applied along the pseudo-cubic [001] direction, so that we can put $P_1 = P_2 = 0$ and $P_3 = P$. Eq. (IV-13) can thus be simplified as follows [see also Eq. (II-1)]:

$$A(P) = \tfrac{1}{2}\chi^X P^2 + \tfrac{1}{4}\xi_{11}^X P^4 + \tfrac{1}{6}\zeta_{111}^X P^6. \qquad \text{(IV-14)}$$

The coefficient $\chi^X$ has already been determined from the temperature behavior of the low-field dielectric constant of the free crystal: $\chi^X = (4\pi/C)(T - T_0)$. The coefficients $\xi_{11}^X$ and $\zeta_{111}^X$ can be determined with a number of experiments, some of which we already know from the previous chapters and which are, of course, related to each other. A list of these experiments follows.

(i) Measurement of the values of the spontaneous polarization and the reciprocal dielectric susceptibility at the Curie temperature. Calling $P_0$ and $\chi_0^X$ these two values, respectively, we derived, in Chapter I, two equations [Eqs. (I-13) and (I-14)]:

$$P_0^2 = \frac{3}{4}\left(-\frac{\xi_{11}^X}{\zeta_{111}^X}\right); \quad \chi_0^X = \frac{3}{16}\frac{(\xi_{11}^X)^2}{\zeta_{111}^X} \qquad \text{(IV-15)}$$

for the determination of the coefficients $\xi_{11}^X$ and $\zeta_{111}^X$.

(ii) Measurement of the polarization dependence of the dielectric constant above the Curie temperature. We have described this method in detail in Section II-3 for the case of tri-glycine sulfate. Calling $\varepsilon_0(T)$ and $\varepsilon_P(T)$ the dielectric constants of the non-polarized and the polarized sample, respectively, we obtain from Eq. (II-3):

$$\frac{4\pi}{\varepsilon_P(T)} - \frac{4\pi}{\varepsilon_0(T)} = 3\xi_{11}^X P^2 + 5\zeta_{111}^X P^4. \qquad \text{(IV-16)}$$

For small values of $P$, the quantity on the left-hand side is a linear function of $P^2$ and the slope of the straight line has the value $3\xi_{11}^X$. This experiment was carried out on single crystals of $BaTiO_3$ by Drougard et al. (D 6). The advantage of this method of determination of the coefficient $\xi_{11}^X$ is that it allows measurements at different temperatures. Drougard et al. (D 6) carried out the measurements between 120 °C and 150 °C and found, in contrast with the assumption made in Section I-5, that $\xi_{11}^X$ is linearly dependent upon temperature in this range (Fig. IV-20). Extrapolation of the linear behavior at higher temperatures reveals that $\xi_{11}^X$ should vanish in the vicinity of 175 °C.

(iii) Measurement of the field dependence of the dielectric constant. This dependence can, of course, be computed from the effect described above under (ii), as done by Triebwasser (T 3) and depicted in Fig. II-8, together with the corresponding effect in tri-glycine sulfate. The comparison between the latter salt and $BaTiO_3$ is particularly instructive, in this case, in showing the difference in the expected field dependence of $\varepsilon$ above a second-order and a first-order transition. Since the coefficient $\xi_{11}^X$ is positive in the former case, the dielectric constant is decreased by a d.c. biasing field, whereas it should be increased in the latter case, in which $\xi_{11}^X$ is negative.

The direct experiment would involve, of course, measuring $\varepsilon$ with a small a.c. field superimposed on a large d.c. biasing field. The pertinent formula for this experiment can be derived in the following way. From:

$$\frac{\partial A}{\partial P} = E = \chi^X P + \xi^X_{11} P^3 + \zeta^X_{111} P^5, \qquad \text{(IV-17)}$$

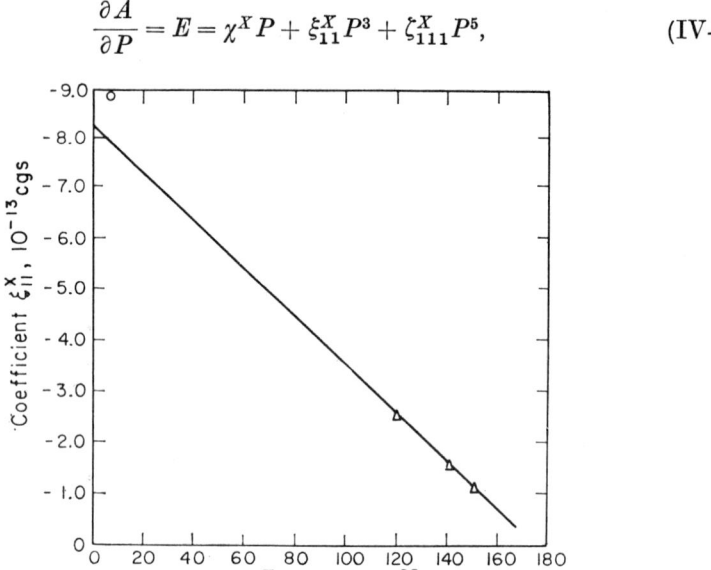

FIG. IV-20. The coefficient $\xi^X_{11}$ of the free energy expansion of BaTiO$_3$ as a function of temperature. The triangles refer to measurements in the non-polar phase by Drougard et al. (D 6), the open circle to the measurement in the orthorhombic phase by Huibregtse and Young (H 1).

we can derive an equation expressing $P$ in terms of powers of $E$ by successive approximations, writing first:

$$P = \frac{1}{\chi^X} (E - \xi^X_{11} P^3 - \zeta^X_{111} P^5), \qquad \text{(IV-18)}$$

and substituting, in the first approximation, $P = E/\chi^X$ in (IV-18). In this way, writing $\varepsilon_0(T)$ and $\varepsilon_E(T)$ for the dielectric constant measured without and with d.c. bias $E$ at the temperature $T$, we obtain, for not too large fields:

$$\varepsilon_E(T) - \varepsilon_0(T) = -\frac{3\,\varepsilon_0^4\,\xi^X_{11}}{(4\pi)^3}\,E^2. \qquad \text{(IV-19)}$$

This formula shows directly that when $\xi^X_{11} < 0$, the difference $(\varepsilon_E - \varepsilon_0)$ is positive when $\xi^X_{11} > 0$, the difference is negative.

The evaluation of this experiment, as well as that described under (ii), should take into consideration the electrocaloric effect (K 1), (M 2), (D 6), i.e. the difference between adiabatic and isothermal dielectric constant, as it was done in the case of tri-glycine sulfate and KH$_2$PO$_4$. The adiabatic correction term of the dielectric constant was computed in Eq. (II-10) as $4\pi T/C^2 c_P$. Insertion of the numerical values pertinent to $BaTiO_3$ shows, however, that this term is negligibly small.

The effect of a d.c. bias on the dielectric behavior of $BaTiO_3$ was investigated by Känzig and Maikoff (K 1) but the crystals used for this investigation were not sufficiently good to allow an analysis of the results on the basis of Eq. (IV-19). Reliable data were obtained, however, for the shift of the transition temperature, $\Delta T_c$, due to the biasing field. This shift can be expressed in terms of Curie constant $C$ and the coefficients $\xi_{11}^X$ and $\zeta_{111}^X$ in the following way (D 5):

$$\frac{\Delta T_c}{\Delta E} = \frac{C}{4\pi} \left( -\frac{4}{3} \frac{\zeta_{111}^X}{\xi_{11}^X} \right)^{1/2}. \tag{IV-20}$$

Insertion of the numerical values leads to:

$$\frac{\Delta T_c}{\Delta E} = 1.6 \times 10^{-3} \,°C/V\,cm^{-1},$$

while the experimental results of Känzig and Maikoff (K 1) yield $1.2 \times 10^{-3}$ °C/ $V\,cm^{-1}$ and those of Merz (M 2), to be described under (iv), yield $1.4 \times 10^{-3}$ °C/ $V\,cm^{-1}$, in good agreement with each other and with the value predicted by Devonshire's thermodynamic theory. It should be noted that the quantity $\Delta T_c/\Delta E$ can also be given by a Clausius–Clapeyron-like expression (D 5), thus:

$$\frac{\Delta T_c}{\Delta E} = -\frac{\Delta P}{\Delta S}, \tag{IV-21}$$

where $\Delta P$ is the discontinuous change in polarization (equal to $-P_0$) and $\Delta S$ is the entropy change at the Curie point. Evaluation of Eq. (IV-21) yields the value $1.4 \times 10^{-3}$ °C/V cm$^{-1}$, also in excellent agreement with the results reported above.

The field dependence of the dielectric constant was recently studied by Triebwasser (T 10) on good single crystals of $BaTiO_3$ in the temperature region just above the Curie point. The results of this investigation are depicted in Fig. IV-21, where the dashed curve represents the behavior expected on the basis of the thermodynamic theory, and the full curve represents the observed behavior. It is seen that the experimental data fit the theoretical curve only in a narrow range of small applied fields, but otherwise deviate drastically from the theoretical expectation. The reason for this discrepancy is that the application of d.c. fields gives rise to space charge layers in the vicinity of the electrodes (see Section 9), and this, in turn, causes a decrease of the apparent dielectric constant measured at equilibrium. Actually, the experimental data indicate that immediately after the application of the d.c. field, the dielectric constant changes to the value predicted by the thermodynamic theory, but as space charges appear near the electrodes, the field in the bulk is reduced drastically to a vanishing value and the dielectric constant of the bulk is therefore almost restored to its unbiased value. The *apparent* dielectric constant, however, is smaller than the unbiased value, because the crystal can now be regarded as a system including one or two surface-layer capacitors in series with that represented by the bulk.

Although this implies a reduction of the effective bulk thickness (reduction which is negligible unless the sample is extremely thin), the apparent dielectric constant is lowered because the intrinsic dielectric constant of the space-charge layer is smaller than that of the bulk material. Since the thickness of such a layer is a function of the applied field, the apparent dielectric constant at equilibrium is dependent upon this field (T 10). We will discuss the nature of these space-charge layers of $BaTiO_3$ in Section 9.

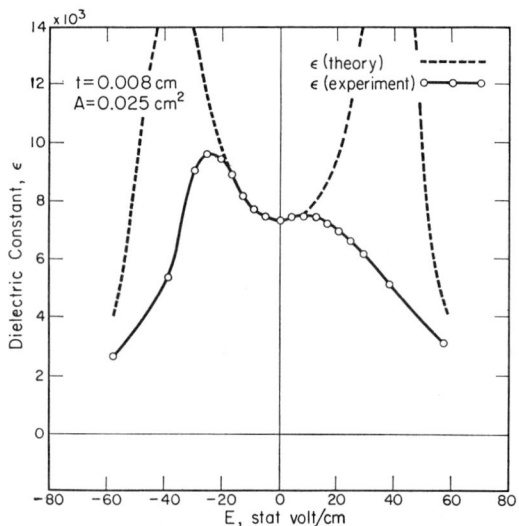

FIG. IV-21. Field dependence of the dielectric constant $\varepsilon_c$ of $BaTiO_3$ a few degrees above the Curie temperature (according to Triebwasser (T 10)).

(iv) Measurement of the $P$–$E$ curves. By this we mean the study of the curves representing the relation between polarization $P$ and field $E$ as displayed, for example, on the screen of an oscilloscope by means of the Sawyer and Tower circuit. Merz (M 2) carried out a detailed study of the $P$–$E$ curves of $BaTiO_3$ crystals above the Curie temperature. A similar study was done independently by Cross (C 3) and, later, repeated by Drougard (D 7). Figure IV-22 shows the observed curves for a crystal whose Curie temperature is 107 °C. Above this temperature, for zero applied field, the crystal is in the non-polar state. With increasing field strength, the polarization increases linearly, as one would expect in an ordinary dielectric material. At a certain critical field strength, however, a discontinuous change in polarization occurs. This critical field strength is that necessary to induce the polarized state at the temperature of observation (see Fig. IV-19). One observes a "double hysteresis loop" which is characteristic of the $P$–$E$ relationship above a transition of the first order.

Such double loops are predicted by the thermodynamic theory, in that they are the graphical representation of the $P$–$E$ relationship expressed by Eq. (IV-17). Such a graphical representation, as computed from Devonshire's theory, is given in Fig. IV-23, where the quantities $p$, $e$, $t$ are normalized polarization, field and

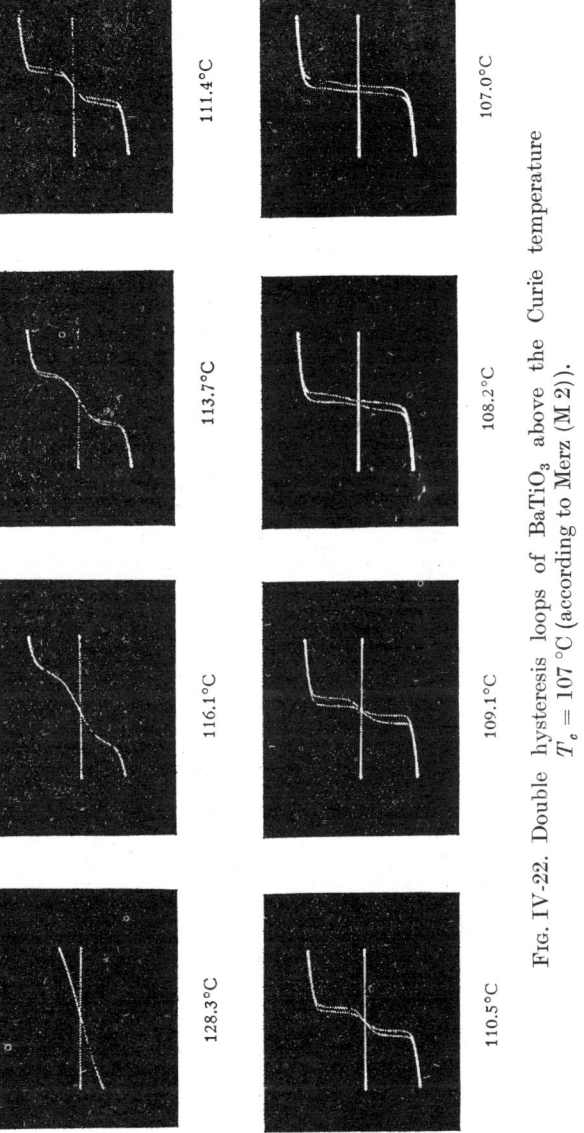

FIG. IV-22. Double hysteresis loops of BaTiO$_3$ above the Curie temperature $T_c = 107\ ^\circ$C (according to Merz (M 2)).

temperature, respectively, and are linearly related to $P$, $E$, $T$ (M 2). It is seen that the slope of the curve at the origin is positive if $T > T_0$, negative when $T < T_0$. Whenever the slope $dP/dE$ becomes negative, which corresponds to an unstable condition, the state of the crystal jumps discontinuously along a

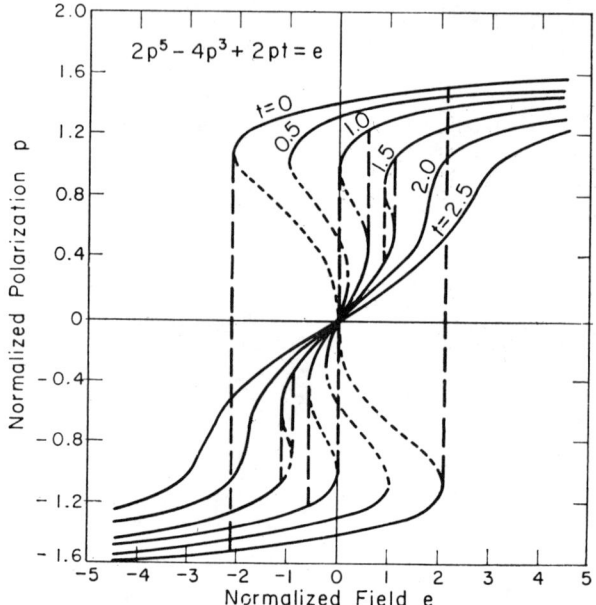

FIG. IV-23. $P$–$E$ curves of BaTiO$_3$ in the vicinity of the Curie temperature. $p$, $e$ and $t$ are normalized quantities linearly related to $P$, $E$ and $T$, respectively (according to Merz (M 2)).

vertical line into a state where $dP/dE$ is positive again. The condition for the slope $dE/dP$ to be zero is given by:

$$\frac{dE}{dP} = \chi^X + 3\xi^X_{11} P^2 + 5\zeta^X_{111} P^4 = 0. \qquad (\text{IV-22})$$

This equation has two roots in $P^2$ when

$$9(\xi^X_{11})^2 - 20\zeta^X_{111}\chi^X > 0, \qquad (\text{IV-23})$$

and if $\chi^X > 0$, both roots are positive, if $\chi^X < 0$, only one root is positive, the other is negative. Condition (IV-23) determines a temperature $T_1$, above which Eq. (IV-22) has no real solution and, therefore, the double loops can no longer be observed. This temperature, $T_1$, lies about 8° above the Curie temperature (M 2).

The field-induced transition revealed by the double-loop experiment and the measurements of the dielectric properties of a crystal biased with a d.c. field were confirmed by other measurements. Chynoweth (C 1) investigated the field-induced pyroelectric effect at temperatures above the Curie point (the crystals used in this investigation had a Curie temperature of 101 °C), with the results depicted in Fig. IV-24(a). At temperatures slightly above the Curie point, the pyroelectric current increases rapidly with increasing field, then drops discontinuously to some lower value (induced transition) after which it decreases slowly and smoothly. With increasing temperature, the discontinuous drop occurs at increasing fields and, also, decreases in magnitude until it no longer occurs. Above the latter temperature $(T_1)$, the curves show a smooth rather than an abrupt peak. The experimental results are again explained by the thermodynamic theory. Differentiation of Eq. (IV-17) with respect to temperature yields an expression for the pyroelectric coefficient, $dP/dT$, which can be plotted as a function of $E$ and compared with the experimental curves. The theoretical plot is shown in Fig. IV-24(b) in terms of normalized quantities.

Meyerhofer (M 4) investigated the dependence of birefringence on electric fields in the [100] direction for various temperatures in the vicinity of the Curie point. A similar investigation was done earlier, qualitatively, by Caspari and

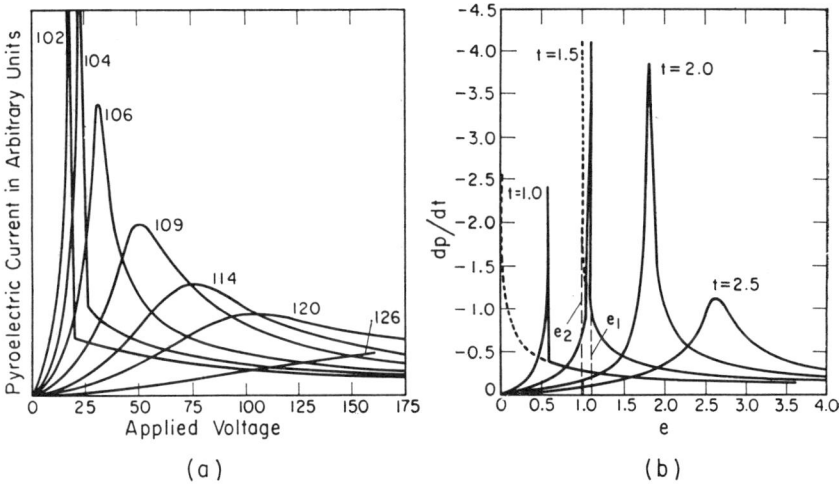

FIG. IV-24. The pyroelectric effect of BaTiO₃. (a) Variation of the pyroelectric current with applied field for various temperatures above the Curie point. In this case, $T_c = 101$ °C. (b) Theoretical plots of the normalized pyroelectric coefficient $(dp/dt)$ vs. the normalized field $e$ for different values of the normalized temperature $t$ (according to Chynoweth (C 1)).

Merz (C 4). A study of Fig. IV-25(a) shows that the birefringence caused by the field in the non-polar phase can be expressed as:

$$\Delta n = \alpha_1 E^2 + \alpha_2 E^4 + \ldots, \qquad (IV-24)$$

in accordance to the quadratic electro-optic effect discussed in Section 2. The coefficients $\alpha_1$ and $\alpha_2$ are of the order of $10^{-10}$ cm² V⁻² and $10^{-18}$ cm⁴ V⁻⁴, respectively, about 2 °C above the Curie point, and decrease rapidly for increasing temperature. The ferroelectric phase is induced for a critical value of the field, in analogy with the other experiments discussed above.

An interesting effect is observed when the field is applied along the cubic [110] direction. In the non-polar phase, the induced birefringence is parallel to the field and its magnitude is given by an equation similar to (IV-24). Figure IV-25 (b) shows that a ferroelectric phase is induced, at higher field strengths, which has a smaller birefringence. Optical measurements indicate that this phase is orthorhombic, but *different* from the phase which is stable below 5 °C. More work is needed to elucidate this point.

The effect of electric fields on the transition from tetragonal into orthorhombic states at 5 °C was investigated by Huibregtse and Young (H 1) and reviewed

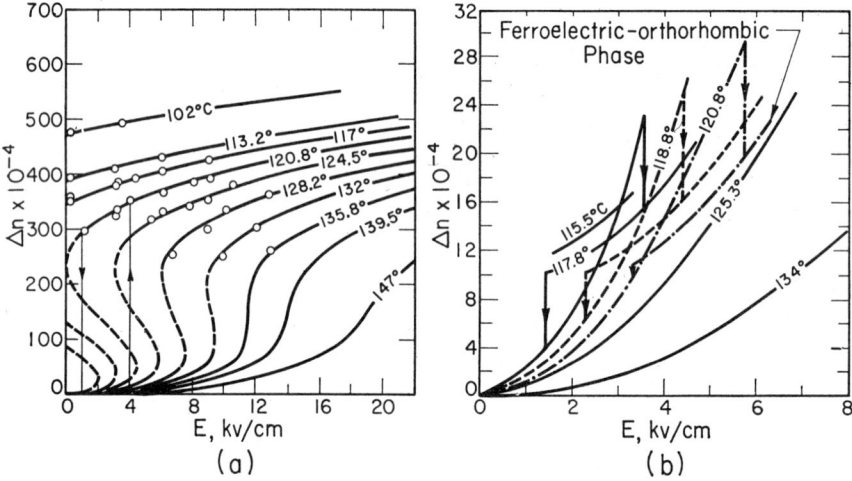

Fig. IV-25. Induced birefringence of BaTiO$_3$ in the vicinity of the Curie point as a function of the electric field applied along: (a) the [100] direction (the points are experimental and the solid curves theoretical); (b) the [110] direction (according to Meyerhofer (M 4)).

by Drougard and Huibregtse (D 8). The transition occurring at 5 °C differs from that occurring at 120 °C in that it is a transition from a ferroelectric state into another ferroelectric state, involving not only a change in magnitude but also in direction of the spontaneous polarization.

The effect of a d.c. biasing field applied along [001] on the dielectric constant in this transition region is depicted in Fig. IV-26 (a). For zero bias, the transition occurs at 6 °C upon cooling and at 12 °C upon heating. For a finite bias, both transitions are lowered, according to expectation, as the applied field is in the direction which favors the tetragonal state. For a field of $2.6 \times 10^4$ V/cm, the orthorhombic–tetragonal transition occurs at a lower temperature than does the tetragonal–orthorhombic transition for zero field. This result suggests the possibility that at an appropriate temperature below 6 °C it is possible to "switch" reversibly between the two states by varying the field applied to the crystal (H 1).

Thus, a large a.c. field applied along [001] just below 6 °C, permits the observation of "triple hysteresis loops" (Fig. IV-26b). At zero and low fields, the

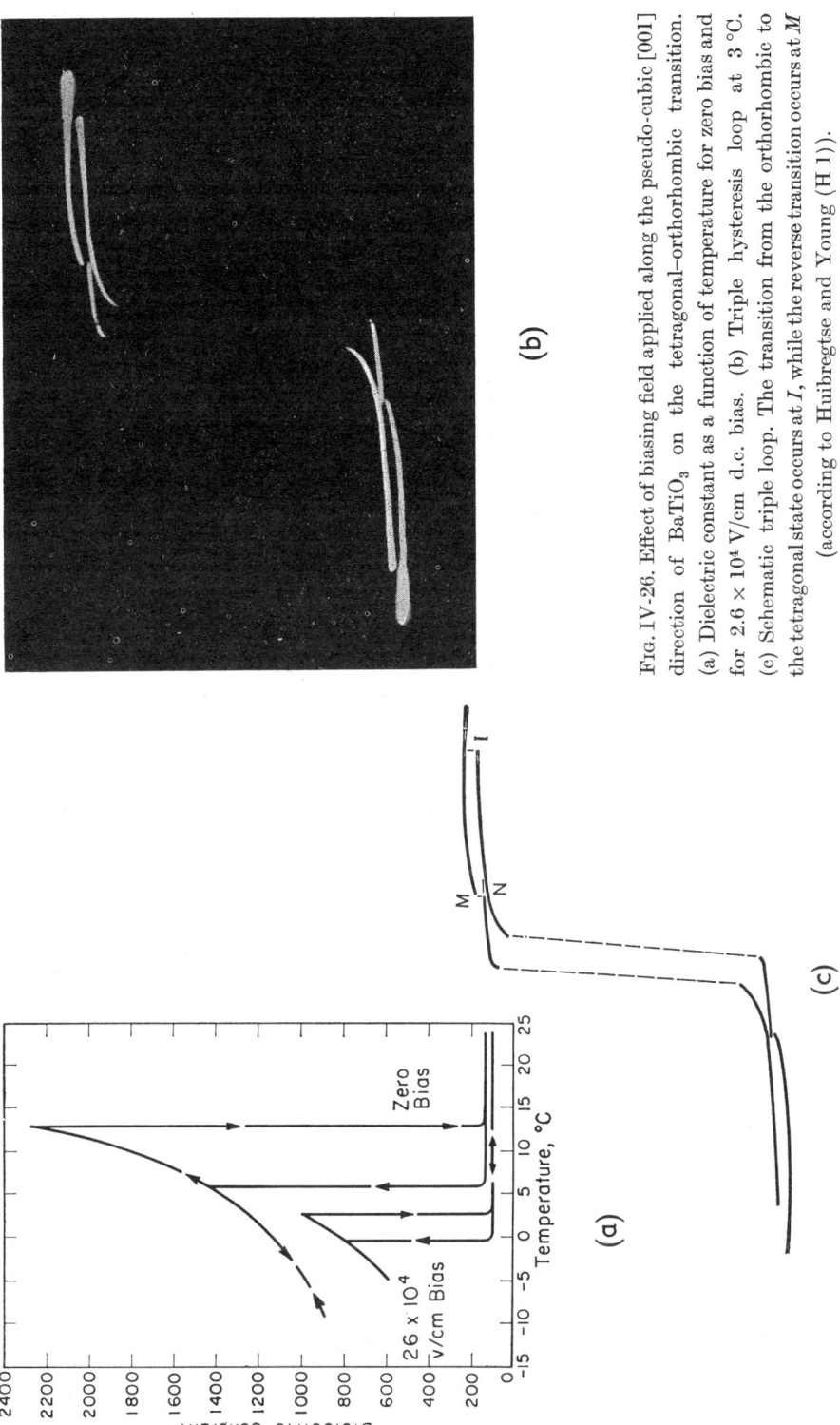

Fig. IV-26. Effect of biasing field applied along the pseudo-cubic [001] direction of BaTiO$_3$ on the tetragonal-orthorhombic transition. (a) Dielectric constant as a function of temperature for zero bias and for $2.6 \times 10^4$ V/cm d.c. bias. (b) Triple hysteresis loop at 3 °C. (c) Schematic triple loop. The transition from the orthorhombic to the tetragonal state occurs at $I$, while the reverse transition occurs at $M$ (according to Huibregtse and Young (H 1)).

orthorhombic state is stable, but at larger fields (point $I$ in Fig. IV-26 c), the tetragonal state is induced. As the polarization in this state is slightly larger than the [001] component of the polarization in the orthorhombic state, a discontinuous increase in polarization is observed at $I$. The reverse transition occurs at the much smaller field indicated by $M$ in Fig. IV-26 (c). These repeated transitions between the two states produce severe shocks to the crystal, and the samples shatter when the frequency of the applied field is 60 c/s. Only at much lower frequencies, around 5 c/s, is it possible to observe triple hysteresis loops without damaging the crystals.

The triple-loop experiment can be explained by the thermodynamic theory. In the orthorhombic phase, only one component of the polarization vector is equal to zero. Choosing this to be $P_2 = 0$, the free energy function (IV-13) becomes:

$$A(P) = \tfrac{1}{2}\chi^X(P_1^2 + P_3^2) + \tfrac{1}{4}\xi_{11}^X(P_1^4 + P_3^4) + \tfrac{1}{2}\xi_{12}^X P_1^2 P_3^2$$
$$+ \tfrac{1}{6}\zeta_{111}^X(P_1^6 + P_3^6) + \tfrac{1}{2}\zeta_{112}^X(P_1^2 P_3^4 + P_3^2 P_1^4) + \cdots \cdots \tag{IV-25}$$

From this expression, one can derive the relationship between $P_3$ and $E_3 = \partial A/\partial P_3$ and compare it with the observed triple loop. The correlation between theoretical and experimental loops allows the determination of the coefficients appearing in the polynomial (IV-25). For the pertinent formula, the reader is referred to the original paper of Huibregtse and Young (H 1). The results are consistent with those reported above: the value of $\xi_{11}^X$ determined at 6 °C fits nicely on the extrapolated line representing the temperature dependence of $\xi_{11}^X$ (open circle in Fig. IV-20). It should be recalled that such a temperature dependence was established in the non-polar phase, i.e. in a temperature region more than 100° higher, with a completely different type of experiment (D 6) (see above). Table IV-4 lists the values of the coefficients of the polynomial (IV-25) as determined with the experiments described above.

TABLE IV-4.  VALUES OF THE COEFFICIENTS OF THE FREE ENERGY EXPANSION (IV-25)
FOR BaTiO$_3$ (ACCORDING TO DROUGARD AND HUIBREGTSE (D 8))

| | |
|---|---|
| $\chi^X$ | $= 7.4 \times 10^{-5}\,(T - T_0)$ c.g.s., where $T_0 = 110$ °C |
| $\xi_{11}^X$ | $= 18.0 \times 10^{-15}\,(T - T_2)$ c.g.s., where $T_2 = 175$ °C |
| $\xi_{12}^X$ | $= 12 \times 10^{-13}$ c.g.s. |
| $\zeta_{111}^X$ | $= 54 \times 10^{-23}$ c.g.s. |
| $\zeta_{112}^X$ | $= 8 \times 10^{-23}$ c.g.s. |

### The Effect of Mechanical Stresses

Turning now to the effects of mechanical stresses on the transition temperature of BaTiO$_3$, it should be noted, at first (K 7), that the present case differs, e.g. from KH$_2$PO$_4$ insofar as no linear correspondence exists between stress and field. In KH$_2$PO$_4$, owing to the piezoelectric properties of the non-polar phase, a shear stress in the (001) plane is essentially equivalent to an electric field applied along the direction of the polar axis. In BaTiO$_3$ such a situation cannot be realized. It is true, however, that a two-dimensional pressure applied in the (001) plane of the tetragonal phase ought to favor the tetragonality and thus increase the Curie temperature. Experiments of this sort were carried out by Forsbergh (F 1). The original experimental data suggested that the increase of Curie temperature is proportional to the square of the two-dimensional pressure, but these data were later considered unreliable, new experiments exhibiting a definite initial shift of the Curie point under two-dimensional pressure (F 2).

The theoretical treatment of this effect can be done on the basis of the thermo-dynamic theory (see also Section II-6). The effect of pressure on the Curie temperature $T_c$ is difficult to compute from the theory but the effect on the Curie–Weiss temperature $T_0$ can be determined easily (F 1). Within the usual limitations imposed upon the temperature dependence of the coefficients of the free energy expansion, the theory predicts a linear dependence of the Curie–Weiss temperature upon two-dimensional pressure. The value of the temperature shift can be computed from the knowledge of the elastic and electrostrictive constants of the crystal (see Section 5).

A similar result is predicted by the theory for the effect of *hydrostatic* pressure on the Curie–Weiss temperature (F 1). The experiment was carried out by Merz (M 6) on single crystals of $BaTiO_3$, with the result reported in Table IV-3. The Curie temperature decreases linearly with pressure, in accordance with the value computed from the Clausius–Clapeyron equation on the basis of thermal expansion and specific heat data (see Section 3). Good agreement is found also between theoretical and experimental values for the effect of pressure on the tetragonal–orthorhombic transition (Table IV-3). This transition temperature is found to decrease with pressure up to about 1500 atm and then increase with about the same slope. It would be interesting to extend this investigation beyond the limit of 3000 atm reached by Merz.

The results of Shirane and Sato (S 5) concerning the effects of pressure are also worth mentioning, although they were obtained with polycrystalline $(Ba, Sr)TiO_3$. The behavior of the dielectric constant under pressure indicates that the Curie constant remains unaffected and the shift of the Curie temperature is in agreement with that measured by Merz. The effect of a *single* compressional stress (T 7) on the other hand, be it parallel or perpendicular to the electroded surface of the ceramic sample, is to raise the temperature of the dielectric constant peak at the rate of $4.0 \times 10^{-3}$ °C cm$^2$/kg.

The measurements of the effect of stresses on the dielectric constant above the Curie point allow the determination of the electrostrictive coefficients, which we are going to discuss in the next section.

## 5. Electrostrictive, Piezoelectric and Elastic Properties

### Electrostrictive and Piezoelectric Effect

In contrast to $KH_2PO_4$, and similarly to tri-glycine sulfate, $BaTiO_3$ is not piezoelectric in the non-polar phase. Owing to the high symmetry of this phase, the discussion of the electrostrictive effect of $BaTiO_3$ is particularly simple;

hence it is worth devoting some time to it. Detailed treatments of this effect can be found in the papers of Devonshire (D 5), Kay (K 8) and Forsbergh (F 2). As mentioned in Section I-5, the electrostriction is a quadratic effect and occurs in all crystals. When an electric field is applied to the crystal, it produces strains proportional to the square of the field.

The fundamental equations governing this effect can be derived from the free energy functions $A(x, P)$ (Eq. I-1) and $A(X, P)$; or $A(x, E)$ and $A(X, E)$, neglecting higher order terms, thus:

$$X_{ij} = -c^P_{ijkl} x_{kl} + q_{ijkl} P_k P_l, \tag{IV-26}$$

$$x_{ij} = -s^P_{ijkl} X_{kl} + Q_{ijkl} P_k P_l, \tag{IV-27}$$

or:

$$X_{ij} = -c^E_{ijkl} x_{kl} + m_{ijkl} E_k E_l, \tag{IV-28}$$

$$x_{ij} = -s^E_{ijkl} X_{kl} + M_{ijkl} E_k E_l. \tag{IV-29}$$

These equations define the electrostrictive coefficients $q, Q, m$ and $M$. Substitution of (IV-27) into (IV-26) and (IV-29) into (IV-28) shows that:

$$\left.\begin{aligned} Q_{ijkl} &= s^P_{ijhn} q_{hnkl} \\ q_{ijkl} &= c^P_{ijhn} Q_{hnkl} \end{aligned}\right\} \tag{IV-30}$$

and:

$$\left.\begin{aligned} M_{ijkl} &= s^E_{ijhn} m_{hnkl} \\ m_{ijkl} &= c^E_{ijhn} M_{hnkl} \end{aligned}\right\} \tag{IV-31}$$

Using the Voigt notation with two suffixes instead of four (see, e.g. Forsbergh (F 2)), the matrix of the electrostrictive constants for cubic crystals is, considering for example the coefficients $Q$:

$$\begin{matrix} Q_{11} & Q_{12} & Q_{12} & 0 & 0 & 0 \\ Q_{12} & Q_{11} & Q_{12} & 0 & 0 & 0 \\ Q_{12} & Q_{12} & Q_{11} & 0 & 0 & 0 \\ 0 & 0 & 0 & Q_{44} & 0 & 0 \\ 0 & 0 & 0 & 0 & Q_{44} & 0 \\ 0 & 0 & 0 & 0 & 0 & Q_{44} \end{matrix} \tag{IV-32}$$

For experimental purposes, it is convenient to consider the equations (IV-27) and (IV-29) and apply them to the stress-free crystal ($X = 0$). The fundamental equations of a stress-free BaTiO$_3$ crystal in the cubic phase are thus:

$$\left.\begin{aligned} x_1 &= Q_{11} P^2_1 + Q_{12} P^2_2 + Q_{12} P^2_3 \\ x_2 &= Q_{12} P^2_1 + Q_{11} P^2_2 + Q_{12} P^2_3 \\ x_3 &= Q_{12} P^2_1 + Q_{12} P^2_2 + Q_{11} P^2_3 \\ x_4 &= Q_{44} P_2 P_3 \\ x_5 &= Q_{44} P_3 P_1 \\ x_6 &= Q_{44} P_1 P_2 \end{aligned}\right\} \tag{IV-33}$$

A similar set of equations is obtained from Eq. (IV-29), relating the strains with the field components.

These equations indicate one way of determining the electrostrictive coefficients: by comparing the field or the polarization with the corresponding strain. The polarization can, of course, be either induced or spontaneous, if the induced or the spontaneous strain is considered, respectively. Thus, knowing the values of the spontaneous strain at room temperature, $x_1 = x_2 = -0.0034$, $x_3 = 0.0075$ (K 5) and the value of the spontaneous polarization at the same temperature, $P_3 = 26 \times 10^{-6}$ C/cm$^2$ (M 2), we obtain (see also Eq. IV-9):

$$Q_{11} = \frac{x_3}{P_3^2} = 1.23 \times 10^{-12} \text{ c.g.s.}; \quad Q_{12} = \frac{x_1}{P_3^2} = -0.56 \times 10^{-12} \text{ c.g.s.} \quad \text{(IV-34)}$$

The values of these two coefficients can also be determined *above* the Curie point from the ratio between electrostrictive strain and induced polarization. This was done by Schmidt (S 6), with results which agree well with (IV-34).

The coefficient $Q_{44}$ can be determined in a similar way from the corresponding values of the spontaneous strain and polarization in the orthorhombic phase. At 0 °C, $x_4 \cong 0.0029$, and the value of the spontaneous polarization is $31 \times 10^{-6}$ C/cm$^2$, as reported by Meyerhofer (M 4), or $29.4 \times 10^{-6}$ C/cm$^2$, as computed from the data reported by Huibregtse and Young (H 1). We obtain, thus:

$$Q_{44} = \frac{x_4}{P_2 P_3} = \begin{cases} 0.67 \times 10^{-12} \text{ c.g.s.,} \\ 0.73 \times 10^{-12} \text{ c.g.s.,} \end{cases} \quad \text{(IV-35)}$$

respectively.

A second way of determining the electrostrictive coefficients is obtained from the following argument (D 5). From the third equation in (IV-33) we obtain

$$\frac{\partial^2 x_3}{\partial P_1^2} = 2Q_{12}; \quad \frac{\partial^2 x_3}{\partial P_3^2} = 2Q_{11}. \quad \text{(IV-36)}$$

On the other hand, from the free energy function $dA = -S\,dT + E\,dP + x\,dX$, we have, for constant $T$, the Maxwell relations:

$$\left(\frac{\partial x_3}{\partial P_1}\right)_X = \left(\frac{\partial E_1}{\partial X_3}\right)_P \quad \text{and} \quad \left(\frac{\partial x_3}{\partial P_3}\right)_X = \left(\frac{\partial E_3}{\partial X_3}\right)_P \quad \text{(IV-37)}$$

and hence:

$$2Q_{12} = \left(\frac{\partial^2 E_1}{\partial X_3 \partial P_1}\right)_P = \left(\frac{\partial \chi_1}{\partial X_3}\right)_P,$$

$$2Q_{11} = \left(\frac{\partial^2 E_3}{\partial X_3 \partial P_3}\right)_P = \left(\frac{\partial \chi_3}{\partial X_3}\right)_P, \quad \text{(IV-38)}$$

so that the electrostrictive constants are a measure of the rate of variation of the inverse dielectric susceptibility with stress. Similarly, we can show that:

$$2M_{12} = \left(\frac{\partial^2 x_3}{\partial E_1^2}\right)_X = \left(\frac{\partial k_1}{\partial X_3}\right)_E;$$

$$2M_{11} = \left(\frac{\partial^2 x_3}{\partial E_3^2}\right)_X = \left(\frac{\partial k_3}{\partial X_3}\right)_E, \quad \text{(IV-39)}$$

where $k_1$, $k_3$ are the dielectric susceptibilities.

We know that $\chi$ is linearly dependent upon temperature, passing through zero at a temperature near the Curie point, and stress causes a linear shift of the Curie–Weiss line up or down the temperature scale (see Section 4, and also Section II-6). The rate of change of $\chi$ with stress is, therefore, independent of temperature, and thus, from (IV-38), the electrostrictive constant $Q$ is independent of temperature: it is a "true" constant of the crystal, in the sense defined in Section III-4. On the other hand, the quantity appearing in Eq. (IV-39) becomes infinite near the Curie temperature, since stress displaces the transition. Thus, the electrostrictive coefficient $M$ has the same sort of anomaly as the dielectric constant when the transition is approached from the high-temperature side.

Information about the electrostrictive constants can be obtained from the measurement of Shirane and Sato (S 5), already discussed in the last section, which were carried out above the Curie temperature on polycristalline BaTiO$_3$ and a mixture of BaTiO$_3$ and SrTiO$_3$ in the molar proportions 60 : 40. From the results obtained with the latter compound, Devonshire (D 5) computed, using Eq. (IV-38):

$$Q_{11} = 0.82 \times 10^{-12} \text{ c.g.s.}; \quad Q_{12} = -0.19 \times 10^{-12} \text{ c.g.s.} \tag{IV-40}$$

These values are smaller than those reported under (IV-34) but it should be kept in mind that they are pertinent to polycrystalline material. The physical constants of ceramics can be obtained from those of a single crystal by a process of averaging indicated by Devonshire (D 4). From the values (IV-34) and (IV-35), and using the averaging formulae given by Devonshire, we obtain:

$$\bar{Q}_{11} = 0.7 \times 10^{-12} \text{ c.g.s.}; \quad \bar{Q}_{12} = -0.2 \times 10^{-12} \text{ c.g.s.,} \tag{IV-41}$$

in good agreement with the values (IV-40).

The values of the electrostrictive constants for single-domain crystals can also be checked by using Merz's measurements of the shift of transition temperature $T_c$ with hydrostatic pressure (M 6) (see Section 4). Namely, it can be shown (D 5) that this shift is:

$$\frac{\partial T_c}{\partial p} = -(Q_{11} + 2Q_{12})\bigg/\frac{\partial \chi}{\partial T}. \tag{IV-42}$$

Since $\partial \chi / \partial T = 4\pi/C = 7.4 \times 10^{-5} \text{ °C}^{-1}$, we obtain from (IV-34) (IV-35), $\partial T_c/\partial p \cong -3 \times 10^{-9} \text{ °C cm}^2/\text{dyn}$, in reasonable agreement, within the experimental accuracy, with the experimental value $-5.7 \times 10^{-9} \text{ °C cm}^2/\text{dyn}$. The agreement would be better if we would take into account the temperature dependence of the higher-order coefficients in the free energy expansion.

A third way of determining the electrostrictive constants is arrived at by considering that, from (IV-27) and (IV-29):

$$\frac{\partial x_{ij}}{\partial P_k} = 2Q_{ijkl}P_l; \quad \frac{\partial x_{ij}}{\partial E_k} = 2M_{ijkl}E_l. \tag{IV-43}$$

Now, the left-hand sides of these equations define the piezoelectric coefficients $b$ and $d$, respectively, according to Eqs. (I-4b) and (I-5b), respectively. This means that a substance which normally has zero piezoelectric coefficients, as BaTiO$_3$ in the cubic phase, will have finite ones if it is given a polarization, e.g. by applying a biasing field. The piezoelectric coefficients will then be proportional to the polarization.

Below the Curie point, the polarization is spontaneous and this explains the piezoelectric effect of the tetragonal phase. We have discussed a similar situation in Section III-4, when we were considering the occurrence of new coefficients due to the lowering of the symmetry, coefficients which are proportional to the polarization. It is justified to speak of a linear piezoelectric effect of the tetragonal phase of BaTiO$_3$, if the polarization components $P_k$ and $P_l$ appearing in (IV-27) are to include the spontaneous polarization $P_s$ and the field-induced polarization $P_{\text{ind}}$ (C 4). For a field applied, e.g. in the $z$ direction, the third equation in (IV-33) can be written as:

$$x_3 = Q_{11}(P_s + P_{\text{ind}})^2 \cong Q_{11}P_s^2 + (2Q_{11}P_s)\,P_{\text{ind}}, \qquad \text{(IV-44)}$$

showing the linearity of the effect with respect to $P_{\text{ind}}$, the piezoelectric constant being of the type $2\,Q\,P_s$.

There are three non-zero piezoelectric coefficients in the tetragonal phase of BaTiO$_3$, namely $b_{15}, b_{31}$ and $b_{33}$ (similarly for the moduli $d_{ik}$). Piezoelectric measurements were carried out by Caspari and Merz (C 4) and, more recently, by Berlincourt and Jaffe (B 21) and Huibregtse et al. (H 3). From the paper by Berlincourt and Jaffe, we take the following values at 25 °C:

$$b_{15} = 5.07 \times 10^{-8} \text{ c.g.s.}; \quad b_{31} = -7.67 \times 10^{-8} \text{ c.g.s.}; \quad b_{33} = 19.17 \times 10^{-8} \text{ c.g.s.}$$

and thus we compute:

$$Q_{11} = \frac{b_{33}}{2\,P_s} = 1.23 \times 10^{-12} \text{ c.g.s.},$$

$$Q_{12} = \frac{b_{31}}{2\,P_s} = -0.49 \times 10^{-12} \text{ c.g.s.}, \qquad \text{(IV-45)}$$

$$Q_{44} = \frac{b_{15}}{P_s} = 0.65 \times 10^{-12} \text{ c.g.s.},$$

in excellent agreement with the values (IV-34) and (IV-35).

Similarly, we can derive the electrostrictive coefficients from the observed values of the piezoelectric moduli $d_{ik}$ using the formula (C 4):

$$d_{ik} = Q_{ik}P_s\frac{\varepsilon_c}{2\,\pi} \qquad \text{(IV-46)}$$

where $\varepsilon_c$ is the dielectric constant along the tetragonal $c$ axis. As $Q_{ik}$ is practically temperature independent, Eq. (IV-46) shows that the moduli $d_{31}$ and $d_{33}$ behave anomalously, as a function of temperature, as $\varepsilon_c$, and drop to zero above the Curie temperature. The thermal behavior of $Q_{12}$ is depicted in Fig. IV-27 (a) and that of $d_{31}$ in Fig. IV-27 (b) from the paper by Huibregtse et al. (H 3). The values of the piezoelectric moduli $d_{ik}$ at 25 °C are given in Table IV-5.

So far, our discussion has related mainly to the constants of a single-domain crystal. The next step is to transfer these properties to single crystals containing many domains, and finally, to ceramic materials. Detailed discussions pertinent

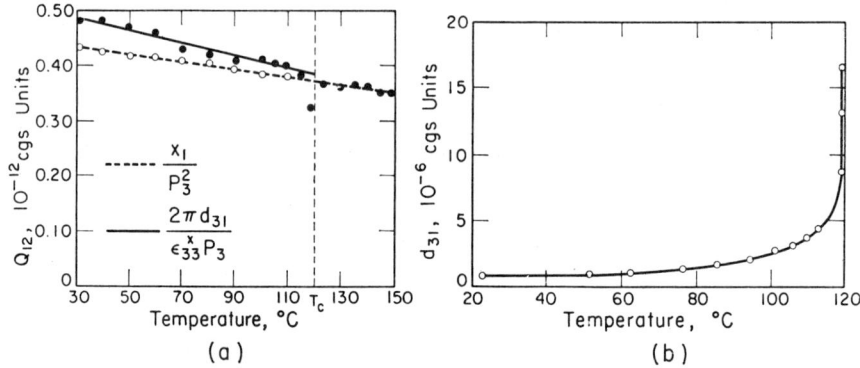

FIG. IV-27. Electrostriction and piezoelectricity of $BaTiO_3$.
(a) The electrostrictive coefficient $Q_{12}$ as a function of temperature.
(b) The piezoelectric coefficient $d_{31}$ as a function of temperature
(according to Huibregtse *et al*.. (H 3)).

to this point were given by Devonshire (D 4) and Kay (K 8). When the crystal consists of an equal number of positive and negative domains, no piezoelectric effect can be observed and the spontaneous polarization cannot be reversed by a mechanical stress. However, an electric field can reverse the spontaneous

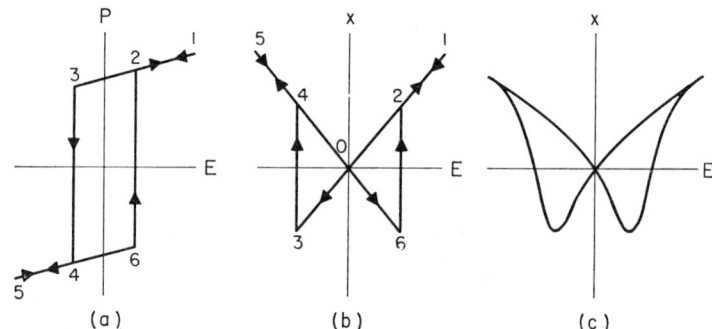

FIG. IV-28. Schematic description of the converse piezoelectric effect in $BaTiO_3$.
(a) Hysteresis loop.
(b) Quadratic hysteresis loop of strain vs. field.
(c) Actual "butterfly" loop.

polarization and thus the piezoelectric polarity. The relationship between strain and field is characterized by a quadratic hysteresis loop (C 4), as indicated in Fig. IV-28(b). Such a loop is best understood by following the normal *P–E* hysteresis loop, reproduced, for convenience, in Fig. IV-28(a). The straight portions 1–3 and 5–6 represent the true piezoelectric component of the strain.

The slope of these lines, in Fig. IV-28(b), represents the piezoelectric modulus $d_{33}$. Between 0 and 3, and 0 and 6, the field opposes the spontaneous polarization, but is not large enough to switch it. At 3 and 6, the coercive field is reached, the spontaneous polarization is reversed and the piezoelectric effect changes sign. In actual crystals, different portions of the sample alter orientation at different parts of the cycle and a smooth curve results, which is called a "butterfly loop", depicted in Fig. IV-28(c).

TABLE IV-5.   ELASTIC AND PIEZOELECTRIC COEFFICIENTS OF TETRAGONAL AND CUBIC BaTiO$_2$ (ACCORDING TO BERLINCOURT AND JAFFE (B 21))

| Elastic compliances at 25 °C ($10^{-13}$ cm²/dyn) | | | Piezoelectric coefficients at 25 °C $d_{ik}$ in $10^{-6}$ c.g.s. units $b_{ik}$ in $10^{-8}$ c.g.s. units | | |
|---|---|---|---|---|---|
| | Single crystals | Ceramics | | Single crystals | Ceramics |
| $s_{11}^{E}$ | 8.05 | 8.55 | | | |
| $s_{33}^{E}$ | 15.7 | 8.93 | $d_{15}$ | 11.76 | 8.10 |
| $s_{12}^{E}$ | $-2.35$ | $-2.61$ | $d_{31}$ | $-1.04$ | $-2.37$ |
| $s_{13}^{E}$ | $-5.24$ | $-2.85$ | $d_{33}$ | 2.57 | 5.73 |
| $s_{44}^{E}$ | 18.4 | 23.3 | $b_{15}$ | 5.07 | 6.27 |
| $s_{66}$ | 8.84 | 22.3 | $b_{31}$ | $-7.67$ | $-1.57$ |
| $s_{11}^{D}$ | 7.25 | 8.18 | $b_{33}$ | 19.17 | 3.80 |
| $s_{33}^{D}$ | 10.8 | 6.76 | Elastic compliances at 150 °C ($10^{-13}$ cm²/dyn) | | |
| $s_{12}^{D}$ | $-3.15$ | $-2.98$ | | | |
| $s_{13}^{D}$ | $-3.26$ | $-1.95$ | | Single crystals | |
| $s_{44}^{D}$ | 12.4 | 18.3 | | | |
| | | | $s_{11}$ | 8.33 | |
| | | | $s_{12}$ | $-2.68$ | |
| | | | $s_{44}$ | 9.24 | |

An interesting consequence of the piezoelectric deformations of individual domains in a multidomain crystal is the effect of "domain clamping" described by Drougard and Young (D 9). Consider a BaTiO$_3$ crystal consisting of a number of positive and negative domains as depicted schematically in Fig. IV-29. If a small electric field, not sufficient to reverse the polarity, is applied to such a crystal (as we do, e.g., when we measure the low-field dielectric constant), the domains with positive polarization will want to expand along the field direction, the domains with negative polarization will want to contract. Thus, the piezoelectric deformations of individual domains are not free, as in a single-domain crystal. The domains effectively clamp one another, in this geometry, and the dielectric constant measured under these conditions is a "thickness-clamped" dielectric constant.

This situation is evidenced by the experimental results depicted in Fig. IV-30(a): this is a plot of dielectric constant (measured with a small a.c. field) vs. the d.c.

biasing field applied to the crystal. The crystal is initially depolarized (point $A$) and becomes a single domain at sufficiently large d.c. fields (point $C$). Upon reversing the d.c. field, the dielectric constant goes through a minimum (point $D$) when the applied field reaches a value comparable with the coercive field and, in

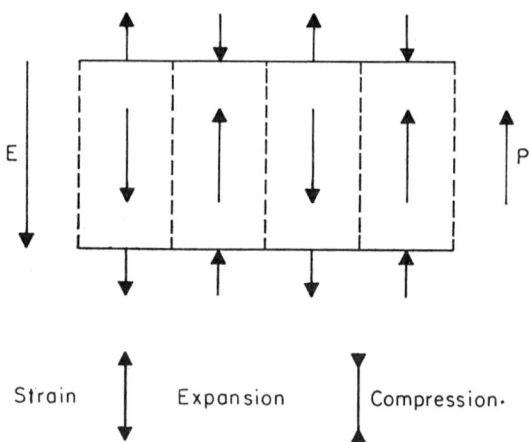

Fɪɢ. IV-29. Opposite piezoelectric deformations of antiparallel domains in response to the same increment of the field (schematic).

the process of reversing the polarity, gives rise to an anti-parallel domain configuration (this field is 675 V/cm for the experiment related to Fig. IV-30a). The same phenomenon occurs by reversing the field again (minimum of $\varepsilon$ at point $F$). These results were later confirmed by measurements of Meitzler and Stadler (M 13) which were carried out in a slightly different way. The dielectric constant was measured as a function of polarization at frequencies below thickness

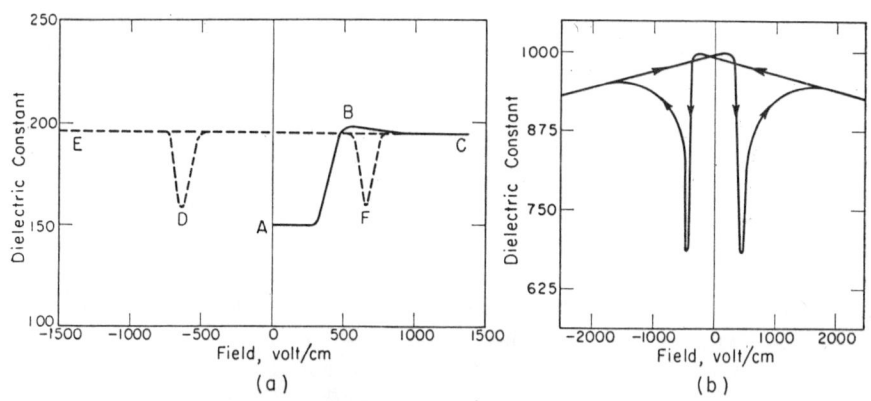

Fɪɢ. IV-30. Domain clamping effect in BaTiO₃.
(a) Dielectric constant as a function of field in the tetragonal phase (according to Drougard and Young (D 9)).
(b) Dielectric constant along pseudo-cubic [001] as a function of field in the ortho-rhombic phase (according to Huibregtse and Young (H 1)).

resonance. A minimum occurred at zero polarization, which is due to domain-clamping effects. At frequencies above the fundamental thickness resonance, on the other hand, the dielectric constant shows a maximum at zero polarization and a minimum at maximum polarization. The reason for this polarization dependence at high frequency is not well understood.

Huibregtse and Young (H 1) have proven that the same domain-clamping effect occurs in the orthorhombic phase of $BaTiO_3$, as shown in Fig. IV-30 (b). The "d.c. coercive field" measured in this way is 435 V/cm, thus about 1.6 times smaller than that of the tetragonal phase, in good agreement with the ratio of coercivities obtained from hysteresis loops measurements.

### The Elastic Properties

The elastic behavior of $BaTiO_3$ has been investigated by several authors (B 1), (B 21), (H 2), (H 3), (M 7) but only Berlincourt and Jaffe (B 21) report a complete list of the elastic compliances for the tetragonal state at 25 °C and for the cubic state at 150 °C, while Huibregtse et al. (H 3) report information about the temperature dependence of some of the elastic compliances. The temperature dependence of $s_{11}^E$ and $s_{11}^P$ is depicted in Fig. IV-31 (a), as measured by Huibregtse et al. (H 3) with the piezoelectric resonance technique. In the cubic phase, no distinction between constant-field and constant-polarization compliance is needed, as the crystal is not piezoelectric. In the tetragonal phase, the compliance $s_{11}^E$ reflects the behavior of the dielectric constant, as expected, while the constant-polarization compliance $s_{11}^P$, contrary to expectations, shows a small anomaly in the vicinity of the Curie temperature. This anomaly is not understood, but is not improbable that the experimental results reported here were affected by the presence of domains with different orientations, a situation which arises often in the vicinity of the Curie temperature in spite of the presence of a polarizing d.c. field. The quantity $(2s_{12}^P + s_{66}^P)$, on the other hand, responds to the expectations in that it is practically temperature independent, while the same quantity at constant electric field, $(2 s_{12}^E + s_{66}^E)$ is markedly anomalous at the Curie point (Fig. IV-31 b). Thus, as in the previously discussed case of $KH_2PO_4$, the elastic

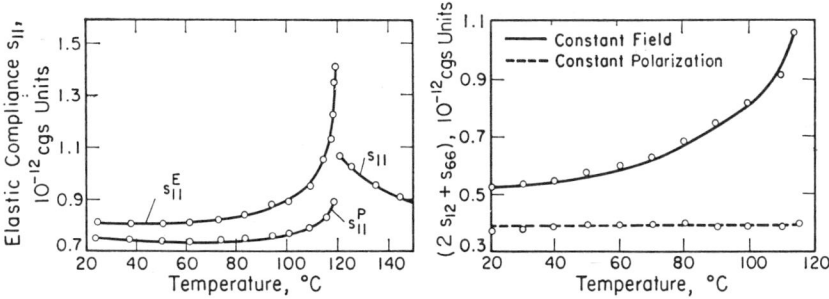

FIG. IV-31. Elastic behavior of $BaTiO_3$.
(a) The compliance coefficients $s_{11}^E$ and $s_{11}^P$ as functions of temperature.
(b) The quantities $(2s_{12} + s_{66})^E$ and $(2s_{12} + s_{66})^P$ as functions of temperature (according to Huibregtse et al. (H 3)).

behavior of $BaTiO_3$ reflects the piezoelectric and, in the final instance, the di-electric behavior.

For reference purposes, Table IV-5 summarizes the numerical values of the elastic compliances and piezoelectric coefficients of $BaTiO_3$ single-crystals and ceramics at 25 °C, as well as the elastic compliances of cubic $BaTiO_3$ at 150 °C, according to Berlincourt and Jaffe (B 21).

### 6. Structural Characteristics

The structure of $BaTiO_3$ above the Curie temperature is the ideal perovskite structure depicted in Fig. IV-1. The space group is $Pm3m$, with one formula-unit per unit cell. The co-ordinates of the atoms in this structure are given by the following special positions:

Ba at $(0, 0, 0)$,

Ti at $(1/2, 1/2, 1/2)$,

$3\,O$ at $(1/2, 1/2, 0)$;  $(1/2, 0, 1/2)$;  $(0, 1/2, 1/2)$,

where the cell edge is about 4 Å (M 5).

It is sometimes convenient to visualize the perovskite structure from a slightly different point of view, namely by taking the origin at the Ti atom. In this way, the continuity of the structure is better visualized (Fig. IV-32). Each Ti atom is at the center of 6 O atoms, arranged at the corners of a regular octahedron. The octahedra are linked by their corners into a three-dimensional framework, enclosing large holes which are occupied by the Ba atoms.

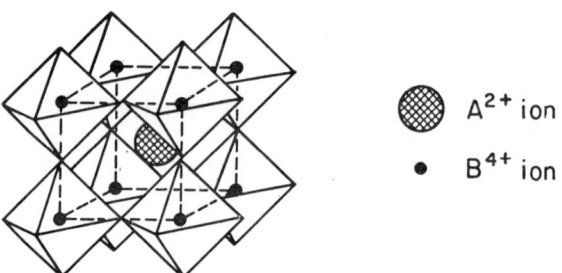

$A^{2+}$ ion

$B^{4+}$ ion

FIG. IV-32. Perovskite structure $ABO_3$ considered as a three-dimensional frame-work of $BO_6$ octahedra.

From the crystal–chemical point of view, on the other hand, the perovskite structure is better visualized as a face-centered–cubic close-packed arrangement of $Ba^{2+}$ and $O^{2-}$ ions. In such close-packed arrangements there are two types of interstitial positions, which are tetrahedrally and octahedrally coordinated, respectively. In the cubic phase of $BaTiO_3$, every available octahedral hole between six $O^{2-}$ ions is filled with a $Ti^{4+}$ ion, and there are no discrete $TiO_6$ groups.

### The Tetragonal Phase

Below the Curie point, the atoms are displaced from their original symmetric positions along one of the $\langle 001 \rangle$ axes. The structure becomes tetragonal, with the cell dimensions (R 4):

$$a = 3.992 \text{ Å}, \quad c = 4.036 \text{ Å},$$

space group $P\,4mm$ and one formula-unit per unit cell. The co-ordinates of the atoms in this structure can be expressed in terms of three parameters, $\delta z_{Ti}$, $\delta z_{OI}$, $\delta z_{OII}$, representing the shifts of the atoms from the symmetrical position, thus:

Ba at $(0, 0, 0)$,

Ti at $(1/2, 1/2, 1/2 + \delta z_{Ti})$,

OI at $(1/2, 1/2, \delta z_{OI})$

2 OII at $(1/2, 0, 1/2 + \delta z_{OII})$; $(0, 1/2, 1/2 + \delta z_{OII})$

These parameters are indicated schematically in Fig. IV-33(b). In addition to the three co-ordinate parameters, the structural analysis involves nine anisotropic temperature parameters in order to describe the thermal vibrations of the atoms (two each for Ba, Ti and OI, three for OII).

The structural problem was first attacked by Känzig, using X-ray diffraction techniques (K 9). Känzig made a careful study of the change of the integrated reflections of the type $(00l)$ and $(0k0)$ through the Curie temperature. In particular, the high-order reflections of the type $(00l)$ were used to determine the change of the structure factors with temperature, thus obtaining an estimate of the atomic displacements $\delta z$. It was *assumed* that $\delta z_{Ti}$ and $\delta z_{OI}$ would have opposite signs and that $\delta z_{OII} = 0$. The latter assumption is actually incorrect (see below), but the value of Känzig's investigation lies less in the determination of the atomic parameters than in the valuable study of the X-ray extinction

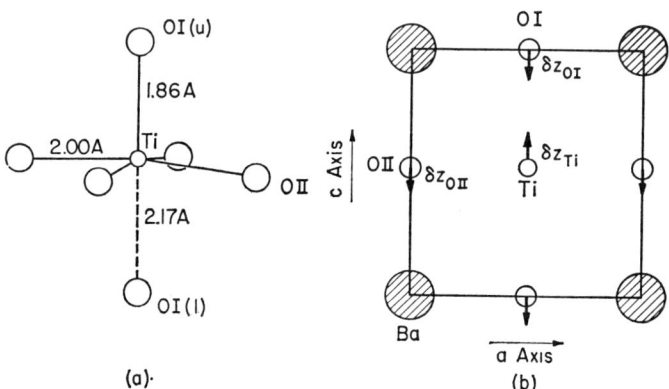

FIG. IV-33. Structure of tetragonal BaTiO$_3$.
(a) Distortion of the TiO$_6$ octahedron.
(b) Schematic projection on (010). The oxygen OII superimposed upon Ti is omitted.

phenomena in the vicinity of the Curie temperature. During the transition, the strain associated with the sharp change in spontaneous polarization increases the mosaic spread, thus diminishing the extinction effects. This leads to very sharp peaks of the extinction-sensitive reflections.

The structural problem was then taken up by Evans (E 1) using conventional methods of X-ray diffraction analysis. Evans avoided the assumption $\delta z_{\text{OII}} = 0$. After a very careful study, he was forced to conclude that the structure of tetragonal $BaTiO_3$ could not be solved uniquely by X-ray methods owing to interactions between co-ordinate and temperature parameters along the tetragonal axis (see Chapter X). He proposed two solutions of the structural problem, both of which yield very good agreement between calculated and observed X-ray intensities. The temperature parameters, however, and particularly the anisotropy of the thermal vibrations are physically unreasonable. This may be the result of the parameter interactions mentioned above, which make an exact solution of the structure impossible by X-rays.

The conditions which prevent a unique solution of the $BaTiO_3$ problem by X-ray methods have been discussed in detail by Shirane et al. (S 2). These conditions arise from the overwhelming scattering power of the barium atoms with respect to the lighter Ti and O atoms, and from the inherent difficulty to separate the effects of small co-ordinate shifts and temperature vibrations of the lighter atoms in a pseudo-centrosymmetrical structure of this kind (see Chapter X). The situation is more favorable if the problem is attacked via neutron diffraction. In this case, the ratios of scattering factors of barium, titanium and oxygen are entirely favorable for a structure determination.

A neutron diffraction analysis of single-domain $BaTiO_3$ crystals at room temperature was carried out by Frazer et al. (F 3). The following parameters were obtained, at 18 °C (D 15):

$$\delta z_{\text{Ti}} = +0.013, \quad \delta z_{\text{OI}} = -0.023, \quad \delta z_{\text{OII}} = -0.013.$$

As for the thermal vibrations of the atoms, it may be recalled that these vibrations are generally described in terms of temperature factors $B_i$, which are related to the mean square displacement $\overline{u_i^2}$ of the $i$th atom in the following way:

$$B_i = 8\pi^2 \overline{u_i^2}.$$

The scattering factor $f_i$ of this vibrating atom is therefore usually written as

$$f_j = f_i^0 \exp\left(-B_i \frac{\sin^2\theta}{\lambda^2}\right)$$

where $f_i^0$ is the scattering of the atom at rest, $\theta$ the Bragg angle, and $\lambda$ the wavelength of the radiation used. When the thermal vibrations are anisotropic, the values of $B_i$ must be specified for each of the principal crystallographic directions. In tetragonal $BaTiO_3$, we call $B_c$ the temperature factor along the c axis, $B_b$ that along the direction approximately parallel to the Ti—OII bond, and $B_a$ that along the direction perpendicular to the other two. With this notation, the anisotropic thermal vibrations provided by the neutron diffraction analysis are (in units of $10^{-16}$ cm²) (D 15):

$$(B_a, B_c)_{Ba} = (0.31, 0.37), \quad (B_a, B_c)_{Ti} = (0.67, 0.42),$$
$$(B_a, B_c)_{OI} = (0.45, 0.40), \quad (B_a, B_b, B_c)_{OII} = (0.62, 0.45, 0.42).$$

Using the cell dimensions reported above, the actual atomic shifts are computed as:

$$\delta z_{Ti} = +0.05 \text{ Å}, \quad \delta z_{OI} = -0.09 \text{ Å}, \quad \delta z_{OII} = -0.05 \text{ Å},$$

thus confirming Känzig's first assumption about the relative sign of $\delta z_{Ti}$ and $\delta z_{OI}$.

Before we proceed with the discussion of the structure, we should emphasize that the choice of the origin of the unit cell is quite arbitrary along the [001] axis. This is a situation that we have already encountered in the discussion of the KDP structure (see Section III-7), and has general validity. Once the results of a structure analysis are obtained, it may occasionally be convenient to choose another origin, which gives a better physical picture of the distorted lattice. In the case of $BaTiO_3$, choosing the origin at the $z$ level of OII, we see that both Ti and Ba shift in the same direction, by 0.10 Å and 0.05 Å, respectively, while OI is displaced slightly in the opposite direction by the amount 0.04 Å. This implies a slight distortion of the oxygen octahedron surrounding the Ti atom.

TABLE IV-6. ATOMIC SHIFTS AND BOND DISTANCES IN TETRAGONAL $BaTiO_3$
(Oxygen OII is placed at $c/2$.)

| Atomic shifts (Å) | Bond distances (Å) | |
|---|---|---|
| $\delta z_{Ba} = +0.05$ | | |
| $\delta z_{Ti} = +0.10$ | Ti – OI $\begin{cases} 1.86 \\ 2.17 \end{cases}$ | Ba – OII $\begin{cases} 2.80 \\ 2.88 \end{cases}$ |
| $\delta z_{OI} = -0.04$ | Ti – OII  2.00 | Ba – OI  2.82 |

Table IV-6 summarizes the results concerning the atomic shifts and the bond distances. The environment of Ti is depicted schematically in Fig. IV-33 (a). It is seen that the bond length Ti−OI is markedly altered by the Ti shift, whereas no large change occurs in the Ba−O bond length. Assuming completely ionic character of the crystal, the positional parameters allow an estimate of the maximum possible contribution of the ionic polarization. The result is $17 \times 10^{-6}$ C/cm², implying that even the minimum contribution of the electronic polarization ($9 \times 10^{-6}$ C/cm²) is considerable.

### The Orthorhombic Phase

The lattice parameters of orthorhombic $BaTiO_3$ are, at $-10$ °C:

$$a = 5.682 \text{ Å}; \quad b = 5.669 \text{ Å}; \quad c = 3.990 \text{ Å}.$$

The $a$ and $b$ axes are parallel to the pseudo-cubic face diagonal, the $c$ axis is parallel to the pseudo-cubic edge. The polar axis is parallel to the axis which has been called $a$, i.e. the *long* diagonal of the pseudo-cubic face. This was confirmed by a number of authors (F 4), (K 5), (S 7), (J 1), (C 5) with optical, electrical and X-ray methods.

The unit cell is thus approximately twice as large as the simple cubic cell and contains two formula-units. The space group, with the above choice of axes, is $C\,2mm$. The atomic positions can again be described in terms of displacements of the atoms from the symmetrical positions in the cubic phase. The structural problem involves then four positional parameters, namely $\delta x_{Ti}$, $\delta x_{OI}$, $\delta x_{OII}$ and $\delta y_{OII}$, as indicated schematically in Fig. IV-34(a).

The problem was solved by a neutron diffraction study of single-domain crystals by Shirane *et al.* (S 7). The resulting parameters are:

$$\delta x_{Ti} = +0.010; \quad \delta x_{OI} = -0.010; \quad \delta x_{OII} = -0.013; \quad \delta y_{OII} = +0.003.$$

This means (Fig. IV-34a) that the Ti atom is displaced from its symmetrical position by 0.06 Å in the positive $x$ direction. The OII atoms are shifted by 0.07 Å in the negative $x$ direction and by less than 0.02 Å in the $y$ direction, so that their resultant displacement occurs toward the Ti atom. The OI atoms, on the other hand, are shifted in the negative $x$ direction.

Choosing the origin such that the shifts of the OII atoms in the $x$ direction are zero, the structure can be described in terms of the displacements of Ti and Ba relative to the oxygen octahedra. These displacements are 0.13 Å and 0.07 Å, respectively, in the positive $x$ direction. These results are reported in Table IV-7 and may be compared with the relative atomic shifts in the tetragonal phase (Table IV-6). The shifts are definitely comparable in the two phases, although they occur along different directions. In the tetragonal phase, Ti has one nearest neighbor (OI) at a distance 1.86 Å, while in the orthorhombic phase Ti has two nearest neighbors (OII) at 1.90 Å. The environment of Ti in the latter phase is depicted in Fig. IV-34(b).

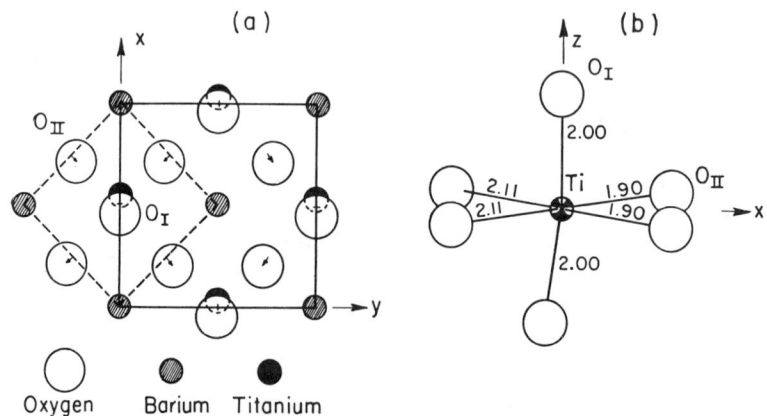

FIG. IV-34. Structure of orthorhombic BaTiO₃.
(a) Schematic projection on orthorhombic (001) plane.
(b) Distortion of the TiO₆ octahedron. Interatomic distances given in Ångström units (according to Shirane *et al.* (S 7)).

It is evident that in both the tetragonal and the orthorhombic phases, the distortion of the oxygen octahedra is very slight. In both phases, this slight distortion can be looked upon as being the result of the displacements of Ti and Ba along the polar axis. Thus, the result of the structure analyses clearly emphasize the essential role played by the Ti atoms in the ferroelectric pheno- menon. In the orthorhombic phase the maximum contribution of the ionic polarization to the total polarization is computed, under the assumption of a completely ionic crystal, as $16 \times 10^{-6}$ C/cm², to be compared with the experimental value of about $30 \times 10^{-6}$ C/cm² for the total polarization.

TABLE IV-7. ATOMIC SHIFTS AND Ti–O BOND DISTANCES IN ORTHORHOMBIC BaTiO₃
(Polar axis along $x$, origin chosen to give $\delta x_{OII} = 0$ (according to Shirane et al. (S 7))

| Atomic shifts (Å) | Ti – O bond distances (Å) |
|---|---|
| $\delta x_{Ba} = +0.07$ | |
| $\delta x_{Ti} = +0.13$ | Ti – OII $\begin{cases} 1.90 \\ 2.11 \end{cases}$ |
| $\delta x_{OI} = +0.02$ | Ti – OI  2.00 |

### The Rhombohedral Phase

The study of this phase is complicated by three experimental difficulties, viz. the low temperatures at which it occurs, the fact that all samples are heavily twinned in this phase and, finally, the fact that plates perpendicular, and bars parallel to the pseudo-cubic [111] direction are not easily available. The symmetry of this phase was established as rhombohedral (space group $R3m$, lattice para- meter $a = 3.998$ Å at $-100$ °C) by Kay and Vousden (K 5) on the basis of optical and X-ray investigations. Later, Jona and Pepinsky (J 1) questioned the assign- ment of rhombohedral symmetry to the low-temperature phase of BaTiO₃. They found that the optical behavior of pseudo-cubic (100) plates under applied fields and, particularly, the dielectric behavior of these plates were not quite consistent with the assumption of rhombohedral symmetry (see also Section 2). These inconsistencies were cleared by a recent careful study of Sawaguchi and Charters (S 8), carried out on small plates perpendicular to the pseudo-cubic [111] direction. The optical anomalies of biased plates were attributed to non-uniform strains. The different values obtained for the dielectric constant of rhombohedral BaTiO₃ along the pseudo-cubic [100] direction (an inconsistency already pointed out in connection with Fig. IV-3) were attributed to domain clamping effects, similar to those observed in the tetragonal and orthorhombic phases (Fig. IV-30).

A structural analysis of the rhombohedral phase of BaTiO₃ was not carried out to date. Neutron diffraction techniques, which would again be necessary in this case, require large specimens extended along the pseudocubic [111] direction.

### Lattice Vibrations and Electronic Structure

Information about the character of lattice vibrations and their change with temperature can be obtained from infrared and optical absorption studies. Both single crystals and powder samples of $BaTiO_3$ were investigated (M 8), (H 4), (T 4), (C 6), (L 2) in the near and far infrared, as well as in the optical range. The most complete infrared-absorption study is due to Last (L 2) who investigated $BaTiO_3$ and a number of perovskite-type compounds.

As there are five atoms in the cubic unit cell of $BaTiO_3$, there are fifteen degrees of freedom, of which three are related to translational and three to torsional motion of the unit cell. The remaining nine modes are associated with vibrational degrees of freedom. The latter can be classified into three vibrations of Ba against the $TiO_3$ group (treated as a single atom) and six internal $TiO_3$ vibrations. The interactions between these motions depend upon the masses and restoring forces of the vibrating atoms and are expected to be small in the case of $BaTiO_3$. The treatment of the $Ba-(TiO)_3$ vibrations leads to a triply degenerate vibration (as there are three equivalent axes in the cubic phase) expected in the vicinity of $225 \ cm^{-1}$. As for the internal $TiO_3$ vibrations, the requirement that atoms in equivalent positions in neighboring unit cells must perform the same vibrational motion reduces the set of six normal vibrations to two infrared-active vibrations, depicted schematically in Fig. IV-35. The absorption band observed at $495 \ cm^{-1}$ (Fig. IV-35 c) is attributed to the "stretching" vibration of the Ti−OI band (Fig. IV-35 a), while the band observed at approximately $340 \ cm^{-1}$ is attributed to the Ti−OII "bending" vibration (Fig. IV-35 b).

The effect of temperature on the "stretching" vibration is depicted in Fig. IV-35 c. In the tetragonal phase, the band is doubly rather than triply degenerate and thus appears slightly broader. However, the band intensity does not vary with respect to the cubic phase, indicating that the frequency difference between vibrations parallel and perpendicular to the polar axis is small. In the ortho-rhombic phase, the band should be tripled. However, two of the lattice parameters are very close to each other and the slight frequency difference between these vibrations cannot be resolved. The rhombohedral band has the same double degeneracy as the tetragonal band, and a double absorption band is observed, as expected.

These results, together with the fact that a number of other perovskite compounds show spectra similar to $BaTiO_3$, lead to the conclusion that the observed infrared spectrum of $BaTiO_3$ can be explained purely on the grounds of crystal symmetry and atomic positions in the cubic phase. No effect is observed which can be related directly to the occurrence of a spontaneous polarization and the ferroelectric activity (L 2).

The optical absorption edge of $BaTiO_3$ lies in the vicinity of $3800 \ Å$ and is only slightly affected by the Curie transition at 120 °C. This result might indicate that the electronic structure of $BaTiO_3$ does not undergo radical changes at the Curie point (K 7).

The X-ray $K$-absorption and emission spectra of Ti in $BaTiO_3$ and related compounds were investigated by the Russian workers (B 6), (B 7), (V 5), (V 6), (V 7). The results of these investigations give additional evidence for the partial covalency of $BaTiO_3$. A marked anomaly of the $K$-absorption spectrum of Ti is observed at the Curie point.

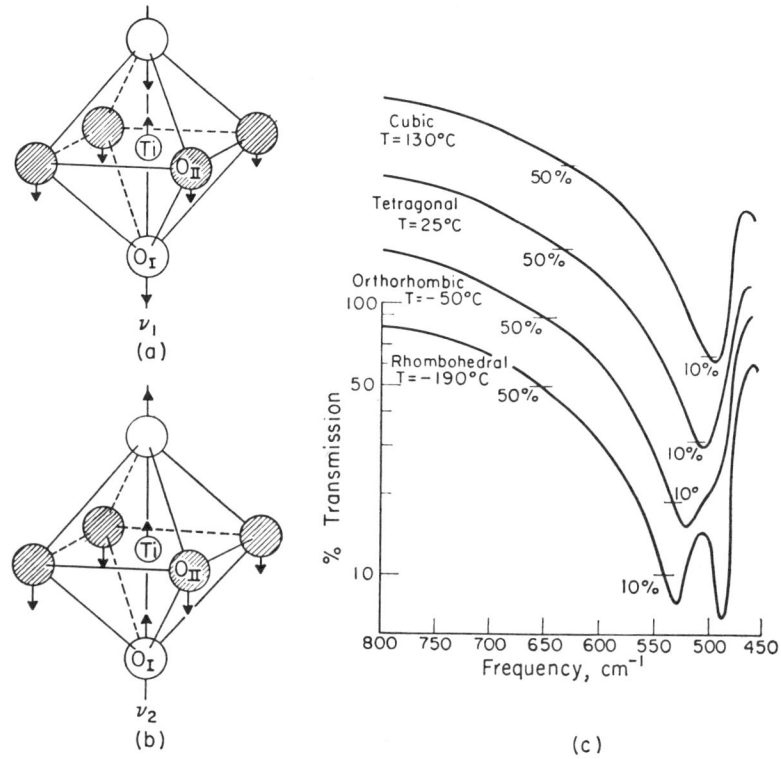

FIG. IV-35. Infrared absorption in $BaTiO_3$.
(a) Higher frequency "stretching" vibration of the $TiO_6$ octahedron (schematic).
(b) Lower-frequency "bending" vibration of the $TiO_6$ octahedron (schematic).
(c) Infrared absorption spectrum in the four phases (according to Last (L 2)).

Information about the effect of the internal field on the degeneracy of the spin levels can be obtained from the study of paramagnetic resonance in $BaTiO_3$ single crystals. Pure crystals, however, do not contain paramagnetic ions, and therefore the effect was studied on $BaTiO_3$ crystals containing small amounts of paramagnetic impurities. The main difficulty of this type of experiment is the identification of the impurity ion responsible for the spectrum observed. In the early stages of these investigations, considerable discrepancies existed among the interpretations of different workers (H 6), (H 7), (L 3), (L 4). A very careful analysis of the experimental data, by Hornig et al. (H 13), has finally established that the electron resonances stem from ferric ions located at the titanium sites. $Fe^{3+}$ is in the $^6S_{5/2}$ state: in the crystalline field, the six levels split into three doublets. When a magnetic field is applied, six energy levels are created and, at frequency larger than the zero-field splittings, five strong transitions will occur. These five transitions are, in fact, observed in the resonance spectrum, providing the strongest argument for the fact that the resonances result from $Fe^{3+}$ ions. The zero-field splittings for this ion in the tetragonal phase of $BaTiO_3$ were determined to be 0.390 cm$^{-1}$ (11.7 kMc/s) and 0.163 cm$^{-1}$ (4.9 kMc/s) at room temperature. The corresponding value in the cubic phase is approximately 0.90 kMc/s. The temperature dependence of the zero-field splittings through the Curie point reflects the behavior of the spontaneous polarization. The final goal of these investigations is, of course, to gain information about local crystalline fields in $BaTiO_3$. Unfortunately, the task of calculating the splittings in terms of crystalline field parameters is quite involved and very little information is available on this subject to date.

## 7. Domains and Domain Walls

As mentioned in Section 1, much more complicated domain patterns are possible in $BaTiO_3$ than in $KH_2PO_4$ and tri-glycine sulfate, because the polar direction can lie along a number of axes which are equivalent in the cubic phase. In the following, we shall be concerned with the geometry of ferroelectric domains in $BaTiO_3$ and with some quantitative estimates of the thickness and the energy of the domain walls. The tetragonal phase, of course, has been investigated much more extensively than the others, owing to the fact that it is stable at room temperature. Moreover, in contrast to the previously discussed ferroelectrics, the direct (optical) observation of domains using a polarizing microscope has played a major role in establishing the characteristic properties of the domain structure. In order to understand the reasons that make the polarizing microscope one of the most valuable tools in this type in research, it may be helpful to recall some of the fundamentals of crystal optics.

The optical properties of crystals are best described in terms of the index ellipsoid or indicatrix (see Section III-5). In the cubic phase of $BaTiO_3$, the indicatrix is a sphere with radius equal to the refractive index $n$. In the tetragonal phase, the indicatrix is an ellipsoid of revolution, obtained from the sphere by a slight compression along the direction of the polar axis. The crystal then is optically uniaxial exhibiting birefringent properties, and it is important to notice that the birefringence has increased from the value zero, in the cubic phase, to a finite value $(n_c - n_a)$. If we observe a plate of tetragonal $BaTiO_3$ with pseudo-cubic habit between crossed nicols, what we see will depend on the angle between the propagation direction of the light beam and the polar (or optic) axis. If this angle is zero (light parallel to the optic axis) the crystal plate will appear dark, and if this angle is 90° (light perpendicular to the optic axis) the plate will appear bright except for two (extinction) positions in which the polar axis is parallel to either one of the polarization planes of the nicols. Thus, in the latter case, one speaks of "parallel" (or straight) extinction, as the pseudo-cubic edges of the crystal plate lie parallel to the polarization planes of the nicols. It is clear that, if two domains are present in the crystal plate, one with the polar axis parallel, the other with the polar axis perpendicular to the light beam, they are going to be distinguished easily from each other by rotating the plate on the microscope stage.

In the orthorhombic phase, the indicatrix is rotated by 45° with respect to the tetragonal indicatrix, and, moreover, it is no longer a rotational ellipsoid: the crystal is optically biaxial. If the polar axis lies *within* the plane of the $BaTiO_3$ plate with pseudo-cubic habit, it can only lie at 45° with respect to the pseudo-cubic edges. Thus, the plate will extinguish only when the pseudo-cubic edges lie symmetrically with respect to the polarization planes of the nicols: one then speaks of "symmetrical" extinction. If the polar axis does *not* lie within the plane of the plate, it can only be inclined at 45° with respect to this plane and the pseudo-cubic edges. Thus, the position of the indicatrix is such that a "parallel" extinction results (see Fig. IV-13).

In the rhombohedral phase, the indicatrix is again an ellipsoid of revolution, and the optic axis lies parallel to the pseudo-cubic [111] direction. Thus, observation of a single-domain plate with pseudo-cubic habit in this phase will reveal that the plate can only exhibit "symmetrical" extinction.

We see then that the observation of the extinction positions of a crystal plate under the polarizing microscope gives us information about the orientation of the polar axis. Obviously, if different regions of the plate extinguish at different positions, this can only happen because these different regions are domains with different orientations of the polar axis.

Summarizing the methods with which a *direct* observation of ferroelectric domains is possible, we have the following list:

(i) Examination of the crystal under the polarizing microscope.

(ii) Examination of the crystal surface after it has been subjected to etching agents.

(iii) Examination of the crystal surface on which electrostatically charged powders have been deposited.

We have already met methods (ii) and (iii) in the study of tri-glycine sulfate domains. These methods permit, of course, only the study of the *surface* of a crystal. On the other hand, method (i) permits only the study of the orientation of the polar axis in different domains, but cannot distinguish between positive and negative end of this axis, which methods (ii) and (iii) can do.

### The Geometry of Domains

The outstanding work in establishing the static properties of domains in $BaTiO_3$ was done by Kay and Vousden (K 5), Forsbergh (F 4), Merz (M 10), (M 11), and Little (L 5). In the tetragonal phase, it is possible for adjacent domains to be polarized at 90° to each other. A pseudo-cubic (001) plate may exhibit the configurations depicted in Fig. IV-36, which are easily identified under the polarizing microscope (B 8), (M 9), (C 6). When the polar axis is perpendicular to the plane of the plate, the corresponding domain is called a "$c$" domain, and when it lies *within* the plane of the plate, the corresponding domain is called an "$a$" domain. Figure IV-36(a) shows an "$a$" domain (parallel extinction) between two "$c$" domains (optically isotropic). Figure IV-36(b) shows a number of "$a$" domains. In order to understand how the latter configuration can be identified under the polarizing microscope, let us introduce some remarks about the nature of the domain walls appearing in Fig. IV-36.

This type of domain wall separates domains polarized perpendicularly to each other and is, therefore, called a "90° wall". The first condition for such a wall to be stable in a (perfectly insulating) crystal is that no charge may be

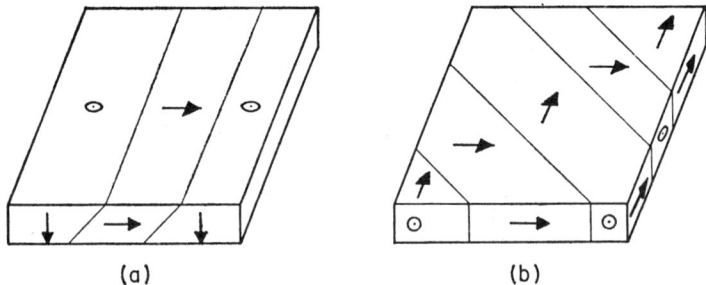

(a)                                    (b)

FIG. IV-36. Domain arrangements in an (001) plate of tetragonal BaTiO$_3$. Arrows represent the direction of the polar axis. The plate surface shows: (a) "$a$" domain between two "$c$" domains, (b) "$a$" domains only.

accumulated on the wall. This requires a "head-to-tail" arrangement of the polar axes of adjacent domains. In actual crystals, "head-to-head" and "tail-to-tail" arrangements have also been observed (L5), but in this case the charge on the wall had been compensated by conduction, and the wall was "locked" in place. The second condition for a stable wall is the continuity and the matching of the lattice at the wall. Such condition is fulfilled by the geometry depicted schematically in Fig. IV-37(a). As the tetragonality of BaTiO$_3$ at room temperature is about 1%, the angle between polar axes of adjacent 90° domains is not exactly 90° but differs from 90° by the angle: 2 arc tangent $c/a \simeq 36'$. This implies that two such 90° domains do not extinguish simultaneously, when the crystal plate is rotated on the microscope stage, and also that the directions of minimum refractive index (optic axes) are almost perpendicular to one another in adjacent domains. They can therefore be distinguished easily by using any kind of retardation plate, such as a quarter-wave plate. Finally, it is apparent from Fig. IV-37(a) that the 90° walls contain unit cells with anomalous strain, being kite-shaped rather than rectangular. Thus, strain birefringence, as well as total reflection of the light on the boundary, makes the 90° wall visible in polarized light. The observed 90° walls in tetragonal BaTiO$_3$ are pseudo-cubic {110} planes, which satisfy both conditions mentioned above.

Adjacent domains can, of course, also be polarized antiparallel to each other, as we know from the study of the ferroelectrics described in Chapters II and III.

Such domains are called 180° domains, and the wall separating them is called "180° wall", indicated by $B-B'$ in Fig. IV-37(a).

Under ideal conditions, there is no way by which antiparallel domains and 180° walls can be directly observed with the polarizing microscope. The spontaneous strain should be the same, in antiparallel domains, as it is proportional to the square of the polarization (see Section 5). The 180° wall is very thin (see

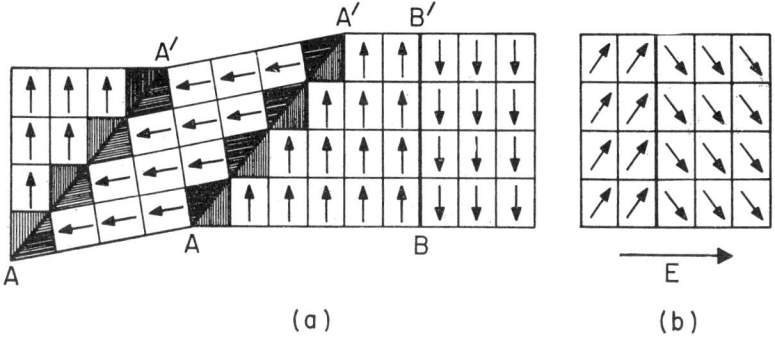

Fig. IV-37. Domain walls in tetragonal BaTiO₃.
(a) $A-A'$ are 90° walls, $B-B'$ a 180° wall. The tetragonality is exaggerated (schematic).
(b) Effect of an electric field perpendicular to the polar axis of antiparallel domains (schematic).

below): the spontaneous polarization can reverse its sign within one unit-cell distance, and thus the periodicity of the lattice is practically undisturbed. Thus, when antiparallel domains are viewed in the direction of the optic axis they all appear optically isotropic, when they are viewed in the direction perpendicular to the optic axis, they all have the same extinction positions. The only way by which the extinction positions of antiparallel domains can be made to differ slightly from one another is to apply an electric field perpendicular to the polar axis. The field should be such as to exert opposite torques on the polar axes of opposite domains, but not large enough as to permanently rotate the polarization. This situation is depicted schematically in Fig. IV-37(b). A similar effect is obtained by applying a suitable stress to the plate, as the piezoelectric effect has different sign in antiparallel domains (see Section 5).

By straining the crystal plates in the way mentioned above, Merz (M 10), (M 11) was able to see the antiparallel domains by looking edgewise at a "c"-domain plate. Little (L 5) achieved the same result by applying a suitable electric field along the [110] direction of an "a"-domain plate: in this case, one component of the applied field makes the antiparallel domains visible, the other component makes them grow. Using the straining technique, Hooton and Merz (H 8) were able to show that very often 180° and 90° walls are present simultaneously, giving rise to complicated domain patterns as visible in Fig. IV-38.

It has been shown recently by Miller and Savage (M 12), however, that direct observation of 180° domains is possible, under the polarizing microscope, with no need for electric fields perpendicular to the polar axis. Figure IV-39(a) is a photomicrograph of a "*c*"-domain plate with one approximately square domain antiparallel to the surrounding region, as observed directly through semi-transparent metal electrodes. The origin of the effects which allow this direct observation is not clear but may conceivably be sought in some kind of clamping between antiparallel domains, clamping which may affect the linear dimensions

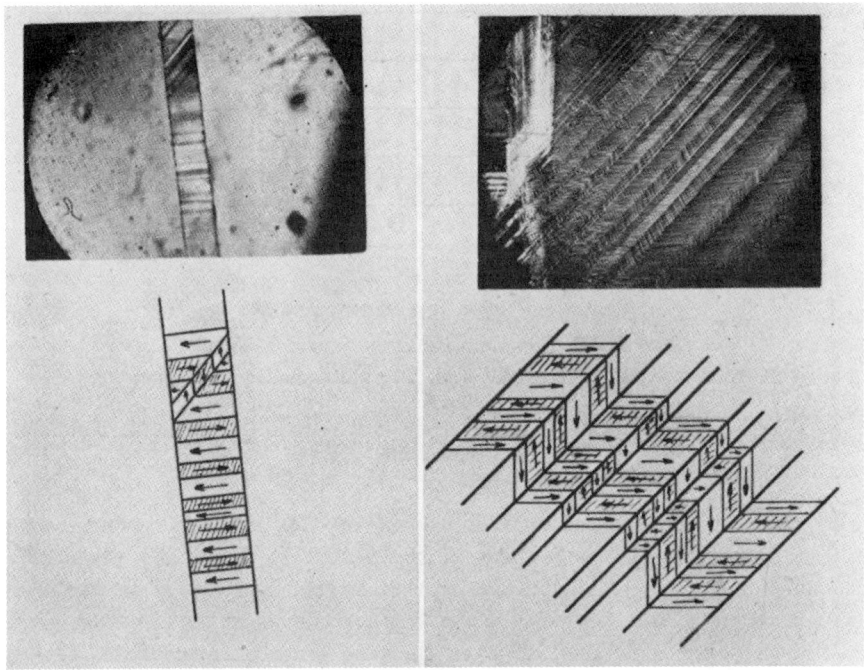

FIG. IV-38. Photomicrograph of a strained BaTiO$_3$ plate in the tetragonal phase, showing 90° and 180° domains (according to Hooton and Merz (H 8)).

and the refractive indices of these domains. This direct observation of 180° domains in unstrained plates is invaluable to the study of the polarization reversal mechanism in BaTiO$_3$ (see Section 8).

The etching technique is, of course, also helpful in revealing antiparallel domains. It was first found by Hooton and Merz (H 8) that the positive end of the polar axis etches faster than the negative end, while the etch rate of the "sides" of the polar axis ("*a*"-domains) is intermediate between that of the positive and that of the negative end. Campbell and Stirland (C 7), (C 8) reported that the positive end of the polar axis etches twice as fast as the negative end and faster than the "sides" in the ratio 4 : 3.

Particularly interesting are the "reflections" of "*c*"-domains by way of "*a*"-domains, as revealed by the etching technique. In Fig. IV-40, the etch marks *a* and *b* and *c* and *d*, respectively, are mirror images of each other and

a)

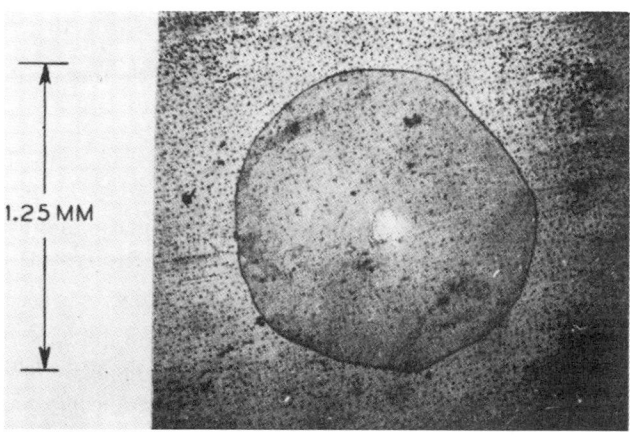

b)

FIG. IV-39 (a) and (b). Antiparallel domains in tetragonal $BaTiO_3$.

(a) Photomicrograph of an approximately square domain polarized oppositely to the surrounding region, as observed directly through semitransparent gold electrodes between crossed nicols (according to Miller and Savage (M 12)).

(b) Antiparallel domain revealed by etching in HF (according to Pearson and Feldmann (P 2)).

represent "$c$"-domains reflected along the surface by the mechanism indicated in the schematic cross-section of the plate. For the details of this mechanism the reader is referred to the original paper by Hooton and Merz (H 8).

The width of antiparallel domains, as determined optically or with the etching technique, is of the order of $10^{-4}$ cm in heavily twinned crystals, but much larger domains, from $10^{-2}$ to $10^{-1}$ cm in linear dimensions, can be made to grow

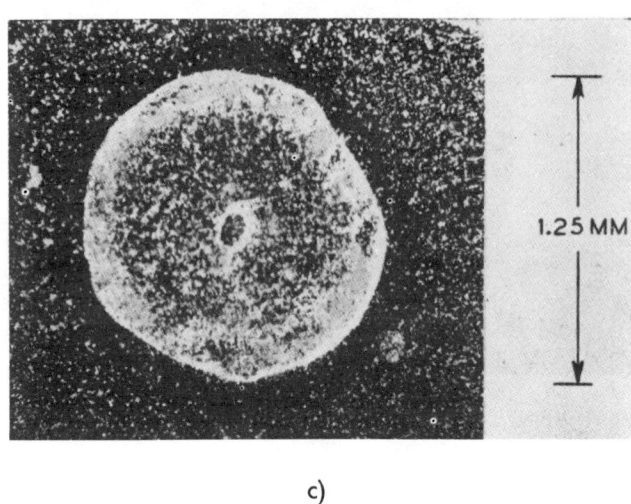

1.25 MM

c)

Fig. IV-39 (c). Antiparallel domains in tetragonal BaTiO$_3$.

(c) Same domain as in (b), revealed by the technique of electrostatically charged powders (according to Pearson and Feldmann (P 2)).

in unstrained crystals. Figure IV-39 shows, in addition, the perfect correlation between the etching technique and that employing electrostatically charged powders (P 2).

The geometry of ferroelectric domains in the orthorhombic and rhombohedral phases of BaTiO$_3$ was studied by Kay and Vousden (K 5) and by Forsbergh (F 4). Figure IV-41 summarizes schematically the possible cases and shows where parallel and symmetrical extinctions are expected. In the orthorhombic phase, $90°$ walls are parallel to pseudo-cubic {001} planes [equal to orthorhombic (110) planes], and $60°$ walls appear also, which are parallel to pseudo-cubic {011} planes [equal to orthorhombic (111) planes]. Domains of $180°$ in the orthorhombic phase were demonstrated by Cameron (C 9), using the etching technique at $0\ °C$ (Fig. IV-42). In the rhombohedral phase, the polar axes of different domains are inclined at approximately $70°$ to each other. The domain walls are pseudo-cubic (100) planes.

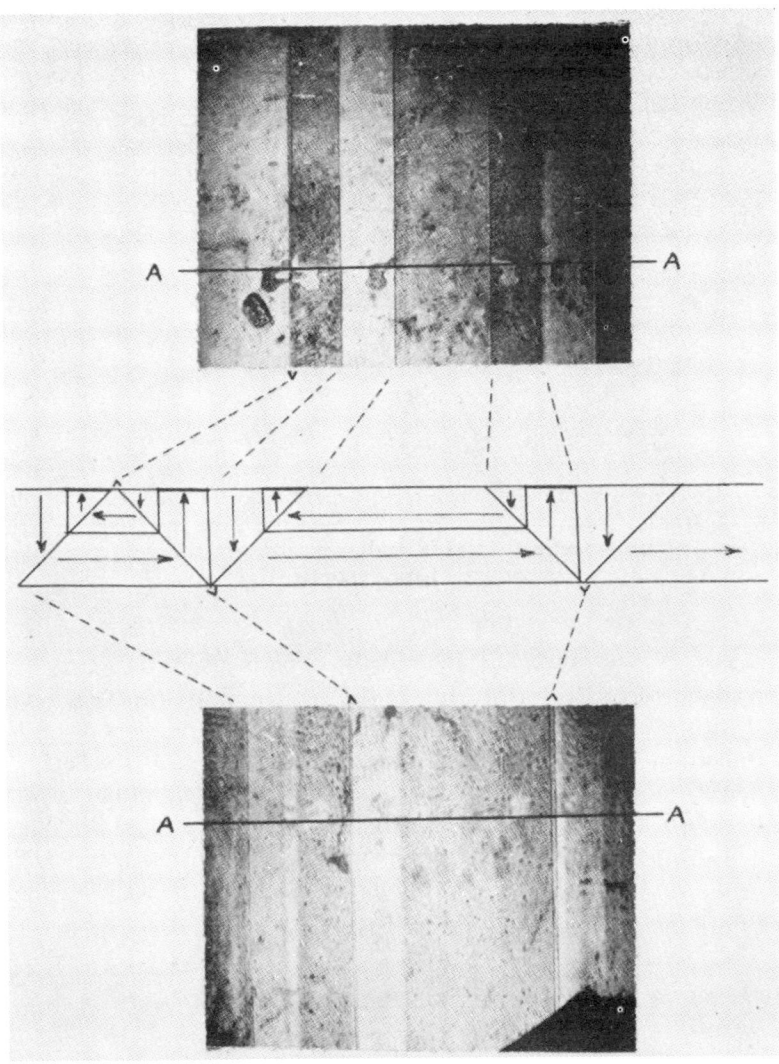

FIG. IV-40. Top and bottom view of etched BaTiO$_3$ plate in the tetragonal phase. Domain "reflections" are visible at $a$ and $b$, and $c$ and $d$, respectively (according to Hooton and Merz (H 8)).

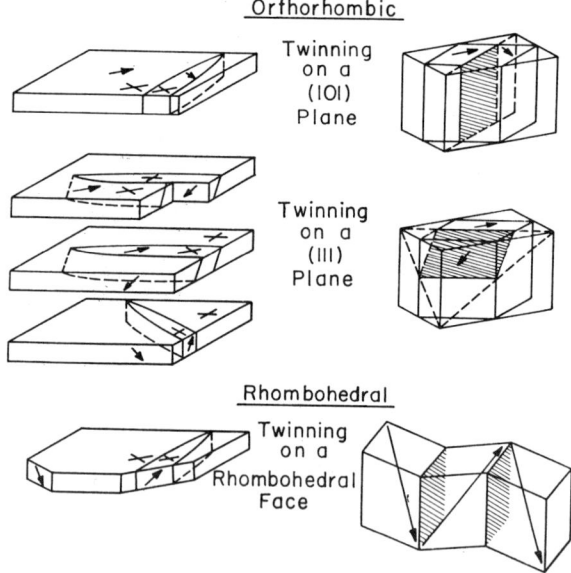

FIG. IV-41. Geometry of the domains in the orthorhombic and rhombohedral phases of $BaTiO_3$. The arrows represent the directions of the polar axis. The plus signs and the crosses refer to parallel and symmetrical extinction, respectively (according to Forsbergh (F 4)).

### Energy and Thickness of the Domain Walls

The available theoretical treatments of the domain walls in $BaTiO_3$ are mainly concerned with 180° walls of the tetragonal phase, as the motions of these walls play an important role in the process of polarization reversal (see Section 8).

Merz (M 11) modified the well-known model of the Bloch wall in ferromagnetics for the case of $BaTiO_3$. The total wall energy was taken to be the sum of the energy due to dipole interactions (corresponding to the exchange energy in the ferromagnetic case) and the anisotropy energy. Calling $t$ the wall thickness, the equation of the wall energy $E_{wall}$ was written as:

$$E_{wall} = \frac{A}{t} + Bt \qquad (IV\text{-}47)$$

where $A$ and $B$ are functions of the lattice constant, the spontaneous strain and the elastic constants. By minimizing Eq. (IV-47) with respect to $t$, Merz estimated $t$ to be of the order of *one* lattice constant, thus very much smaller than in the Bloch case. The corresponding wall energy was estimated to be of the order of 7 ergs/cm².

Kinase and Takahashi (K 11) approached the same problem from a different point of view. They assumed a model for the 180° wall in which the *magnitude* (rather than the direction) of the spontaneous polarization varies while approaching the domain wall, goes through zero at the wall and increases in the

FIG. IV-42. Three directions of polarization in orthorhombic BaTiO₃ as revealed by the etching technique at 0 °C. In the smoothest areas, the negative ends of the dipoles are at the surface; in the roughest areas, the positive ends of the dipoles are at the surface. In the areas of intermediate roughness, the dipoles are parallel to the surface (according to Cameron (C 9)).

opposite direction on the other side. This model is quite reasonable in view of the fact that the large anisotropy within the wall makes a rotation of the dipoles highly improbable. Kinase (K 10) carried out detailed calculations of the local field at the Ti sites for a few cases of parallel and antiparallel dipole arrays result-ing from displacements of Ti only. The local field was found to be of the same order of magnitude in all cases, but slightly larger in the case of a parallel dipole array, in contrast to the results of Cohen (C 10). The fact that the energies of the two states are of the same order of magnitude makes it possible to consider the antipolar state as the limiting case of a polar state in which each domain is only one lattice constant thick. Accordingly, the 180° wall involves an abrupt change of polarization. Kinase's calculations of the local fields at the Ti sites (K 10) are useful for the estimate of this field at the domain boundary. Using these results, Kinase and Takahashi (K 11) computed the lattice spacing and the ionic shift at the boundary with the same method outlined in Section 3 for the calculation of the lattice distortion in a single-domain crystal. The calculations show that the changes in lattice spacing and ionic shift at the domain wall are only 0.002% and 0.5%, respectively. Accordingly, the 180° wall involves no strain of the lattice and has practically no thickness. The energy required to form a domain wall was computed from the difference between the total energy of the lattice at the boundary and that of a single domain crystal. The resulting wall energy was $E_{wall} = 1.40$ erg/cm$^2$, thus much smaller than Merz's value reported above.

These results were used by Kinase and Takahashi to estimate the width of antiparallel domains, using the condition of equilibrium between wall energy and electrostatic energy caused by free charges at the crystal surface. Three basic domain configurations were considered: the stripe type, the checker-board type and the cylindrical type (Fig. IV-43). As the electrostatic energy is a volume energy, the calculated domain width is a function of the thickness of the crystal plate considered. For all three configurations and for a plate thickness of 0.1 cm, the domain width was computed as $0.6 - 1.5 \times 10^{-4}$ cm agreeing in order of magni-tude with the observed value (see above). This estimate applies, of course, only to the virgin state of a free, well-annealed crystal, cooled from above the Curie point to room temperature. The domains created during the process of polari-zation reversal, on the other hand, cannot be treated with equilibrium con-siderations (see Section 8).

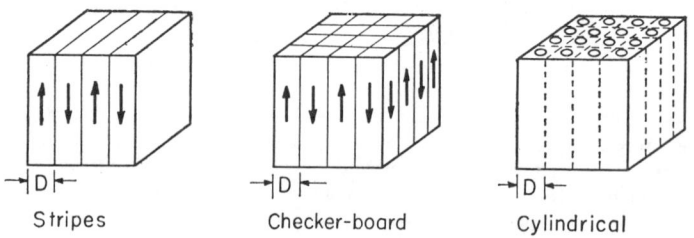

Stripes          Checker-board          Cylindrical

FIG. IV-43. Three domain configurations of tetragonal BaTiO$_3$ assumed for the calculation of the domain width D (according to Kinase and Takahashi (K 11)).

The values reported above for the thickness and the energy of the 180° wall are only valid at room temperature and under the assumption that only the titanium atoms are displaced from their symmetrical positions. It was stated that when both Ti and OI shifts are considered, the calculated wall thickness has a value of two lattice constants. If the displacements of the other oxygen atoms are also considered, the problem becomes more complicated, and the thickness of the 180° wall increases slightly (K 10). Of particular interest, with regard to the growth process of 180° domains under the influence of applied fields, is the study of the front wall of a wedge-shaped domain with antiparallel polarization with respect to its surroundings. Theoretical considerations indicate that the domain wall must be thicker at the tip than on the sides of the wedge because the dipole field acting on the Ti ions is quite different in the two cases. Kinase (K 13) estimated that the thickness of the front wall may be equivalent to four lattice spacings ($\sim$16 Å), and its energy of the order of 80 ergs/cm$^2$.

Zhirnov (Z 1) developed a phenomenological treatment of the ferroelectric domain wall by modifying the method suggested by Landau and Lifshitz (L 6) for the study of the domain structure in ferromagnetics. The equilibrium variation of the polarization vector in the boundary region was determined from the condition of a minimum in the thermodynamic potential. Explicit expressions were obtained for the thickness and the energy of the domain walls. The numerical values obtained for the 180° wall are closer to those of Merz than to those of Kinase and Takahashi, as expected from the similarity of the treatments. Zhirnov estimated a thickness of from 5 to 20 Å and an energy of about 10 ergs/cm$^2$ for the 180° wall; a thickness of from 50 to 100 Å and an energy of from 2 to 4 ergs/cm$^2$ for the 90° wall.

A molecular treatment of the 90° wall is not available to date. The problem is much more complicated than for the 180° wall because the lattice distortion at the wall involves a noticeable contribution of elastic energy (see Fig. IV-37). This contribution is inversely proportional to the wall thickness. Moreover, the spontaneous polarization may change its direction gradually. Little estimated from direct optical observation that the thickness of the 90° wall is approximately 0.4 $\mu$. However, such a visual estimate may easily be affected by total reflection of the light at the boundary. At any rate, the 90° wall is expected to be much thicker than the 180° wall. Kinase (K 10) has suggested an interesting model for the 90° wall, which involves the consideration of an antiferroelectric array of the PbZrO$_3$ type. This treatment was, however, not developed in detail. The *origin* of the 90° domains is, on the other hand, more easily understood than that of the antiparallel domains. Owing to the strain involved at a 90° boundary, a crystal which is mechanically clamped must split into domains in order to relax the elastic stresses.

## 8. The Problem of Polarization Reversal

We have repeatedly pointed out that the theoretical value of the field strength necessary to reverse the polarization of a ferroelectric crystal is orders of magnitude larger than the experimental value of the coercive field. Landauer *et al.* (L 1) estimated from the free energy expression (IV-14) a value of about $2 \times 10^5$ V/cm for the coercive field of BaTiO$_3$ at room temperature. Kinase and Takahashi (K 11) calculated that the theoretical value of the field strength required to cause a simultaneous reversal of the polarization of a whole domain is $5.9 \times 10^6$ V/cm. On the other hand, the minimum field strength required to move a 180° wall was computed to be $5.8 \times 10^6$ V/cm, so that, according to the theoretical predictions, the motion of a 180° wall is no easier than the reversal of the polarization of a whole domain, in sharp contrast to the situation in the ferromagnetic case. The experimental values of the coercive field of BaTiO$_3$ at room temperature, on the other hand, vary only within the range of $500-2000$ V/cm.

It is evident that a satisfactory explanation for this discrepancy between theoretical and experimental values of the coercive field can only be given if the mechanism of polarization reversal in BaTiO$_3$ is known in every detail. In recent years, a large amount of experimental and theoretical effort has been directed toward the solution of this problem. In seeking this solution, the first question that one would like to have answered is the following. When we apply a given electric field to a single-domain crystal in the direction opposite to the spontaneous polarization, what is the amount of polarization that we can reverse and how long does it take for this process to be completed? It is, of course, expected that the answer to this question will be dependent on the magnitude of the field applied, and it is exactly this dependence that we now seek.

Merz was the first to perform experiments with this purpose (M 10), (M 11), (M 3). He did this by employing a pulsing technique, which was already described in Section II-7. In this technique, a train of rectangular voltage pulses of successive opposite directions is applied to the crystal, and quantities like the switching time $t_s$ and the maximum current $i_{max}$ are measured. These quantities are defined in the switching transient depicted in Fig. II-15, which ought to be self-explanatory.

The results of these measurements are shown in Fig. IV-44 (a) which is a plot of reciprocal switching time vs. applied voltage. Within the range of fields considered in this graph (a few hundred volts per centimeter), it appears that the switching process can be accurately described by the following exponential law:

$$i_{max} \sim \frac{1}{t_s} = \frac{1}{t_\infty} e^{-\alpha/E}, \tag{IV-48}$$

where the activation field $\alpha$ (of the order of $10^4$ V/cm) depends on temperature and sample thickness. This important exponential law, which was first established by Merz, was later confirmed by the measurements of Little (L 5) and Wieder (W 4). The latter author studied the reversal process at fields as low as 210 V/cm, where periods of several hours are required. We have already discussed in Section 2 how Landauer *et al.* (L 1) were able to show that the frequency and field dependence of the coercive field is fully consistent with the switching law (IV-48). The same law was recently studied in greater detail by Drougard (D 10), who determined how the *instantaneous*, rather than the maximum, value of the switching current depends on the state of net polarization of the crystal. According to Drougard's results, the switching process can be described fairly accurately by the law:

$$i_s = Q_s \beta \left[ 1 - \left( \frac{P}{P_s} \right)^2 \right] e^{-\alpha/E}, \tag{IV-49}$$

where $i_s$ is the actual switching current at any instant of time, $P$ the net polarization of the sample $(-P_s \leqq P \leqq +P_s)$ and $Q_s$ the total charge switched during

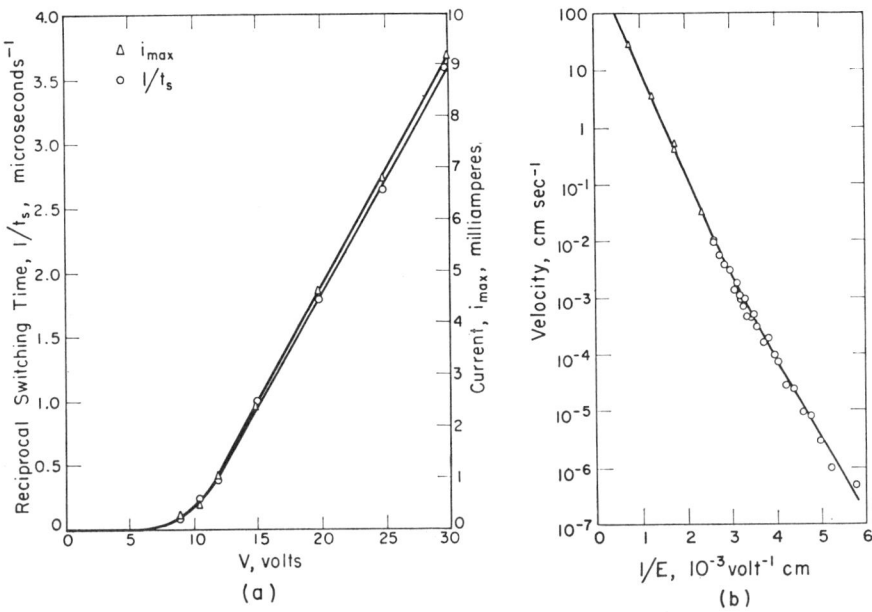

Fig. IV-44. Polarization reversal in tetragonal BaTiO$_3$.

(a) Switching current $i_{max}$ and reciprocal switching time, $1/t_s$, vs. applied field (according to Merz (M 3)). The thickness of the sample was specified only within the limits of 2.5 to 35 × 10$^{-3}$ cm.

(b) Semilogarithmic plot of sidewise velocity of 180° walls vs. reciprocal field (according to Miller and Savage (M 25)).

a complete reversal of the polarization. The coefficient $\beta$ was found to vary from 0.4 to 2.3 × 10$^{-7}$ sec$^{-1}$ in different crystals. $\alpha$, on the other hand, is a temperature-dependent constant (for a given crystal) which is independent of the state of polarization of the sample. There are a few important conclusions that can be drawn from the form of the law (IV-49) but we defer their discussion to the end of this section.

Now that the phenomenological relationship between the rate of polarization reversal and the applied field has been established, we may ask the second question, most important to the problem considered in this section: what is the mechanism that brings about the experimental laws (IV-48) and (IV-49)?

Many workers have tried to give an answer to this question by employing different experimental techniques. Most of these techniques involved the direct observation of the domain motions under the polarizing microscope. Merz (M 11) was the first to observe the motion of 180° walls during polarization reversal: he did this by using the straining technique described in Section 7, i.e. by applying the driving field along the direction of the polar axis and the straining field perpendicular to it. He concluded from this observations that the mechanism of polarization reversal involves the nucleation of antiparallel domains and their subsequent growth in the forward direction, rather than a sideway expansion of the domain walls. Little (L 5) employed a different geometry by applying a field at 45° to the polar axis, acting both as the driving and as the straining field, and concluded also that the sidewise motion of 180° walls should be negligible.* There is reason to believe, however, that the validity of Merz's and Little's results may be limited to the experimental techniques employed. Both authors made use of a straining field. The exponential law (IV-48) and Fig. IV-44(a), on the other hand, were obtained by applying only a field parallel to the ferroelectrics axis. It is difficult, therefore, to try and interpret the observations done with a straining field in terms of the reversal expressed by the exponential law (IV-48).

A different approach has been followed by Miller and Savage in a series of papers (M 12), (M 14), (M 15), (M 25), (M 35). The technique employed by these authors does not make use of a straining field. The domains created by a given field in a partially switched crystal are revealed *after* the reversal by the etching techniques (see Section 7). Also, the method mostly involves the use of liquid electrodes, rather than the metal electrodes employed in all other investigations described above. The most striking result of Miller and Savage's investigations is that extensive sidewise motion of 180° walls was found to occur, at least with applied fields of the order of a few hundreds up to a thousand volts per centimeter. The motion is called "extensive" when the domain wall moves sidewise through distances which are a significant fraction of the diameter of the electroded area. In some cases, the entire electroded area (of the order of 2.5 mm in diameter) can be reversed by a single growing domain, expanding through sidewise 180° domain-wall motion. Having established this fact, the main purpose of the investigations was the measurement of the velocity of sidewise wall motion as a function of the applied field and, finally, the correlation of the results with the exponential law (IV-48). Before we discuss these points, however, it may be worthwhile to summarize a number of other interesting observations made by Miller and Savage.

In the first place, it was found that the sideway velocity is several orders of magnitude smaller, in metal-electroded crystals, than that found for the same field in liquid-electroded crystals (M 12), (M 35). This evidence could partially explain the previous results of Merz and Little, who used exclusively metal-electroded samples. Furthermore, it was found that after a crystal has been completely polarized in one direction and the field removed from the crystal,

---

* Little (L 5) was also able to observe the motion of 90° walls under the influence of electric fields. Similar studied were carried out by Fousek and Brezina (F 5).

a few small reversed domains may form, a phenomenon called "back-switching". This result seems to be confirmed by high-resolution photographs of etch patterns of switched crystals taken by Cameron (C9). On this basis, it cannot be decided, of course, whether actually back-switching occurred or rather some regions having the original direction of polarization never did disappear. But for the purpose of the present discussion this point is immaterial. The important point is to realize that, when the field is reversed, these left-over or back-switched domains start expanding immediately, a phenomenon which may reduce the requirements of the number of new domains to be nucleated.

Finally, the shapes of the expanding domains were found to be both size and field dependent. The smallest domains observed are always circular (i.e. when the diameter is smaller than about $10^{-3}$ cm in diameter); once the domains grow larger, their shapes are determined principally by the magnitude of the field applied. For fields of a few hundred volts per centimeter, the domains are approximately square with the sides at about $45°$ to the crystalline $a$

Fig. IV-45(a). Field and size dependence of the shape of switched domains in BaTiO$_3$. Photomicrograph of a portion of the $c$ domain configuration of a specimen that had been partially switched and etched five times. The small circular region in the approximate center of the four concentric squares represents a "back-switched" domain. The squares represent four partial polarization reversals made with successively decreased pulse length, each of the same amplitude (695 V/cm) (according to Miller and Savage (M25)).

axes (Fig. IV-39a). With fields of about 1000 V/cm the domains become octagonal (Fig. IV-45(a) and (b)) and finally, with larger fields, approximately square again, but now with the long sides parallel to the crystalline $a$ axes (M25). There are therefore two types of 180° domain boundaries, namely (110) and (100) walls (cf. also Husimi (H14)).

Let us now discuss the results of Miller and Savage in terms of the sideway wall velocity in liquid-electroded crystals. For fields ranging up to about

1000 V/cm, the velocity of the 180° domain walls can be expressed by:

$$v = v_\infty \, e^{-\delta/E} \tag{IV-50}$$

where $E$ is the applied electric field. The quantities $\delta$ and $v_\infty$ are essentially field independent for fields lower than about 350 V/cm, i.e. over a range of about four decades in velocity, $10^{-6}$ cm sec$^{-1}$ to $10^{-2}$ cm sec$^{-1}$. In this range, $\delta$ is of the order of $2 \times 10^3$ V/cm. Figure IV-44(b) shows, however, that when the range of

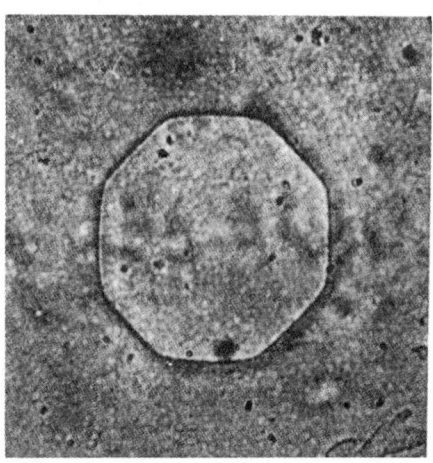

FIG. IV-45 (b). Field and size dependence of the shape of switched domains in BaTiO$_3$.
Photomicrograph of an octagonal shaped reversed domain. The crystal was partially switched with a field of 700 V/cm (according to Miller and Savage (M 25)).

velocity is extended to cover eight orders of magnitude, then a small variation of $v_\infty$ and $\delta$ with field is detected. In the higher-field range, $\delta$ is of the order of $5 \times 10^3$ V/cm. The effect of temperature on the wall velocity can unfortunately not be studied extensively with a technique employing liquid electrodes, but the data presented indicate that this effect is quite pronounced. The temperature dependence lies primarily in $\delta$ and not in $v_\infty$ (see also more recent results by Savage and Miller (S 23)). Also, the numerical values of these constants are somewhat affected by the impurity content of the material investigated.

Let us now summarize the mechanisms that have been proposed for explaining the process of polarization reversal in BaTiO$_3$. We may conveniently classify these mechanisms into three types:

(i) nucleation of antiparallel domains;

(ii) forward motion of the 180° domain wall, in which case the wall moves in the direction of the polar axis (what we mean here is the forward motion of the tip of an antiparallel spike that does not simultaneously grow in cross-section at its base); and

(iii) sidewise motion of the 180° wall, in which case the wall moves in a direction perpendicular to the polar axis.

What seems difficult to do, however, is to estimate just what fractions of the reversal process are due to which of these three mechanisms. Can one say, for example, that the field dependence of the sideway velocity of the domain walls, as determined by Miller and Savage, is alone sufficient to explain the empirical switching law (IV-48)? Miller and Savage (M 15) have shown that, by making certain assumptions regarding the rate of nucleation of reversed domains, Eq. (IV-48) follows from Eq. (IV-50), and the quantities $\alpha$ and $\delta$ are related to one another by rather simple expressions.

Such expressions, however, do not provide much quantitative information, because a direct comparison between the numerical value of $\alpha$ and that of $\delta$ is hardly possible. In fact, $\delta$ was determined from wall velocity measurements on liquid-electroded crystals whereas $\alpha$ was obtained from switching transients on metal-electroded crystals. Moreover, the experimental values of $\alpha$ and $\delta$ were mostly measured on different samples and their difference lies within the range of fluctuations caused by different impurity contents. Drougard, for example, reported values of $\alpha$ ranging from 5 to 10 kV/cm (D 10), which are almost equal to the values of $\delta$ reported by Miller and Savage (M 25).

Only little can be said, on the other hand, about the role played by the nucleation of *new* domains. There is, in fact, no definite experimental evidence for such a nucleation process. Merz's direct observation of the reversal process (M 11) would seem to provide such evidence but the presence of a straining field in the actual experiment, which was carried out under conditions of stroboscopic illumination, may cast some doubt about the weight of this observation. The study of the Barkhausen pulses, on the other hand, usually related each pulse to an individual nucleation (N 2), (K 12), (C 11), (C 12). But Miller has shown that Barkhausen pulses can also be generated when two domain walls in sideways motion approach each other closely (M 14). In this interpretation, the pulses would then be attributed to the merging of two expanding domains, rather than to the individual nucleation of a completely new domain. From the theoretical point of view, it is also quite difficult to understand how the nucleation of new domains can occur. Merz (M 11) assumed that the appearance of nuclei is due to the thermal agitation. But Landauer (L 7) has shown that this thermal nucleation process has such a low probability as to be virtually impossible (see also Prutton (P 3)).

We may therefore conclude that the nucleation of new antiparallel domains plays a minor role, if any, in the process of polarization reversal, and that this process is mainly governed by domain-wall motions. This, of course, is only part of the problem, as we have yet to understand the mechanism by which such a motion is made possible. It is generally agreed that a 180° wall is not very likely to move as a unit parallel to itself on an atomic scale, at least with the fields that have been used experimentally (M 11), (K 11). There is, however, a certain amount of experimental evidence pointing toward a sidewise wall motion controlled by the nucleation of reversed domains *at* an existing 180° wall (L 7), (M 15). A strong support for this idea is given by the very fact that the wall velocity has been found to depend exponentially upon the field. It should also be noticed that such a nucleation mechanism is energetically quite different from that required by nuclei appearing in regions remote from existing walls, as the nucleation adjacent to a wall requires the creation of less domain wall than the nucleation in completely unswitched environments (Fig. IV-46a).

Theoretical models for the growth of $180°$ domains in $BaTiO_3$ have been advanced by Abe (A 2) and by Burfoot (B 23), (B 24). However, the most satisfactory picture of this phenomenon is that inferred from the ideas suggested by Drougard (D 10) in order to explain the polarization dependence of the instantaneous switching current (Eq. IV-49), picture that was ·later confirmed by Fatuzzo and Merz (F 10) through their study of nucleus–domain interactions (see Section II-7), and more recently analyzed in greater detail by Miller and Weinreich (M 36). Let us assume, at first, that a poled crystal always contains a certain number of left-over or back-switched domains. It may be pointed out, in this connection, that an interesting explanation of this fact has recently been advanced by Janovec (J 6). The existence of space-charge layers at the surface of the crystal (see Section 9) gives rise to a strong electric field directed perpendicular to the surface. Although it cannot as yet be decided whether this field is directed into or out of the bulk of the crystal, there is always one surface of the crystal on which this field is opposite to the spontaneous polarization of the bulk. The formation of antiparallel domains in this particular surface region is thus energetically advantageous. When we apply an electric field to the crystal in the direction opposite to its polarization, each existing antiparallel domain starts growing immediately by way of wall motion. This motion, in turn, occurs by way of nucleation of a protuberance at the existing wall, which expands then throughout the thickness of the crystal very rapidly. The nucleation of the protuberance would have a probability proportional to the perimeter of the existing wall and its expansion would cause a change in polarization also proportional to this perimeter. In this way, Drougard could explain why the instantaneous switching current is proportional to the quantity $(1 - P^2/P_s^2)$, where $P$ is the net existing polarization (D 10). This simple model would require, however, that the time rate of change in polarization for a single domain be proportional to the area of the domain, which is not compatible with the fact that the wall velocity is independent of domain size, as proved experimentally by Miller and Savage (M 25).

This difficulty was overcome by Miller and Weinreich (M 36) through the assumption that many protuberances are nucleated almost simultaneously on the wall of each domain. The sequence of events may tentatively be conceived of in the way indicated schematically in Fig. IV-46(b). As the nucleation probability increases with the area of the wall, the switching process would be governed by this nucleation at all times, independently of domain size. The nucleated protuberances were assumed to be triangular slabs along the wall and about one lattice constant thick, as depicted schematically in Fig. IV-46(a). The energy change consequent on the formation of such a nucleus can be written, in a general way, as

$$\Delta U = -2P_s E V + \sigma_w A + U_d \qquad (\text{IV-51})$$

where $\sigma_w$ is the wall energy per unit area, $U_d$ the depolarization energy, $V$ the volume and $A$ the wall area of the domain nucleus. For a triangular slab such as depicted in Fig. IV-46(a), the only additional domain walls which must be created during the nucleation event are the two edges, i.e. twice the area $ABED$. The depolarization energy of such a triangular nucleus cannot be calculated in the usual manner in which a depolarization factor is introduced since the shape of the nucleus is not one of the family of general ellipsoids. Consequently, the depolarization energy was computed with the formula (M 36):

$$U_d = \tfrac{1}{2} \int\int \frac{\varrho_1 \varrho_2}{\varepsilon r_{12}} \, dv_1 \, dv_2, \qquad (IV\text{-}52)$$

where $\varrho_1$ and $\varrho_2$ are the charge densities in volumes $dv_1$ and $dv_2$ separated by a distance $r_{12}$, and $\varepsilon$ is the mean dielectric constant of the medium. The dimensions of the critical nucleus and the activation energy can then be determined by the conditions for a minimum of $\Delta U$. For a field of 300 V/cm, the critical nucleus was calculated to be $7 \times 10^{-6}$ cm wide (along the electroded crystal surface) and $16 \times 10^{-6}$ cm high (along the ferroelectric axis). For limited ranges of field, the model gives a wall velocity dependence upon the field of $v = $ constant $\times \exp(-\delta/E)$, which agrees with experiment. The magnitude of the calculated activation field $\delta$ agrees with experiment if the energy of the additional wall consequent on a nucleation is set equal to 0.4 erg/cm$^2$, which, in turn, agrees in order of magnitude with theoretical estimates of the wall energy (see Section 7). The approximately square domains observed in the low field region are consistent with the model, and the change in shape of the domains observed at higher fields (Fig. IV-45(a, b)) can likewise be explained if slightly different wall energies are assumed for the edges of the nucleated steps on the two different types of 180° domain walls (M 36) (i.e. the (110) and (100) walls, respectively) (cf. also Nakamura (N 6)). This nucleation model is therefore capable of explaining much of the experimental data. Once the nucleus has grown throughout the thickness of the crystal in the direction of the polar axis, it is not inconceivable that some sideway expansion could occur by way of new nucleation events at the edges of the grown slab, as indicated in step 3 of Fig. IV-46(b).

Another model for the motion of 180° walls was suggested by Nakamura (N1). A screw dislocation running perpendicularly to the wall would imply the presence of a step in the wall, and the spiral climb of such a step would effectively cause a parallel shift of the domain wall. This model was investigated in more detail by Miller and Weinreich (M 36), but it was found to give the correct dependence of the wall velocity $v$ upon field only with certain unrealistic restrictions on the dislocation density or the wall mobility. It was suggested, however, that this mechanism may contribute to the wall motion with fields of a few thousand volts per centimeter or higher.

At higher fields, in fact, the phenomenological dependence of the switching time upon fields is no longer exponential. For fields of several thousand volts per centimeter, Merz (M 3) found that a BaTiO$_3$ crystal behaves like an ohmic resistance in series with a voltage bias. Switching time $t_s$ and maximum current $i_{\max}$

are linearly related to the applied field:

$$i_{\max} \propto \frac{1}{t_s} = \frac{E - E_0}{\gamma d},\qquad\text{(IV-53)}$$

where $E_0$ is a limiting field similar to the coercive field, $d$ the thickness of the sample and $\gamma$ a constant which depends upon temperature. At fields as high as

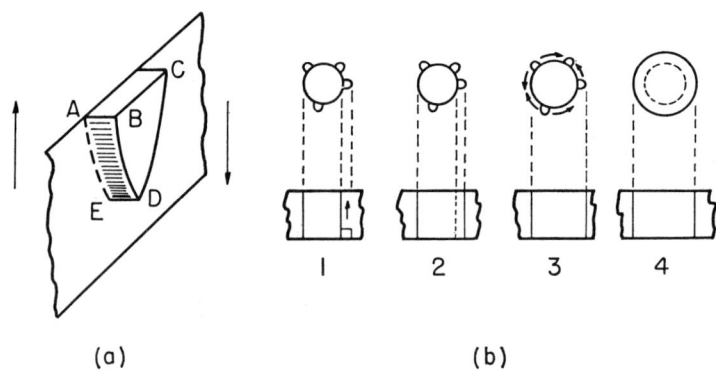

(a)                              (b)

FIG. IV-46. Schematic model of nucleation-controlled domain-wall motion.
(a) Nucleation of a protuberance at an existing 180° wall. The arrows on the right and on the left of the wall indicate the directions of the spontaneous polarization in the respective regions. Only the area $ABED$ (and the corresponding area on the opposite side) must be newly created. The area $BCD$ existed already before nucleation of the protuberance.
(b) Succession of events during domain growth. In step (1), protuberances are created at the (cylindrical) wall of the domain. In step (2), these protuberances have grown very rapidly throughout the thickness of the sample. In step (3), some sidewise motion may occur. In step (4), one stage of the process is completed.

about $10^5$ V/cm, Stadler (S 9) found that the relationship is no longer linear:

$$\frac{1}{t_s} \cong \text{constant} \times E^{3/2}.\qquad\text{(IV-54)}$$

No more information is available, at this stage, about high-field switching, and practically nothing is known about the mechanism governing it. All we can say is that the motion of the domain walls must occur by way of some phenomenon other than nucleation. It has been inferred (M 3), (S 9) that the switching time could be no shorter than the time taken for an elastic strain wave to traverse the crystal thickness, so that the limiting velocity of switching would be the velocity of sound. It is not clear, however, why and how acoustical modes would have to be connected with the phenomenon of polarization reversal. It would seem more logical to think that, at very high fields, all the unit cells of the crystal would reverse their polarization simultaneously, so that for infinitely large fields the crystal would switch infinitely fast. This would occur when one of the two minima of the potential well characterizing, say, the Ti ions is raised to such a level that the energy barrier vanishes. At what value of the applied field this would be the case we do not know with certainty. We can expect, however, that this critical field is approximately equal to the value of the coercive field estimated from the expansion of the free energy function for the free crystal (the fields applied by Stadler (S 9) were still somewhat smaller than this value at room temperature) or, better yet, to the same quantity for the clamped crystal.

### 9. The Surface of BaTiO₃ Crystals

A large variety of anomalies in the behavior of $BaTiO_3$ crystals have led a number of investigators to invoke the existence of various sorts of layers at the surface of these crystals. The detailed nature of these layers is not well understood, at the present stage, but their existence seems to be indisputably confirmed by a large number of experiments. In the following, we will first list and discuss the most important among these experiments, and then we will try to classify the various types of proposed surface layers into two main groups.

The first suggestion about the existence of a surface layer on $BaTiO_3$ crystals was made by Känzig (K 4) on the basis of experiments with very small particles of $BaTiO_3$ (A 1). These experiments were done with the same purpose as that which inspired the investigation on $KH_2PO_4$ (see Section III-9). The results obtained in that case, however, cannot be expected to be the same for the case of $BaTiO_3$ because of two reasons:

(i) In $KH_2PO_4$, the spontaneous polarization can only lie along the tetragonal $c$ axis, whereas in $BaTiO_3$ it can occur along three directions perpendicular to each other. Thus, in $BaTiO_3$ domain configurations with "closed flux" are possible.

(ii) In the case of $KH_2PO_4$, the ferroelectric phase occurs at such low temperatures that both the crystal and the medium in which the small particles are imbedded can be considered to be perfect insulators. This is no longer the case for $BaTiO_3$ particles.

However, an effect due to the small particle size of $BaTiO_3$ was found to exist on the basis of X-ray and electron diffraction experiments (A 1). At room temperature, the spontaneous strain was found to decrease with decreasing particle size. Also, with decreasing particle size, the experiments revealed that the Curie transition becomes smeared out over an increasing temperature range. The spontaneous strain of the tetragonal phase does not vanish at the Curie point of the macroscopic crystal. Particles with an average diameter of 1000 Å still exhibit a tetragonal distortion of the order of 0.2% at 500 °C. Electron diffraction experiments indicate the existence of a surface layer about $10^{-6}$ cm thick which exhibits a larger strain than the bulk of the crystal. The distortion of this surface layer is practically independent of temperature.

Further evidence for the existence of such a surface layer was given by Merz (M 3) on the basis of the dependence of the activation field $\alpha$ (Eq. IV-51) and the coercive field $E_c$ (Fig. IV-10) upon the thickness of the sample. $\alpha$ depends on the crystal thickness in the following fashion:

$$\alpha = \alpha_\infty \left(1 + \frac{d_0}{d}\right) \qquad \text{(IV-55)}$$

where $\alpha_\infty$ is the value of $\alpha$ for very thick crystals, $d$ is the crystal thickness and $d_0$ is approximately equal to $10^{-2}$ cm. Merz's explanation of this thickness dependence consists in assuming that, in the bulk of the crystal, switching proceeds as $\exp(-\alpha_\infty/E_b)$, where $E_b$ is the field actually existing in the bulk of the crystal. $E_b$ is smaller than the average applied field $E$ because the dielectric constant of the surface layer is assumed to be small ($\sim 5$) and thus an appreciable portion of the applied voltage lies across the surface layer. The thickness of this layer was estimated by Merz as of the order of $10^{-4}-10^{-5}$ cm.

As we have pointed out in the preceding section, the measurements of the sideway velocity of domain walls established an exponential dependence of this velocity upon the applied field in the form: $v = v_\infty \exp(-\delta/E)$, where the value of $\delta$, at least for metal-electroded crystals, was found to be nearly equal to the value of the activation field $\alpha$ discussed above. The interpendence between $\delta$ and $\alpha$ makes it logical, therefore, to expect that these two quantities may exhibit approximately the same behavior. Miller and Savage (M 35) have established, in fact, that $\delta$ depends upon the thickness of the crystal in the following way:

$$\delta = \delta_\infty \left(1 + \frac{d_0}{d}\right), \qquad \text{(IV-55a)}$$

where $d_0 = 5 \times 10^{-3}$ cm, or just about half the value obtained by Merz from the plot of $\alpha$ vs. $d^{-1}$ (see Eq. IV-55). The closeness between these results may become particularly significant when we think that the fields used by Merz in establishing Eq. (IV-55) were considerably higher than those used Miller and Savage in deriving Eq. (IV-55a), as proved by the fact that the longest switching time measured by Merz was about 10 msec, while the shortest switching time measured by Miller and Savage was about 1 sec.

Above the Curie point, a space-charge layer at the surface of $BaTiO_3$ crystals was detected directly by means of pyroelectric measurements. Since, above the Curie temperature, the polarizability is very high and strongly temperature dependent, it is reasonable to expect that any spontaneous polarization in the space-charge layer at the surface should give rise to an appreciable pyroelectric effect. This was proven experimentally by Chynoweth (C 13) by means of the dynamic method described in Section 2 (C 1). A non-zero pyroelectric current was measured above the Curie point, the direction of this current being independent of the direction in which the crystal was originally polarized at room temperature. This proves the existence of an appreciable polarization above the Curie point, even though no external fields are applied. This residual polarization must be ascribed to space-charge fields inside the crystal. Chynoweth developed an equivalent circuit involving a barrier at each surface of the crystal plate, the two barriers being polarized either in the same or in the opposite direction. From the experimental data, the effective d.c. barrier resistance was estimated as $1.6 \times 10^5 \, \Omega$ and, upon assuming a barrier conductivity similar to that of the bulk crystal ($3 \times 10^{-9}$ mho/cm) the barrier thickness was estimated as about $3 \times 10^{-5}$ cm. In accordance with the results of Anliker et al. (A 1), Chynoweth found from annealing tests that the surface layer decreases with increasing temperature. Measurements of the photovoltaic effect as a function of temperature above the Curie point showed that this effect behaves similarly to the pyroelectric effect, indicating a direct association between the photovoltaic effect and the barriers responsible for the pyroelectric signals. The detailed correlating mechanism, however, was not discussed.

It was originally concluded (C 13), on the basis of pyroelectric tests below the Curie point, that the surface layer is responsible for the fact that the crystal shows a bias for the direction in which it polarizes. Thus, the asymmetry often observed in hysteresis loops was generally attributed to the surface space-charge layers on the crystal. Also, the ultrasonic method used by Husimi et al. (H 9), (H 10), (K 14) for the measurement of the polarization (see Section 2) indicated that the absolute values of the remanent polarization may be quite different for the two orientations of the polarization in c-domain plates. A recent investigation of Miller and Savage (M 16), however, showed no evidence for a room-temperature polarization bias due to surface space-charge layers. This investigation made use of Chynoweth's dynamic pyroelectric techniques. It was shown that an apparent polarization bias may occur, under certain electroding conditions, because of electrode edge effects. When these effects were eliminated, the pyroelectric hysteresis loops appeared symmetric with respect to the polarization. Miller and Savage's data strongly suggest that the asymmetry of the remanent polarization reported in the piezoelectric and pyroelectric investigations cited above may have been due to edge effects. However, the residual pyroelectric signals observed by Chynoweth at temperatures *above* the Curie point do not result from fringe effects and are consistent with Känzig's hypothesis of a polarized layer at the surface of $BaTiO_3$ crystals.

Chynoweth's pyroelectric experiments were repeated under somewhat different conditions by Wieder and White (W 5) in order to determine some of the properties of the space-charge layer at room temperature. The light flashes were made to fall either directly upon the electrode area ("longitudinal pyroelectric excitation") or perpendicular to it ("transversal pyroelectric excitation"). The pyroelectric transients were essentially the same in the two cases. The effects were explained qualitatively in terms of a localized space-charge layer consisting in trapped charge carriers. However, very little information can be gathered from these experiments on the nature of the potential barrier at the surface. It cannot be said, therefore, whether the barriers arise from impurity ion distributions or from trapped electrons or positive holes.

More information on this subject was provided by Triebwasser (T 10) with a number of interesting experiments involving the application of d.c. fields to $BaTiO_3$ crystals. In particular, the study of the birefringence induced by an applied field above the Curie temperature clearly demonstrated the building up, in time, of space-charge layers near the electroded surfaces of the crystal. These space charges result in a reduction of the actual field existing in the bulk of the sample, and give rise to a lowering of the apparent dielectric constant as discussed in Section 4. The important result of these experiments is that the thickness of the space-charge layer was found to be dependent upon the magnitude and duration of the applied field, as evidenced by the data plotted in Fig. IV-21. The model proposed by Triebwasser for the surface layer is that of Schottky exhaustion barriers with a donor concentration of about $10^{19}$ cm$^{-3}$ and a dielectric constant of the order of 200, which leads to an estimated thickness of the barrier of $5 \times 10^{-6}$ cm. When an external field is applied to the crystal, there is an additional voltage drop across the barrier, which is expected to increase with the external field, causing the barrier at one of the contacts to broaden to the point where the capacitive impedance of the barrier becomes comparable with the bulk impedance. When this happens, the measured impedance will be higher, and the measured capacitance will be lower than expected from the phenomenological theory. Similar conclusions were reached also by Drougard and Schlosser (D 16) who examined the frequency and thickness dependence of the dielectric constant and loss factor of single crystals above the Curie temperature. Again, the results were explained on the basis of surface layers with comparatively low permittivity near the electrodes. The relaxation time of this layer at 120 °C was found to be larger than $10^{-4}$ sec, so that the impedance of the layer would be expected to affect the switching properties even at low audio frequencies.

Recently, Drougard and Landauer (D 11) have proposed a different approach to the problem of surface layers in ferroelectrics. They have shown that the thickness dependence of the activation field $\alpha$ found by Merz (and that of the quantity $\delta$ measured by Miller and Savage) could be explained by assuming a very thin and lossy surface layer. This is supposedly not a space-charge layer of the kind detected in many of the experiments described above, and produced by an electric field, but rather a thin layer which is a permanent part of the crystal and electrode structure, and could well be a result of the electrode evaporation process (Miller and Savage (M 35) have found that the surface layer is less effective, although still present, with liquid electrodes than with metal electrodes). The effect of such a layer upon the switching characteristics was assumed to be due to a surface density of magnitude $2P_s$ present immediately behind an advancing 180° domain wall, on the interface between the surface layer and the bulk material. The layer was assumed to be lossy in order for charge distributions due to previous polarization reversals to be neutralized. Assuming a dielectric constant of 5 for this layer, Drougard and Landauer calculated a layer thickness of about 4 Å, i.e. only about one lattice constant thick. Miller and Savage (M 35) have pointed out that the average field in a layer of this sort would be of the order of $10^8$ V/cm, and have modified this picture of a lossy layer by assuming a dielectric constant of about 100. With this assumption, the layer thickness was computed to be of the order of 100 Å, and the field within the layer about $3 \times 10^6$ V/cm. It was pointed out that with such high fields, electron field emission would be expected to occur, and field emission may in turn manifest itself in electroluminescence. These processes would contribute to the reduction of the high fields in the surface layer behind a rapidly advancing wall. For low fields in the surface layer, field emission would become unimportant and dielectric relaxation was proposed as the mechanism which further reduces the field in the layer. The switching characteristics were also found to be a function of the impurity content of the crystal, the electrode material and the humidity of the surroundings, all of which probably alters the properties of the surface layer (see also the effects of ambient atmosphere on the shape of the hysteresis loops of BaTiO$_3$ (A 4)).

Electroluminescent phenomena during polarization reversal were in fact detected by Harman (H 11). The measurements were made with high-frequency excitation ($\gtrsim$ 500 kc/s) and low average field strength (< 1000 V/cm). It was shown that the light emission must be the result of a high r.f. field across a thin surface barrier. The dependence of the emitted light intensity $I$ upon the applied voltage $V$ may be approximated by the following law:

$$I = a V^b, \qquad (IV\text{-}56)$$

where $a$ and $b$ are constants which depend on the electrode materials. Values of $b$ range from about 4 to 6. The intensity of the emitted light varies with temperature similarly as the dielectric constant and thus reaches large values at the Curie point. This effect can be explained by assuming that the surface layer responsible for the light emission has a low dielectric constant which remains essentially unaffected through the ferroelectric transition. When the bulk dielectric constant increases, more of the total r.f. voltage applied appears across the surface layer and the electroluminescence increases according to Eq. (IV-56). The effect persists far above the Curie point to at least 300 °C. The overall efficiency of the light emission is very low and a tentative value of $10^{-6}\%$ was obtained.

The model proposed by Harman in order to explain the phenomena observed consists in assuming that BaTiO$_3$ contains a small amount of oxygen vacancies. These centers may serve as electron traps and would be ionized by the large r.f. fields existing in the electrode–crystal interface. The resulting free electrons would then be accelerated sufficiently to produce electron–hole pairs by collision. The remaining ionized centers would be isolated point charges which could produce field emission from the electrodes (G 3), if located within a few atomic diameters from the surface. These point charges near the crystal surface and the surface properties of the electrodes may then be influenced by the ambient atmosphere. The experiments showed, in fact, that the electroluminescence is affected by the surrounding atmosphere. The analysis of the experimental data allowed an estimate of the upper limit for the ratio of dielectric constant to thickness in the surface layer. This ratio was found to have a value of $\leq 1.4 \times 10^5$ cm$^{-1}$. Assuming, together with Merz (M 3), a value of 5 for the dielectric constant of the barrier layer, Harman arrived at a barrier thickness of about $3.5 \times 10^{-5}$ cm, in agreement with Chynoweth (C 13). It is interesting to note that Harman's investigations of the electroluminescence yielded qualitatively similar results when extended to perovskite materials other than BaTiO$_3$, namely ferroelectric KNbO$_3$, antiferroelectric PbZrO$_3$ and non-ferroelectric CaTiO$_3$ and SrTiO$_3$. The study of high-frequency induced electroluminescence of SrTiO$_3$ in the temperature interval ranging from $-40$ °C to $+300$ °C led to an estimated thickness of the surface layer about four times as wide as that in BaTiO$_3$.

The existence of large electric fields within a thin layer near the surface would of course lead one to expect a larger concentration of $c$ domains at the surface of $BaTiO_3$ crystals. Subbarao *et al.* (S 16) have studied the intensities of suitably chosen X-ray reflections in order to gain information about the relative ratio of $c$ and $a$ domains on the surface of ceramic $BaTiO_3$ samples. The number of $c$ domains present was in fact found to be larger than that expected from a random distribution law. Moreover, the percentage of $c$ domains at the surface could be affected by a number of operations such as grinding, etching and irradiation with ultraviolet light, all of which is again expected to alter the properties and the thickness of the surface layer.

In conclusion, we may now summarize the large number of experimental evidences for the existence of surface layers in $BaTiO_3$ as follows:

(1) Observation of a tetragonal surface layer above the Curie point.

(2) Thickness dependence of the switching characteristics, in particular, the coercive field, the activation field $\alpha$ and the sideway wall velocity of domains.

(3) Dependence of the dielectric constant upon field, frequency and thickness above the Curie point.

(4) Residual pyroelectric effects above the Curie temperature.

(5) Inhomogeneous distribution of the birefringence induced by d.c. fields above the Curie point.

Unraveling this large number of anomalous effects has not yet led to any consistent picture of the surface layer except in qualitative terms. We may tentatively classify the various types of surface layers that have been proposed into two main groups (C 20):

(a) Space-charge layers due either to surface ionic vacancies as originally suggested by Känzig (K 4) or to exhaustion barriers as suggested by Triebwasser (T 10).

(b) Chemically or mechanically distorted layers which take no part in the polarization reversal process but will give rise to interface charges, as suggested by Drougard and Landauer (D 11) and further discussed by Miller and Savage (M 35).

## 10. Effects of Radiation and Decay

There are a number of indirect evidences that the character of the surface layer of $BaTiO_3$ crystals may be affected by various types of radiation. A qualitative study of the photosensitivity of $BaTiO_3$ was carried out by Arend (A 3) on very small $(1-5\,\mu)$ particles. It was observed that prolonged exposure to daylight affects visibly the coloration of the crystallites, an effect that was attributed to photochemical processes within the surface layer. Similar coloration effects were observed in single crystals of $BaTiO_3$ by Lefkowitz and Mitsui (L 8) as consequences of $\gamma$-ray and neutron irradiations.

Quantitative measurements of radiation damage effects were carried out by various authors with respect to changes in dielectric and crystallographic properties. Rogers (R 5) studied the effect of prolonged pile irradiation on the dielectric constant of ceramic $BaTiO_3$. The samples were exposed to neutron radiation to the extend of $nvt = 2.1 \times 10^{20}$ c.g.s. units. The dielectric constant measured at 1000 c/s was drastically reduced, at room temperature, to about one third of its original value and remained constant over the temperature range from 30 to 140 °C, showing no anomaly at the Curie point. These results are consistent with the study of fast neutron effects on the crystal structure of tetragonal $BaTiO_3$, carried out by Wittels and Sherill (W 6). These authors found that a fast neutron flux of $1.8 \times 10^{20}$ n/cm² impinging on a (tetragonal) crystal at 100 °C transforms it into cubic. This transformation involves an anisotropic expansion of both the $a$ and the $c$ lattice parameters by 2.26% and 1.17%, respectively. The cubic phase thus obtained exhibits no transition upon cooling and is still stable at 78 °K.

Lefkowitz (L 11), on the other hand, has found that integrated pile dosages of $1 \times 10^{18}$ $nvt$ depress the peak value of the dielectric constant of "pure" ceramic $BaTiO_3$ at the Curie point but do not affect the Curie temperature. Little modification of the room-temperature dielectric constant was observed. For the same dosages, ceramics of $BaTiO_3$ made with additives (4% of $PbTiO_3$) showed a shift of the Curie point (up to 8°) as well as a depression of the peak in dielectric constant. The same ceramics, however, would not exhibit hysteresis loops, showing that the amount of reversible polarization had been practically reduced to zero.

Lefkowitz and Mitsui (L 8) investigated the effects of $\gamma$-ray and neutron irradiation on the coercive field of single crystals of $BaTiO_3$. For $\gamma$-ray dosages up to $1.2 \times 10^8$ r and neutron dosages up to $2.8 \times 10^{16}$ $nvt$ the room-temperature coercive field was found to decrease and the spontaneous polarization to increase with exposure. A possible explanation for these effects is that the defects produced by the radiation act as domain nucleation sites. Another possibility is that the stress field around existing sites may be modified by the radiation effects to make the polarization reversal easier. Double hysteresis loops were observed for fast neutron dosages of the order of $10^{17}$ $nvt$, similar to those observed by Chynoweth in tri-glycine sulfate bombarded with X-rays (see Section II-6). These double hysteresis loops are apparently not due to induced transitions but rather to a biasing effect of the incident radiation. However, the detailed mechanism of the radiation damage effects in $BaTiO_3$ is not known at the present stage.

Another, as yet unexplained, effect which may be related to the properties of the surface layer is the fatigue or decay effect observed after repeated pulsing of a given crystal. This effect consists in a gradual reduction of the charge switched or loss of response after a few million cycles of switching (M 17). This effect is of great importance in the field of possible applications of $BaTiO_3$ crystals as matrix memories for digital storage in computer and switching systems. Generally, the reduced switching charge can be restored to its initial value by switching the crystal during a few minutes with a 60 c/s sine-wave voltage. It was established by Stadler (S 10) that the decay effect does not occur at $-195$ °C. Apparently, decay is not primarily dependent on the crystal phase, and the fact that $BaTiO_3$ is rhombohedral at $-195$ °C plays no role in the decay phenomenon, as some crystals may decay at $-100$ °C but others do not at $-20$ °C. A great deal more experimentation is needed before light can be shed on this subject.

## 11. Theoretical Treatments

### Phenomenological Theory of Devonshire

The fundamentals of the phenomenological theory of Devonshire (D 3), (D 4), (D 5) have already been described in some detail in Section I-5 and applied to the case of the tetragonal phase of $BaTiO_3$ in Section IV-4. There are a few important points of Devonshire's theory which we have not discussed previously and to which it is worth devoting some time at this stage.

We recall that Devonshire's approach consists in expanding the free energy of the cubic crystal in terms of the strains and polarization (Eq. I-1) and using certain properties of the crystal to determine the coefficients. Other properties are then predicted. Results obtained in this way are, of course, independent of any atomic model. The success of Devonshire's theory lies in the fact that it is capable of explaining the dielectric, piezoelectric and elastic behavior of the crystal at any temperature by means of a *single* free energy polynomial involving a *limited* number of terms. By assuming reasonable values for the coefficients of these terms one can explain the successive transitions through the cubic, tetragonal, orthorhombic and rhombohedral phases. In determining the coefficients, one makes use only of the following experimental data:

(i) the Curie temperature $T_c$ and the temperature at which any one of the lower transitions occurs, e.g. the tetragonal–orthorhombic transition;

(ii) the Curie constant $C$ and the Curie–Weiss temperature $T_0$ describing the behavior of the dielectric constant in the cubic region; and

(iii) the spontaneous polarization and strain at any single temperature in the tetragonal region.

It is easy to see, in a purely qualitative way, that the expression for the free energy of the stress-free crystal (Eq. IV-13) allows four sets of solutions which may correspond to minima of the free energy, namely (D 3):

$$\left.\begin{array}{l} P_1 = P_2 = P_3 = 0; \\ P_1 = P_2 = 0,\ P_3 \neq 0; \\ P_3 = 0,\ P_1 = P_2 \neq 0; \\ P_1 = P_2 = P_3 \neq 0. \end{array}\right\} \qquad \text{(IV-57)}$$

The first equation represents the cubic phase, in which the polarization is zero, the remaining three equations represent the tetragonal, orthorhombic (referred to monoclinic axes) and rhombohedral phases, in this order, where the polarization will in turn point along a cube edge, a face diagonal and a body diagonal. The relative depths of the minima of the free energy function change with the coefficient $\chi^X$. If this decreases steadily with decreasing temperature and suitable constant values are chosen for the coefficients $\xi_{11}^X$, $\xi_{12}^X$ and $\zeta_{111}^X$ of Eq. (IV-13) (neglecting $\zeta_{112}^X$ and $\zeta_{123}^X$), the sequence of, and the temperatures at which the transitions occur are those actually observed in $BaTiO_3$.

Knowing the values of the coefficients, Devonshire was able to derive theoretical curves for the free energy, the spontaneous polarization and the principal dielectric constants. Comparison between the theoretical curves in Fig. IV-47 and the experimental curves in Fig. IV-3 shows that the quantitative agreement is not exact but the qualitative agreement is very satisfactory.

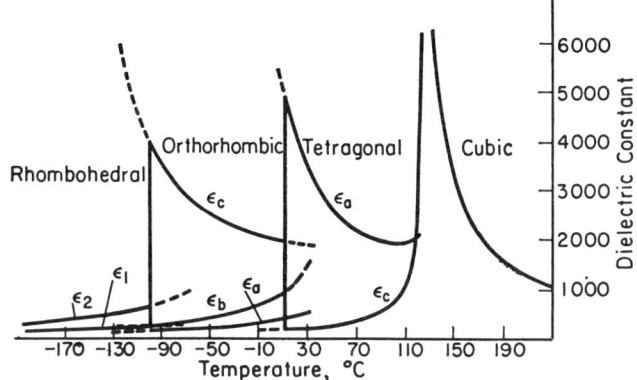

FIG. IV-47. Theoretical behavior of the principal dielectric constants of BaTiO₃ (according to Devonshire (D 3)).

The spontaneous strain can be calculated from the free energy function (I-1) of polarization and strain. Upon differentiating the free energy with respect to the strains and equating the resulting expressions to zero (i.e. the stresses are assumed to be zero) one obtains six relations between strain and polarization involving the elastic and the electrostrictive constants. Thus, knowing the numerical values of the latter constants, one can evaluate the variation of the strains with temperature over the whole range of interest. The results are depicted in Fig. IV-48 as obtained by Devonshire (D 3), showing again that the qualitative agreement with the experimental data shown in Fig. IV-15 is quite satisfactory.

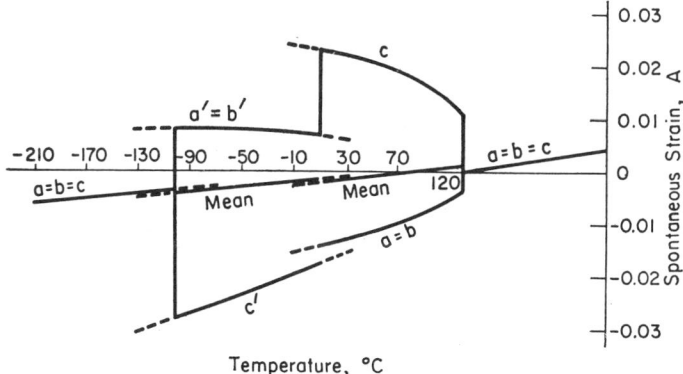

FIG. IV-48. Theoretical behavior of the spontaneous strains of BaTiO₃ relative to the value at 120 °C (according to Devonshire (D 3)).

The effect of clamping on the nature of the transition was investigated by Devonshire by eliminating the strains from Eq. (I-1) and comparing the resulting expression with that of the free energy (IV-13) in terms of polarization for zero stress. This comparison yields the relations between the coefficients $\chi^X$, $\xi^X_{11}$, $\xi^X_{12}$ of the free crystal and the coefficients $\chi^x$, $\xi^x_{11}$, $\xi^x_{12}$ of the clamped crystal (D 3). It turns out that $\chi^x = \chi^X$, as expected, because the non-polar phase is not piezo-electric and thus no distinction is made between free and clamped dielectric constant. The relations between $\xi^X_{11}$ and $\xi^x_{11}$, and $\xi^X_{12}$ and $\xi^x_{12}$ involve the elastic and the electrostrictive constants. Using the values of these constants available to him, Devonshire was able to compute the numerical value of $\xi^x_{11}$ and $\xi^x_{12}$ (D 4) and thus conclude that, although $\xi^X_{11}$ is negative, $\xi^x_{11}$ is positive. Since the co-efficients of higher powers in polarization are probably also positive, this means that the transition of the clamped crystal would be of the second order, although that of the free crystal is of the first order. This point is very important when atomic models for $BaTiO_3$ are considered, as the most useful information in setting up a statistical atomic theory is the dielectric behavior of the crystal which is not allowed to assume a spontaneous deformation. The importance of Devon-shire's result about the sign of $\xi^x_{11}$ is shown by the fact that Slater's model theory of $BaTiO_3$ leads to a transition of the second order (see below).

It should be mentioned, before closing this discussion of Devonshire's theory, that a general thermodynamic treatment of the properties of ferroelectric crystals was developed earlier by Ginzburg (G 4) and applied by him to the Curie transition of $BaTiO_3$. However, only a limited amount of experimental evidence was available when Ginzburg's paper was published, so that a fair evaluation of the theory was not possible. The thermodynamic treatment of the $BaTiO_3$ problem was subsequently expanded by Ginzburg (G 5) himself, Smolenskii and Pasynkov (S 19) and Khodolenko (K 22).

### The Lorentz Correction of Slater

Slater (S 15) has assumed, following Devonshire (D 3), that the ferroelectric behavior of $BaTiO_3$ arises because of the Lorentz correction, leading to a vanishing term in the denominator of the expression for the dielectric constant. In order to understand Slater's approach, it may be useful to recall briefly some of the ideas expressed in Section I-2 in connection with the nature of the spontaneous polarization and the "$4\pi/3$ catastrophe". There are in general two types of polarization: the ionic polarization, arising from the displacement of ions from their equilibrium positions, and the electronic polarization, arising from the distortion of the electronic clouds of the ions. The latter contribution can be estimated from the optical dielectric constant, i.e. the dielectric constant found by squaring the value of the index of refraction in the visible spectrum. The optical dielectric constant of $BaTiO_3$ is no larger than one would expect from the electronic polarizability of its constituents, being about $(2.4)^2 = 5.76$. Thus, it is logical to expect that the anomaly of the low-frequency dielectric constant is a conse-quence of ionic displacements. Slater assumes that only the Ti ion undergoes a displacement and that it is this displacement which is responsible for the effect.

The problem is now to find out just how big a contribution we need from the ionic polarization in order to reach a spontaneously polarized state. To get an idea of the orders of magnitude involved, let us first consider a hypothetical case in which there would be no electronic polarization at all, and, also, the "Lorentz" correction would be exactly $(4\pi/3)P$. In this case, the dielectric constant can be expressed by the formula of Clausius–Mosotti (see Section I-2):

$$\varepsilon = 1 + \frac{4\pi\alpha_{ion}/v}{1 - 4\pi\alpha_{ion}/3v}, \qquad (IV\text{-}58)$$

where $\alpha_{ion}$ is the ionic polarizability of Ti and $v$ is the volume of the unit cell. We know that in this case the dielectric constant becomes infinite (Curie point) when:

$$\frac{4\pi}{3} \frac{\alpha_{ion}}{v} = 1. \qquad (IV\text{-}59)$$

If we next relieve only the first of the above assumptions, i.e. we allow a contribution from the electronic polarization, but we still keep the Lorentz correction as $(4\pi/3)P$, then the formula for the dielectric constant becomes:

$$\varepsilon = 1 + \frac{4\pi \sum (\alpha_{el}/v + \alpha_{ion}/v)}{1 - (4\pi/3) \sum (\alpha_{el}/v + \alpha_{ion}/v)}, \qquad (IV\text{-}60)$$

leading to a Curie point when:

$$\frac{4\pi}{3} \sum \left( \frac{\alpha_{el}}{v} + \frac{\alpha_{ion}}{v} \right) = 1. \qquad (IV\text{-}61)$$

Here, $\alpha_{el}$ represents the electronic polarizability of Ba, Ti and O. For Ba and Ti Slater used the values of $\alpha_{el}$ computed by Anderson and Shockley (reported in Slater's paper (S 15)), while for the value of $\alpha_{el}$ of oxygen he chose a figure which would give the electronic polarizability of the crystal as $(2.4)^2 = 5.76$. In this way, Slater found that

$$\frac{4\pi}{3} \sum \frac{\alpha_{el}}{v} = 0.61. \qquad (IV\text{-}62)$$

Thus, according to Eq. (IV-61), a further contribution of only:

$$\frac{4\pi}{3} \sum \frac{\alpha_{ion}}{v} = 0.39 \qquad (IV\text{-}63)$$

ought to be made by the ionic polarizability of Ti in order to produce ferroelectricity.

This effect is yet much more drastically enhanced if we relieve also the second assumption made originally about the value of the Lorentz correction. Slater has shown that when the Lorentz correction is applied properly, one does not need nearly as great an ionic polarizability on the part of the Ti ion as one would otherwise suppose. This is done in the following way.

In a $BaTiO_3$, each unit cell contains five ions. Assuming that the crystal is polarized along the $z$ axis, symmetry demands that each of the ions be polarized along the $z$ axis. The total polarization of the crystal can then be written as the sum of five terms: the polarizations arising from each type of ion. Each of these types of ions forms a simple cubic lattice and a $BaTiO_3$ crystal may be described as consisting of five interpenetrating simple cubic lattices. We know that the field of a simple cubic lattice of dipoles at one of its lattice points is $(4\pi/3)P$, where $P$ is the polarization resulting from this lattice alone. However, the field of a simple cubic lattice of dipoles is not equal to $(4\pi/3)P$ at most points within the unit cell. Given a point with co-ordinates $(u, v, z)$ (in fractions of the cell edge) within the unit cell, the field acting at this point may be given by the general formula:

$$F(u, v, z) = E + \left[\frac{4\pi}{3} + S(u, v, z)\right] P, \qquad (\text{IV-64})$$

where the term in the brackets represents the factor applying to the proper Lorentz correction. The quantity $S(u, v, z)$ was computed by McKeehan (M 21) and by Luttinger and Tisza (L 10) for a few special points within the unit cell, with the following results:

$$\left.\begin{aligned}
S(0, 0, 0) &= S(\tfrac{1}{2}, \tfrac{1}{2}, \tfrac{1}{2}) = 0 \\
S(0, \tfrac{1}{2}, \tfrac{1}{2}) &= 4.334 \\
S(\tfrac{1}{2}, \tfrac{1}{2}, 0) &= -8.668 \\
S(\tfrac{1}{2}, 0, 0) &= -15.040 \\
S(0, 0, \tfrac{1}{2}) &= 30.080 .
\end{aligned}\right\} \qquad (\text{IV-65})$$

These results show that the ordinary Lorentz correction $(4\pi/3)P$ is accurate not only at the lattice points but also at a point in the center of the unit cell. Taking Ti at the origin (Fig. IV-32), this means that the ordinary Lorentz correction is valid for the action of the Ti and Ba ions on themselves as well as on each other. On the other hand, if we consider the action of the Ti ions on the oxygens OI (Fig. IV-33) which lie along the same line parallel to the polarization, Eq. (IV-65) tells us that the Lorentz correction is really $(4\pi/3 + 30.080)P$, or approximately 8.2 times the ordinary value $(4\pi/3)P$. In a general way, we can compute the local fields $F^k$ acting on the $k$ ion, from (IV-65), as follows:

$$
\begin{pmatrix} F^{\text{Ti}} \\ F^{\text{Ba}} \\ F^{\text{OI}} \\ F^{\text{OII}} \\ F^{\text{OII}} \end{pmatrix}
= E + 4\pi
\begin{pmatrix}
\frac{1}{3}, & \frac{1}{3}, & \frac{1}{3}+q, & \frac{1}{3}-\frac{1}{2}q, & \frac{1}{3}-\frac{1}{2}q \\
\frac{1}{3}, & \frac{1}{3}, & \frac{1}{3}-p, & \frac{1}{3}+\frac{1}{2}p, & \frac{1}{3}+\frac{1}{2}p \\
\frac{1}{3}+q, & \frac{1}{3}-p, & \frac{1}{3}, & \frac{1}{3}+\frac{1}{2}p, & \frac{1}{3}+\frac{1}{2}p \\
\frac{1}{3}-\frac{1}{2}q, & \frac{1}{3}+\frac{1}{2}p, & \frac{1}{3}+\frac{1}{2}p, & \frac{1}{3}, & \frac{1}{3}-p \\
\frac{1}{3}-\frac{1}{2}q, & \frac{1}{3}+\frac{1}{2}p, & \frac{1}{3}+\frac{1}{2}p, & \frac{1}{3}-p, & \frac{1}{3}
\end{pmatrix}
\begin{pmatrix} P^{\text{Ti}} \\ P^{\text{Ba}} \\ P^{\text{OI}} \\ P^{\text{OII}} \\ P^{\text{OII}} \end{pmatrix}
\qquad (\text{IV-66})
$$

where $p = 8.668$, $q = 30.080$; $P^{\text{Ti}}$, $P^{\text{Ba}}$, $P^{\text{OI}}$ and $P^{\text{OII}}$ are the polarizations of Ti, Ba, OI and OII, respectively. These polarizations $P^k$, in turn, are related to the polarizabilities $\alpha^k$ of the $k$ ions by the general formula:

$$P^k = \frac{\alpha^k}{v} F^k. \tag{IV-67}$$

In the case of Ba, OI and OII the polarizabilities are only electronic ($\alpha_{el}^k$), in the case of Ti, the polarizability is assumed to be both electronic and ionic ($\alpha^{Ti} = \alpha_{el}^{Ti} + \alpha_{ion}^{Ti}$)

In this way, Slater found that the static dielectric constant can be written in the form:

$$\varepsilon = \frac{c_2}{c_4} + \frac{(c_1/c_3) - (c_2/c_4)}{1 + (c_4/c_3)\,(4\pi\alpha_{ion}/v)} \tag{IV-68}$$

where $c_1$, $c_2$, $c_3$, $c_4$ are functions of the electronic polarizabilities $\alpha_{el}$ of Ba, Ti and O. Inserting numerical values, Slater re-wrote Eq. (IV-68) as follows:

$$\varepsilon = 3.84 + \frac{1.93}{1 - 5.39\,(4\pi\alpha_{ion}/v)}\,, \tag{IV-69}$$

showing that ferroelectricity obtains when:

$$\frac{4\pi}{3}\,\frac{\alpha_{ion}}{v} = 0.062\,. \tag{IV-70}$$

Comparing (IV-70) with (IV-63), we see that the ionic polarizability of Ti is only one sixth of what it would need to be if the Lorentz correction were only $(4\pi/3)\,P$. Thus, the effect of the ionic displacement of the Ti ions is enormously enhanced by the electronic polarization of the other ions when the Lorentz correction is applied properly.

In examining the relative polarizations of the various types of ions, Slater concluded that the polarization is almost entirely contributed by the Ti and the OI ions. The Ba and OII ions are hardly polarized at all. Taking into account the polarizabilities, the Ti ions contribute about 37% of the total polarization (of which 31% comes from ionic shift, 6% from electronic polarization), the oxygen OI about 59%, the oxygen OII about 6%, and the Ba ions about 2% in the reverse direction. The effect is thus due to strong interaction between Ti and OI, which polarize one another and build linear chains of dipoles all pointing in the same direction.

In evaluating the contribution of Slater's theory to the problem of ferroelectricity in BaTiO$_3$ it should be kept in mind that the model used was only intended as an approximation and as such it is highly successful. It is true that there is no real justification for assigning the whole ionic polarization to Ti (M 22), as this depends on the choice of the origin of the unit cell and this choice is quite arbitrary. However, we do know, from the values of conventional ionic radii, that the Ti ions are rather loosely held in the center of the oxygen octahedra, the restoring forces being small. Moreover, the significance of Slater's results does not lie in the particular assignment to Ti of the only ionic polarizability allowed by the theory, but rather in revealing the enhancement of the field along the Ti—OI chains and thus the importance of these dipole chains oriented in the direction of the polar axis.

A particularly elegant way for the treatment of the ionic polarization has been devised by Takagi (T 5). Every ionic crystal may be described as consisting of a number of suitable sublattices $j$, $k$, etc., interpenetrating each other. The mean displacements of the $j$, $k$, etc. ..., ions from the sites of the corresponding $j$, $k$, etc. ..., sublattices are denoted by $z_j$, $z_k$, ..., respectively. Assuming that these mean displacements are all parallel to the average electric field, it can be shown that the electrostatic energy of the crystal contains only the mean relative displacements $(z_j - z_k)$, ..., but not the displacement of each sublattice separately. The local field at the mean position of the $j$ ions is given by an expression of the following type:

$$F_j = E + \sum_k \left[ \mu_k + q_k (z_k - z_j) \right] N_k \gamma_{jk}, \qquad \text{(IV-70a)}$$

where $\mu_k$ is the electronic moment of the $k$ ions, $q_k$ their ionic charge, $N_k$ their number per unit volume and $\gamma_{jk}$ the Lorentz factor for the field on the $j$ sublattice arising from the ions on the $k$ sublattice. The quantity $q_k(z_k - z_j) N_k$ appearing in Eq. (IV-70a) shows that, as far as the field on the $j$ sublattice is concerned, the $k$ sublattice appears to have an ionic polarization $q_k(z_k - z_j) N_k$. However, if one would consider the local field at another sublattice, the same $k$ sublattice would appear to have a different ionic polarization, depending on the mean relative displacements of the two sublattices considered. Thus the ionic polarization of each sublattice loses its unique meaning. It is evident that Takagi's approach allows the treatment of the total ionic polarization with no need for the obscure concept of ionic polarization of a given sublattice. This approach was applied by Takagi himself to the original Kängig's model of $BaTiO_3$ in which only displacements of Ti and OI were assumed. The relative stability of ferroelectricity vs. antiferroelectricity was discussed in terms of ionic and electronic polarizabilities.

It has also been pointed out (M 22), in connection with Slater's theory, that the treatment assumes parallel rather than antiparallel displacements in adjacent Ti-OI chains, while Cohen's calculations (C 10) would indicate that the antipolar state is favored, in $BaTiO_3$, with respect to the polar state. Cohen's calculations were done by taking into account the local fields at the location of the displaced ions. Only dipole–dipole interactions between Ti and OI were considered, while the polarizations of Ba and OII were neglected. However, Takagi (T 5) has obtained the opposite result: the driving forces acting on Ti and OI favor the polar rather than the antipolar state. The same result was obtained by Kinase (K 10), using a slightly different approach (see also Section IV-7). Therefore, Cohen's result should be considered incorrect most probably because of the complete neglectfulness of the polarizations of Ba and OII.

Slater's theory was later extended by Triebwasser (T 6), who took into account the actual displacements of all ions, rather than that of Ti alone. The ionic polarizations were computed on the basis of the displacements relative to the Ba ion, taken as the origin. From the viewpoint expressed above, the results of this extension of Slater's approximation should be interpreted with caution, particularly as far as the real physical meaning of the ionic polarizations of the individual ions is concerned.

### The Model Theory of Slater–Devonshire

The basic idea of the theory of $BaTiO_3$, developed by Devonshire (D 3) and considerably extended and improved by Slater (S 15), is to calculate the free energy of the crystal from a model and compare it with the expression postulated phenomenologically for the same function in (I-1). The model, as described above, is that involving electronic polarizabilities of all the ions and ionic polarizability of Ti alone, taking into account, of course, the proper Lorentz correction.

Assume that only the ionic polarizability of Ti is different from zero, all other polarizabilities being equal to zero, and that the Ti ions can be treated independently from each other. Let the potential energy of a Ti ion at position $(x, y, z)$ (measured from its equilibrium position) be $\Phi(x, y, z)$ in the absence of an external field. Let there also be a local electric field $F$ of components $(F_x, F_y, F_z)$ acting on the ion, and let $q$ be the charge of the ion. Then, the total potential energy of the ion can be written as:

$$U_E = \Phi(x, y, z) - q\,F \cdot r,\qquad\text{(IV-71)}$$

where $r$ is the radius vector with components $(x, y, z)$. Straightforward use of the methods of statistical mechanics allows one to compute the partition function $Z$ of the crystal containing $N$ identical ions:

$$Z = \left[\frac{e}{Nh^3}(2\pi m k T)^{3/2}\right]^N w^N,\qquad\text{(IV-72)}$$

where $m$ is the mass of the ion and

$$w = \int \exp\left[\frac{-\Phi(x, y, z) + q\,F \cdot r}{kT}\right] dv.\qquad\text{(IV-72a)}$$

The free energy expressed in terms of $F$ and $T$, $A(F, T)$ is then given by:

$$A(F, T) = -kT \ln Z,\qquad\text{(IV-73)}$$

and from this we can compute the ionic polarization of Ti, $P^{ion}(\mathrm{Ti})$ (with components $P_x^{ion}(\mathrm{Ti})$, $P_y^{ion}(\mathrm{Ti})$, $P_z^{ion}(\mathrm{Ti})$), as follows:

$$P_x^{ion}(\mathrm{Ti}) = -\left[\frac{\partial A(F, T)}{\partial F_x}\right]_T = NkT\left[\frac{\partial \ln w}{\partial F_x}\right]_T,\ \text{etc.}\qquad\text{(IV-74)}$$

However, we are rather interested in the free energy expressed in terms of polarization and temperature, $A(P^{ion}, T)$, which can be obtained from $A(F, T)$ by way of the formula:

$$A(P^{ion}, T) = A(F, T) + F \cdot P.\qquad\text{(IV-75)}$$

In this way, we have the necessary statistical background for understanding Slater's calculation of the free energy.

The potential energy of a displaced ion in the clamped crystal is assumed by Slater to have cubic symmetry, thus:

$$\begin{aligned}\Phi(x, y, z) = {}& a(x^2 + y^2 + z^2) + b_1(x^4 + y^4 + z^4) + \\ & + 2b_2(y^2 z^2 + z^2 x^2 + x^2 y^2).\end{aligned}\qquad\text{(IV-76)}$$

With this assumption, and using series expansion methods, Slater computes the free energy $A(P^{\text{ion}}, T)$, and from it, by differentiation with respect to the polarization components, he derives the components of the local field polarizing the Ti ion. The $x$ component of this field can thus be written as follows:

$$F_x(\text{Ti}) = \frac{2\,a\,P_x^{\text{ion}}(\text{Ti})}{N\,q^2}\left[1 + \frac{kT}{a^2}(3\,b_1 + 2\,b_2)\right] +$$
$$+ \frac{4\,P_x^{\text{ion}}(\text{Ti})}{N^3 q^4}\left\{b_1[P_x^{\text{ion}}(\text{Ti})]^2 + b_2([P_y^{\text{ion}}(\text{Ti})]^2 + [P_z^{\text{ion}}(\text{Ti})]^2)\right\}. \tag{IV-77}$$

Now, if we allow all ions to have electronic polarizabilities, we know that the local fields are given by Eqs. (IV-66), where we now must distinguish between electronic and ionic contributions to the polarization of Ti. We do this by replacing, in (IV-66), $P^{\text{Ti}}$ with $[P^{\text{el}}(\text{Ti}) + P^{\text{ion}}(\text{Ti})]$. Combining (IV-77) with (IV-66) we obtain one equation (for each component) and combining (IV-66) with (IV-67) we obtain four more independent equations which we can solve for the five parts of the $x$-component of polarization

$$P_x^{\text{ion}}(\text{Ti}),\ P_x^{\text{el}}(\text{Ti}),\ P_x^{\text{el}}(\text{Ba}),\ P_x^{\text{el}}(\text{OI})\ P_x^{\text{el}}(\text{OII}),$$

the sum of which is the $x$-component of the macroscopic polarization $P$. It is then possible, with further mathematical manipulations, to write down expressions for the components $E_x$, $E_y$, $E_z$ of the macroscopic field in terms of the components $P_x$, $P_y$, $P_z$ of the total macroscopic polarization. Integration of these expressions yields then finally the free energy in terms of the components of total polarization for the clamped crystal:

$$A(P_x, P_y, P_z, T) = -NkT\ln\left[\frac{e}{Nh^3}(\pi kT)^3\left(\frac{2m}{a}\right)^{3/2}\right] +$$
$$N\left(\frac{kT}{a}\right)^2\frac{3}{4}(3\,b_1 + 2\,b_2) +$$
$$\left(\frac{4\pi c_4}{c_5}\right)^2\frac{Nq^2}{4a^3}k(T-T_0)(3\,b_1 + 2\,b_2)(P_x^2 + P_y^2 + P_z^2) + \tag{IV-78}$$
$$\left(\frac{4\pi c_4}{c_5}\right)\frac{Nq^2}{16\,a^4}[b_1(P_x^4 + P_y^4 + P_z^4) + 2\,b_2(P_y^2 P_z^2 + P_z^2 P_x^2 + P_x^2 P_y^2)].$$

Comparing this expression with the free energy polynomial in terms of polarization and strain postulated by Devonshire's phenomenological theory (see above) Slater obtains:

$$\left.\begin{aligned}
\chi^x &= \left(\frac{4\pi c_4}{c_5}\right)^2\frac{Nq^2}{2a^3}(3\,b_1 + 2\,b_2)\,k(T-T_0), \\[2mm]
\xi_{11}^x &= \left(\frac{4\pi c_4}{c_5}\right)^2\frac{Nq^4}{4a^4}b_1, \\[2mm]
\xi_{12}^x &= \left(\frac{4\pi c_4}{c_5}\right)^4\frac{Nq^4}{4a^4}b_2.
\end{aligned}\right\} \tag{IV-79}$$

and for the $x$ component of polarization:

$$P_x^2 = \frac{c_5^2}{c_3 c_4} N k \frac{(3b_1 + 2b_2)}{b_1} (T - T_0). \qquad \text{(IV-80)}$$

As pointed out by Slater, a comparison between the numerical predictions of the theory and the experiment can be done in two ways. First, the observed Curie temperature allows an evaluation of $q^2/a$, and the temperature dependence of the dielectric constant yields $(3b_1 + 2b_2)/a^2$. Assuming that $q$ equals $n$ electronic charges, one finds that the constants are of reasonable order of magnitude if $n$ is chosen to be equal to unity rather than 4 (see also Kinase and Takahashi (K 6), (K 11) and Section IV-3). On the other hand, the behavior of the polarization below the Curie point, Eq. (IV-80), yields information about the fourth-power term. The polarization of the clamped crystal increases much more slowly below the Curie temperature than that observed on the free crystal. This is in agreement with Devonshire's result that the transition of the clamped crystal is of the second-order.

Slater's method for deriving the free energy from an ionic potential function was later extended by Triebwasser to include terms of the sixth power in the displacement of the ions (T 6).

### Theory of Mason and Matthias

The model introduced by Mason and Mattias (M 23) involves the assumption that the Ti ions are randomly distributed among six equilibrium positions within the oxygen octahedra. This corresponds to the assumption of permanent pseudo-dipoles which are capable of six orientations. The interaction between these dipoles is taken care of by a Lorentz internal field.

We have already seen in Section I-4 that this type of treatment leads to a Curie constant which is of the order of magnitude of the Curie temperature. In order to arrive at a larger Curie constant, Mason and Matthias find it necessary to introduce an artificially small Lorentz factor (0.14, rather than of the order of $4\pi/3$). The theory can be criticized on a number of grounds. In the first place (J 3), (F 2) the assumption of rotating permanent dipoles with six directional states leads to an entropy change of $k \ln 6$ per unit cell, i.e. 3.6 cal/mole °C, in contrast with the observed value which is of the order of 0.1 cal/mole °C (see Table IV-1). Secondly, the assumption of permanent dipoles involves again the doubtful concept of ionic polarizability of a given type of ions in the lattice. Finally, the nuclear-resonance experiments performed on $KNbO_3$ (which is isomorphous to $BaTiO_3$) by Cotts and Knight (C 15) strongly indicate that the equilibrium position of Nb above the Curie point is in the center of the oxygen octahedron, an evidence which speaks against the existence of permanent dipoles in the perovskite lattice.

### The Oxygen-displacement Model of Jaynes

An interesting model for $BaTiO_3$ was treated qualitatively by Jaynes (J 4), who considered that displacement of the oxygen ions is at least as effective a cause of polarization as is Ti displacement. The argument goes as follows: The mismatch of ionic sizes of Ba and O causes the Ba−O contacts to be too tight and, as a consequence, the O atoms are squeezed out. Upon cooling, first one group of oxygen is squeezed out of their symmetrical positions, causing the symmetry to become tetragonal. Contraction in the directions at right angles to the polarization occurs, as the Ba ions which were pressing against the displaced oxygens can now come together by a small amount. Upon further cooling, a second group of oxygens is squeezed out of their positions, the crystal distorting to orthorhombic symmetry. Finally, at yet lower temperatures, the third group of oxygens is squeezed out of place, causing the symmetry to become rhombohedral. This simple model is capable of giving a qualitative understanding of two features of $BaTiO_3$, viz. the existence of three phase transitions, involving changes in direction of spontaneous polarization, and the very large electromechanical coupling.

### The Electronic Theory of Jaynes

All the models described above for the solution of the $BaTiO_3$ problem have a common feature in that the electronic polarizability is assumed to be constant, independent of temperature and field. With this assumption, in order to get a Curie point and a finite polarization it is mostly necessary to invoke the coupling of units such as rotating dipoles, ions with more than one equilibrium position, or anharmonically bound ions (F 2). A theory which essentially results in a temperature-dependent electronic polarizability and does not require the attribution of dipole moments to the atomic displacements was put forward by Jaynes (J 3). In this theory, the electronic states of the entire $TiO_6$ octahedra are considered, rather than the electronic polarizabilities of the individual ions. The octahedra are assumed to be independent from one another and their interactions are described by a field of the Lorentz type. The abnormally large polarizability is attributed to the existence of thermally excited states which are coupled to the ground state by the dipole-moment operator. The theory predicts an entropy change which is in agreement with that found experimentally. It also predicts a strong infrared absorption line in the vicinity of $\sim 10\,\mu$. No such line was detected by the spectroscopic research (L 9), (M 8).

### The Static Model of Hagedorn

A completely different approach to the $BaTiO_3$ problem was attempted by Hagedorn (H 12) by introducing a static model. The word static is referred to the fact that the thermal motion is neglected and, instead, the time average of the center of gravity of each atomic charge is replaced by a static point charge to which a static dipole is superimposed. In the non-polar state, the Coulomb field vanishes at the equilibrium positions of each ion and the ions have no dipole moment. When the ions are displaced from their symmetrical positions they are located at points where the Coulomb field no longer vanishes and they become polarized. The field generated by these dipoles is superimposed on the Coulomb field and modifies the induced dipole moments. The interactions between charges and dipoles, as well as the internal fields, are described by a system of equations which allow one to compute the dipole moments, the internal fields and the dielectric constant, provided that the ionic charges, the polarizabilities and the external field are known. Upon assuming a purely ionic crystal, i.e. attributing the charges $+2$, $+4$ and $-2$ to Ba, Ti and O, respectively, Hagedorn obtained a spontaneous polarization about fifty times larger than that observed, and a dielectric constant about 20% smaller than the experimental value. This shows again, in a very drastic way, that the ionic character of $BaTiO_3$ is very small. An estimate of the ionic nature of $BaTiO_3$ can be arrived at by introducing "effective" charges and expressing the polarizabilities in terms of these charges. By an iteration method, Hagedorn was able to determine the values of the effective charges which yield the correct values of the spontaneous polarization and dielectric constant. These effective charges turned out to be of the order of $0.1-0.6$, which is considered reasonable for a definitely non-ionic crystal such as $BaTiO_3$. The model allows a qualitative estimate of the optical dielectric constant in agreement with the experimental results.

### The Importance of Homopolar Forces

A purely qualitative interpretation of the ferroelectric phenomenon in $BaTiO_3$ and other crystals was discussed by Megaw (M 24), (M 22), who greatly emphasized the importance of covalent bond character in the occurrence of ferroelectricity. According to this picture, the oxygen octahedra are linked by their corners in such a way as to satisfy as well as possible both the bond-angle requirements of the oxygen and the packing requirements of the other cations. The occurrence of spontaneous polarization is connected with a change in the covalent bond system of the crystal.

There is no doubt that the bond character of the constituent atoms plays an important role in the ferroelectric phenomenon, in addition to the dipole–dipole interactions described, e.g., by Slater. The difficulty with Megaw's idea, however, is that the origin of ferroelectricity is sought in "abrupt" changes of the bond character at each transition (S 7). No reasonable physical explanation for such a phenomenon seems possible at the present stage, especially because the phenomenological theory implies a gradual and continuous change of the pertinent physical quantities through the transitions.

It may be mentioned that the importance of covalent bondings in perovskite-type ferroelectrics was also emphasized by Smolenskii (S 20). A concise survey of the model theories of $BaTiO_3$ proposed by the Russian workers can be found in a review article of Smolenskii (S 21).

### 12. Growth and Preparation of Single Crystals of $BaTiO_3$

A considerable amount of work has been done in the past several years on the problem of growing large crystals of $BaTiO_3$. Since most of the methods developed for this purpose involve the use of the polycrystalline compound

$BaTiO_3$ as a starting material, it is worth devoting a few words to the problem of its preparation.

An extensive and useful study of the phase-diagram in the system $BaO–TiO_2$ was carried out by Rase and Roy (R 6), revising the previous results of Statton (S 11). Probably the most important information obtained from the phase diagram of Rase and Roy is the fact that $BaTiO_3$ transforms from the cubic to a hexagonal phase at 1460 °C, before melting at 1618 °C (see also Wood (W 7)). The phase change at 1460 °C is very sluggish and thus the hexagonal phase can be quenched to room temperature by rapid cooling through 1460 °C. This fact explains the occurrence of hexagonal crystals in some of the crystal-growing attempts reported in the literature and should be kept in mind when crystals are grown from the melt. The hexagonal modification of $BaTiO_3$ is not ferroelectric. Its structure was studied by Burbank and Evans (B 9), who revealed a framework of $TiO_6$ octahedra linked in quite a different way then in cubic $BaTiO_3$.

The best method for the synthesis of fine-crystalline $BaTiO_3$ involves a solid state reaction between stoichiometric quantities of BaO and $TiO_2$ at temperatures lower than 1460 °C. It should also be pointed out that an aqueous synthesis of $BaTiO_3$ was reported by Flaschen (F 7), who obtained a very fine powder (1–5 $\mu$ particle size) from a mixture of a dilute propyl alcohol solution of a titanium ester and a basic solution of barium hydroxide. Reaction above 80 °C yielded directly a crystalline product, reaction below this temperature required aging of the precipitate for a few hours.

Let us now briefly review the methods and techniques that have been employed for the growth of $BaTiO_3$ single crystals:

(i) Growth from the melt (V 8), (K 15). Both the Stockbarger and the flame-fusion process were applied with limited success. Large single crystals could indeed be obtained but they were generally badly twinned and highly strained.

(ii) Hydrothermal synthesis (R 7). In spite of numerous (mostly unpublished) attempts to obtain large single crystals of $BaTiO_3$ under hydrothermal conditions, no satisfactory degree of success has yet been attained with this method.

(iii) Growth from binary (or ternary) melts. Generally speaking, this is the most satisfactory method of growth to date. A number of salts have been found to be good solvents for $BaTiO_3$ in their liquid phases. Since some of these salts have fairly low melting points, crystals of $BaTiO_3$ can be obtained from these melts at comparatively low temperatures, in much the same way as water soluble crystals are grown from aqueous solutions. The solvent is often referred to as the "flux". For this method, it is important to know the phase diagram of the system of $BaTiO_3$ and the flux to be used. A very extensive investigation of melting point diagrams involving $BaTiO_3$ and twenty-six different fluxes was carried out by Sholokhovich and Belyaev (S 12), (S 13). The most successful fluxes are, in order of increasing usefulness, sodium and potassium carbonates, barium chloride and potassium fluoride. A fairly complete critical survey of the methods involving carbonates and barium chloride has been published by Benes et al. (B 10) (See also the more general review article by Novosil'tsev et al. (N 5)).

The use of various mixtures of $Na_2CO_3$ and $K_2CO_3$ led only occasionally to limited success (B 11), (B 12), (B 13), (N 3). $BaTiO_3$ crystals with linear dimensions ranging from 0.01 to 2 mm were obtained by Kay (K 16) starting from reacted $BaTiO_3$ powder. Results were not satisfactory when a mixture of the constituents (BaO or $BaCO_3$ and $TiO_2$) was used in place of the compound $BaTiO_3$. The choice of crucibles has some bearing on the results owing to possible inclusion of crucible material as an impurity in the crystals obtained. The quality of the end product may also be affected by the surrounding atmosphere during the growth process (K 17). It can be stated, in general, that the process of growing $BaTiO_3$ crystal from melts containing sodium and potassium carbonate is very slow (order of days) and poorly reproducible. The crystals obtained are rather small and often impure.

The use of barium chloride as a flux presents a number of advantages with respect to the carbonates mentioned above. The growing process is somewhat faster than that described above (order of hours), and larger, more uniform crystals can be obtained than in the previous case. The first successful growth of $BaTiO_3$ crystals was accomplished by Blattner et al. (B 5) in 1947 by dissolving $BaCO_3$ and $TiO_2$ in $BaCl_2$ and slowly cooling from above 1000 °C. It was ascertained that $TiO_2$ dissolves well in $BaCl_2$ only when $BaCO_3$ is present, which led to the use of an excess of $BaCO_3$ over $TiO_2$ in the melt (M 18), (M 19), (M 20). Crystals up to 5 mm in linear dimensions were occasionally grown in this way (M 18) but more often the largest dimension of the crystals obtained in earlier attempts was of the order of 2 mm (S 14), (B 14). An important consideration to take into account when growing crystals of $BaTiO_3$ from barium chloride (or generally from a flux) is that the flux evaporates at high temperatures, so that the molar ratios of the components in the melt change toward lower values of the flux content. The most successful recipe for the growth of $BaTiO_3$ crystals from melts of $BaCl_2$ and excess $BaCO_3$ is that reported by Cherepanov (C 14). A mixture of $BaCl_2$, $BaCO_3$ and $TiO_2$ in molar ratios 1 : 1.4 : 1 is heated in an open corundum crucible to 1300–1450 °C in 4–5 hr, soaked for 0.5–3 hr, and cooled at a rate of 20–100 °C/hr to approximately 1000–1200 °C, at which temperature the melt is decanted. The crystals remaining in the crucible are then cooled to 800 °C at the rate of 50 °/hr. According to Cherepanov, this procedure yields thin triangular plates with the hypotenuse as long as 8 mm. It should be pointed out here that one of the most essential points in the above procedure is the decanting of the liquid flux at high temperatures in order to avoid cracking or strains in the crystals grown. This technique must be credited to Remeika, who first applied it to his most successful method to be described below. A final remark about the use of $BaCl_2$ as a flux is that the latter is best used in conjunction with the constituents of $BaTiO_3$ but it can also be used with the reacted polycrystalline compound. In this case, the method is rather slow (the soaking periods are of the order of 20 hr at 1150 °C and the cooling cycles may involve several days) but crystals of about 2 mm in linear dimensions can be obtained in this way (K 16). The phase diagram of the system $BaTiO_3$–$BaCl_2$, useful for this type of growth, was investigated by Rase and Roy (R 8).

By far the most successful growing technique for $BaTiO_3$ crystals is that developed by Remeika (R 9) in 1954, involving the use of potassium fluoride as a flux. A typical procedure consists in starting from a mixture containing approximately 30% of reacted $BaTiO_3$, 0.02% $Fe_2O_3$ and the balance of KF, placed in a platinum crucible in the above order. The covered crucible is introduced into a furnace at 1150–1200 °C, soaked for approximately 8 hr and then cooled at a slow rate to 850–900 °C. The still liquid flux is poured off at this temperature and the crystals are annealed by slowly cooling down to room temperature. The small addition of $Fe_2O_3$, which is not essential to the success of the growing procedure, is provided in order to reduce the conductivity of the crystals due to loss of oxygen at high temperatures. The essential points in Remeika's method are the use of KF as the flux and the trick of decanting the liquid flux before it solidifies.

The procedure described above generally yields crystal plates of triangular shape with hypotenuse as long as 2–3 cm and thickness varying between a few-tenths of a millimeter to a few millimeters, depending on the cooling rate. The hypothenuse of these triangular plates is parallel to a pseudo-cubic $\langle 110 \rangle$ direction and the sides are parallel to the pseudocubic $\langle 100 \rangle$ directions. A characteristic feature of the crystals grown by Remeika's method is that the triangular plates often occur in pairs which are joined along the common hypotenuse at an acute angle of approximately 39°. These are twinning configurations which are normally termed "butterfly wings". The crystallographic features of these butterfly twin crystals were studied by White (W 8) and by Curien and Le Corre (C 16). Usually, a narrow well-developed face is present along the junction of the wing plates. This face is a (100) face which is connected to the plates through two (111) twinning planes. Occasionally, crystals are found which have the plates joined at an obtuse angle (about 109°). Crystals of the latter type possess a sharp edge along the junction and have only a single (111) twinning plane (which is essentially a hexagonal stacking fault (D 17)). These butterfly twins conform to the established twinning laws for cubic crystals, but the reasons for the precise form of twins observed are not clear.

The phase diagram of the system $BaTiO_3$–KF was investigated by Karan and Skinner (K 18). The effect of temperature on the habit of the crystals obtained by Remeika's method was studied by Eustache (E 2). The results of Eustache's experiments show that when crystallization starts above 1000 °C, one obtains the butterfly twins described above; when crystallization starts between 1000 °C and 950 °C, one obtains small square plates; when crystallization starts below 950 °C, one obtains transparent cubes and parallelepipeds (see also Novosil'tsev and Khodakov (N 4)). A very detailed study of the peculiarities of Remeika's method and its results has been published recently by De Vries (D 17), with particular attention to the dentritic forms often obtained with this technique.

In conclusion, it should be emphasized that the most important characteristic of the method as described by Remeika is that one directly obtains plates, whose thickness can be controlled by varying the cooling rate in the growing process. In many respects this is an advantage, as it would otherwise be quite laborious to cut thin plates out of a larger single crystal of $BaTiO_3$. When very thin plates are desired, the technique described by Last (L 9) can be applied, which consists in etching the plates in phosphoric acid at 130 °C at the rate of 1 $\mu$/min.

The plates as grown possess rather large $c$ and $a$ domains, if care is taken to anneal the crystal properly. More often, however, they are multidomain crystals. They can be converted into a single $c$ domain by a poling process in which the plates are first etched in phosphoric acid at about 140 °C, then washed in water and immersed in glycerol between two plate electrodes. The system is then heated above the Curie temperature and a d.c. field applied and maintained while decreasing the temperature through the Curie point. Another successful poling technique which does not require heating of the crystal has been described by Campbell (C 21). Fang *et al.* (F 8) have reported that the same domain conversion can be accomplished if, instead of being heated above the Curie point, the plate is cooled below 0 °C and the temperature increased through the orthorhombic-tetragonal transition. Furthermore, either $c$ or $a$ domains can be obtained depending on the liquid in which the crystals are immersed. Liquids with large dielectric constants (or the order of 40) lead to complete $c$ domain conversion, and liquids with small dielectric constants (of the order of 5) lead to complete $a$ domain conversion. The influence of the dielectric constant of the imbedding liquid is understood by considering the direction of the lines of force of the dielectric displacement vector entering the $BaTiO_3$ plate as a consequence of the applied d.c. field.

### 13. Physical Properties of Ceramic $BaTiO_3$

In the preceding discussion of $BaTiO_3$ we have been concerned primarily with the properties of single crystals, and we have referred only very briefly to some results obtained with ceramic materials. Clearly, the measurements of many physical quantities are more meaningful and give more reliable results when carried out on single crystals. In particular, the anisotropic character of many quantities in the polar phases can only be detected on single-domain specimens.

Nevertheless, a very considerable amount of work has been and is being done on polycrystalline specimens ("ceramic") of $BaTiO_3$ and its modifications. One reason for this fact is historical, owing to the difficulties encountered earlier in the preparation of single crystals. The predominant reason for these studies, however, stems from the many applications of ceramic $BaTiO_3$ in various fields of engineering. A hard ceramic body can be easily prepared by way of standard sintering procedures at temperatures of the order of 1350–1450 °C and the shape of the ceramic body can be easily modified to suit a number of practical requirements. Under suitable conditions, the ceramics can be "polarized" in any given direction by the application of a strong d.c. field. These ceramics can then be used as electromechanical transducers in a number of applications.

In the present section, we will focus our attention on the relationships between the physical properties of (non-polarized) ceramics and those of single crystals of $BaTiO_3$. We will be concerned with such problems as the calculation of the dielectric, elastic and piezoelectric properties of the ceramic material from the corresponding properties of the single crystal. The characteristics of some solid solutions containing $BaTiO_3$ will be described in the following chapter. A good survey of the work done prior to 1956 on ceramic $BaTiO_3$ and some of its solid solutions can be found in a book by Sachse (S 1).

In comparing the properties of ceramics with those of single crystals we will meet some which are, obviously, almost identical in the two cases and some which are not. Among the former we might include the transition temperatures, the specific heat, the lattice constants and, generally, most of the physical properties measured in the cubic phase, such as the Curie constant (see Section 2). It should be noted, however, that most of the result obtained with ceramic samples must be corrected for porosity before they can be considered representative of the material investigated. It is evident, for example, that the dielectric constant actually measured on a ceramic specimen will be a function of the number, the shapes and the dielectric constant of the pores. Correction formulae have been worked out, and tested experimentally, in this case, by Rushman and Strivens (R 10). Roberts (R 11) has shown, for the cases of $BaTiO_3$ and $PbZrO_3$, that the Curie constant is also a function of porosity.

The fundamental problem is then that of describing the properties of $BaTiO_3$ ceramics below the Curie temperature in terms of the physical constants of single crystals. This is not an easy task, because we know that the latter constants are highly anisotropic in the polar phases. The ceramic bodies are aggregates of crystallites which, in turn, may consist of many domains. We have little knowledge of the forces that are at work among these crystallites and of the mechanical strains created around the boundaries. Although there is no doubt that the properties of ceramic materials reflect some kind of averaging of the constituents crystallites, the averaging process can be made in many ways and, moreover, is complicated by the nature of the interactions among crystallites. Further complications occur when polarized ceramics are being considered, because such ceramics show anisotropic character of the dielectric, elastic and piezoelectric properties.

### Physical Constants of Non-polarized Ceramics

Let us first consider the dielectric constant of unpolarized ceramics in the tetragonal phase. Formulae for the dielectric constant of an aggregate of non-piezoelectric crystallites were reported by Bruggeman (B 15). Various conditions of aggregation were assumed, and it was shown that both the assumptions of spherical and laminar crystallites lead to the same average value of the dielectric constant. Using Bruggeman's formula, and assuming the values $\varepsilon_a = 4500$ and $\varepsilon_c = 250$ for the dielectric constants of single crystals, Marutake (M 26) computed the value $\varepsilon = 2500$ for the dielectric constant of ceramic $BaTiO_3$. This is definitely larger than the experimental values, which range from 1400 to 2000 (see, e.g., Sachse (S 1)). The reason for this discrepancy must be sought in the fact that $BaTiO_3$ is piezoelectric below the Curie point and hence the physical constants of ceramics cannot be computed properly without taking into account the piezoelectric interactions.

A detailed treatment of this problem was given by Marutake (M 26), who made the following simplifying assumptions for the calculations: (i) the shape of each crystallite is spherical, and (ii) each crystallite consists of a single domain, so that the motions of the domain walls can be neglected.

A given spherical crystallite is considered to be immersed in an isotropic medium which has the same dielectric constant as that of the ceramic. As the dielectric constant of the crystallite is highly anisotropic, the electric field acting in the interior of the little sphere is different from that acting in the surrounding medium, which is equal to the applied field. The relation between these fields is computed in terms of the polarization present inside and outside the crystallite, where both the dielectric and the piezoelectric contributions to the polarizations are considered. Knowing the above relations and using the fundamental piezo-electric equations of tetragonal $BaTiO_3$, Marutake was able to compute the average dielectric constant of the ceramic with the result: $\varepsilon = 1500$, which is in excellent agreement with the observed values (see Table IV-8). It was emphasized that the large value of the piezoelectric modulus $d_{15}$ of single crystals plays the major role in reducing the value of the dielectric constant of ceramic materials.

The agreement between observed and calculated values should, however, be considered with some reserve. In the first place, the observed value of the dielectric constant of ceramic $BaTiO_3$ is not a very well defined quantity, as it depends upon a number of factors, such as the pressure applied before firing, the sintering temperature, etc. In the second place, the assumption of spherical crystallites appears to be quite unrealistic, although any other assumption might have made the calculations very difficult. The domain structure at the surface of $BaTiO_3$ ceramics was studied by Kulcsar (K 19), Cook (C 17), De Vries and Burke (D 12) and Tennery and Anderson (T 8). The polished and etched surfaces of dense ceramic bodies were observed optically and with an electron microscope. The domain structure was found to consist of thin laminar layers of 90° domains with thickness ranging from 0.5 to 2.0 $\mu$, and irregularly shaped 180° domains. Although the domain structure observed is often extremely complicated, it seems to suggest that the assumption of thin laminar crystallites may be closer to reality than that of spherical grains.

It appears, nevertheless, that Marutake's approach is a very important step toward the success of the problem mentioned above. A procedure similar to that followed for the calculation of the dielectric constant was also applied to the treatment of elastic constants of the ceramic by Marutake (M 26). The theoretical results are in satisfactory agreement with the observed values when the porosity is vanishingly small, but the dependence of the physical properties on the porosity shows much less satisfactory agreement with the experimental results. This is certainly due to the assumptions that had to be made about the shapes of the pores.

There are also other characteristics of ceramic $BaTiO_3$ which, at present, cannot be explained properly with Marutake's approach. It is known, for example, that $BaTiO_3$ ceramics exhibit anomalously large values of the dielectric constant (about 3000) when the grain size is made smaller than about 1 $\mu$ (K 20), (E 3). Also, in this case, the hysteresis curve observed at room temperature appears to be less non-linear, and elastic anomalies are expected as well (M 27). In this case the explanation must be sought in the anomalies exhibited by the dielectric and elastic properties of the crystallites themselves when their size is reduced, as suggested by the work of Känzig and co-workers (A 1), (K 4) (see Section 9).

### Properties of Polarized Ceramics

For many applications, $BaTiO_3$ ceramics are generally used after having been polarized by a d.c. biasing field of suitable magnitude and direction. It was first reported by Roberts (R 12) and later confirmed by Mason (M 28), (M 7) that such polarized ceramics exhibit a marked piezoelectric effect. Subsequently, a number of other investigators studied the dielectric and piezoelectric properties of $BaTiO_3$ ceramics either under the direct influence of an electric d.c. field or after removal of the field (M 29), (B 16), (P 4). While the non-polarized ceramic is isotropic and, therefore, non-piezoelectric, the polarized ceramic has, in the Schönflies notation, the symmetry $C_n$ with $n = \infty$ (M 7), and thus is expected to exhibit anisotropy of the dielectric and elastic properties as well as a piezoelectric effect.*

These properties were confirmed experimentally by a number of authors. Mason (M 7) demonstrated the hysteresis effects of the resonance and antiresonance frequencies as functions of the applied d.c. field (see Fig. IV-49). A similar hysteresis effect was observed by Takagi and Sawaguchi (T 9) in the dimensional change caused by the application of the d.c. field. It was also found that the d.c. bias causes a marked decrease of the dielectric constant measured with small a.c. field (R 12), (M 28). After removal of the d.c bias, the dielectric constant was found to become slightly larger than in the unpolarized state. A similar effect was also observed (M 29) in the behavior of the elastic compliance $s^E$.

These phenomena were treated as essentially electrostrictive in character by Mason, who considered the non-biased ceramic as a uniform non-piezoelectric medium in which the polarization is induced by the applied field. Baerwald (B 17) developed a thermodynamic continuum theory along similar lines. This approach is, no doubt, very convenient for the description of the observed phenomena: the analogy with the magnetostrictive effect is quite obvious. This type of treatment leads generally to results which are good approximations to the experimental

---

* A detailed study of the crystallographic and physical properties of piezoelectric ceramics has been published by Shubnikov (S 22). The point group of polarized ceramics such as those of $BaTiO_3$ was defined as $(\infty \cdot m)$, implying the existence of an $\infty$-fold rotation axis and of an infinite number of longitudinal mirror planes parallel to this axis. Choosing a co-ordinate system such that the $x_3$ axis is identified with the $\infty$-fold rotation axis, the piezoelectric tensor of such a material was calculated as:

$$\begin{matrix} 0 & 0 & 0 & 0 & d_{15} & 0 \\ 0 & 0 & 0 & d_{15} & 0 & 0 \\ d_{31} & d_{31} & d_{33} & 0 & 0 & 0 \end{matrix}.$$

It may also be noted that such a material should exhibit a piezoelectric polarization $P_3$ when subjected to a hydrostatic pressure $p$:

$$P_3 = (2d_{31} + d_{33})p.$$

facts. However, it should be kept in mind that the thermodynamic variables describing the ceramic are not necessarily unique, as different domain configurations may exhibit the same polarization, but otherwise different properties.

An excellent characterization of the "piezoelectric ceramics" can be found in a concise article by Jaffe (J 5). The polarization of such ceramics can be measured, e.g. ballistically upon heating them through the Curie temperature. The results range between 5 and 8 × $10^{-6}$ C/cm², i.e. remarkably smaller than the value of

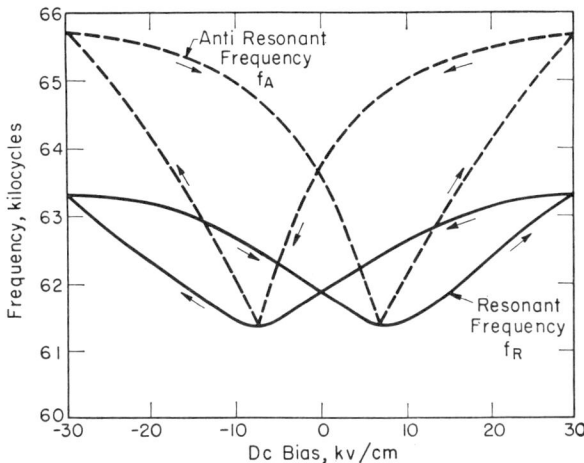

FIG. IV-49. Resonant and antiresonant frequencies for a circular disk of ceramic BaTiO₃ in radial vibration (according to Mason (M 7)).

single crystals at room temperature ($P_s = 26 \times 10^{-6}$ C/cm²). This implies that, although the poling process definitely alters the relative orientations of domains, only a comparatively small number of them remain aligned thereafter. A straight averaging process would lead one to expect the value $\langle P_s \rangle_{\mathrm{av.}} = 0.84\,P_s$, if both 180° and 90° reorientations were allowed, and the value $\langle P_s \rangle_{\mathrm{av.}} = 0.50\,P_s$, if only 180° reversals are permitted. The spontaneous polarization of ceramic BaTiO₃, on the other hand, turns out to be only about 5 × $10^{-6}$ C/cm² when measured at 60 c/s from the hysteresis loops. This is not too surprising, as we know that even single crystals, when of inferior quality, used to exhibit rather small spontaneous polarizations (15–18 × $10^{-6}$ C/cm²).

The values of the piezoelectric moduli of polarized ceramics are, on the other hand, quite large. As shown in Table IV-8, the moduli $d_{31}$ and $d_{33}$ have the same order of magnitude as the corresponding moduli of single crystals. The anisotropy of the dielectric and elastic properties was measured by Bechmann (B 18), Moseley (M 30) and Marutake and Ikeda (M 29) and was generally found to be rather small. For the dielectric constants, for example:

$$\varepsilon_3/\varepsilon_0 = 1.12; \quad \varepsilon_1/\varepsilon_0 = 1.03,$$

where $\varepsilon_3$ is the value measured in the direction of the poling field, $\varepsilon_1$ that perpendicular to it, and $\varepsilon_0$ is the value of the unpolarized ceramic.

We conclude from this that the dielectric constant of polarized ceramics is from 10% to 15% larger than that of unpolarized samples. This is, of course, not the change that one would expect if the effect of the poling process were merely that of aligning more $c$ domains parallel to the field. In fact, in such a case, the dielectric constant should decrease. However, if we assume that the main effect of the poling field, in reversing a number of 180° domains, is that of relieving at least part of the clamping to which the crystallites are subjected, then an increase in dielectric constant is expected. Such an effect is reminiscent of the domain clamping effect demonstrated by Drougard and Young (D 9) and discussed in Section 5.

Let us now study the quantitative aspects of the problem. The first question to consider is this: what is the distribution of domain orientations under the effect of the poling field? Mason (M 28) estimated that approximately 10% of the domains undergo a 90° rotation under the influence of a field of 30 kV/cm. Subbarao *et al.* (S 16) reached the figure of 12% for the same quantity. Berlincourt and Krueger (B 22) estimated that 90° reorientation amounts to 17% during poling and 12% after removal of the field. This figure should be considered more reliable than the large fraction of 80% reported by Danielson (D 13) on the basis of X-ray intensity data. Thus, the effect of the poling field is mainly to cause 180° reversals, while most of the 90° domains remain unaffected. These conclusions were also qualitatively confirmed by the direct observation of domain structures of polarized ceramics by De Vries and Burke (D 12).

The first step toward finding the relationship between the properties of ceramics and those of single crystals consists again in an averaging procedure which assumes a certain distribution of domain orientations but disregards the interactions between the crystallites. Calculations of this kind were carried out by a number of Russian workers for the dielectric, piezoelectric and elastic properties of ceramic $BaTiO_3$ in all three ferroelectric phases (B 19), (B 20), (K 21), (S 17). We have already seen in the case of unpolarized ceramics that, although reasonable success may occasionally be achieved, the assumption of an ideal polycrystal may be quite inadequate when the piezoelectric interactions cannot be ignored, as is the case in materials such as $BaTiO_3$.

Marutake and Ikeda (M 26), (M 29) were able to prove that the anisotropy of polarized ceramics is, in fact, mainly due to the piezoelectric interactions among the crystallites. The latter were again assumed to be spherical in shape, and permanent orientation of 90° domains by the poling field was fully neglected. In spite of these oversimplified assumptions, the calculated values of the physical constants of ceramics turn out to be in reasonably good agreement with the observed ones. Table IV-8 shows the results for the piezoelectric moduli, where $\langle d \rangle_{\mathrm{av.}}$ represents the result of the straight averaging process. Here again, as in the case of the dielectric constant, the comparatively large values of $d_{11}$ and $d_{33}$ are due mainly to the large value of $d_{15}$ in single crystals. The anisotropy of the dielectric constant was calculated as:

$$\varepsilon_3/\varepsilon_0 = 1.10; \quad \varepsilon_1/\varepsilon_0 = 1.04,$$

in excellent agreement with the experimental data.

In conclusion, we can again say that Marutake's approach can explain several of the effects observed in ceramic $BaTiO_3$, in spite of the fact that the basic assumptions are rather crude and the agreement with the experiment is not always quantitative. It should be noted that both Marutake's and the Russian authors' concepts are in contrast to Mason's idea, in that the former ascribe to the ceramic "average properties" of the single crystal while the latter considers the ceramic as a uniform electrostrictive medium.

TABLE IV-8. DIELECTRIC AND PIEZOELECTRIC PROPERTIES OF POLYCRYSTALLINE $BaTiO_3$
(ACCORDING TO MARUTAKE (M 26))

| Dielectric constant, unpolarized ceramic | | | | |
|---|---|---|---|---|
| $\langle \varepsilon \rangle_{av.}$ | $1/\langle 1/\varepsilon \rangle_{av.}$ | Bruggeman's formula (B 15) | Marutake (M 26) | Observed |
| 3100 | 680 | 2500 | 1500 | 1600 |

Piezoelectric moduli ($10^{-6}$ c.g.s. units), polarized ceramic
The values in parentheses are quoted by Jaffe (J 5)

| | $\langle d \rangle_{av.}$ | Marutake (M 26) | Observed | Single crystal value used for calculations (M 26) |
|---|---|---|---|---|
| $d_{31}$ | $-3.8$ | $-2.1$ | $-2.0\,(-2.3)$ | $-2.0$ |
| $d_{33}$ | $8.5$ | $5.1$ | $5.4\,(5.7)$ | $6.0$ |
| $d_{15}$ | $9.5$ | $6.0$ | $8.0$ | $30.0$ |

It should also be mentioned that McQuarrie (M 31) arrived at a picture of the ceramic similar to that of Marutake by an independent method. In McQuarrie's picture, a small contribution of the 90° domains to the small-signal dielectric constant is allowed. The poling process is visualized as the removal of the 180° walls and a temporary orientation of the 90° domains. Most of the latter return to the original configuration after removal of the field, but it is probable that a small fraction remains oriented in the direction of the field. This seems to be substantiated by the dilatometric experiments of Subbarao et al. (S 16). These authors measured the thermal expansion of polarized and unpolarized ceramics through the Curie temperature and found that the polarized specimens exhibit a larger contraction at the transition than the unpolarized ones. However, the amount of 90° orientation may not be more than about 5%.

## The Aging Effect

It has been shown that the piezoelectric moduli, the dielectric and the elastic constants of polarized ceramics decrease gradually with time over a period of years. This so-called aging effect occurs also, to a lesser extent, in the dielectric properties of unpolarized ceramics. The first systematic investigations of these effects were carried out by Mason (M 32) and by McQuarrie and Buessem (M 33). It was clearly demonstrated that aging occurs only in the ferroelectric phase.

This suggests that processes of domain reorientation play a major role in the aging phenomenon. Mason attributed the effect to a slow temperature-induced motion of domain walls. McQuarrie (M 31) emphasized the diminishing contribution of 90° walls in time. Marutake and Ikeda (M 29), on the other hand, called upon the changes in piezoelectric interactions between crystallites to explain the changes in dielectric and elastic properties. The piezoelectric interactions may account for part of the aging effect, but other causes must also be invoked in the case of unpolarized ceramics. It is conceivable, for example, that the sample may be inhomogeneously strained when it is first cooled through the Curie temperature, and aging may be due to a slow relief of the strains with time. More experimentation is needed in order to clear up this problem.

Sawaguchi and Charter (S 18) studied the aging effect in ceramic samples of $(Pb_{0.5}, Ca_{0.5})TiO_3$, whose Curie point lies in the vicinity of room temperature. They found that the effect could be explained by the slow separation of the solid solution into two phases in the course of time. This, however, is only a special case of the aging problem. Aging was also observed in thin films (about $1 \mu$ thick) by Feldman (F 9). In this case, again, domain reorientations were considered responsible for most of the observed effects.

The above discussion of ceramic $BaTiO_3$ is, of course, not intended to be complete. There are many important problems in this field that we did not treat here. For the most recent survey of the properties and application of ferroelectric ceramics, the reader is referred to an article by Mason (M 34).

## BIBLIOGRAPHY

(A 1)  ANLIKER, M., BRUGGER, H. R. and KÄNZIG, W., *Helv. Phys. Acta* **27**, 99 (1954).
(A 2)  ABE, R., *J. Phys. Soc. Japan* **14**, 633 (1959).
(A 3)  AREND, H., *Czechoslov. J. Phys.* **9**, 124 (1959).
(A 4)  ANDERSON, J. R., BRADY, G. W., MERZ, W. J. and REMEIKA, J. P., *J. Appl. Phys.* **26**, 1387 (1955).
(A 5)  ABE, R., *J. Phys. Soc. Japan* **15**, 795 (1960).
(B 1)  BOND, W. L., MASON, W. P. and McSKIMIN, H. J., *Phys. Rev.* **82**, 442 (1951).
(B 2)  BENEDICT, T. S. and DURAND, J. L., *Phys. Rev.* **109**, 1091 (1958).
(B 3)  BUSCH, G., FLURY, H. and MERZ, W., *Helv. Phys. Acta* **21**, 212 (1948).
(B 4)  BLATTNER, H., KÄNZIG, W. and MERZ, W., *Helv. Phys. Acta* **22**, 35 (1949).
(B 5)  BLATTNER, H., MATTHIAS, B. and MERZ, W., *Helv. Phys. Acta* **20**, 225 (1947).
(B 6)  BLOKHIN, M. A., *Doklady Akad. Nauk S.S.S.R.* **95**, 965 (1954).
(B 7)  BLOKHIN, M. A., *Doklady Akad. Nauk S.S.S.R.* **95**, 1165 (1954).
(B 8)  BLATTNER, H., KÄNZIG, W., MERZ, W. and SUTTER, H., *Helv. Phys. Acta* **21**, 207 (1948).
(B 9)  BURBANK, R. D. and EVANS, H. T., JR., *Acta Cryst.* **1**, 330 (1948).
(B 10) BENES. J., BEDNAROVA, V. and SAFRATA, S., *Czechoslov. J. Phys.* **6**, 56 (1956).
(B 11) BELYAEV, I. N., NOVOSIL'TSEV, N. S., KHODAKOV, A. L. and FESENKO, E. G., *Doklady Akad. Nauk S.S.S.R.* **78**, 875 (1951).
(B 12) BELYAEV, I. N. and SHOLOKHOVICH, M. L., *Doklady Akad. Nauk S.S.S.R.* **77**, 51 (1951).
(B 13) BELYAEV, I. N., NOVOSIL'TSEV, N. S., FESENKO, E. G. and KHODAKOV, A. L., *Zhur. Eksptl. i Teoret. Fiz.* **23**, 211 (1952).

(B 14) BELYAEV, I. N. and SHOLOKHOVICH, M. L., *Zhur. Priklad. Khim.* **25**, 818 (1952).
(B 15) BRUGGEMAN, D. A. G., *Z. Physik* **92**, 561 (1934).
(B 16) BERLINCOURT, D. and KRUEGER, H. H. A., *Phys. Rev.* **105**, 56 (1957).
(B 17) BAERWALD, H. G., *Phys. Rev.* **105**, 480 (1957).
(B 18) BECHMANN, R., *J. Acoust. Soc. Am.* **28**, 347 (1956).
(B 19) BOGDANOV, S. V., WUL, B. M. and TIMONIN, M. M., *Izvest. Akad. Nauk S.S.S.R., Ser. Fiz.* **21**, 374 (1957).
(B 20) BOGDANOV, S. V., WUL, B. M. and RAZBASH, R., YA. *Kristallografiya* **2**, 115 (1957).
(B 21) BERLINCOURT, D. and JAFFE, H., *Phys. Rev.* **111**, 143 (1958).
(B 22) BERLINCOURT, D. and KRUEGER, H. A., *J. Appl. Phys.* **30**, 1804 (1959).
(B 23) BURFOOT, J. C., *Proc. Phys. Soc. (London)* **73**, 641 (1959).
(B 24) BURFOOT, J. C. and PEACOCK, R. V., *Proc. Phys. Soc. (London)* **73**, 973 (1959).
(C 1) CHYNOWETH, A. G., *J. Appl. Phys.* **27**, 78 (1956).
(C 2) CADY, W. G., *Piezoelectricity*, McGraw-Hill, New York (1946).
(C 3) CROSS, L. E., *Phil. Mag.* **44**, 1161 (1953).
(C 4) CASPARI, M. E. and MERZ, W. J., *Phys. Rev.* **80**, 1082 (1950).
(C 5) COOK, W. R., JR., cited by D. BERLINCOURT and H. JAFFE, *Phys. Rev.* **111**, 143 (1958).
(C 6) CROSS, L. E., DENNISON, A. T. and NICHOLSON, M. M., *Proc. Leeds Phil. Lit. Soc., Sci. Sect.* **5**, 199 (1949).
(C 7) CAMPBELL, D. S., *Phil. Mag.* **46**, 1261 (1955).
(C 8) CAMPBELL, D. S. and STIRLAND, D. J., *Brit. J. Appl. Phys.* **7**, 62 (1956).
(C 9) CAMERON, D. P., *IBM J. Res. Developm.* **1**, 2 (1957).
(C 10) COHEN, M. H., *Phys. Rev.* **84**, 369 (1951); **84**, 368 (1951).
(C 11) CHYNOWETH, A. G., *Phys. Rev.* **110**, 1316 (1958).
(C 12) CHYNOWETH, A. G., *J. Appl. Phys.* **30**, 280 (1959).
(C 13) CHYNOWETH, A. G., *Phys. Rev.* **102**, 705 (1956).
(C 14) CHEREPANOV, A. M., *J. Tech. Phys. U.S.S.R.* **27**, 2280 (1957).
(C 15) COTTS, R. M. and KNIGHT, W. D., *Phys. Rev.* **96**, 1285 (1954).
(C 16) CURIEN, H. and CORRE, Y. LE, *Bull. Soc. Franç. Minéral. et Crist.* **78**, 604 (1955).
(C 17) COOK, W. R., JR., *J. Am. Ceram. Soc.* **39**, 17 (1956).
(C 18) CAMPBELL, D. S., *J. Electronics and Control* **3**, 330 (1957).
(C 19) CASELLA, R. C. and KELLER, S. P., *Phys. Rev.* **116**, 1469 (1959).
(C 20) CHYNOWETH, A. G., *Phys. Rev.* **117**, 1235 (1960).
(C 21) CAMPBELL, D. S. J., *Brit. Inst. Radio Engrs.* **17**, 385 (1957).
(D 1) DROUGARD, M. E. and YOUNG, D. R., *Phys. Rev.* **95**, 1152 (1954).
(D 2) DANIELSON, G. C., MATTHIAS, B. T. and RICHARDSON, J. M., *Phys. Rev.* **74**, 986 (1948).
(D 3) DEVONSHIRE, A. F., *Phil. Mag.* **40**, 1040 (1949).
(D 4) DEVONSHIRE, A. F., *Phil. Mag.* **42**, 1065 (1951).
(D 5) DEVONSHIRE, A. F., Theory of Ferroelectrics, *Phil. Mag. Suppl.* **3**, 85 (1954).
(D 6) DROUGARD, M. E., LANDAUER, R. and YOUNG, D. R., *Phys. Rev.* **98**, 1010 (1955).
(D 7) DROUGARD, M. E., *J. Appl. Phys.* **27**, 1559 (1956).
(D 8) DROUGARD, M. E. and HUIBREGTSE, E. J., *IBM J. Res. Developm.* **1**, 318 (1957).
(D 9) DROUGARD, M. E. and YOUNG, D. R., *Phys. Rev.* **94**, 1561 (1954).
(D 10) DROUGARD, M. E., *J. Appl. Phys.* **31**, 352 (1960).
(D 11) DROUGARD, M. E. and LANDAUER, R., *J. Appl. Phys.* **30**, 1663 (1959).
(D 12) VRIES, R. C. DE and BURKE, J. E., *J. Am. Ceram. Soc.* **40**, 200 (1957).
(D 13) DANIELSON, G. C., *Acta Cryst.* **2**, 90 (1949).
(D 14) DROUGARD, M. E., FUNK, H. L. and YOUNG, D. R., *J. Appl. Phys.* **25**, 1166 (1954).
(D 15) DANNER, H. R., FRAZER, B. C. and PEPINSKY, R., Private communication (June, 1960).
(D 16) DROUGARD, M. E. and SCHLOSSER, H., *J. Appl. Phys.* **32**, 1227 (1961).
(D 17) VRIES, R. C. DE, *J. Am. Ceram. Soc.* **42**, 547 (1959).
(E 1) EVANS, H. T., Techn. Report No. 58, Laboratory for Insulation Research, Massachusetts Institute of Technology (1953).
(E 2) EUSTACHE, H., *Compt. rend.* **244**, 1029 (1957).
(E 3) EGERTON, L. and KOONCE, S. E., *J. Am. Ceram. Soc.* **38**, 412 (1955).

(F 1) FORSBERGH, P. W., JR., *Phys. Rev.* **93**, 686 (1954).
(F 2) FORSBERGH, P. W., JR., *Piezoelectricity, Electrostriction and Ferroelectricity, Handbuch der Physik*, vol. 17, pp. 264–392, Springer-Verlag, Berlin (1956).
(F 3) FRAZER, B. C., DANNER, H. and PEPINSKY, R., *Phys. Rev.* **100**, 745 (1955).
(F 4) FORSBERGH, P. W., JR., *Phys. Rev.* **76**, 1187 (1949).
(F 5) FOUSEK, J. and BREZINA, B., *Czechoslov. J. Phys.* **9**, 265 (1959).
(F 6) FOUSEK, J., *Czechoslov. J. Phys.* **8**, 254 (1958).
(F 7) FLASCHEN, S. S., *J. Am. Chem. Soc.* **77**, 6194 (1955).
(F 8) FANG, P. H., MARZULLO, S. and BROWER, W. S., *Phys. Rev.* **108**, 242 (1957).
(F 9) FELDMANN, C., *Rev. Sci. Instr.* **26**, 463 (1955); *J. Appl. Phys.* **27**, 870 (1956).
(F 10) FATUZZO, E. and MERZ, W. J., *Phys. Rev.* **116**, 61 (1959).
(G 1) GRÄNICHER, H., *Helv. Phys. Acta* **22**, 395 (1949).
(G 2) GRÄNICHER, H., Phys. Inst. E.T.H., Zürich, Private communication (March 1960).
(G 3) GELLER, M., *Phys. Rev.* **101**, 1685 (1956).
(G 4) GINZBURG, V. L., *Zhur. Eksptl. i Teoret. Fiz.* **15**, 739 (1945); *J. Phys. U.S.S.R.* **10**, 107 (1946).
(G 5) GINZBURG, V. L., *Zhur. Eksptl. i Teoret. Fiz.* **19**, 36 (1949).
(H 1) HUIBREGTSE, E. J. and YOUNG, D. R., *Phys. Rev.* **103**, 1705 (1956).
(H 2) HUIBREGTSE, E. J., DROUGARD, M. E. and YOUNG, D. R., *Phys. Rev.* **98**, 1562 (1956).
(H 3) HUIBREGTSE, E. J., BESSEY, W. H. and DROUGARD, M. E., *J. Appl. Phys.* **30**, 899 (1959).
(H 4) HILSUM, C., *J. Opt. Soc. Am.* **45**, 771 (1955).
(H 5) HORIE, T., KAWABE, K. and SAWADA, S., *J. Phys. Soc. Japan* **8**, 823 (1954).
(H 6) HORNIG, A. W., JAYNES, E. T. and WEAVER, H. E., *Phys. Rev.* **96**, 1703 (1954).
(H 7) HORNIG, A. W., REMPEL, R. C. and WAEVER, H. E., *Phys. Rev. Letters* **1**, 284 (1958).
(H 8) HOOTON, J. A. and MERZ, W. J., *Phys. Rev.* **98**, 409 (1955).
(H 9) HUSIMI, K. and KATAOKA, K., *J. Appl. Phys.* **29**, 1247 (1958); *Rev. Sci. Instr.* **31**, 418 (1960).
(H 10) HUSIMI, K., *J. Appl. Phys.* **29**, 1379 (1958).
(H 11) HARMAN, G. G., *Phys. Rev.* **111**, 27 (1958).
(H 12) HAGEDORN, R., *Z. Physik* **133**, 394 (1952).
(H 13) HORNIG, A. W., REMPEL, R. C. and WEAVER, H. E., *J. Phys. Chem. Solids* **10**, 1 (1959).
(H 14) HUSIMI, K., *J. Phys. Soc. Japan* **15**, 731 (1960).
(J 1) JONA, F. and PEPINSKY, R., *Phys. Rev.* **105**, 861 (1957).
(J 2) JANOVEC, V., *Czechoslov. J. Phys.* **8**, 3 (1958).
(J 3) JAYNES, E. T., *Ferroelectricity*, Princeton University Press (1953).
(J 4) JAYNES, E. T., *Phys. Rev.* **79**, 1008 (1950).
(J 5) JAFFE, H., *J. Am. Ceram. Soc.* **41**, 494 (1958).
(J 6) JANOVEC, V., *Czechoslov. J. Phys.* **9**, 468 (1959).
(K 1) KÄNZIG, W. and MAIKOFF, N., *Helv. Phys. Acta* **24**, 343 (1951).
(K 2) KITTEL, C., *Phys. Rev.* **83**, 458 (1951).
(K 3) KÄNZIG, W. and MEIER, R., *Helv. Phys. Acta* **21**, 585 (1949).
(K 4) KÄNZIG, W., *Phys. Rev.* **98**, 549 (1955).
(K 5) KAY, H. F. and VOUSDEN, P., *Phil. Mag.* **40**, 1019 (1949).
(K 6) KINASE, W. and TAKAHASHI, H., *J. Phys. Soc. Japan* **10**, 942 (1955).
(K 7) KÄNZIG, W., *Ferroelectrics and Antiferroelectrics, Solid State Physics*. vol. 4, pp.1–197, Academic Press, New York (1957).
(K 8) KAY, H. F., *Rep. Progr. in Phys.* **43**, 230 (1955).
(K 9) KÄNZIG, W., *Helv. Phys. Acta* **24**, 175 (1951).
(K 10) KINASE, W., *Progr. Theoret. Phys. (Kyoto)* **13**, 529 (1955).
(K 11) KINASE, W. and TAKAHASHI, H., *J. Phys. Soc. Japan* **12**, 464 (1957).
(K 12) KIBBLEWHITE, A. C., *Proc. Inst. Elec. Engrs. (London)* **B 102**, 59 (1955).
(K 13) KINASE, W., *Busseiron-Kenkyu* **4**, 721 (1958).
(K 14) KIYASU, Z., HUSIMI, K. and KATAOKA, K. J., *Phys. Soc. Japan* **12**, 432 (1957).
(K 15) KREMERS, H. C., WILLIAMS, H. D. and HACSKAYLO, M., Harshaw Chem. Co., Contract No. DA-36-039-sc-15493, Final Report (1 July, 1946–31 October, 1954).

(K 16) KAY, H. F., *Acta Cryst.* **1**, 229 (1948).
(K 17) KAWABE, K. and SAWADA, S., *J. Phys. Soc. Japan* **12**, 218 (1957).
(K 18) KARAN, C. and SKINNER, B., *J. Chem. Phys.* **21**, 2225 (1953); **22**, 957 (1954).
(K 19) KULCSAR, F., *J. Am. Ceram. Soc.* **39**, 13 (1956).
(K 20) KNIEPKAMP, H. and HEYWANG, H., *Z. angew. Phys.* **6**, 385 (1954).
(K 21) KHOLODENKO, L. P. and SHIROBOKOV, M. IA., *Zhur. Tekh. Fiz.* **27**, 929 (1957).
(K 22) KHOLODENKO, L. P., *Zhur. Eksptl. i Teoret. Fiz.* **31**, 244, 1034 (1956).
(K 23) KINASE, W., KOBAYASHI, J. and YAMADA, N., *Phys. Rev.* **116**, 348 (1959).
(K 24) KINASE, W., *Progr. Theoret. Phys.* (*Kyoto*) **22**, 736 (1959).
(L 1) LANDAUER, R., YOUNG, D. R. and DROUGARD, M. E., *J. Appl. Phys.* **27**, 752 (1956).
(L 2) LAST, J. T., *Phys. Rev.* **105**, 1740 (1957).
(L 3) LOW, W. and SHALTIEL, D., *Phys. Rev. Letters* **1**, 51 (1958).
(L 4) LOW, W. and SHALTIEL, D., *Phys. Rev. Letters* **1**, 286 (1958).
(L 5) LITTLE, E. A., *Phys. Rev.* **98**, 978 (1955).
(L 6) LANDAU, L. D. and LIFSHITZ, E. M., *Physik. Z. Sowjetunion* **8**, 153 (1935).
(L 7) LANDAUER, R., *J. Appl. Phys* **28**, 227 (1957).
(L 8) LEFKOWITZ, I. and MITSUI, T., *J. Appl. Phys.* **30**, 269 (1959).
(L 9) LAST, J. T., *Rev. Sci. Instr.* **28**, 720 (1957).
(L 10) LUTTINGER, J. M. and TISZA, L., *Phys. Rev.* **70**, 954 (1946); **72**, 257 (1947).
(L 11) LEFKOWITZ, I., *J. Phys. Chem. Solids* **10**, 169 (1959).
(L 12) LURIO, A. and STERN, E., *J. Appl. Phys.* **31**, 1805 (1960).
(M 1) MERZ, W. J., *Phys. Rev.* **76** , 1221 (1949).
(M 2) MERZ, W. J., *Phys. Rev.* **91**, 513 (1953).
(M 3) MERZ, W. J., *J. Appl. Phys.* **27**, 938 (1956).
(M 4) MEYERHOFER, D., *Phys. Rev.* **112**, 413 (1958).
(M 5) MEGAW, H. D., *Proc. Roy. Soc.* (*London*) **189**, 261 (1947).
(M 6) MERZ, W. J., *Phys. Rev.* **78**, 52 (1950).
(M 7) MASON, W. P., *Piezoelectric Crystals and Their Application to Ultrasonics,* Van Nostrand, New York (1950).
(M 8) MARA, R. T., SUTHERLAND, G. B. B. M. and TYRELL, H. V., *Phys. Rev.* **96**, 801 (1954).
(M 9) MATTHIAS, B. T. and HIPPEL, A. VON, *Phys. Rev.* **73**, 1378 (1948).
(M 10) MERZ, W. J., *Phys. Rev.* **88**, 421 (1952).
(M 11) MERZ, W. J., *Phys. Rev.* **95**, 690 (1954).
(M 12) MILLER, R. C. and SAVAGE, A., *Phys. Rev. Letters* **2**, 294 (1959).
(M 13) MEITZLER, A. H. and STADLER, H. L., *Bell System Tech. J.* **37**, 719 (1958).
(M 14) MILLER, R. C., *Phys. Rev.* **111**, 736 (1958).
(M 15) MILLER, R. C. and SAVAGE, A. *Phys. Rev.* **112**, 755 (1958).
(M 16) MILLER, R. C. and SAVAGE, A., *J. Appl. Phys.* **30**, 808 (1959).
(M 17) MERZ, W. J. and ANDERSON, J. R., *Bell Labs. Record* **33**, 335 (1955).
(M 18) MATTHIAS, B. T., BRECKENRIDGE, R. G. and BEAUMONT, D. W., *Phys. Rev.* **72**, 532 (1947).
(M 19) MATTHIAS, B. T., *Phys. Rev.* **73**, 808 (1948).
(M 20) MATTHIAS, B. T., *Nature* **161**, 325 (1948).
(M 21) MCKEEHAN, L. W., *Phys. Rev.* **43**, 913 (1933); **72**, 78 (1947).
(M 22) MEGAW, H. D., *Ferroelectricity in Crystals*, Methuen, London (1957).
(M 23) MASON, W. P. and MATTHIAS, B. T., *Phys. Rev.* **74**, 1622 (1948).
(M 24) MEGAW, H. D., *Acta Cryst.* **5**, 739 (1952); 7, 187 (1954).
(M 25) MILLER, R. C. and SAVAGE, A., *Phys. Rev.* **115**, 1176 (1959).
(M 26) MARUTAKE, M., *J. Phys. Soc. Japan* **11**, 807 (1956).
(M 27) MARUTAKE, M. and IKEDA, T., *J. Phys. Soc. Japan* **11**, 814 (1956).
(M 28) MASON, W. P., *Phys. Rev.* **74**, 1134 (1948).
(M 29) MARUTAKE, M. and IKEDA, T., *J. Phys. Soc. Japan* **12**, 233 (1957).
(M 30) MOSELEY, D. S., *J. Acoust. Soc. Am.* **27**, 947 (1955).
(M 31) QUARRIE, M. MC, *J. Am. Ceram. Soc.* **39**, 54 (1956)
(M 32) MASON, W. P., *J. Acoust. Soc. Am.* **27**, 73 (1955).

(M 33) QUARRIE, M. C. MC and BUESSEM, W. R. *J., Am. Ceram. Soc.* **34**, 402 (1955).

(M 34) MASON, W. P., Piezoelectric and Ferroelectric Devices, *Molecular Science and Molecular Engineering* (Edited by VON HIPPEL, A.), ch. 17, John Wiley, New York (1959).

(M 35) MILLER, R. C. and SAVAGE, A. J., *Appl. Phys.* **31**, 662 (1960).

(M 36) MILLER, R. C. and WEINREICH, G., *Phys. Rev.* **117**, 1460 (1960).

(N 1) NAKAMURA, T., *J. Phys. Soc. Japan* **9**, 425 (1954).

(N 2) NEWTON, R. R., AHEARN, A. J. and KAY, K. G. MC, *Phys. Rev.* **75**, 103 (1949).

(N 3) NOVOSIL'TSEV, N. S. and KHODAKOV, A. L., *Zhur. Eksptl. i Teoret. Fiz.* **27**, 94 (1954).

(N 4) NOVOSIL'TSEV, N. S. and KHODAKOV, A. L., *Doklady Akad. Nauk S.S.S.R.* **85**, 1263 (1952).

(N 5) NOVOSIL'TSEV, N. S., KHODAKOV, A. L., SHOLOKHOVICH, M. L., FESENKO, E. G. and KRAMEROV, O. P., *Kristallografiya* **4**, 101 (1959).

(N 6) NAKAMURA, T., *J. Phys. Soc. Japan* **15**, 1379 (1960).

(N 7) NAKAMURA, E. and FURNICHI, J., *J. Phys. Soc. Japan* **15**, 1955 (1960).

(P 1) POWLES, J. G. and JACKSON, W., *Proc. Inst. Elec. Engrs. (London)* **96**, No. 3, 383 (1949).

(P 2) PEARSON, G. L. and FELDMAN, W. L., *J. Phys. Chem. Solids* **9**, 28 (1959).

(P 3) PRUTTON, M., *Proc. Phys. Soc. (London)* **72**, 307 (1958).

(P 4) PIPES, L. A., *J. Appl. Phys.* **23**, 818 (1952).

(R 1) ROBERTS, S., *Phys. Rev.* **75**, 989 (1949).

(R 2) RABENHORST, H. and MELICHERCIK, J., *Ann. Physik* (7) **1**, 261 (1958).

(R 3) ROBERTS, S., *Phys. Rev.* **85**, 925 (1952).

(R 4) RHODES, R. G., *Acta Cryst.* **4**, 105 (1951).

(R 5) ROGERS, E. T., JR., *J. Appl. Phys.* **27**, 1066 (1956).

(R 6) RASE, D. E. and ROY, R., *J. Am. Ceram. Soc.* **38**, 102 (1955).

(R 7) ROY, R., The Pennsylvania State University, 1st and 4th Quarterly Reports, U.S.A. Signal Corps Contract DA-36-039-sc-5594 (1951-52).

(R 8) RASE, D. E. and ROY, R., *J. Phys. Chem.* **61**, 744 (1957).

(R 9) REMEIKA, J. P., *J. Am. Chem. Soc.* **76**, 940 (1954); *U.S. Pat.* 2 852 400 (16 September, 1958).

(R 10) RUSHMAN D. F., and STRIVENS, M. A., *Proc. Phys. Soc. (London)* **59**, 1011 (1947).

(R 11) ROBERTS, S., *J. Am. Ceram. Soc.* **33**, 63 (1950).

(R 12) ROBERTS, S., *Phys. Rev.* **71**, 890 (1947).

(S 1) SACHSE, H., *Ferroelektrika*, Springer-Verlag, Berlin (1956).

(S 2) SHIRANE, G., JONA, F. and PEPINSKY, R., *Proc. I.R.E.* **43**, 1738 (1955).

(S 3) SCHMITT, H. J., *Z. angew. Phys.* **9**, 107 (1957).

(S 4) SHIRANE, G. and TAKEDA, A., *J. Phys. Soc. Japan* **7**, 1 (1952).

(S 5) SHIRANE, G. and SATO, K., *J Phys. Soc. Japan* **6**, 20 (1951).

(S 6) SCHMIDT, G., *Naturwiss.* **45**, 8 (1958).

(S 7) SHIRANE, G., DANNER, H. and PEPINSKY, R., *Phys. Rev.* **105**, 856 (1957).

(S 8) SAWAGUCHI, E. and CHARTERS, M. L., *Phys. Rev.* **117**, 465 (1960).

(S 9) STADLER, H. L., *J. Appl. Phys.* **29**, 1485 (1958).

(S 10) STADLER, H. L., *J. Appl. Phys.* **29**, 743 (1958).

(S 11) STATTON, W. O., *J. Chem. Phys.* **19**, 33 (1951).

(S 12) SHOLOKHOVICH, M. L. and BELYAEV, I. N., *Zhur. Obshchei Khim.* **24**, 218 (1954).

(S 13) SHOLOKHOVICH, M. L. and BELYAEV, I. N., *Zhur. Obshchei Khim.* **24**, 1118 (1954).

(S 14) SAFRATA, S., BEDNAROVA, V. and BENES, J., *Czechoslov. J. Phys.* **6**, 185 (1956).

(S 15) SLATER, J. C., *Phys. Rev.* **78**, 748 (1950).

(S 16) SUBBARAO, E. C., QUARRIE, M. C. MC and BUESSEM, W. R., *J. Appl. Phys.* **28**, 1194 (1957).

(S 17) SHUVALOV, L. A., *Kristallografiya* **2**, 119 (1957).

(S 18) SAWAGUCHI, E. and CHARTERS, M. L., *J. Am. Ceram. Soc.* **42**, 157 (1959).

(S 19) SMOLENSKII, G. A. and PASYNKOV, R. E., *Zhur. Eksptl. i Teoret. Fiz.* **24**, 69 (1953).

(S 20) SMOLENSKII, G. A., *Zhur. Tekh. Fiz.* **27**, 1778 (1957).

(S 21) SMOLENSKII, G. A., *Izvest. Akad. Nauk S.S.S.R., Ser. Fiz.* **21**, 233 (1957).

(S 22) SHUBNIKOV, A. V., *Issledovanie pezoelektricheskich tekstur* (*Investigation of Piezoelectric Textures*), Izdatel'stvo Akad. Nauk SSSR, Moscow (1955); see also: SHUBNIKOV, A. V., ZHELUDEV, I. S., KONSTANTINOVA, V. P. and SILVESTROVA, I. M., *Etude des textures piézoélectriques* (French translation by DAKNOFF, A.), Dunod, Paris (1958).

(S 23) SAVAGE, A. and MILLER, R. C., *J. Appl. Phys.* **31**, 1546 (1960).

(T 1) TROMBORULLO, R. F., cited by BENEDICT, T. S. and DURAND, J. L., *Phys. Rev.* **109**, 1091 (1958).

(T 2) TODD, S. S. and LORENSON, R. E., *J. Am. Chem. Soc.* **74**, 2043 (1952).

(T 3) TRIEBWASSER, S., *IBM J. Research Develop.* **2**, 212 (1958).

(T 4) TURLIER, P., EYRAUD, L. and EYRAUD, C., *Compt. rend.* **243**, 659 (1956).

(T 5) TAKAGI, Y., *Proc. Internat. Conference Theor. Phys.* (*Kyoto and Tokyo*), September, 1953, p. 824.

(T 6) TRIEBWASSER, S., *J. Phys. Chem. Solids* **3**, 53 (1957).

(T 7) TAKAGI, Y., SAWAGUCHI, E. and AKIOKI, T., *J. Phys. Soc. Japan* **3**, 270 (1948).

(T 8) TENNERY, V. J. and ANDERSON, F. R., *J. Appl. Phys.* **29**, 755 (1958).

(T 9) TAKAGI, Y. and SAWAGUCHI, E., *J. Phys. Soc. Japan* **4**, 363 (1949).

(T 10) TRIEBWASSER, S., *Phys. Rev.* **118**, 100 (1960).

(V 1) VON HIPPEL, A., BRECKENRIDGE, R. G., CHESLEY, F. G. and TISZA, L., *Ind. Eng. Chem.* **38**, 1097 (1946).

(V 2) SANTEN, J. H. VAN and JONKER, G. H., *Nature* **156**, 717 (1945).

(V 3) HIPPEL, A. VON, *Revs. Modern Phys.* **22**, 221 (1950).

(V 4) VOLGER, J., *Philips Research Repts.* **7**, 21 (1952).

(V 5) VAINSHTEIN, E. E., STARYI, I. B. and BRIL, M. N., *Izvest. Akad. Nauk S.S.S.R., Ser. Fiz.* **20**, 784 (1956).

(V 6) VAINSHTEIN, E. E. and VASIL'EV, IN. N., *Soviet Physics Doklady* **2**, 207 (1957).

(V 7) VAINSHTEIN, E. E., BRIL, M. N. and STARYI, I. B., *Doklady Akad Nauk S.S.S.R.* **117**, 597 (1957).

(V 8) VON HIPPEL, A., Massachusetts Institute of Technology, Tech. Rep. 51 (1950).

(V 9) VON HIPPEL, A., *Z. Physik.* **133**, 158 (1952).

(V 10) VON HIPPEL, A., *Dielectric and Waves*, John Wiley, New York (1954).

(W 1) WUL, B. and GOLDMAN, I. M., *Compt. rend. Acad. sci. U.R.S.S.* **46**, 139 (1945); **49**, 177 (1945); **51**, 21 (1946).

(W 2) WIEDER, H. H., *J. Appl. Phys.* **26**, 1479 (1955).

(W 3) WIEDER, H. H., *J. Appl. Phys.* **28**, 367 (1957).

(W 4) WIEDER, H. H., *J. Appl. Phys.* **27**, 413 (1956).

(W 5) WIEDER, H. H. and WHITE, D. J., NOL Corona, Tech. Memo No. 42-28, May (1959).

(W 6) WITTELS, M. C. and SHERILL, F. A., *J. Appl. Phys.* **28**, 606 (1957).

(W 7) WOOD, E. A., *J. Chem. Phys.* **19**, 976 (1951).

(W 8) WHITE, E. A. D., *Acta Cryst.* **8**, 845 (1955).

(Z 1) ZHIRNOV, V. A., *Zhur. Eksp. i Teoret. Fiz.* **35**, 1175 (1958).

# CHAPTER V

# PEROVSKITE-TYPE OXIDES

## 1. Introduction

Barium titanate, which we have discussed in detail in the preceding chapter, is only one member of a large group of compounds which is called the perovskite family, of which the parent member, as mentioned in Section IV-1, is the mineral $CaTiO_3$, called perovskite. The general formula of the compounds belonging to this family is $ABO_3$, where $A$ is a monovalent, divalent or trivalent metal and $B$ a pentavalent, tetravalent or trivalent element, respectively. We include in the perovskite family not only the compounds having the ideal cubic perovskite structure, but also all the compounds with structures which can be derived from the ideal one by way of small lattice distortions or omission of some atoms. The general formula $ABO_3$ implies that the compounds encompassed by it are double oxides, but it should be pointed out that a large number of double fluorides, such as $KMgF_3$, $KZnF_3$ and others, also belong to the perovskite family. No ferroelectric material has been found, to date, in the subgroup of double fluorides, and for this reason we will limit our discussion, in the present chapter, to the perovskite-type double oxides. The crystal–chemical characteristics of perovskite-type compounds in general have been studied by Naray-Szabo (N 1), Megaw (M 1) and Roth (R 1).

Owing to the extraordinarily structure-sensitive character of the phenomenon of ferroelectricity, it was only logical that, after the discovery of ferroelectricity in $BaTiO_3$, considerable effort would be expended on a search for other ferroelectric materials within the same family. This search was definitely fruitful, as it led to the discovery of ferroelectric activity in $PbTiO_3$, $KNbO_3$ and $KTaO_3$. As can be seen from Table V-1, these compounds have Curie points scattered over a wide range of temperatures, while the room-temperature symmetry is characterized by very different distortions from the original perovskite cubic lattice. Table V-1 lists also a number of other $ABO_3$ compounds with interesting dielectric properties, such as the antiferroelectrics $PbZrO_3$ and $NaNbO_3$, which are the only antiferroelectric crystals that we shall discuss in detail in this book. There are also several distorted perovskite compounds, such as $CaTiO_3$ and $CaZrO_3$, which exhibit phase changes at certain temperatures but no large anomaly of the dielectric constant at these temperatures. These compounds are characterized by the rather small size of the $A$ ions; the distortion occurring below the transition temperature can be attributed mainly to steric effects. The nature of these transitions is thus quite different from that of the transitions occurring in $BaTiO_3$. There are, of course, limiting cases in which the transition seems to be caused by both the steric effect and the co operative phenomenon characteristic of $BaTiO_3$. The compounds undergoing these transitions may be interesting, from the viewpoint of dielectric theory, purely because they do *not* exhibit ferroelectric properties. However, we are not going to discuss this type of compound; the reader interested in the structural characteristics of the various perovskite-type oxides is referred to the detailed description given in Megaw's book (M 2).

TABLE V-1.  REPRESENTATIVES OF PEROVSKITE-TYPE COMPOUNDS

$C$ = cubic; $T$ = tetragonal; $O$ = orthorhombic. A star in front of the values of the lattice parameters denotes a multiple unit cell. The orthorhombic distortion is described in terms of monoclinic axes. ($F$) in the last column indicates that the phase transition to the cubic phase coincides with the ferroelectric Curie point.

| Compound | Structure at 20 °C | | | | Transition temperature to cubic phase (°C) |
|---|---|---|---|---|---|
| | Symmetry | Lattice parameters | | | |
| | | $a = b$ | $c/a$ | $\gamma$ | |
| $BaTiO_3$ | $T$ | 3.992 | 1.010 | | 120 $(F)$ |
| $SrTiO_3$ | $C$ | 3.905 | — | | $-220$ |
| $CaTiO_3$ | $O$ | *3.827 | 0.999 | 90° 40′ | 1260 |
| $PbTiO_3$ | $T$ | 3.905 | 1.063 | | 490 $(F)$ |
| $CdTiO_3$ | $O$ | *3.791 | 1.004 | 91° 10′ | — |
| $PbZrO_3$ | $O$ | *4.159 | 0.988 | 90° | 230 |
| $PbHfO_3$ | $O$ | *4.136 | 0.991 | 90° | 215 |
| $KNbO_3$ | $O$ | 4.038 | 0.983 | 90° 15′ | 435 $(F)$ |
| $NaNbO_3$ | $O$ | *3.914 | 0.991 | 90° 41′ | 640 |
| $AgNbO_3$ | $O$ | *3.944 | 0.993 | 90° 34′ | 550 |
| $KTaO_3$ | $C$ | 3.989 | — | | $-260$ $(F)$ |
| $NaTaO_3$ | $O$ | *3.890 | 0.998 | ·90° 29′ | 470 |
| $AgTaO_3$ | $O$ | *3.931 | 0.992 | 90° 21′ | 485 |

In looking for ferroelectrics other than $BaTiO_3$ within the perovskite family, there are two important questions that should be considered in the course of a systematic study. The first is related to the likelihood that a given compound $ABO_3$ will crystallize with a perovskite-type lattice — what conditions must be satisfied by the $A$ and $B$ ions for this to be the case? The second question inquires about the particular factors that play the major role in establishing ferroelectricity in the perovskite lattice.

The answer to the first question can be given, to a certain extent, by considerations of ionic size. We have summarized in Table V-2 the ionic radii of the $A$ and $B$ ions which are likely to form perovskite-type structures. We have limited ourselves to compounds of the type $A^{2+}B^{4+}O_3$ and $A^{1+}B^{5+}O_3$. As mentioned above, the perovskite family also numbers many compounds having the general formula $A^{3+}B^{3+}O_3$, but no ferroelectrics have been discovered, to date, in this particular group. It should also be pointed out that not all the possible $ABO_3$ combinations considered in Table V-2 actually result in a perovskite-type structure. In fact, some of the compounds reported earlier by Naray—Szabo (N 1) as being perovskites could not be prepared as such by other authors — Roth (R 1) reported that "$PbSnO_3$" and "$PbCeO_3$" cannot be obtained as perovskites using the normal firing procedures. In the cases of all the other compounds which do crystallize with a perovskite-type lattice, it may be seen from Table V-2 that the ionic radii of both the $A$ and the $B$ ions can be varied within a wide range.

TABLE V-2. IONIC RADII OF A FEW OF THE $A$
AND $B$ IONS IN PEROVSKITE-TYPE COMPOUNDS
$ABO_3$, ACCORDING TO AHRENS (SEE ROTH(R1))

| $A$ ion | Radius (Å) | $B$ ion | Radius (Å) |
|---------|------------|---------|------------|
| $Na^{1+}$ | 0.94 | $V^{5+}$ | 0.59 |
| $Ag^{1+}$ | 1.26 | $Nb^{5+}$ | 0.69 |
| $K^{1+}$ | 1.33 | $Ta^{5+}$ | 0.68 |
| $Rb^{1+}$ | 1.47 | | |
| $Cs^{1+}$ | 1.67 | | |
| $Cd^{2+}$ | 0.97 | $Ti^{4+}$ | 0.68 |
| $Ca^{2+}$ | 0.99 | $Sm^{4+}$ | 0.71 |
| $Sr^{2+}$ | 1.12 | $Hf^{4+}$ | 0.78 |
| $Pb^{2+}$ | 1.20 | $Zr^{4+}$ | 0.79 |
| $Ba^{2+}$ | 1.34 | $Ce^{4+}$ | 0.94 |
| | | $Th^{4+}$ | 1.02 |

Table V-1, on the other hand, shows that the resulting dielectric properties are quite different. This observation is already related, in part, to the second question mentioned above, in that it implies an inquiry about the role played by the individual ions in the dielectric behavior of $BaTiO_3$-type ferroelectrics.

The geometrical packing of the ions in the lattice is certainly an important factor determining the structure type. We have already pointed out in Section IV-6 that the perovskite structure is essentially a three-dimensional framework of $BO_6$ octahedra (see also Fig. IV-32), but it can also be regarded as a cubic close-packed arrangement of $A$ and O ions, with the $B$ ions filling the interstitial positions. The packing of this structure can be characterized by a "tolerance factor" $t$ which is defined in the following way:

$$R_A + R_O = t \sqrt{(2)}(R_B + R_O), \qquad (V-1)$$

where $R_A$, $R_B$ and $R_O$ are the ionic radii of the $A$, $B$ and O ions, respectively. When $t$ is exactly equal to unity, the packing is "ideal". When $t$ is greater than one, there is too large a space available for the $B$ ion, and this ion can therefore "rattle" inside its octahedron. Table V-3 shows that this is the case in $BaTiO_3$, for which, using the Goldschmidt ionic radii, the tolerance factor $t$ assumes the value of 1.02.

In the early stages of research, it was considered that a "rattling" Ti ion could be the main cause of ferroelectricity in $BaTiO_3$. This idea was supported by the fact that partial substitution of Ba with Sr causes the Curie temperature, as well as the lower transition temperatures, to decrease with increasing Sr concentration. Sr is smaller than Ba and consequently it provides less "rattling" space for Ti. Similar effects were found by replacing Ti in $BaTiO_3$ by Zr or Sn. Both Zr and Sn are larger than Ti, so that the replacement decreases the effective volume available to the "rattling" $B$ ion: the Curie temperature again decreases with increasing concentration of Zr or Sn.

TABLE V-3. VALUES OF TOLERANCE FACTOR $t$ FOR VARIOUS PEROVSKITE-TYPE COMPOUNDS OF THE TYPE $A^{2+}B^{4+}O_3$

$R_A$ and $R_B$ are the Goldschmidt radii of the $A$ and $B$ ions, respectively, in angström units. $t$ was calculated by using $R_A$ corrected for the co-ordination number $(CN)12$. The parentheses indicate that the compound is not known as a perovskite. $R_O = 1.32$ Å (according to Megaw (M 1)).

| | | $B$ ion $\longrightarrow$ | $Ti^{4+}$ | $Sn^{4+}$ | $Zr^{4+}$ | $Th^{4+}$ |
|---|---|---|---|---|---|---|
| | | $R_B$ $CN=6$ | 0.64 | 0.74 | 0.77 | 1.10 |
| $A$ ion | $R_A$ $CN=6$ | $R_A$ $CN=12$ | Tolerance factors $t$ | | | |
| $Ca^{2+}$ | 1.06 | 1.16 | 0.89 | 0.85 | 0.84 | (0.72) |
| $Sr^{2+}$ | 1.27 | 1.37 | 0.97 | 0.92 | 0.91 | (0.78) |
| $Ba^{2+}$ | 1.43 | 1.52 | 1.02 | 0.97 | 0.96 | 0.83 |
| $Pb^{2+}$ | 1.32 | 1.40 | 0.98 | (0.92) | 0.93 | (0.79) |
| $Cd^{2+}$ | 1.03 | 1.11 | 0.88 | (0.83) | (0.82) | (0.71) |

It soon became clear, however, that packing alone cannot explain the results obtained in the solid solution system $(Ba, Pb)TiO_3$. Pb is also smaller than Ba, as is Sr, but the Curie point was found to increase with increasing concentration of Pb. Obviously, the role of the various ions in the structure cannot be expressed solely by means of their ionic radii. Thus, a classification of the $ABO_3$-type compounds cannot be successfully made by the use of only these radii (see Wood (W 1)). Other factors, such as polarizability and the character of the bonding must also be taken into account. A more consistent picture for the classification of $A^{2+}B^{4+}O_3$ compounds has been obtained by Roth (R 1) upon considering the polarizabilities of the $A^{2+}$ ions and the radii of the $B^{4+}$ ions (see Fig. V-1). The values of the polarizability of the $A^{2+}$ ions were computed by Roberts (R 2), in much the same way as that followed in the various compilations of ionic radii of the various elements. Roberts was interested in computing the *total* polari-izability (i.e. ionic plus electronic) of each set of ions in such a way that the sum over all ions in a given compound would in all cases represent the observed polarizability of the compound. A similar approach was made by Tessman *et al.* (T4) for the electronic polarizabilities. Roberts was able to show, in fact, that each set of ions has very nearly the same total polarizability in different crystals of the perovskite type. Gränicher and Jakits (G 1) extended Roberts' approach to a whole series of solid solution systems, in an attempt to explain their dielectric properties. It turned out, however, that the experimental results could not be explained quantitatively in this fashion.

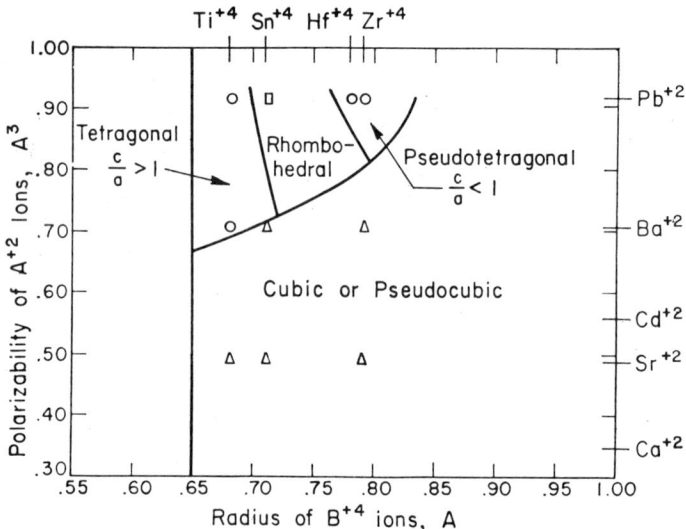

FIG. V-1. Classification of perovskite-type compounds $A^{2+}B^{4+}O_3$ (according to Roth (R 1)).

Megaw (M 5), (M 6) took quite a different approach in emphasizing the role played by the covalent bonds in double oxides of this kind. As the bonding character and the bond angles depend upon the individual ions considered, it should not be surprising that two ions with the same polarizability may cause two compounds to exhibit very different dielectric properties. A similar idea was put forward by Orgel (O 1) in a more general discussion of the crystal chemistry of oxides containing ions of transition metals. This kind of argument, however, does not allow, at the present stage, a quantitative analysis of the ferroelectric problem, although it is quite clear that the importance of covalent bonding cannot be denied.

It should also be pointed out that, independently of the particular model adopted, a very essential role is played, in the phenomenon of ferroelectricity in these compounds, by the framework of $BO_6$ octahedra. This role was first recognized by Matthias (M 3) and also, independently, by Smolenskii and Kozhevnikova (S 1). Matthias advanced an empirical rule for the occurrence of ferroelectricity in perovskite-type compounds, namely that the $B$ ion must have the electronic configuration of a noble gas, a rule that is obeyed by ions such as Ti,

Zr, Nb and Ta. This rule turned out to be highly effective and led, in fact, to the discovery of a number of perovskite ferroelectrics. Also, it appears that the framework of $BO_6$ octahedra is essential to the ferroelectric activity of other, non-perovskite, oxides, such as $Cd_2Nb_2O_7$ and $PbNb_2O_6$ (see Chapter VI). The general problem of occurrence of spontaneous polarization in perovskite-type compounds was discussed by several Russian investigators and reviewed by Smolenskii (S 32).

In the present chapter, we are going to discuss the properties of typical perovskite ferroelectrics, such as $KNbO_3$ and $PbTiO_3$, as well as some of the related binary solid solutions. Particularly in the latter case, the pertinent literature is exceedingly voluminous, owing to the importance of many ferro-electric compounds in the field of technical applications. A unique classification of the various solid solution system is practically impossible: we have arranged them in the following sections in such a way that a given binary system is dis-cussed in the section identified with the end member having the higher Curie temperature. For example, solid solutions between $BaTiO_3$ and $PbTiO_3$ are discuss-ed in the section of $PbTiO_3$. Exceptionally, however, $(K, Na)NbO_3$ solutions are discussed in the section of $NaNbO_3$. This arrangement, of course, is mainly a matter of convenience, and may be only justified by the fact that the study of solid solution systems usually starts from the high-temperature side, except for the case of the system $BaTiO_3$–$PbTiO_3$. As far as antiferroelectrics are concerned, we are going to discuss in detail $PbZrO_3$ and $NaNbO_3$, and some of their solid solutions with corresponding ferroelectrics. Some of the material covered in the present chapter has been discussed, from a somewhat different point of view, by Smolenskii (S 2), (S 3) and by McQuarrie (M 4).

As regards the conventions followed for the crystallographic axes of the various phases of perovskite-like compounds, it may be useful to recall (see Section IV-3) that in the orthorhombic pseudo-cubic distortion the $a_o$ axis is chosen along the polar axis and the $c_o$ axis along the original cube edge. This implies that:

$$a_o \geqq b_o \approx a_{\text{cubic}} \times \sqrt{2}; \quad c_o \approx n \times a_{\text{cubic}},$$

where $n$ is an integer. When the orthorhombic phase is discussed in terms of monoclinic axes, the orthorhombic $c_o$ axis is chosen as the monoclinic unique axis $c_m$, while $a_m = b_m \sim a_{\text{cubic}}$. The subscripts $o$ for orthorhombic and $m$ for monoclinic will often be omitted when no confusion can arise.

## 2. Potassium Niobate, $KNbO_3$, Potassium Tantalate, $KTaO_3$, and Their Solid Solutions

### Potassium Niobate, $KNbO_3$

Ferroelectricity in $KNbO_3$ was first discovered by Matthias (M 7) in 1949. A large amount of attention has subsequently been given to this compound, in the form of dielectric, optical and X-ray investigations (M 8), (W 1), (S 4), (S 5), (T 1). The Curie point of $KNbO_3$ lies at 435 °C, and the crystal exhibits, upon cooling, the same sequence of phase transitions as is found in $BaTiO_3$. The symmetry of the polar phase below 435 °C is tetragonal, it becomes orthorhombic at 225 °C and finally rhombohedral at − 10 °C (S 4). All these phase transitions are of the first order and connected with detectable temperature hystereses (see Fig. V-2).

Single crystals of $KNbO_3$ were first produced from binary melts using an excess of $K_2CO_3$ as a flux (W 1). These crystals, however, were mostly small and had,

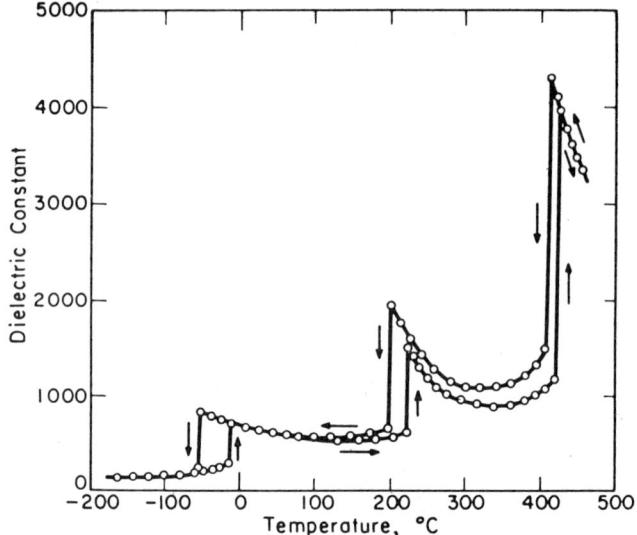

Fɪɢ. V-2. Temperature dependence of the dielectric constant of multi-domain single crystals of $KNbO_3$ (according to Shirane, Newnham and Pepir sky (S 5)).

in addition, rather large electrical conductivity, owing to the slight chemical reduction which occurred during the growth process. Large, flawless single crystals up to one inch in linear dimensions have since been grown by Miller (M 9) by means of a technique involving the pulling of a seed from a molten mixture of 1.2 mole $K_2CO_3$ and 1.0 mole $Nb_2O_5$. To date, no technique for growing large $KNbO_3$ single crystals with plate-like habit is known.

Most of the work done on $KNbO_3$ has employed polycrystalline samples. It is not easy to prepare well sintered ceramic specimens of $KNbO_3$, in contrast to $BaTiO_3$, mainly because the melting point of $KNbO_3$ is rather low (1050 °C). The phase diagram of the system $K_2CO_3$ and $Nb_2O_5$, very useful in connection with the preparation of polycrystalline $KNbO_3$ as well as the growth of single crystals, has been investigated by Reisman and Holtzberg (R 3), (R 4).

The dielectric constant of multi-domain crystals of $KNbO_3$ is depicted in Fig. V-2 as a function of temperature. This curve shows a marked resemblance to the higher branch of Merz's curve for $BaTiO_3$ (Fig. IV-3), indicating that the domain pattern of the particular crystal investigated consisted mainly of $a$ domains, in the tetragonal phase. The temperature dependence of the spontaneous polarization is shown in Fig. V-3, as measured by Triebwasser (T 1). The value of $P_s$ at the Curie point is $26 \times 10^{-6}$ C/cm². The theoretical values plotted on the same curve represent the results of calculations of the free energy function determined from data obtained above and just at the Curie temperature.

The temperature dependence of the lattice parameters of $KNbO_3$ was studied first by Wood (W 1) and later, in more detail, by Shirane et al. (S 5). The general appearance of the curve is again quite similar to the corresponding curve of

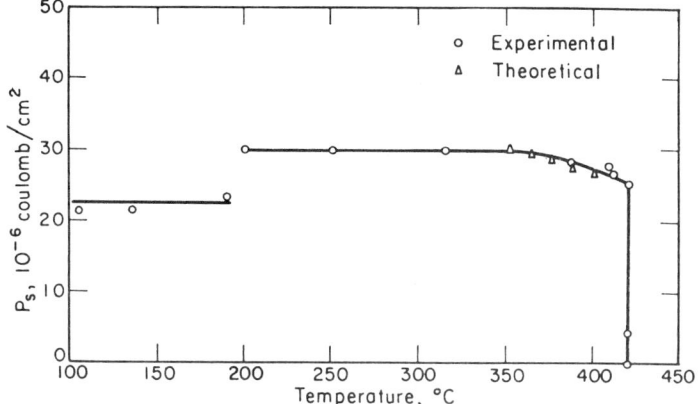

FIG. V-3. Spontaneous polarization of $KNbO_3$ as a function of temperature. The theoretical points were calculated from the free-energy expansion using the values of coefficients determined above the Curie temperature (according to Triebwasser (T 1)).

$BaTiO_3$, but the spontaneous strain in $KNbO_3$ is larger in all three phases. This is not unexpected, owing to the fact that the Curie temperature of $KNbO_3$ is higher than that of $BaTiO_3$, and a higher Curie temperature is generally indicative of stronger interactions. The maximum tetragonality of $KNbO_3$, for example, is $c/a = 1.017$, to be compared with the value of $c/a = 1.010$ for $BaTiO_3$. Accordingly, the transition energies are considerably larger for $KNbO_3$ than for $BaTiO_3$ at each of the corresponding phase changes (S 5), (T 2).

It is worth mentioning that $KNbO_3$ is the *only* ferroelectric crystal that exhibits the same phase symmetries and the same sequence of transitions as $BaTiO_3$. A comparison between some of the pertinent data of the two crystals is made in Table V-4. It would be tempting to try and explain the similarities between the two compounds on the basis of factors such as size and polarizability of the constituent atoms, but unfortunately such a comparison cannot be made so simply. Phenomenologically, of course, the properties of $KNbO_3$ can be described by the free energy expansion (I-1) and (IV-13) just as well as those of $BaTiO_3$, if the proper values are given to the coefficients. These values have been determined, for $KNbO_3$, by Triebwasser (T 1), and are also listed in Table V-4.

The nuclear quadrupole resonance of $Nb^{93}$ has been studied in all four phases of $KNbO_3$ by Cotts and Knight (C 1). The results of this experiment yield information about the field gradient and its asymmetry at the position of the Nb

TABLE V-4. COMPARISON BETWEEN $KNbO_3$ AND $BaTiO_3$

Most of the data on $KNbO_3$ originate from Triebwasser's paper (T 1), with the exception of the transition energies (S 5). Triebwasser (T 2) (H 4) reported smaller values (110–135 cal/mol) than that listed below for the transition energy of $KNbO_3$ at the Curie point.

| | $KNbO_3$ | $BaTiO_3$ |
|---|---|---|
| Transition temperatures (°C) | 435, 225, − 10 | 120, 5, − 90 |
| Transition energies (cal/mole) | 190, 85, 32 | 49, 21, 11 |
| Maximum tetragonal distortion $c/a$ | 1.017 | 1.010 |
| Spontaneous polarization at the Curie point ($10^{-6}$ C/cm$^2$) | 26 | 18 |
| Curie constant $C$ (°K) as defined from: $$\varepsilon = \frac{C}{T - T_0}$$ | $2.4 \times 10^5$ | $1.7 \times 10^5$ |
| $T_c - T_0$ (°C) | 58 | 11 |
| Coefficients $\xi_{11}^X$ and $\zeta_{111}^X$ of the free energy expansion: $$A = \frac{1}{2}\,\chi^X P^2 + \frac{1}{4}\,\xi_{11}^X P^4 + \frac{1}{6}\,\zeta_{111}^X P^6$$ | $\xi_{11}^X = -10 \times 10^{-13}$ c.g.s. and $\zeta_{111}^X = 54 \times 10^{-23}$ c.g.s. | $-20 \times 10^{-13}$ c.g.s. and $25 \times 10^{-23}$ c.g.s. |

nucleus. A quantitative interpretation of the results in terms of the parameters of the crystalline field is, however, very difficult. Qualitatively, on the other hand, the symmetry of the resonance lines can be used as a very sensitive detector of the different phase transitions (see Fig. V-4).*

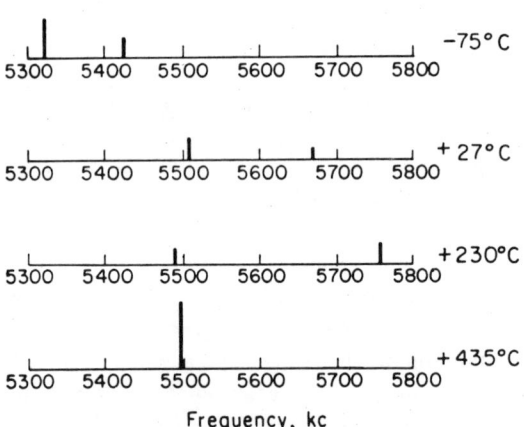

FIG. V-4. Nuclear quadrupole resonance of Nb$^{93}$ in $KNbO_3$ at different temperatures. The magnetic spectra show the effect of changes in crystal structure upon the quadrupole splitting (according to Cotts and Knight (C 1)).

* A more recent study of the temperature dependence of the electric field gradient at the Nb$^{93}$ site is that of R. R. Hewitt (*Phys. Rev.* **121**, 45 (1961)).

## Potassium Tantalate, KTaO₃

The ferroelectric properties of this compound were discovered by Matthias (M 7). The symmetry of the room temperature phase is cubic and the lattice parameter is almost identical to that of $KNbO_3$ in its centrosymmetric phase, i.e. above 435 °C. In spite of this similarity, the Curie temperature of $KTaO_3$ is 13 °K, one of the two lowest ferroelectric transition temperatures known to date (the other is that of lithium thallium tartrate monohydrate, to be discussed in Chapter IX).

The temperature dependence of the dielectric constant has been measured down to 1.3 °K by Hulm *et al.* (H 1). The results show that the dielectric constant obeys the Curie–Weiss law only down to 52 °K, below which temperature it still increases, but less rapidly than required by that law. A flat maximum is finally reached at the Curie point, below which the curve levels off in a smooth fashion. No information is available, to date, about the symmetry of the polar phase. The deviation from the Curie–Weiss law at low temperatures is a phenomenon characteristic of ferroelectrics with very low Curie points. It was shown by Barrett (B 1) that quantum effects play an important role in these low temperature regions and that these effects can explain the deviation from the Curie–Weiss law observed experimentally.

## Solid Solutions of KNbO₃ and KTaO₃

The system of solid solutions K(Nb, Ta)O₃ was first investigated by Reisman *et al.* (R 5), (R 6) with thermal and X-ray measurements. More recently, a detailed study of this system was carried out by Triebwasser T( 3), with particular attention to its dielectric properties. The measurements were made on single crystals of solid solutions with different compositions. The results show that the Curie temperature varies almost linearly between the two end-members of the system (see Fig. V-5). An interesting feature revealed by the dielectric study is

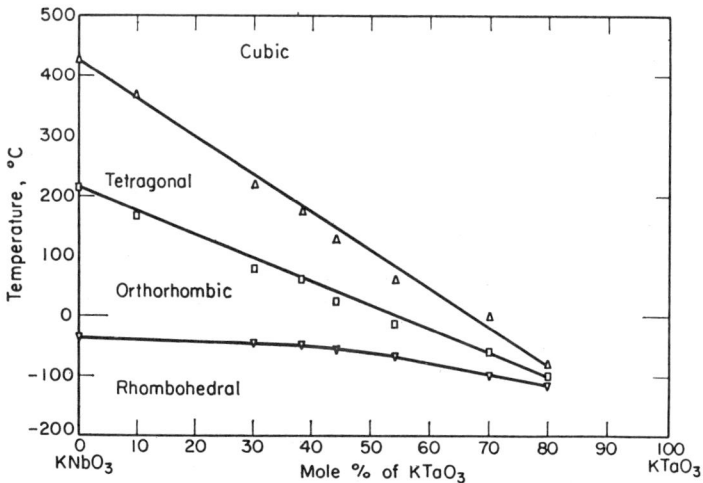

FIG. V-5. Phase diagram of the system KNbO₃–KTaO₃ (according to Triebwasser (T 3)).

that the ferroelectric transition, which is of the first order in $KNbO_3$, becomes of the second order when the Ta concentration exceeds 55%.

The experimental results of the dielectric measurements were analyzed by Triebwasser in terms of Slater's theory of $BaTiO_3$ (see Section IV-11). It was concluded that the dependence of the Lorentz correction upon volume (which is changing very little in this solid solution system) plays a major role in the onset of ferroelectricity. It is a well-known fact that the volume may be a very important quantity in the occurrence of a spontaneous polarization (see Section I-2). However, it is questionable whether Slater's theory can be applied to the treatment of these solid solutions as it stands, i.e. without further refinements. This theory is certainly the most successful treatment of the ferroelectric effect in $BaTiO_3$, at the present time; its basic assumption limiting ionic polarizability to Ti only turns out to be a fairly good approximation in explaining the dielectric properties of $BaTiO_3$. It seems rather doubtful, however, if the same assumption can explain quantitatively the *difference* between the behaviors of two compounds, in this case, $KNbO_3$ and $KTaO_3$.

### 3. Sodium Niobate, NaNbO₃, Sodium Tantalate, NaTO₃, and Solid Solutions

#### Sodium Niobate, NaNbO₃

The dielectric and crystallographic properties of $NaNbO_3$ are quite complex, in sharp contrast with the well-defined situation encountered in $KNbO_3$. Originally, $NaNbO_3$ was reported to be ferroelectric by Matthias and Remeika (M 8), but later the room-temperature structure of this compound was found to be non-polar by Vousden (V 1), a fact which, of course, is not compatible with ferroelectricity. This inconsistency was subsequently explained through a careful study of the dielectric properties by Cross and Nicholson (C 2). The room-temperature phase is non-polar when no electric field is applied, but a ferroelectric state can be induced by the application of strong fields. Thus, at room temperature $NaNbO_3$ is truly antiferroelectric.

Table V-5 summarizes the sequence of transitions encountered in this compound upon decreasing the temperature. Above 640 °C, the symmetry is cubic (perovskite-type, phase $\alpha$); below this temperature, it becomes tetragonal, or pseudo-tetragonal, with an axial ratio $c/a = 1.0015$ (phase $\beta$). At 562 °C, another transition occurs, below which the symmetry is only pseudo-tetragonal with axial ratio $c/a = 1.0023$, but truly orthorhombic (phase $\gamma$). A third transition is found at 354 °C, the symmetry remaining orthorhombic and the phase ($\delta$) becoming antiferroelectric. Finally, at $-200$ °C a fourth transition transforms the lattice into a polar, ferroelectric phase ($\varepsilon$) with monoclinic symmetry.

The room-temperature structure has been the object of detailed crystallographic investigations. The space group is $P222_1$, with eight formula units per cell, and the lattice parameters are (W 1), (V 2):

$$a = 5.568 \text{ Å}, \quad b = 5.505 \text{ Å}, \quad c = 15.518 \text{ Å}.$$

TABLE V-5. SEQUENCE OF PHASE TRANSITIONS IN $NaNbO_3$
WITH DECREASING TEMPERATURE

| Phase | α | β | γ | δ | ε |
|---|---|---|---|---|---|
| Temperature (°C) | 640 | 562 | 354 | −200 | |
| Symmetry | Cubic | Tetragonal $c/a = 1.0015$ | Pseudo-tetragonal $c/a = 1.0023$ | Orthorhombic | Monoclinic |
| Properties | Non-polar | Non-polar | Non-polar | Anti-ferroelectric | Ferroelectric |

The $a$ and $b$ axes lie along the original cubic $\langle 110 \rangle$ directions and are therefore approximately equal to $a_c \times \sqrt{2}$, where $a_c$ is the cubic lattice constant. The $c$ spacing is approximately four times larger than $a_c$. Vousden's analysis revealed the existence of antiparallel shifts (of 0.11 Å) of the Nb atoms along the direction of the $a$ axis (see Fig. V-6). This structure was recently re—examined by Megaw and Wells (M 10); the space group assignment was questioned but the essential features of Vousden's structure were confirmed.

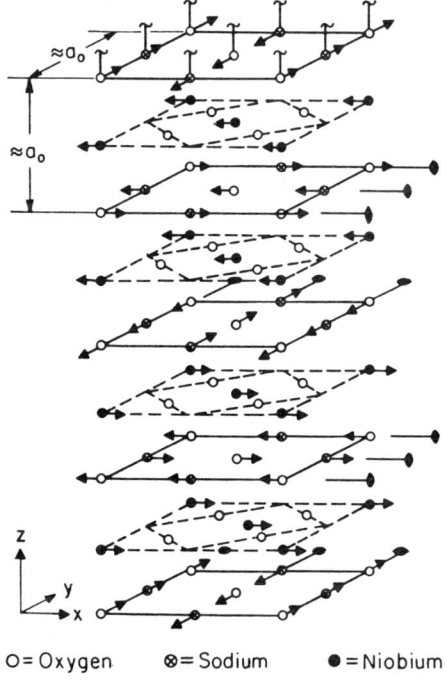

O= Oxygen    ⊗= Sodium    ●=Niobium

FIG. V-6. Schematic drawing of the structure of $NaNbO_3$. The arrows indicate the directions of atomic displacements from the centrosymmetrical positions (according to Vousden (V 1)). In this figure, $a_0$ corresponds to the pseudo-cubic spacing.

The phase transitions in NaNbO$_3$ have been studied by a number of authors using dielectric, optical and X-ray techniques. In the early stages of the problem, some confusion existed in relation to the symmetries of the various phases (W 1). Subsequently, the situation has become more clear, at least in part, thanks to a series of careful X-ray and optical studies (S 5), (C 2), (F 1) and detailed investigations of the phase diagram by Reisman *et al.* (R 7) and by Shafer and Roy (S 13). The three transition temperatures of 354 °C, 562 °C and 640 °C listed in Table V-5 were in fact reported by Reisman *et al.*, correcting an inaccuracy in the earlier papers which gave the temperature of the transition between the $\beta$ and the $\gamma$ phase as 480 °C. Reisman *et al.* were able to show that this transition temperature is very sensitive to the state of strain in the samples, so that only after repeated annealing is it possible to observe reproducibly a sharp transition at 562 °C. The structures of the $\beta$ and the $\gamma$ phases, however, are not known: in both cases, the symmetry is apparently pseudo-tetragonal but truly ortho-

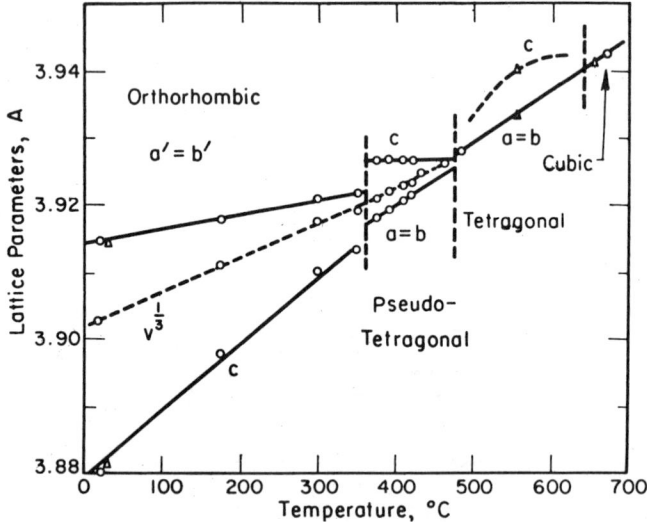

FIG. V-7. Temperature dependence of the lattice parameters of NaNbO$_3$ (circles according to Shirane *et al.* (S 5), triangles according to Francombe (F 1)).

rhombic (see Fig. V-11 b). The temperature dependence of the lattice parameters above room temperature is depicted in Fig. V-7, that below room temperature in Fig. V-8.

The structure of the $\varepsilon$ phase was studied by Johns *et al.* (J 2) by means of X-ray Weissenberg photographs of single crystals. The transition between the room-temperature $\delta$ phase and the lowest phase, which occurs at $-200$ °C upon cooling, was found to exhibit a marked temperature hysteresis (C 2), (M 8); the reverse transition was found to be as high as $-10$ °C. The $\varepsilon$ phase is monoclinic and the lattice parameters (at $-160$ °C, for increasing temperature) are:

$$a = 5.564 \text{ Å}, \quad b = 5.548 \text{ Å}, \quad c = 2 \times 3.906 \text{ Å}, \quad \beta = 91° 09'$$

where $b$ is the unique monoclinic axis. The spacing along the $c$ axis is doubled with respect to the pseudo-cubic parameter, but if this fact is disregarded it turns out that the unit cell has a rhombohedral geometry (see Fig. V-8b). The lattice parameters of this pseudo-rhombohedral cell are:

$$a_R = 3.906 \text{ Å}, \quad \alpha_R = 89° \, 11'$$

which implies again an elongation in the direction of the body diagonal, as in the low-temperature phases of $BaTiO_3$ and $KNbO_3$.

The most complete investigation of the dielectric properties of $NaNbO_3$ in all its phases was carried out by Cross and Nicholson (C 2) (see Fig. V-9). The single crystal used in this study were grown from binary melts using NaF as a flux. Only the transition at 354 °C was found to be accompanied by a large anomaly of the dielectric constant, while the two upper transitions (at 562 °C and 640 °C) are characterized by rather small dielectric anomalies. This may be considered an evidence for the fact that the phase change at 354 °C has the character of an antiferroelectric transition, while the two other phase changes are

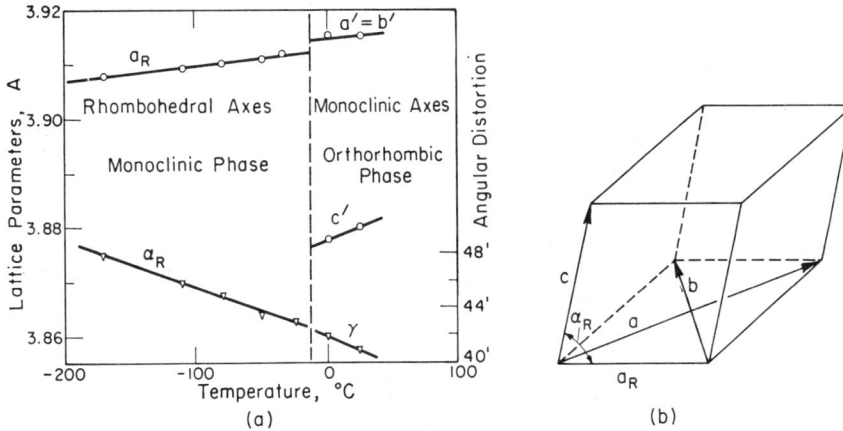

FIG. V-8. Low-temperature phase transition of $NaNbO_3$.

(a) Temperature dependence of the lattice parameters. To facilitate comparison, the room-temperature orthorhombic phase is described in terms of monoclinic axes $a' = b'$, $c'$ and $\gamma$, while the low-temperature monoclinic phase is described in terms of rhombohedral parameters $a_R$ and $\alpha_R$.

(b) Relation between orthorhombic and rhombohedral parameters. (According to Johns et al. (J 2)).

caused by factors other than dielectric. It may be pointed out, in this connection, that $NaNbO_3$ is one of those compounds in which steric effects (due to the small Na ion) may play as important a role as interactions of dielectric origin (see Section 1).

As for the transition into the ferroelectric phase occurring at $-200$ °C, it should be pointed out that it is essentially different from all the ferroelectric phase changes encountered in the other perovskite ferroelectrics. The latter are transitions from non-polar, or polar, to polar states, while in the present case we are considering a transition from an antiferroelectric into a ferroelectric phase. The spontaneous polarization of the $\varepsilon$ phase of NaNbO$_3$ is of the order of $12 \times 10^{-6}$ C/cm$^2$ along the $c$ axis (C 2). When the driving field is applied in a direction perpendicular to the $c$-axis the spontaneous polarization is found to be very small, if it exists at all. This tells us that the $c$ axis cannot be rotated into the direction of the applied field, in contrast to the case of BaTiO$_3$ and KNbO$_3$.

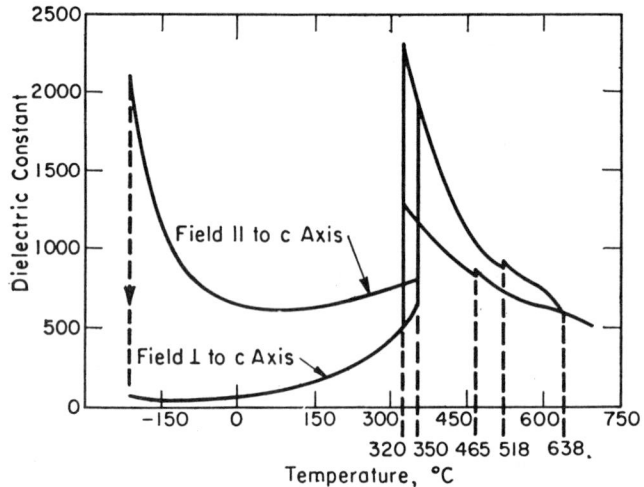

Fig. V-9. Dielectric constant of single crystals of NaNbO$_3$ as a function of temperature (according to Cross and Nicholson (C 2)). Axes are labeled in the same way as in Fig. V-7. The dashed line on the left-hand side indicates the phase transition at low temperatures. The transition temperatures differ somewhat from those reported in Table V-5.

The effect of large electric fields on the dielectric properties of NaNbO$_3$ was investigated by Cross and Nicholson (C 2) with the results depicted schematically in Fig. V-10. In view of what we have mentioned above, the line between region ($A$) and region ($B$), in this figure, merely represents the field dependence of the transition temperature. The fact that hysteresis loops can be observed in region ($B$) once they have been initiated in region ($A$) is equivalent to saying that the thermal hysteresis of this transition is just as large as the width of region ($B$). These phenomena are observed for fields applied along the direction of the $c$ axis.

If, on the other hand, the electric field is applied in a direction perpendicular to the $c$ axis, double hysteresis loops are observed in the room-temperature antiferroelectric phase. These double loops persist, if sufficiently large fields are applied, in the low-temperature region. Double hysteresis loops are generally indicative of the fact that a ferroelectric phase is being induced by the field, but it should again be noted that the present case is different from that encountered in BaTiO$_3$ (see Section IV-4), as here the ferroelectric phase is forced upon an antiferroelectric, rather than a non-polar, state. This field-induced ferroelectric phase must obviously be different from that occurring spontaneously below $-200$ °C, because in the former the field is applied perpendicular to $c$, while in the latter it is applied parallel to $c$. It is conceivable that the field-induced phase is identical to that observed in solid solutions $(Na, K) NbO_3$ at small K concentrations (see below). This supposition arises from the result of observations made in the course of the study of a similar situation in $PbZrO_3$, where more

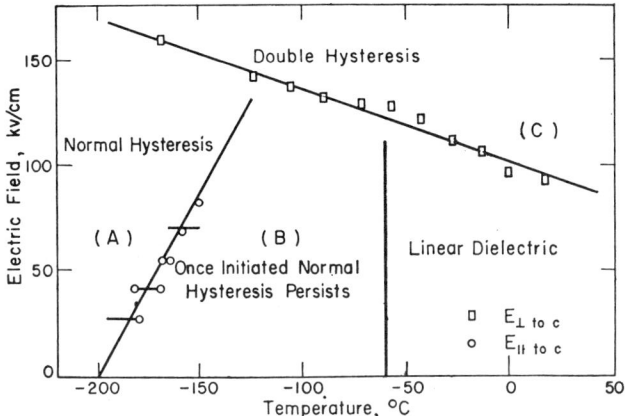

Fig. V-10. Dielectric properties of single crystals of $NaNbO_3$ in large electric fields at different temperatures (according to Cross and Nicholson (C 2)). Axes labeled in the same way as in Fig. V-7.

direct experimental evidence is available (see Section 6). Accordingly, it is expected that the critical field for the induced ferroelectric phase will depend in a very sensitive way upon the presence of impurities and upon temperature.

### Solid Solutions of $KNbO_3$ and $NaNbO_3$: $(K, Na) NbO_3$

The dielectric and crystallographic properties of the system of solid solutions between $KNbO_3$ and $NaNbO_3$ were first studied by Shirane et al. (S 5). In general, we can say that most of the phases in this system are ferroelectric, except those very close to the $NaNbO_3$-side of the diagram. It is interesting to see that, despite the large difference between the ionic radii and the polarizabilities of K and Na, the Curie temperature remains almost constant over the whole system of solid solutions, except of course for the region very near to $NaNbO_3$ (see Fig. V-11).

We observe at first that since the room-temperature modifications of $NaNbO_3$ and $KNbO_3$ are not isostructural, a phase boundary must exist between these two orthorhombic phases (dashed line in Fig. V-11a). To date, it is not well established where exactly this phase boundary lies, within the diagram. The dielectric data indicate that ferroelectricity occurs in solid solutions containing 90% (S 5) and even 98% (E 1) of $NaNbO_3$. The X-ray data, on the other hand, indicate that the weak reflections associated with the multiple unit cell of $NaNbO_3$ at room temperature are already found in solid solutions containing about 75% of $NaNbO_3$ (R 5). As these weak X-ray reflections are usually related to the antiferroelectric character of the room-temperature phase of $NaNbO_3$, there seems to be a conflict between the X-ray and the dielectric data. However, this contradiction may be explained in the same way as the problem of ferroelectricity vs. antiferroelectricity in pure $NaNbO_3$: the solid solutions rich in $NaNbO_3$ are probably antiferroelectric, so that strong electric fields, as applied in the dielectric studies, can induce ferroelectric phases. It is significant, in this respect,

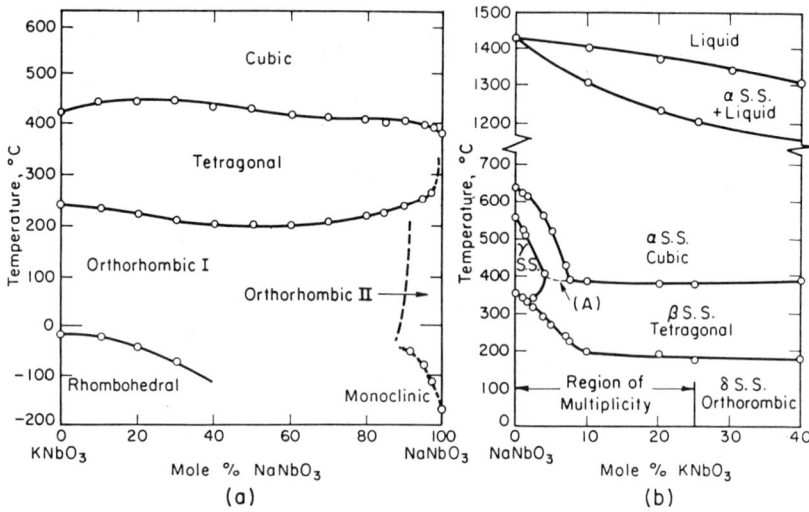

FIG. V-11. Phase diagram of the system $NaNbO_3$–$KNbO_3$.
(a) According to Shirane *et al.* (S 5). The lowest phase line on the $NaNbO_3$ side
of the diagram has been added according to the data of Johns (J 2).
(b) According to Reisman and Banks (R 5). The dashed line at (*A*) does not appear
in the original diagram (see text).

that no miscibility gap was found between the pure $KNbO_3$ and the pure $NaNbO_3$ orthorhombic structures, a fact evidenced by the continuous change in lattice constant (S 5), (R 5) and in density (R 5) over the whole range of compositions.

The problem is more complicated in the immediate vicinity of the $NaNbO_3$ end of the phase diagram. The pictures drawn by different investigations differ somewhat from one another (S 5), (R 5), (C 2). Figure V-11 (b) depicts that proposed by Reisman and Banks (R 5). The dashed line indicated at (*A*) is actually not part of the original diagram drawn by these authors. There are, on the other hand, good reasons to believe that such a line exists. Without it, the diagram implies that the $\beta$ phase of $NaNbO_3$ is the same ferroelectric tetragonal phase observed in $KNbO_3$ between 225 °C and 435 °C. This, however, is in contradiction with the multiple unit cell of this $\beta$ phase observed by Francombe (F 1). Furthermore, the absence of line (*A*) would imply that the transition at 640 °C is a ferroelectric Curie point. It would be logical, in this case, to expect a large anomaly of the dielectric constant at this temperature, which is not what is observed experimentally. The phase diagrams reported by Shirane *et al.* (S 5) and by Cross (C 3) include a phase line drawn approximately at (*A*).

It should be realized, however, that it is not essential, for the discussion of the dielectric properties of this solid solution system, to know exactly the location of every 'phase boundary. The most important characteristic of this system is rather the fact that a ferroelectric phase is found to exist in the region very close to pure $NaNbO_3$.

Experimentally, the dielectric study of this solid solution system is complicated by the difficulty of obtaining dense ceramic samples. This difficulty is partially related to the comparatively low melting points of the end members, and was recently overcome by Egerton and Dillon (E 1) with a critical firing process carried out just below the melting point of the samples. Promising electromechanical coefficients and a rather low dielectric constant were reported for solid solutions in the vicinity of 50% Na.

From the theoretical point of view, this system was studied in detail by Cross (C 4), who applied the thermodynamic treatment developed by Devonshire for $BaTiO_3$ to both $NaNbO_3$ and the system $(Na, K)NbO_3$. The theory can explain fairly well the dielectric properties and the various transition temperatures observed in $NaNbO_3$. The treatment of the antiferroelectric phase, however, requires the knowledge of more experimental parameters than does Devonshire's theory of $BaTiO_3$ because in the former case the value of the spontaneous polarization of the sublattices cannot be determined experimentally.

### Sodium Tantalate, NaTaO₃

$NaTaO_3$ was originally reported by Matthias (M 7) as having ferroelectric properties below a Curie point at approximately 480 °C. Two additional phase changes were detected by Cross (C 5) at 580 °C and 630 °C through optical studies. This suggests that the dielectric properties of this compound may be quite similar to those observed in $NaNbO_3$. However, no anomalies of the dielectric constant and the coefficient of thermal expansion were observed between $-180$ °C and 510 °C by Smolenskii et al. (S 6).

The symmetry of the room-temperature phase is reportedly orthorhombic (V 2). Kay and Miles (K 1) have found that the $c$ spacing of this phase is only twice that of the original cubic cell, in contrast to the case of $NaNbO_3$, whose $c$ spacing at room temperature is four times that of the pseudo-cubic lattice. The parameters of $NaTaO_3$ are:

$$a = 5.513 \text{ Å}, \quad b = 5.494 \text{ Å}, \quad c = 2 \times 3.875 \text{ Å},$$

which implies that $NaTaO_3$ and $NaNbO_3$ are not strictly isomorphous.

### Solid Solutions of NaNbO₃ and NaTaO₃

The system of solid solutions $Na(Nb, Ta)O_3$ has been investigated by Smolenskii et al. (S 6). The transition temperature occurring at 354 °C in pure $NaNbO_3$ (see Table V-5), which is characterized by a large dielectric anomaly, decreases with increasing Ta concentration. The peak of the dielectric constant was found to occur at $-150\,°C$ in the solid solution $Na(Nb_{0.25}Ta_{0.75})O_3$. Ferroelectric hysteresis loops were observed in the same solid solution at $-193\,°C$. Unfortunately, no more information is available, to date, about this system.

## 4. Silver Niobate, $AgNbO_3$, Silver Tantalate, $AgTaO_3$, and Rubidium Tantalate, $RbTaO_3$

### Silver Niobate, $AgNbO_3$

It has been reported by Francombe and Lewis (F 5) that $AgNbO_3$ exhibits an orthorhombic distortion from the perovskite lattice at room temperature. The lattice parameters of this phase are listed in Table V-1, and are quite similar to those of $NaNbO_3$. Their temperature dependence also recalls that encountered in $NaNbO_3$, as a comparison between Figs. V-7 and V-12(b) shows. The X-ray data reveal a transition from orthorhombic to tetragonal symmetry at 325 °C and from tetragonal to cubic symmetry at 550 °C.

The dielectric constant of a ceramic specimen of $AgNbO_3$ is depicted in Fig. V-12(a) as a function of temperature. Above 325 °C, the Curie–Weiss law $\varepsilon = C/(T - T_0)$ is obeyed approximately, the Curie constant $C$ being equal to $1.8 \times 10^5$ degrees. The highest peak of the dielectric constant curve occurs at the transition between the orthorhombic and tetragonal phases, while no detectable dielectric anomaly occurs at the transition from tetragonal to cubic symmetry. This behavior, again, is quite similar to that of $NaNbO_3$. Under moderate electric fields, $AgNbO_3$ seems to exhibit some slight non-linearity in the relationship between polarization and field but no hysteresis loops can be observed. Owing

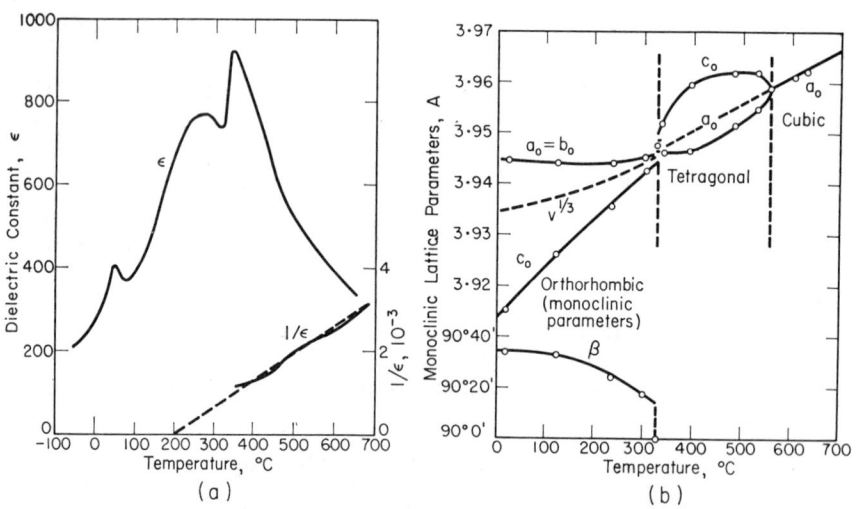

FIG. V-12. Properties of $AgNbO_3$.
(a) Dielectric constant of ceramic samples as a function of temperature.
(b) Lattice parameters as functions of temperature. (According to Francombe and Lewis (F 5)).

to its close similarity to $NaNbO_3$, it seems therefore justifiable to suspect that a ferroelectric phase could be induced by very strong electric fields or by additions of isomorphous compounds, i.e. that $AgNbO_3$ might be antiferroelectric.

Figure V-12(a) shows also that two other anomalies in the dielectric constant curve occur, at 50 °C and 250 °C. The nature of these anomalies, which could not be detected by the X-ray study, is not understood. Reisman and Holtzberg (R 12) carried out a study of the phase diagram of the system $Ag_2O–Nb_2O_5$ and observed anomalies in the differential thermal analysis of $AgNbO_3$ at 292 °C and 578 °C. These anomalies correspond to the two phase changes detected by the X-ray study.

### Silver Tantalate, $AgTaO_3$

The temperature dependence of the lattice parameters of $AgTaO_3$ shows a strong similarity to that of $AgNbO_3$ (F 5). Transitions were observed at 370 °C and 485 °C (see Table V-1). The dielectric properties could not be studied because the samples have a relatively high electrical conductivity.

### Rubidium Tantalate, $RbTaO_3$

It has been reported by Smolenskii and Kozhevnikova (S 1) that $RbTaO_3$ is ferroelectric with a Curie point at 250 °C and has a tetragonal structure at room temperature. No independent confirmation of the fact that this compound has a perovskite type-structure is available. It also appears that this compound cannot be prepared by means of the conventional procedure followed for solid-state reactions. Subsequent review articles by Smolenskii (S 2), (S 3) do not include this compound in the list of known ferroelectrics, so that there is room for doubt about its ferroelectric activity until further evidence is presented.

## 5. Lead Titanate, $PbTiO_3$, and Solid Solutions

### Lead Titanate, $PbTiO_3$

$PbTiO_3$ is ferroelectric at room temperature and has a Curie point at 490 °C, as reported first by Shirane *et al.* (S 9) and, independently, by Smolenskii (S 7), (S 8). Subsequent investigations of the properties of this compound have provided data for an interesting comparison with the behavior of $BaTiO_3$ (S 10), (S 11).

Most of the experimental results were obtained on ceramic samples. Single crystals can be grown with flux methods but not as easily as in the case of $BaTiO_3$, mostly because of the loss of PbO at higher temperatures. Rogers (R 8) obtained crystals with edges of a few millimeters from binary melts in which PbO was used as a flux. More recently, larger, plate-like crystals up to 1 cm in edge were grown by Kobayashi (K 2) using KF as a flux. Successful results were also reported by the Russian workers (F 2), (S 12) using different fluxes, such as $Pb(BO_2)_2$ and $Na_2SiO_3$.

The structure of $PbTiO_3$, at room temperature, involves a tetragonal distortion from the perovskite lattice, and the compound is isomorphous with tetragonal $BaTiO_3$. The lattice parameters are (M 1):

$$a = 3.904 \text{ Å}, \quad c = 4.150 \text{ Å}, \quad c/a = 1.063,$$

showing that the tetragonal distortion is markedly larger than that of $BaTiO_3$ (where $c/a = 1.01$). This large distortion causes rather drastic changes in the crystal lattice at the Curie point, as evidenced by the temperature dependence of the

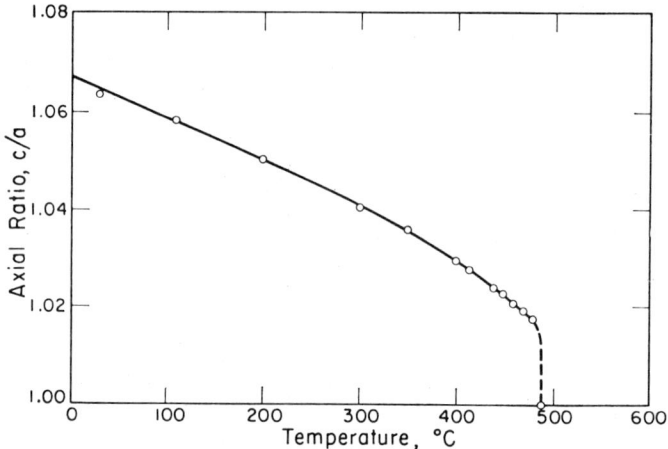

FIG. V-13. Temperature dependence of the axial ratio $c/a$ in $PbTiO_3$ (according to Shirane and Hoshino (S 10)).

axial ratio depicted in Fig. V-13. The transition is of the first order, and the discontinuous change in the ratio $c/a$ at the Curie point is of the order of 2%.

Both the specific heat and the specific volume exhibit large anomalies at the transition (S 10). The specific heat anomaly is spread over a wide temperature interval, ranging from 340 °C to 540 °C, in accordance with the temperature dependence of the axial ratio, and its integrated value is 1150 cal/mole.

The dielectric constant of $PbTiO_3$ has been measured as a function of temperature only with ceramic samples (S 10). The general appearance of the curve obtained is quite similar to that of $BaTiO_3$ in the tetragonal and cubic regions. Above the Curie point, the Curie–Weiss law $\varepsilon = C/(T - T_0)$ is obeyed, the value of the Curie constant $C$ being $1.1 \times 10^5$ degrees. As for the spontaneous polarization $P_s$, no reliable data are available to date, although it is known that the value of $P_s$ is considerably larger than that for $BaTiO_3$. Some information about the temperature dependence of $P_s$ can be obtained from Fig. V-13. In fact, if we write the $c$ spacing as $c = a + \Delta a$, we see that the quantity $(c/a - 1)$ represents the spontaneous strain along the tetragonal axis, which has been shown, in the case of $BaTiO_3$, to be proportional to $P_s^2$. Thus, we are justified in expecting a marked temperature dependence of the spontaneous polarization.

The character of this dependence, however, is apparently in contradiction with that inferred from birefringence measurements. The quantity $\Delta n = n_a - n_c$ ($n_a$ = refractive index along $a$; $n_c$ = refractive index along $c$) is depicted in Fig. V-14 as a function of temperature (S 11), (F 3). The behavior of this curve is in contradiction with the expectation that $\Delta n$ be proportional to $P_s^2$ (spontaneous Kerr effect), a relationship that was experimentally confirmed in BaTiO$_3$. It was suggested by Känzig (K 3) that the anomalous behavior of the birefringence in PbTiO$_3$ may be due to an increasing overlap of ions with increasing polarization.

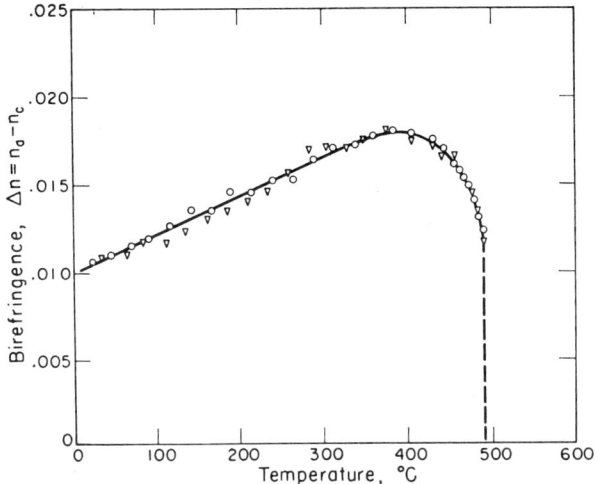

FIG. V-14. Temperature dependence of the birefringence of PbTiO$_3$, measured with Na light (according to Shirane et al. (S 11)).

It may also be pointed out, in this connection, that the birefringence shows a very strong dependence upon the wavelength of the light (R 8).

Optical studies of single crystals revealed that the geometry of the ferroelectric domains is substantially the same as that observed in tetragonal BaTiO$_3$. In most virgin crystals, $c$ domains are visible, but they are usually small and irregular in shape. After heating the crystals above the Curie point, the $c$ domains disappear almost completely and $a$ domains prevail (K 2). The effect of d.c. fields upon the domain structure was studied only briefly (F 3), (F 4), and practically no information is available about the dynamics of domain wall motions.

The atomic positions in the tetragonal lattice of PbTiO$_3$ were determined by Shirane et al. (S 11) using both X-ray and neutron diffraction techniques. The results are expressed in terms of displacements $\delta z$ of the ions from their symmetrical positions, where $\delta z$ is expressed in fraction of the cell edge. Taking the origin at the position of Pb, the results are:

$$\delta z_{Ti} = +0.040, \quad \delta z_{OI} = +0.112, \quad \delta z_{OII} = +0.112,$$

where, as was the case in BaTiO$_3$, OI is the oxygen located above or below Ti along the tetragonal axis and OII represents the other oxygen atoms (see

Fig. IV-1). We notice first that the oxygen shifts are identical for the two sites considered, meaning that the oxygen octahedron suffers no distortion in $PbTiO_3$, and furthermore that the oxygens seem to be shifted in the same direction as Ti, which is in contrast to the situation found in $BaTiO_3$. However, we have already pointed out in Section IV-6 that the structure may be considered from different viewpoints according to the choice of the origin and that this choice is quite arbitrary. Instead of taking the origin at the Pb position, it is more enlightening in this case to choose it at such a level along the tetragonal axis that the oxygen network remains fixed. From this viewpoint, both Ti and Pb appear to be shifted in the same direction, namely, Ti by 0.30 Å and Pb by 0.47 Å. Looked at in this way, the structure of $PbTiO_3$ shows a number of features in common with that of $BaTiO_3$, although there are a few important differences, mainly in connection with the Pb$-$O (Ba$-$O) bond distances. Figure V-15 shows schematically the structure models of $BaTiO_3$ and $PbTiO_3$ from this point of view.

The large ionic displacements observed in $PbTiO_3$ cause marked changes in the bond lengths, as indicated in Table V-6. The distance between Ti and OI($u$), of 1.78 Å, is considerably smaller than the sum of the corresponding Goldschmidt

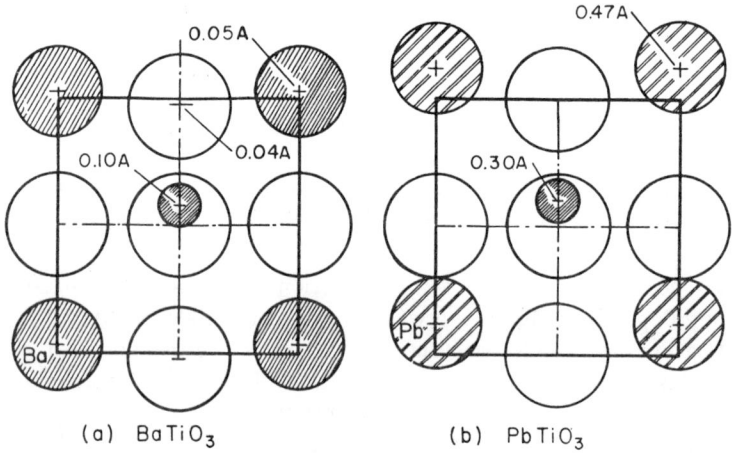

FIG. V-15. Schematic models for the structures of $BaTiO_3$ and $PbTiO_3$. The size of the ions was chosen arbitrarily.

radii, 1.96 Å. The system of Pb$-$O bonds is affected appreciably by the large displacement of Pb along the tetragonal axis. The shortest Pb$-$O distances, 2.53 Å (there are four of these distances) is again much shorter than the sum of the corresponding Goldschmidt radii, 2.72 Å.

The qualitative treatment of ferroelectricity proposed by Megaw, who emphasized the role of covalent bonding as the origin of the phase transition in $BaTiO_3$, was extended, accordingly, to the case of $PbTiO_3$ (M 6). It was pointed out, in particular, that the Pb$-$O bond system may be very important, especially in view of its similarity to the bonding occurring in crystals of lead oxide, PbO, where four oxygens and one lead atom form a flat tetragonal pyramid with lead at the apex and a Pb—O distance of 2.33 Å.

More quantitative treatments of the ferroelectric phenomenon in $PbTiO_3$ have been carried out by the Russian workers. Attention was devoted, in parti- cular, to the fact that, owing to the large atomic displacements occurring in this crystal, the Lorentz correction of the internal field at the displaced positions

TABLE V-6. LATTICE PARAMETERS AND BOND LENGTHS IN
TETRAGONAL AND CUBIC $PbTiO_3$

OI($u$) represents the oxygen atom closer, OI($l$) that farther away from Ti. The middle column shows the bond lengths that would be found in a fictitious centrosymmetrical tetra- gonal structure in which all shifts $\delta z$ are zero. The numbers are given in Ångström units (according to SHIRANE et al. (S 11)).

| | Tetragonal $\delta z_{Ti} = 0.040$ $\delta z_O = 0.112$ | Tetragonal $\delta z = 0$ | Cubic at 490 °C |
|---|---|---|---|
| $a$ | 3.90 | 3.90 | 3.97 |
| $c$ | 4.15 | 4.15 | |
| Ti—OI($u$) | 1.78 ⎱ | 2.08 ⎱ | |
| Ti—OI($l$) | 2.38 ⎰ | 1.95 ⎰ | 1.98 |
| Ti—OII | 1.98 | | |
| Pb—OI | 2.80 | 2.76 ⎱ | |
| Pb—OII($u$) | 2.53 ⎱ | 2.85 ⎰ | 2.80 |
| Pb—OII($l$) | 3.20 ⎰ | | |

may differ considerably from that calculated at the cubic positions. Zhdanov et al. (Z 1) estimated that the Lorentz correction at the displaced Pb position is about 75% larger than that for the cubic case. There is, however, another factor affecting the Lorentz correction, namely the tetragonal distortion of the lattice. Calculations show that the effect of tetragonality on the Lorentz correction is opposite to that of the ionic shifts, so that the two effects compensate one another. Thus, in the case of $PbTiO_3$, despite its considerable tetragonal distortion and large ionic displacements, the difference between the proper Lorentz correction and that which obtains for the cubic case is less than 20%.

Detailed calculations of the electronic and ionic polarizations in $PbTiO_3$ (and $BaTiO_3$) were carried out, more recently, by Venevtsev et al. (V 6), (V 7) following the method of Kozlovskii (K 12). The model employed by these authors is one which assumed point charges and superimposed dipoles fixed at the posi- tions occupied by the ions in the lattice. The origin of the unit cell was taken at the Ti position, so that this ion was given only electronic polarizability, while Pb and O were ascribed both electronic and ionic polarizabilities. The effective charge

on each ion was assumed to be half of that corresponding to a purely ionic structure, as this assumption led to good agreement between the calculated and the
observed values of the spontaneous polarization of $BaTiO_3$. The theory indicates
that the character of the internal field distribution in $PbTiO_3$ is qualitatively
the same as in $BaTiO_3$, but that the fields are stronger in the former case. The
calculated value of the spontaneous polarization of $PbTiO_3$, at room temperature,
is $81 \times 10^{-6}$ C/cm² and is thus about three times larger than for $BaTiO_3$.

Before closing this brief discussion of $PbTiO_3$, it should be mentioned that
a further phase transition was reported to occur, in this crystal, at approximately
$-100$ °C by Kobayashi et al. (K 4), (K 5). This phase change does not occur if
the crystal is cooled rapidly to low temperatures, but only if the cooling rate is
extremely slow. The low-temperature phase appears to be non-ferroelectric and
possesses a multiple tetragonal cell. No further information is available, at
present, about the properties of this low-temperature phase.

### Solid Solutions Containing $PbTiO_3$

$PbTiO_3$ has the highest Curie temperature known among perovskite-type
ferroelectrics. Almost any substitution of Pb or Ti with suitable atoms which are
likely to form a perovskite-type lattice causes a lowering of the Curie point. A
great number of solid solution systems involving $PbTiO_3$ have been investigated:
references to the pertinent publications have been summarized in Table V-7
for the convenience of the reader.

Figure V-16 shows the dependence of the Curie temperature upon composition
when Pb is replaced by Ba, Sr or Ca. In all three cases considered the Curie point
decreases almost linearly with increasing concentration of the substituting
atoms. The axial ratio $c/a$ of the tetragonal solid solutions at room temperature
decreases accordingly. It may be of interest to note that the two lower transition
temperatures of $BaTiO_3$ decrease with increasing concentration of Pb (S 14).
In the system (Pb, Ca)$TiO_3$ the composition $(Pb_{0.5}Ca_{0.5})TiO_3$ has been studied
more recently by Sawaguchi et al. (S 18), (S 19) with particular attention to the
problem of aging of ferroelectric ceramics. The results of these studies have
already been mentioned in Section IV-13.

Another interesting system of solid solutions is that involving $PbTiO_3$ and
$PbZrO_3$. Considerable attention has been devoted to the dielectric properties
of the compounds in this system because of the antiferroelectric character of
$PbZrO_3$ (see the next section). Detailed X-ray and dielectric studies have been
reported by Shirane et al. (S 20), (S 21), (S 22) and by Sawaguchi (S 23). Figure V-17
depicts the results of these studies, which reveal the appearance of a ferroelectric
rhombohedral phase. There is reason to believe that this rhombohedral distortion from the perovskite lattice is isomorphous with the phase occurring in $BaTiO_3$
below $-90$ °C. The rhombohedral angle differs from $90°$ by $20'$ in the solid solution $Pb(Ti_{0.4}Zr_{0.6})O_3$.

Extensive investigations of the piezoelectric properties of poled ceramics
in this system have been carried out by B. Jaffe et al. (J 4) and by Kulcsar (K 9).

TABLE V-7. REFERENCES TO LITERATURE CONCERNING SOLID
SOLUTIONS AMONG COMPOUNDS OF THE PEROVSKITE TYPE·

Most of the investigations reported in the papers cited are concerned with the structure and the transition temperature of the
solid solutions. Single-crystal studies are denoted by a star.

| System | Reference |
|---|---|
| $PbTiO_3$–$BaTiO_3$ | S 14 S 15 S 16 J 3 N 2 F 2 S 12* |
| –$SrTiO_3$ | N 3 N 4 S 8 |
| –$CaTiO_3$ | S 17 S 18 S 19 |
| –$PbZrO_3$ | S 20 S 21 S 22 S 23 J 4 K 9 |
| –"$PbSnO_3$" | N 5 N 23 V 3 |
| $PbZrO_3$–$BaZrO_3$ | R 9 S 33 S 34 |
| –$SrZrO_3$ | S 33 S 34 K 7 |
| –$CaZrO_3$ | S 33 K 6 |
| –"$PbSnO_3$" | V 3 S 24 K 7 |
| $BaTiO_3$–$SrTiO_3$ | R 10 V 4 G 1 S 38 B 4 K 10* R 3 R 11 K 13* |
| –$CaTiO_3$ | D 1 D 2 D 3 M 11 B 2 |
| –$BaZrO_3$ | K 8 M 11 I 1 S 35 S 36 |
| –$BaSnO_3$ | D 2 N 5 S 37 |
| (Ba, Pb, Ca)$TiO_3$ | M 15 |
| (Ba, Sr, Ca)$TiO_3$ | M 14 D 2 |
| (Ba, Pb)(Ti, Zr)$O_3$ | I 1 S 39 P 1 |
| (Ba, Pb)(Ti, Sn)$O_3$ | M 12 S 24 V 5 I 4 |
| (Ba, Sr)(Ti, Sn)$O_3$ | M 13 |
| (Ba, Ca)(Ti, Zr)$O_3$ | M 11 |
| (Pb, Sr)(Ti, Zr)$O_3$ | I 4 |

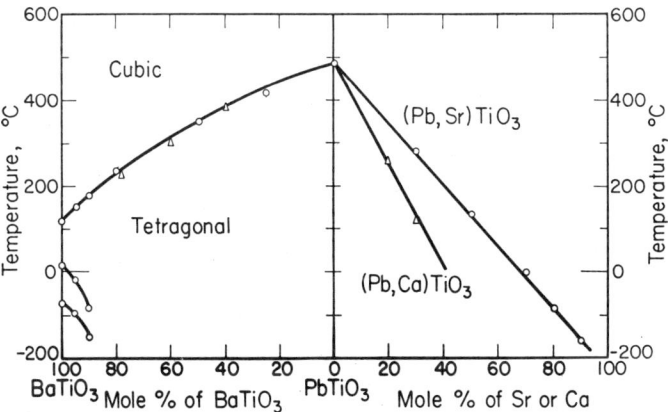

FIG. V-16. Solid solutions of $PbTiO_3$. Left: phase diagram of the system $PbTiO_3$–
$BaTiO_3$ (according to Shirane and Suzuki (S 4), triangles according to Nomura
and Sawada (N 2)). Right: phase diagrams of the systems $PbTiO_3$–$SrTiO_3$
(according to Nomura and Sawada (N 4) and $PbTiO_3$–$CaTiO_3$ (according to
Smolenskii (S 17)).

A very high electromechanical coupling coefficient and a very large piezoelectric modulus $d_{33}$ were reported for solid solutions with compositions close to the phase boundary between the rhombohedral and the tetragonal phases (Fig. V-17). Recently, Berlincourt and Krueger (B 5) carried out a detailed study of the processes of domain reorientation in, $Pb(Zr_{0.43}Ti_{0.57})O_3$, a composition with tetragonal symmetry ($c/a = 1.022$). They showed that 44% of the 90° domains and almost all of the 180° domains remain oriented after removal of the pooling field. For more general information on piezoelectric ceramics, the reader is referred to the excellent review article of H. Jaffe (J 5).

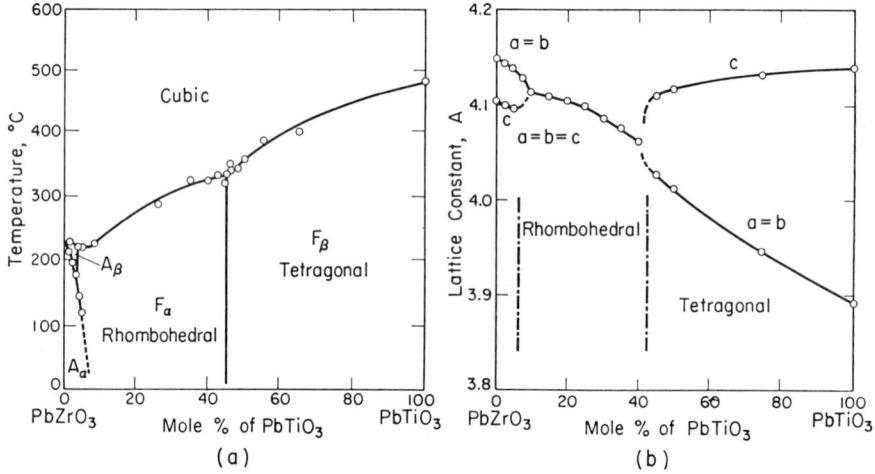

FIG. V-17. Solid solutions of $PbTiO_3$ and $PbZrO_3$.
(a) Phase diagram (according to Sawaguchi (S 23)).
(b) Lattice parameters vs. composition (according to Shirane and Suzuki (S 21)).

Solid solutions $Pb(Ti, Sn)O_3$ were studied by Nomura (N 5), Smolenskii et al. (S 24) and Venevtsev and Zhdanov (V 3). The phase diagram exhibits characteristics that are very similar to those of the system $PbTiO_3$–$PbZrO_3$. A ferroelectric rhombohedral phase was observed in the range of Sn concentration from 60 to 80%. Compositions richer in Sn could not be prepared with a perovskite-type lattice. This is in contradiction with an early study of Naray–Szabo (N 1), who reported the existence of a compound $PbSnO_3$ having a perovskite-type lattice, but is in agreement with a more recent investigations of Roth (R 1), who found that a perovskite-type "$PbSnO_3$" does not exist. The Curie point of $PbTiO_3$ is lowered by the addition of Sn and occurs at approximately 200 °C for the compound $Pb(Ti_{0.2}Sn_{0.8})O_3$, which represents the solubility limit in this system. It may be interesting to recall, in this respect, that the Sn ion does not have a noble gas configuration, in contrast to both Ti and Zr. As mentioned in Section 1, such an electronic configuration for the $B$ ion has been proposed by Matthias (M 3) as a condition essential for the occurrence of ferroelectricity in perovskite-type compounds $ABO_3$.

## 6. Lead Zirconate, PbZrO$_3$, its Solid Solutions and Lead Hafnate, PbHfO$_3$

### Lead Zirconate, PbZrO$_3$

The existence of a marked dielectric anomaly in PbZrO$_3$ was first reported by Roberts (R 9) and by Smolenskii (S 8) on the basis of measurements carried out on ceramic material. Figure V-18 shows that the dielectric constant reaches a

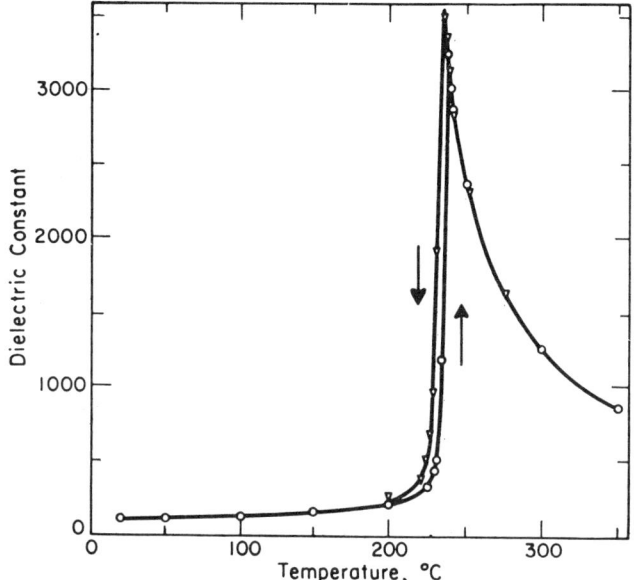

FIG. V-18. Dielectric constant of ceramic PbZrO$_3$ as a function of temperature (according to Roberts (R 9)).

high peak at 230 °C. Above this temperature, the structure is of the cubic perovskite type and the dielectric constant follows the Curie–Weiss law $\varepsilon = C/(T - T_0)$ with $C = 1.6 \times 10^5$ degrees. Although the character of the dielectric anomaly is similar to that observed in BaTiO$_3$, no ferroelectric hysteresis loops were observed below the transition temperature. Another characteristic feature of PbZrO$_3$ is the sign of the volume anomaly, there being a contraction of the volume on cooling (S 25), (S 26), in contrast to BaTiO$_3$ which, upon cooling, exhibits a volume expansion. Subsequent careful studies of the dielectric and structural properties of PbZrO$_3$, by Sawaguchi *et al.* (S 29), (S 30) have established that the phase occurring below 230 °C is antiferroelectric: in the vicinity of the transition temperature, strong electric fields can induce a ferroelectric phase.

The symmetry of the antiferroelectric phase was first reported to be tetragonal, on the basis of X-ray powder diffraction patterns, with the lattice parameters (M 1):

$$a_T = 4.159 \text{ Å} \quad \text{and} \quad c_T/a_T = 0.988$$

Later, however, an X-ray and optical study of small single crystals (S 27) revealed that the true symmetry of this phase is orthorhombic, the $a$ and $b$ axing being oriented along pseudo-cubic $\langle 110 \rangle$ directions. The lattice parameters are: $a = a_T \sqrt{2} = 5.88$ Å; $b = 2a$; $c = 2c_T = 8.20$ Å. The temperature dependence of the lattice parameters (S 28) is depicted in Fig. V-19, where the antiferroelectric phase is described in terms of pseudo-tetragonal axes for the convenience of comparison with the lattice constant of the cubic phase above 230 °C.

The X-ray study of Sawaguchi et al. (S 27) revealed that the structure of the orthorhombic phase is derived from the cubic perovskite-type lattice essen-

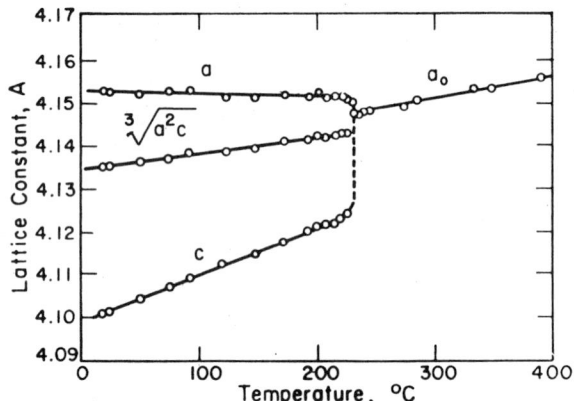

Fig. V-19. Temperature dependence of the lattice parameters of $PbZrO_3$. The antiferroelectric phase is described in terms of pseudo-tetragonal axes $a$ and $c$. $a_0$ = cubic parameter (according to Sawaguchi (S 28)).

tially through antiparallel displacements of the Pb ions along one of the original $\langle 110 \rangle$ directions, which is called the $a$ axis. The character of this distortion is depicted schematically in Fig. V-20. A further detailed investigation of this structure by Jona et al. (J 6), on the basis of X-ray diffraction photographs of single crystals and neutron diffraction patterns of polycrystalline material, has revealed that the oxygen atoms are also displaced antiparallel to each other in the (001) plane. If we compare this structure with that of $NaNbO_3$ (see Fig. V-6), we see that antiparallel atomic shifts within the (001) plane represent a feature common to both lattice distortions. This may explain the fact that the pseudo-tetragonal axial ratio $c/a$ is smaller than unity, since we know, from the case of $BaTiO_3$, that the spontaneous strain perpendicular to the polarization always results in a contraction. The fact that, in $PbZrO_3$, the orthorhombic $b$ spacing is exactly twice as large as the $a$ spacing, which gives rise to the pseudo-tetragonal character of this phase, may only be a coincidence due to fortuitous balancing of the oxygen shifts along $b$ and the lead shifts along $a$.

In contrast to $NaNbO_3$, however, the neutron diffraction analysis has revealed the existence of some uncompensated oxygen displacements along the [001] axis. As a consequence of these displacements, the [001] axis is polar and the true space group of the antiferroelectric phase is *Pba2*. Thus, this phase ought to exhibit a piezoelectric effect, in accordance with the results reported by Roberts (R 13), and it is reasonable to ask why the crystal is not ferroelectric along the $c$ axis. Although this question is not completely settled, the detailed study of the structure seems to suggest that reversal of the polarity by means of external electric fields is likely to be very difficult (J 6).

The dielectric properties of $PbZrO_3$ were investigated in detail by Sawaguchi et al. (S 29), (S 30), (S 31). It was shown that, in contrast to $BaTiO_3$, the application of a d.c. bias causes the Curie temperature to decrease: for 20 kV/cm, this decrease amounts to 1.5 °C. As mentioned above, no hysteresis loops can be

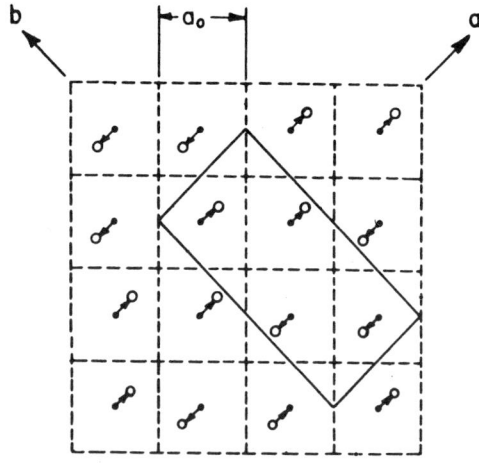

FIG. V-20. Schematic projection on (001) of the antiferroelectric structure of $PbZrO_3$. Arrows indicate the directions of Pb displacements from the centrosymmetrical positions. Dashed lines delineate the original perovskite lattice, the solid line shows the orthorhombic unit cell. $a_0$ is the pseudo-cubic lattice parameter (according to Sawaguchi et al. (S 27)).

observed in the room temperature phase: the relation between polarization $P$ and field $E$ is strictly linear. In the vicinity of the transition temperature, however, a ferroelectric phase is induced when the applied field reaches a critical value. The magnitude of this critical field increases with decreasing temperature, as indicated in Fig. V-21. Thus, double hysteresis loops are observed similar to those obtained with $NaNbO_3$ at room temperature. Analogous results were also reported by Smolenskii (S 2). It may be of historical interest to point out that $PbZrO_3$ is the compound in which double hysteresis loops were observed first. Also, it may not be superfluous to recall that these double loops have quite a different origin from that responsible for the analogous phenomenon in $BaTiO_3$. In the latter crystal, double loops are observed *above* the Curie point and ferroelectricity is induced in a phase that is originally non-polar; in $PbZrO_3$, the double loops are observed *below* the transition temperature and ferroelectricity is induced in an antiferroelectric phase. An X-ray study by Shirane and Hoshino (S 34) has shown that the ferroelectric phase induced in $PbZrO_3$ has rhombohedral symmetry, as one may have expected from the phase diagram of PbZrO$_3$ and PbTiO$_3$ (see Fig. V-17).

The dielectric data indicate that the phase transition in PbZrO$_3$ is of the first order. Calorimetric measurements reveal a large anomaly of the specific heat: the integrated value yields 440 cal/mole for the heat of transition (S 29). Optical investigations of small single crystals were carried out by Jona *et al.* (J 1), who studied the geometry of antiferroelectric domains (twinning rules) and the temperature dependence of the birefringence. The elastic properties of ceramic specimens were investigated by Marutake and Ikeda (M 18), who reported an anomaly of the elastic constant at the transition temperature.

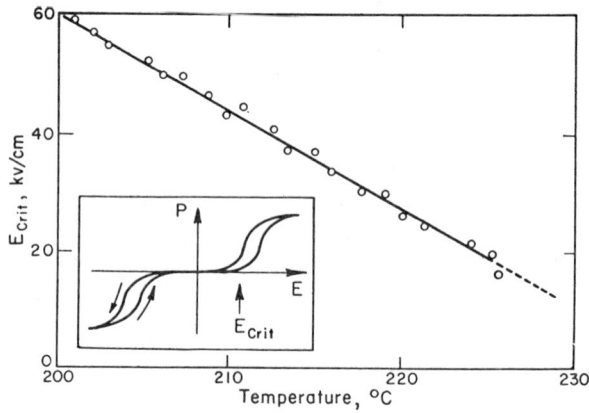

FIG. V-21. Temperature dependence of the threshold field strength E$_{crit}$ of PbZrO$_3$. Insert: Definition of E$_{crit}$ (according to Sawaguchi and Kittaka (S 31)).

### Solid Solutions Containing PbZrO$_3$

Several systems of solid solutions involving PbZrO$_3$ and a number of other perovskite type compounds have been extensively investigated. Particular attention has been devoted to the dielectric properties and the structural distortion of the new phases which are observed.

The system PbZrO$_3$–PbTiO$_3$ has already been discussed in Section 5 (see Fig. V-17). The appearance of a ferroelectric rhombohedral phase on the PbZrO$_3$ side of the phase diagram is in accordance with the rhombohedral symmetry of the ferroelectric phase induced by large electric fields in pure PbZrO$_3$. Investigations of the solid solutions (Pb, Ba)ZrO$_3$ have revealed that the same phase appears when Pb is partially replaced by Ba (see Fig. V-22) (S 33), (S 34), (R 9). This is a further confirmation of the fact that the stable antiferroelectric phase of PbZrO$_3$, below its transition temperature, is energetically very close to the ferroelectric rhombohedral phase, as the latter can be obtained not only by the application of electric fields but also by relatively small changes in composition.

A rather interesting feature of the solid solutions (Pb, Ba) ZrO$_3$ and Pb (Zr, Ti) O$_3$ is the nature of the volume anomaly at the corresponding phase transitions. We have already pointed out that all perovskite-type ferroelectrics, such as BaTiO$_3$, KNbO$_3$ and PbTiO$_3$, are characterized by a volume *expansion* when the temperature is lowered through their Curie points. $PbZrO_3$, on the other hand, exhibits an anomalous volume *contraction* when the temperature is decreased through the antiferroelectric transition point. These phenomena have been referred to as

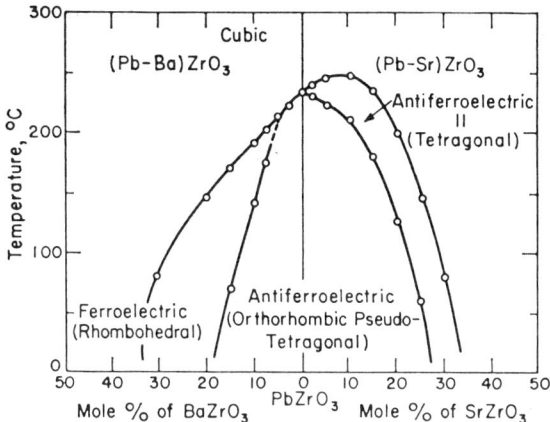

Fig. V-22. Phase diagrams of the systems $(Pb, Ba)ZrO_3$ and $(Pb, Sr)ZrO_3$ (according to Shirane (S 33)).

positive and negative electrostriction, respectively, by Smolenskii (S 26). Thus, the hypothesis was advanced that some solid solutions between $PbZrO_3$ and any one of the ferroelectric perovskites may exhibit, at suitable compositions, no volume anomaly at the transition temperature or, in other words, a phenomenon of zero electrostriction. Figure V-23 shows, however, that this is not the case. The

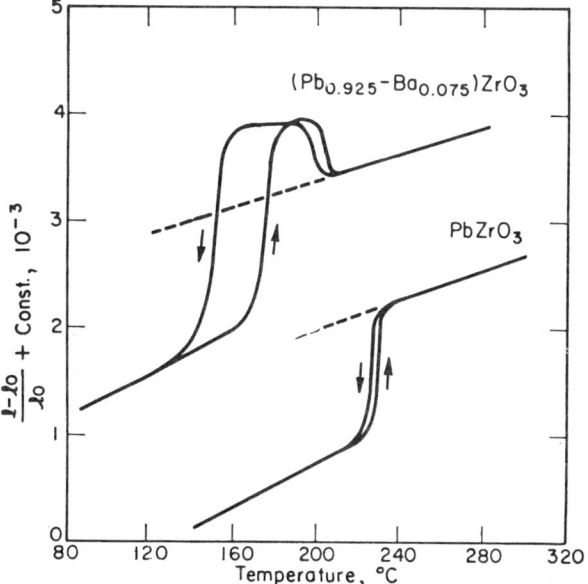

Fig. V-23. Volume anomalies in $PbZrO_3$ and $(Pb, Ba)ZrO_3$ (according to Shirane (S 33)).

solid solutions undergo, in sequence for decreasing temperature, first a volume expansion at the Curie point from the non-polar to the ferroelectric state and then a larger volume contraction at the transition from the ferroelectric into the antiferroelectric phase.

The appearance of a ferroelectric rhombohedral phase upon slight changes in composition from $PbZrO_3$ is, on the other hand, not a rule with general validity. In the systems $PbZrO_3$–$SrZrO_3$ (S 33), (K 7) and $PbZrO_3$–$CaZrO_3$ (S 33), (K 6), for example, another phase does indeed appear but it is a new antiferroelectric phase which has a tetragonal structure, i.e. different from that of pure $PbZrO_3$ (see Fig. V-22). A number of other solid solutions involving $PbZrO_3$ and additions of Sn, Ce or Cd have been investigated by Krainik (K 6), (K 7).

### Lead Hafnate, $PbHfO_3$

The phase transitions and the dielectric properties of $PbHfO_3$ have been studied by Shirane and Pepinsky (S 40) and were found to be very similar to those of $PbZrO_3$. Two transitions were reported, at 163 °C and 215 °C, respectively. X-ray powder diffraction patterns indicate that the structure of the room-temperature phase is isomorphous with that of $PbZrO_3$, suggesting that this phase may also be antiferroelectric. However, no induced ferroelectric phase was observed under the action of electric fields. The intermediate phase, between 163 and 215 °C, is likely to be isomorphous with the tetragonal antiferroelectric phase occurring in solid solutions $(Pb, Sr)ZrO_3$ (see Fig. V-22).

### 7. Solid Solutions Containing $BaTiO_3$

As $BaTiO_3$ was the first, and most important, ferroelectric perovskite to be discovered, a large number of solid solutions containing $BaTiO_3$ as one of the components have been studied in great detail. The addition of $PbTiO_3$ has already been discussed in Section 5. It turned out that the solid solutions $BaTiO_3$–$PbTiO_3$ represent the only known system in which the Curie temperature of $BaTiO_3$ can be raised by isomorphous substitution. Figure V-24 shows that, otherwise, replacement of Ba with Ca or Sr, and Ti with Zr or Sn causes a decrease of the Curie point.

The system $BaTiO_3$–$SrTiO_3$ was the first to be studied systematically (R 10), (G 1). As was already mentioned in Section 1, the Curie point of $BaTiO_3$ decreases linearly with increasing concentration of Sr. Addition of Sr is, in fact, still considered as the standard method to lower the Curie temperature of $BaTiO_3$ for the purpose of practical applications. The volume of the unit cell also exhibits a linear decrease with increasing Sr concentration (R 10), a fact that, in the early stages, seemed to confirm the hypothesis of a "rattling" Ti ion as the origin of ferroelectric activity (see Section 1). Extrapolation of the straight line representing the dependence of the Curie temperature upon Sr concentration would lead one to expect a Curie point for pure $SrTiO_3$ around 40 °K. This, however, does not obtain; the dielectric properties of pure $SrTiO_3$ are, in fact, still an object of controversy and will be discussed in the next section.

An interesting result of the study of these (Ba, Sr)TiO$_3$ solid solutions is the fact that the tetragonal–orthorhombic transition temperature of BaTiO$_3$ is also lowered by the Sr addition but less rapidly than the Curie temperature. As a result, the temperature range of the tetragonal phase is reduced. For example, the solid solutions containing 25% and 50% of Sr exhibit the tetragonal–orthorhombic transition at $-30$ °C and $-100$ °C, respectively. The lowest phase change of BaTiO$_3$ (orthorhombic to rhombohedral), on the other hand, is affected only little by the Sr addition. It is therefore reasonable to expect that the symmetry of the ferroelectric phase occurring in Sr-rich solid solutions is rhombohedral. This is also quite analogous to the situation encountered in the system KNbO$_3$–KTaO$_3$ (see Section 2).

Figure V-24 shows that the system BaTiO$_3$–CaTiO$_3$ behaves somewhat differently than the other systems, in that the Curie temperature of BaTiO$_3$ is hardly affected by the Ca addition. This result is quite unexpected, as Ba, Sr and Ca belong all into the same column of the periodic table of the elements and have ionic radii and polarizabilities decreasing in this order. Ca can replace Ba, at room tempera-

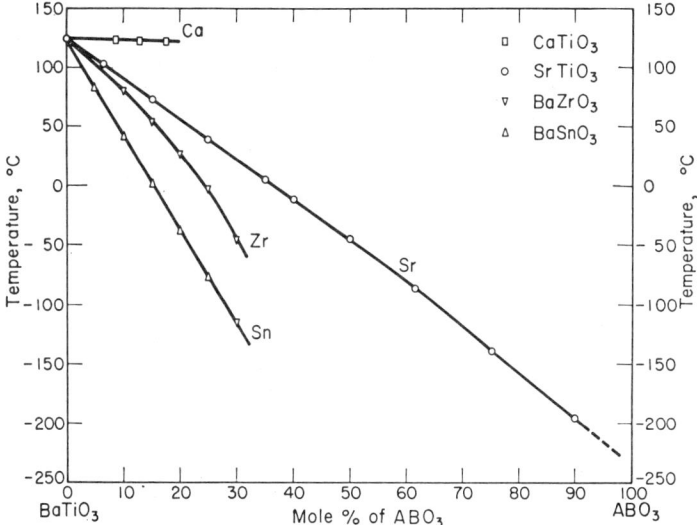

Fig. V-24. Dependence of the Curie temperature of BaTiO$_3$ upon various additions (references are given in Table V-7).

ture, only up to 25 mole %, the lattice parameters decreasing with increasing Ca content (D 1), (M 11). A "dome" of insolubility has been reported by DeVries and Roy (D 3) between the end members of this solid solution system. In contrast to the invariance of the Curie point, the room temperature value of the dielectric constant of BaTiO$_3$ is decreased by the Ca addition. Similarly, the tetragonal–orthorhombic transition temperature is lowered (B 2) and occurs at $-60$ °C for a Ca content of 10 mole %.

Two other interesting systems are those involving the substitution of Ti, in BaTiO$_3$, with Zr or Sn. The Curie temperature is lowered in both cases (see

Fig. V-24), while the two lower transitions are raised in such a way as to converge at the Curie point near room temperature (K 8), (S 35), (S 37).

The references cited in Table V-7 are essentially concerned with structural and dielectric investigations of the phase diagrams. A very large number of other publications have appeared on the subject of dielectric and piezoelectric characteristics of ceramic solid solutions containing $BaTiO_3$. These investigations have been concerned not only with the binary systems discussed above but also with the effect of a variety of additions upon ceramic properties such as porosity, conductivity, internal friction, etc. A review of the pertinent literature up to 1955 has been written by McQuarrie (M 4). More recently, a rapidly increasing number of Russian papers are being published on this general subject (see, for example, Bokov (B 3) and Roi (R 11).)

Need for new and better materials in the field of technical applications has also initiated a series of investigations of ternary and quaternary systems. Some of the pertinent literature is included in Table V-7. The phase diagram of the system $(Ba, Pb)(Ti, Zr)O_3$ reported by Ikeda (I 1) is depicted in Fig. V-25.

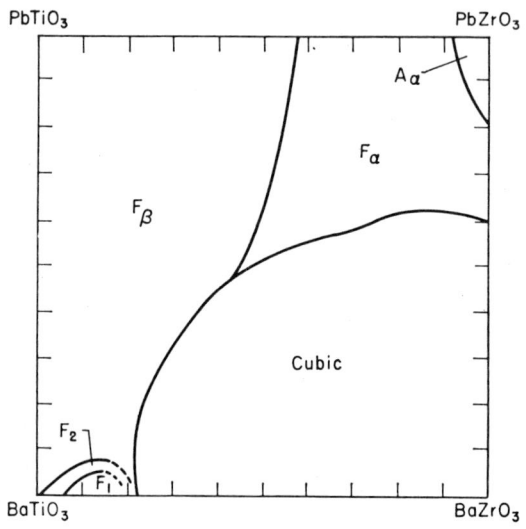

FIG. V-25. Phase diagram of the system $(Ba, Pb)(Ti, Zr)O_3$. $F_\alpha$: rhombohedral, $F_\beta$: tetragonal, $F_1$: rhombohedral, $F_2$: orthorhombic (according to Ikeda (I 1)).

## 8. Strontium Titanate, $SrTiO_3$ and Cadmium Titanate, $CdTiO_3$

### Strontium Titanate, $SrTiO_3$

We have already mentioned $SrTiO_3$ in connection with the solid solution systems $(Ba, Sr)TiO_3$ and $(Pb, Sr)TiO_3$. In both cases, extrapolation of the straight line representing the dependence of the Curie temperature upon composition indicates that pure $SrTiO_3$ should have a Curie point around 40 °K. This extrapolation, however, is not very reliable, as in both systems no solid solution with more than 90 mole % of $SrTiO_3$ was studied. If, on the other hand, the temperature dependence of the dielectric constant of pure $SrTiO_3$ is considered, it turns out that the Curie–Weiss law is obeyed, and extrapolation of this relation also seems to predict a Curie point at low temperature.

A peak in the dielectric constant of ceramic $SrTiO_3$ around 20–30 °K was in fact reported by Smolenskii (S 7), (S 8). The value of the dielectric constant at room temperature was reported to be 250 and that at the peak about 2000. Below the peak, well-defined hysteresis loops were observed. These results, however, are in sharp contrast with those of Hulm (H 2), who observed no maximum in the dielectric constant down to 1.3 °K. In the temperature range between 4 °K and 1.3 °K, the value of the dielectric constant is 1300 and appears to be practically temperature independent. At the lower temperature, it is reportedly also independent of applied electric fields up to field strengths of 25 kV/cm. This is not the behavior that one would expect from a ferroelectric compound. It is

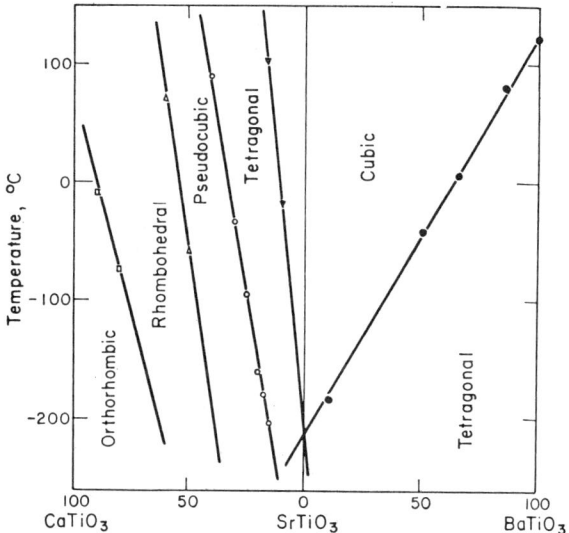

FIG. V-26. Phase diagrams of the systems $SrTiO_3$–$CaTiO_3$ and $SrTiO_3$–$BaTiO_3$ (according to Gränicher and Jakits (G 1)).

possible that these discrepancies may be explained, as suggested by Smolenskii (S 2), by the fact that the dielectric properties of $SrTiO_3$ are very sensitively dependent upon impurities and conditions of sample preparation.

More recent studies of Gränicher and Jakits (G 1), (G 2), however, seem to shed light on this problem from a different point of view. The phase diagrams $SrTiO_3$–$BaTiO_3$ and $SrTiO_3$–$CaTiO_3$, depicted in Fig. V-26, are quite instructive, in this respect. The (Ba, Sr)$TiO_3$ line extrapolates to about 40 °K, whereas the (Ca, Sr)$TiO_3$ line extrapolates to 90 °K. The tetragonal phase on the $CaTiO_3$ side is not ferroelectric and therefore different from the tetragonal phase on the $BaTiO_3$

side, which, of course, is ferroelectric. These extrapolations are admittedly not very accurate but this does not invalidate the supposition that both tetragonal phases must be energetically very close to one another in pure $SrTiO_3$ at low temperatures. Gränicher's dielectric measurements on single crystals indicate that the Curie–Weiss law $\varepsilon = C/(T-T_0)$ is obeyed down to 95 °K, with $C = 8.3 \times 10^4$ degrees and $T_0 = 38$ °K. At lower temperatures, deviations from this law are observed but the dielectric constant keeps increasing monotonously up to a value of 20,000 at 20 °K. Hysteresis loops can be observed below 50 °K and the spontaneous polarization equals $3 \times 10^{-6}$ C/cm² at 4.2 °K, but for applied fields of only a few hundred volts per centimeter, the relation between polarization and field appears to be linear. Thus, it is reasonable to hypothesize that the observed ferroelectric phase is induced by the strong electric fields applied. The results of a recent study of the dielectric properties of single crystals, by Weaver (W 3), are consistent with this picture.

Gränicher (G 3) reported also that a sluggish transformation of the crystal into a non-ferroelectric tetragonal phase occurs below 90 °K. Recent experiments by Dobrov et al. (D 4) and by Müller (M 19) on the paramagnetic resonance of $Fe^{3+}$ in $SrTiO_3$ indicate also a deviation from cubic symmetry at 80 °K.

### Cadmium Titanate, CdTiO₃

$CdTiO_3$ was reported to be ferroelectric below 50–60 °K by Smolenskii (S 7), (S 8), (S 41). The dielectric constant has the value 250 at room temperature and obeys the Curie–Weiss law $\varepsilon = C/(T-T_0)$ with $C = 4.5 \times 10^4$ degrees.

At and above room temperature, however, $CdTiO_3$ is not cubic. The symmetry of the room-temperature phase is orthorhombic and the lattice parameters, as reported by Megaw (M 1), are:

$$a = 10.834 \text{ Å}, \quad b = 10.695 \text{ Å}, \quad c = 7.615 \text{ Å}.$$

The corresponding monoclinic axes (see Section 1) are:

$$a = b = 2 \times 3.791 \text{ Å}, \quad c = 2 \times 3.807 \text{ Å}, \quad \gamma = 91° 10'.$$

The atomic positions in the room-temperature phase were determined by Kay and Miles (K 1), but no structural study was carried out below the Curie temperature.

It is reasonable to expect that the cubic perovskite structure will be stable, at some higher temperature, but the exact transition temperature is not known. It must be mentioned, in this respect, that the ceramic samples of $CdTiO_3$ must be sintered at relatively low temperatures, namely, 1050–1180 °C. When the sintering temperature is lower than 1000 °C, the compound crystallizes with the ilmenite structure; when it is higher, an irreversible inversion into the perovskite structure type occurs (P 2). However, according to Smolenskii, the compound decomposes at higher temperatures, which justifies the relatively low temperature of preparation. The fact that $CdTiO_3$ is not cubic around room temperature is probably only a result of mismatched atomic sizes. The $Cd^{2+}$ ion may namely be too small (radius 1.03 Å) to co-operate with the oxygen ions in the formation of a face-centered-cubic close-packed structure.

This situation is not uncommon in the perovskite family. We have already mentioned that a number of titanates and zirconates crystallize with a distorted perovskite structure. Perovskite itself, $CaTiO_3$, is a typical example of this phenomenon, as it becomes cubic only at 1260 °C (G 1). It ought to be stressed once more that an unfortunate custom has been established, in the literature, in labeling all these non-cubic non-ferroelectric compounds as antiferroelectrics. This habit, however, should not be encouraged, as already discussed in Chapter I, essentially because these types of compounds offer no particular interest as dielectrics until it is proved that their properties are closely related to ferroelectricity.

## 9. Complex Perovskite-type Compounds

We have learned, so far, that perovskite-like compounds of the type $ABO_3$ can be obtained with $A$ and $B$ ions which have ionic radii within fairly wide limits. The same structure type was found in solid solutions of the type $A(B_i, B_j)O_3$ and $(A_i, A_j)BO_3$, and even in quaternary systems of the general kind $(A_i, A_j)(B_i, B_j)O_3$ (see Section 7). Accordingly, one can consider the preparation of perovskite-type compounds with more complicated combinations of atoms than those discussed so far, provided, of course, that charge neutrality is preserved. Gränicher and Jakits (G 1), for example, investigated the compounds:

$$(K_{0.5}, La_{0.5})TiO_3, \quad (K_{0.5}, Nd_{0.5})TiO_3$$

and their solid solutions with $BaTiO_3$.

More recently, Smolenskii and co-workers (S 42–S 45), (S 53) have carried out extensive investigations of compounds prepared according to the same

TABLE V-8. COMPLEX PEROVSKITE-TYPE FERROELECTRICS
Chemical formulae are given according to the terminology used by Smolenskii and coworkers. The relationship to the formula $ABO_3$ is explained in the text.

| Composition | Curie temperature $T_c$ (°C) | Reference |
|---|---|---|
| $Pb_2(FeNb)O_6$ | 112 | S 43 |
| $Pb_2(FeTa)O_6$ | − 30 | S 53 |
| $Pb_2(ScNb)O_6$ | 90 | S 45 I 3 |
| $Pb_2(ScTa)O_6$ | 26 | S 45 I 3 |
| $Pb_2(YNb)O_6$ | 280* | S 43 |
| $Pb_2(MgW)O_6$ | 39* | S 53 |
| $Pb_3(MgNb_2)O_9$ | − 8 | S 42 S 44 |
| $Pb_3(NiNb_2)O_9$ | − 150 | S 42 M 20 |
| $Pb_3(Fe_2W)O_9$ | − 80 † | S 53 |

\* Probably non-ferroelectric.
† Probably also antiferromagnetic.

general idea. Several ferroelectrics have, in fact, been discovered in systems of the following type:

$$\mathrm{Pb}^{2+}\left[(B_i^{3+})_{1/2}\,(B_j^{5+})_{1/2}\right]\mathrm{O}_3,$$

$$\mathrm{Pb}^{2+}\left[(B_i^{2+})_{1/3}\,(B_j^{5+})_{2/3}\right]\mathrm{O}_3.$$

Practical examples are summarized in Table V-8. Although the details of the dielectric and structural characteristics of these compounds are not known, they all seem to exhibit a behavior very similar to that of $\mathrm{BaTiO}_3$. The lattice parameters of $\mathrm{Pb}(\mathrm{Sc}_{0.5},\mathrm{Nb}_{0.5})\mathrm{O}_3$ have recently been reported by Ismailzhade (I 3) as:

$$a = 4.074\ \text{Å}\quad \text{and}\quad c/a = 1.002$$

with a multiple unit cell.

## 10. Perovskite-type Compounds with Non-stoichiometric Composition

It has been established that perovskite-type compounds can not only be obtained with the general formula $ABO_3$ but also with non-stoichiometric compositions of the type $A_{1-x}BO_3$ or $ABO_{3-y}$ (see, for example, Kestigian et al. (K 11)). If we put $x = 1$ in the former formula, we see that we can consider tungsten trioxide, $\mathrm{WO}_3$, as the extreme case of compounds of this type. It is known, in fact, that the tungsten ions assume the same positions, in the lattice, as the titanium ions in $\mathrm{BaTiO}_3$, but the Ba ions (or, generally, the $A$ ions) are missing.

In recent years, research into ferroelectricity has been extended to various systems of solid solutions with non-stoichiometric compositions, i.e. containing lattice vacancies. The crystal chemistry of these compounds is not well established, to date. Dielectric measurements are often hindered by the fact that most of these compounds tend to have a rather high electrical conductivity. In a few cases, however, compounds of this type have turned out to be interesting from the viewpoint of dielectric behavior so that it may be worth devoting some space to them.

### Tungsten Trioxide, $\mathrm{WO}_3$

This compound has been the object of very extensive investigations ever since ferroelectric properties were reported by Matthias in 1949 (M 16). Unfortunately, no reliable dielectric measurements could be carried out at and above room temperature because of high electrical conductivity. Evidence for ferroelectric activity was given only below a phase transition occurring at $-50\ °\mathrm{C}$ (M 16), (M 17).

The symmetry of the room-temperature phase is monoclinic. Three phase transitions were reported to occur above room temperature, namely at 330 °C, 740 °C and 910 °C. The structural characteristics and other physical properties of the various phases are described in detail in the book of Megaw (M 2), the review article of Känzig (K 3), and recent papers by Sawada and Danielson (S 46).

The data available so far seem to indicate that $\mathrm{WO}_3$ is similar to $\mathrm{NaNbO}_3$ in its dielectric characteristics and phase transitions. It appears, however, that the high electrical conductivity of the samples investigated is not an intrinsic characteristic of the compound $WO_3$ but rather a consequence of the variable valence states that can be assumed by the tungsten ion. Thus, the preparation of nonconducting samples of $WO_3$ may be extremely difficult.

### Additions of Nb and Ta to $BaTiO_3$: $Ba(Ti, Nb)O_3$ and $Ba(Ti, Ta)O_3$

These systems of solid solutions were first studied by Smolenskii et al. (S 47) and by Isupov (I 2). The samples were prepared according to the general equation.

$$(1 - x)BaTiO_3 + xBa_{0.5}BO_3 \rightarrow Ba_{1-0.5x}(Ti_{1-x}, B_x)O_3,$$

where $B$ stands for either Nb or Ta. If neither $Ti^{4+}$ nor $Nb^{5+}$, and $Ta^{5+}$, respectively, change their valence states, the chemical equation given above represents correctly the solid state reaction involved. A recent X-ray study by Subbarao and Shirane (S 48) has confirmed that reactions of this type lead indeed to solid solutions. What is not certain is whether the lattice vacancies introduced in this fashion are limited to the $A$ sites of the perovskite lattice or exist also in the $B$ and O sites. At any rate, a rather high electrical conductivity for the samples is expected, owing to the probability of mixed valence states, $Ti^{3+}$ and $Ti^{4+}$, of the titanium ion. Saburi (S 51) on the other hand, has shown that the resistivity of the samples with Nb additions goes through a minimum when the Nb contents is 0.2–0.4 mole % but then increases again with larger Nb concentrations. An explanation for this odd phenomenon is not available at the present time.

The Curie temperature of $BaTiO_3$ is lowered rapidly by the addition of Nb and occurs at 0 °C when the variable $x$ of the formula given above equals 0.05 (see Fig. V-27). Addition of Ta has almost identical effects. The two lower phase

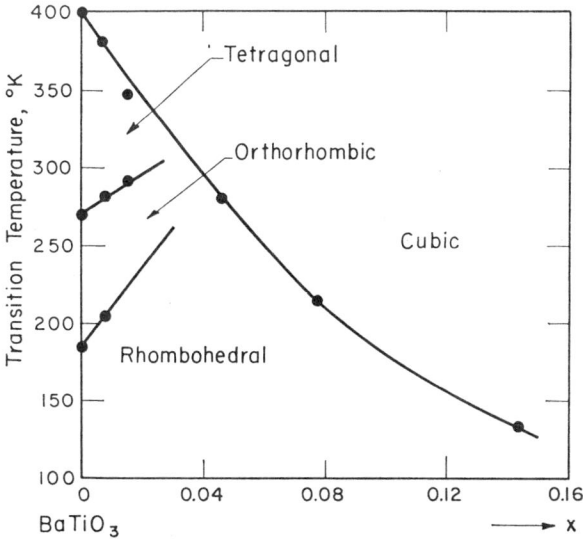

FIG. V-27. Variation of the transition temperatures in the system $Ba_{1-0.5x}(Ti_{1-x}, Nb_x)O_3$ (according to Subbarao and Shirane (S 48)).

transition temperatures of $BaTiO_3$ are raised by the additions, in much the same way as those for solid solutions $Ba(Ti, Zr)O_3$. The limit of solubility is found at about 10%.

Similar results were obtained with solid solutions of the type $Pb(Ti, Nb)TiO_3$ and $Pb(Ti, Ta)O_3$ by Subbarao (S 50). In these systems, the Curie temperature of $PbTiO_3$ is lowered to 465 °C at the solubility limit of 3–5%. Compounds of the type $Pb(Zr, Nb)O_3$ and $Pb(Zr, Ta)O_3$ were studied by Krainik (K 7): the corresponding phase diagrams recall that encountered in the system $(Pb, Ba)ZrO_3$ (see Fig. V-22).

### Additions of $Fe_2O_3$ to $BaTiO_3$

We have mentioned in Section IV-12 that Remeika's method for growing large tabular crystals of $BaTiO_3$ from binary melts with KF usually involves the addition of small amounts of $Fe_2O_3$ (about 0.2 atomic %) in order to reduce the electrical conductivity of the crystals obtained. As a result, the Curie point of these crystals is usually a few degrees lower than that of crystals grown without addition of $Fe_2O_3$. This effect was studied more systematically by Nishioka et al. (N 6) who found that the Curie temperature is lowered linearly to 15 °C by the addition of 5 atomic % of $Fe_2O_3$. The lattice parameters change accordingly. The Fe ions are probably replacing Ti in the states $Fe^{2+}$ or $Fe^{3+}$, thus creating some vacancies in the Ba and O sites. Similar replacements with oxides of other $3d$ transition metals were also studied (S 49).

### Conducting $BaTiO_3$

In recent years, rather interesting behavior of the electrical conductivity of samples of $BaTiO_3$ containing small amounts (0.1–0.3 mole %) of $La_2O_3$ or $Sm_2O_3$ has been reported in the literature. These additions have two striking effects: in the first place, the value of the electrical resistivity at room temperature is decreased drastically from about $10^{10}\ \Omega$ cm to $10 - 10^2\ \Omega$ cm; in the second place, the temperature dependence of the resistivity exhibits a pronounced anomaly in the vicinity of the Curie point of pure $BaTiO_3$ (see Fig. V-28).

These phenomena were first reported by Sauer and Flaschen (S 52) for the case of $La_2O_3$ additions, then by Harman (H 3) for the case of $Sm_2O_3$ additions. Extensive investigations were later carried out by Saburi (S 51), who found that similar phenomena can also be observed by adding to $BaTiO_3$ oxides of other metals such as Bi, Ce, Sb, Nb and Ta, as well as by reducing pure $BaTiO_3$. The chemical formulae of these conducting samples may probably be written in the following way:

$$\left(Ba^{2+}_{1-x}, La^{3+}_x\right)\left(Ti^{4+}_{1-x}, Ti^{3+}_x\right)O_3 \quad \text{and} \quad Ba^{2+}\left(Ti^{4+}_{1-2x} Ti^{3+}_x Ta^{5+}_x\right)O_3.$$

The mechanism of the change in resistivity around the Curie temperature is not well understood at the present stage. There is experimental evidence for the belief that the phenomenon is intimately connected with the ferroelectric phase transition. It was established, namely, by Saburi with an X-ray study of samples of $(Ba, Ce)TiO_3$ that the increase in resistivity at 120 °C is indeed related to the tetragonal–cubic phase change. Moreover, Sauer and Flaschen demonstrated that the critical temperature of the resistivity increase can be lowered by partial replacement of Ba with Sr, which is in accordance with the lowering of the Curie point of $BaTiO_3$ in the non-conducting system (Ba, Sr)$TiO_3$ discussed in Section 7. The piezoresistance of these conducting $BaTiO_3$ compounds was also studied by Sauer *et al.* (S 17).

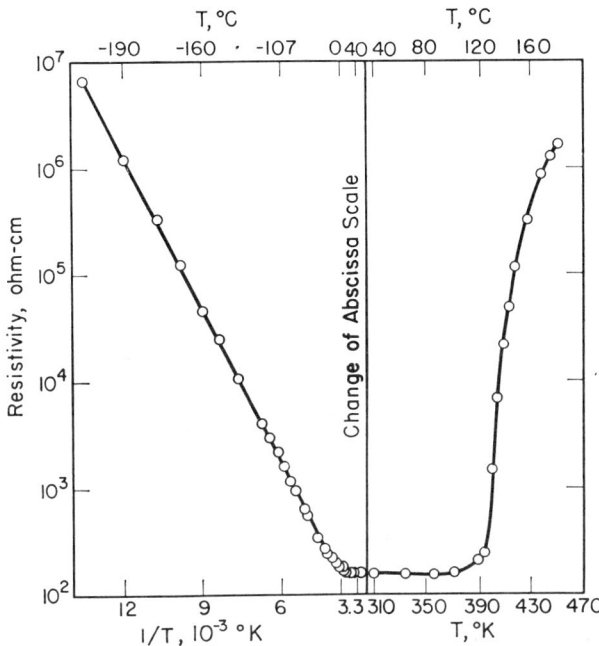

FIG. V-28. Temperature dependence of the resistivity in a sample of $BaTiO_3$ containing 0.1 mole % of $Sm_2O_3$. Note the change in the abscissa axis (according to Harman (H 3)).

It is particularly interesting to notice that the resistivity exhibits a continuous increase after the sharp anomaly at 120 °C. It is conceivable that this behavior is related to the change in polarizability of the crystal in the range where the dielectric constant of non-conducting $BaTiO_3$ decreases hyperbolically. No noticeable change in resistivity was detected, on the other hand, at the two lower transition temperatures of pure $BaTiO_3$.

### Additions of Cd and Pb to $NaNbO_3$: (Na, Cd)$NbO_3$ and (Na, Pb)$NbO_3$

Compounds of this type may be prepared according to the following reaction:

$$(1 - x)NaNbO_3 + xCd_{0.5}NbO_3 \rightarrow (Na_{1-x}Cd_{0.5x})NbO_3$$

and similarly for the case in which Cd is replaced by Pb. Written in this way, these solid solutions appear to be very closely related to Ba(Ti, Nb)$O_3$, in that lattice vacancies are created by. proper chemical substitutions. The system

(Na, Cd)$NbO_3$ was studied by Wainer and Wentworth (W 2) and by Lewis and White (L 1). The system (Na, Pb)$NbO_3$ was investigated by Francombe and Lewis (F 6).

The dielectric and structural properties of both types of compounds are very similar to those of (Na, K)$NbO_3$. Solid solutions with perovskite-type structure were obtained for values of $x$ up to 0.30. They exhibit ferroelectric properties as soon as Na is replaced by Cd or Pb to the extent of a few per cent. The Curie point of these ferroelectric compounds decreases with increasing additions and occurs at 200 °C at the solubility limit ($x = 0.30$). The tetragonal–orthorhombic phase boundary occurring in the phase diagram $NaNbO_3$–$KNbO_3$ is also found to occur in the system discussed here.

## BIBLIOGRAPHY

(B 1)  BARRETT, J. H., *Phys. Rev.* **86**, 118 (1952).
(B 2)  BERLINCOURT, D. A. and KULCSAR, F., *J. Am. Acoust. Soc.* **24**, 709 (1952).
(B 3)  BOKOV, V. A., *Zhur. Tekh. Fiz.* **23**, 77 (1958); *Izvest. Akad. Nauk S.S.S.R., Ser. Fiz.* **21**, 382 (1957).
(B 4)  BASMAJIAN, J. A. and DeVRIES, R. C., *J. Am. Ceram. Soc.* **40**, 373 (1957).
(B 5)  BERLINCOURT, D. and KRUEGER, H. A., *J. Appl. Phys.* **30**, 1804 (1959).
(C 1)  COTTS, R. M. and KNIGHT, W. D., *Phys. Rev.* **96**, 1285 (1954).
(C 2)  CROSS, L. E. and NICHOLSON, B. J., *Phil. Mag.*, Ser. 7, **46**, 453 (1955).
(C 3)  CROSS, L. E., *Nature* **181**, 178 (1958).
(C 4)  CROSS, L. E., *Phil. Mag.* Ser. 8, **1**, 76 (1956).
(C 5)  CROSS, L. E., unpublished. Quoted by CROSS, L. E., *Phil. Mag.* Ser. 8, **1**, 76 (1956).
(D 1)  DURST, G., GROTENHUIS, M. and BARKOV, A. G., *J. Am. Ceram. Soc.* **33**, 133 (1950).
(D 2)  DUNGAN, R. H., KANE, D. F. and BICKFORD, L. R., JR., *J. Am. Ceram. Soc.* **35**, 318 (1952).
(D 3)  DeVRIES, R. C. and ROY, R., *J. Am. Ceram. Soc.* **38**, 142 (1955).
(D 4)  DOBROV, W. I., VIETH, R. F. and BROWNE, M. E., *Phys. Rev.* **115**, 79 (1959).
(E 1)  EGERTON, L. and DILLON, D. M., *J. Am. Ceram. Soc.* **42**, 438 (1959).
(F 1)  FRANCOMBE, M. H., *Acta Cryst.* **9**, 256 (1956).
(F 2)  FESENKO, E. G., KRAMAROV, O. P., KHODAKOV, A. L. and SHOLOKHOVICH, M. L., *Izvest. Akad. Nauk S.S.S.R., Ser. Fiz.* **21**, 305 (1957)
(F 3)  FESENKO, E. G. and KOLESCOVA, R. V., *Kristallografiya* **4**, 62 (1959).
(F 4)  FESENKO, E. G., *Doklady Akad. Nauk S.S.S.R.* **88**, 785 (1953).
(F 5)  FRANCOMBE, M. H. and LEWIS, B., *Acta Cryst.* **11**, 175 (1958).
(F 6)  FRANCOMBE, M. H. and LEWIS, B., *J. Electronics* **2**, 387 (1957).
(G 1)  GRÄNICHER, H. and JAKITS, O., *Nuovo Cimento, Suppl.* **11**, 480 (1954).
(G 2)  GRÄNICHER, H., *Helv. Phys. Acta* **29**, 211 (1956).
(G 3)  GRÄNICHER, H., *Archiv. sci. (Geneva)* **11**, 28 (1958).
(H 1)  HULM, J. K., MATTHIAS, B. T. and LONG, E. A., *Phys. Rev.* **79**, 885 (1949).
(H 2)  HULM, J. K., *Proc. Phys. Soc. (London)* **63**, 1184 (1950).
(H 3)  HARMAN, G., *Phys. Rev.* **106**, 1358 (1957).
(H 4)  HALL, J. J. and TRIEBWASSER, S., *J. Am. Chem. Soc.* **81**, 6394 (1959).
(I 1)  IKEDA, T., *J. Phys. Soc. Japan* **14**, 168 (1959).
(I 2)  ISUPOV, V. A., *Izvest. Akad. Nauk S.S.S.R., Ser. Fiz.* **21**, 402 (1957).
(I 3)  ISMAILZHADE, I. G., *Kristallografiya* **4**, 417 (1959).
(I 4)  IKEDA, T., *J. Phys. Soc. Japan* **14**, 1286 (1959).
(J 1)  JONA, F., SHIRANE, G. and PEPINSKY, R., *Phys. Rev.* **97**, 1584 (1955).

(J 2)   JOHNS, J. F., Master Thesis, The Pennsylvania State University, Graduate School of Physics, 13 May (1957); see also: SHIRANE, G., JOHNS, J. F., JONA, F. and PEPINSKY, R.. *Abstract Annual Meeting Am. Cryst. Assoc.*, 1956.

(J 3)   JONKER, G. H. and VAN SANTEN, J., *Chem. Weekblad* **43**, 672 (1947).

(J 4)   JAFFE, B., ROTH, R. S. and MARZULLO, S., *J. Appl. Phys.* **25**, 809 (1954); *J. Research Natl. Bur. of Standards* **55**, 239 (1955).

(J 5)   JAFFE, H., *J. Am. Ceram. Soc.* **41**, 494 (1958).

(J 6)   JONA, F., SHIRANE, G., MAZZI, F. and PEPINSKY, R., *Phys. Rev.* **105**, 849 (1957).

(K 1)   KAY, H. F. and MILES, J. L., *Acta Cryst.* **10**, 213 (1957).

(K 2)   KOBAYASHI, J., *J. Appl. Phys.* **29**, 866 (1958).

(K 3)   KÄNZIG, W., *Solid State Physics*, vol. 4, p. 96, Academic Press, New York (1957).

(K 4)   KOBAYASHI, J. and UEDA, R., *Phys. Rev.* **99**, 1900 (1955).

(K 5)   KOBAYASHI, J., OKAMOTO, S. and UEDA, R., *Phys. Rev.* **103**, 830 (1956).

(K 6)   KRAINIK, N. N., *Izvest. Akad. Nauk S.S.S.R., Ser. Fiz.* **21**, 411 (1957).

(K 7)   KRAINIK, N. N., *Zhur. Tekh. Fiz.* **28**, 525 (1958).

(K 8)   KELL, R. C. and HELLICAR, N. J., *Acoustica* **6**, 235 (1956).

(K 9)   KULCSAR, F., *J. Am. Ceram. Soc.* **42**, 49, 343 (1959).

(K 10)  KHODAKOV, A. L., SHOLOKHOVICH, M. L., FESENKO, E. G. and KRAMAROV, O. P., *Doklady Akad. Nauk S.S.S.R.* **108**, 825 (1956).

(K 11)  KESTIGIAN, M., DICKINSON, J. G. and WARD, R., *J. Am. Chem. Soc.* **79**, 5598, 5601 (1957).

(K 12)  KOZLOVSKII, V. K., *Zhur. Tekh. Fiz.* **21**, 1388 (1951).

(K 13)  KISAKA, S., IKEGAMI, S. and SASAKI, H., *J. Phys. Soc. Japan* **14**, 1680 (1959).

(L 1)   LEWIS, B. and WHITE, E. A. D., *Acta Cryst.* **8**, 849 (1955); *J. Electronics* **1**, 646 (1956).

(M 1)   MEGAW, H. D., *Proc. Phys. Soc. (London)* **58**, 10 (1946).

(M 2)   MEGAW, H. D., *Ferroelectricity in Crystals*, ch. 5, Methuen, London (1957).

(M 3)   MATTHIAS, B. T., *Science* **113**, 591 (1951).

(M 4)   McQUARRIE, M., *Am. Ceram. Soc. Bull.* **34**, 170, 225, 256, 295, 328 (1955).

(M 5)   MEGAW, H. D., *Acta Cryst.* **5**, 739 (1952).

(M 6)   MEGAW, H. D., *Acta Cryst.* **7**, 187 (1954).

(M 7)   MATTHIAS, B. T., *Phys. Rev.* **75**, 1771 (1949).

(M 8)   MATTHIAS, B. T. and REMEIKA, J. P., *Phys. Rev.* **82**, 727 (1951).

(M 9)   MILLER, C. E., *J. Appl. Phys.* **29**, 233 (1958).

(M 10)  MEGAW, H. D. and WELLS, M., *Acta Cryst.* **11**, 858 (1958).

(M 11)  McQUARRIE, M. C. and BEHNKE, F. W., *J. Am. Ceram. Soc.* **37**, 539 (1954).

(M 12)  MYL'MIKOVA, I. E., *Izvest. Akad. Nauk S.S.S.R., Ser. Fiz.* **21**, 423 (1957).

(M 13)  MEDVOI, A. I., *Zhur. Tekh. Fiz.* **28**, 1006 (1958).

(M 14)  McQUARRIE, M., *J. Am. Ceram. Soc.* **38**, 444 (1955).

(M 15)  McQUARRIE, M., *J. Am. Ceram. Soc.* **40**, 35 (1957).

(M 16)  MATTHIAS, B. T., *Phys. Rev.* **76**, 430 (1949).

(M 17)  MATTHIAS, B. T. and WOOD, E. A., *Phys. Rev.* **84**, 1255 (1951).

(M 18)  MARUTAKE, M. and IKEDA, T., *J. Phys. Soc. Japan* **10**, 424 (1955).

(M 19)  MÜLLER, K. A., *Helv. Phys. Acta* **31**, 173 (1958).

(M 20)  MYL'NIKOVA, I. E. and BOKOV, V. A., *Kristallografiya* **4**, 433 (1959).

(N 1)   NARAY-SZABO, S., *Naturwiss.* **31**, 202, 466 (1943); *Müegyetemi Közlemenyek* **1**, 30 (1947).

(N 2)   NOMURA, S. and SAWADA, S., *J. Phys. Soc. Japan* **6**, 36 (1951).

(N 3)   NOMURA, S. and SAWADA, S., *J. Phys. Soc. Japan* **5**, 270 (1950).

(N 4)   NOMURA, S. and SAWADA, S., *J. Phys. Soc. Japan* **10**, 108 (1955).

(N 5)   NOMURA, S., *J. Phys. Soc. Japan* **10**, 112 (1955).

(N 6)   NISHIOKA, A., SEKIKAWA, K. and OWAKI, M., *J. Phys. Soc. Japan* **11**, 180 (1956).

(O 1)   ORGEL, L. E., *Discussions Faraday Soc.* **26**, 138 (1958).

(P 1)   PORAI-KOSHITS, E. A., KARASIK, N. YA. and GOMON, G. O., *Zhur. Tekh. Fiz.* **25**, 945 (1955).

(P 2)   POSNJAK, E. and BARTH, T. F. W., *Z. Krist.* **88**, 271 (1934).

(R 1)   ROTH, R. S., *J. Research Natl. Bur. Standards* **58**, 75 (1957).

(R 2) ROBERTS, S., *Phys. Rev.* **76**, 1215 (1949).
(R 3) REISMAN, A. and HOLTZBERG, F., *J. Am. Chem. Soc.* **77**, 2115 (1955).
(R 4) REISMAN, A., HOLTZBERG, F., TRIEBWASSER, S. and BERKENBLIT, M., *J. Am. Chem. Soc.* **78**, 719 (1956).
(R 5) REISMAN. A. and BANKS, E., *J. Am. Chem. Soc.* **80**, 1877 (1958).
(R 6) REISMAN, A., TRIEBWASSER, S. and HOLTZBERG, F., *J. Am. Chem. Soc.* **77**, 4228 (1955).
(R 7) REISMAN, A., HOLTZBERG, F. and BANKS, E., *J. Am. Chem. Soc.* **80**, 37 (1958).
(R 8) ROGERS, H. H., Technical Report No. 56, Laboratory for Insulation Research, Massachusetts Institute of Technology (1952).
(R 9) ROBERTS, S., *J. Am. Ceram. Soc.* **33**, 63 (1950).
(R 10) RUSHMAN. D. F. and STRIVENS, M. A., *Trans. Faraday Soc.* **A 42**, 231 (1946).
(R 11) ROI, N. A., *Akaust. Zhur.* **1**, 264, 353 (1955); **2**, 62 (1956).
(R 12) REISMAN, A. and HOLTZBERG, F., *J. Am. Chem. Soc.* **80**, 6503 (1958).
(R 13) ROBERTS, S., *Phys. Rev.* **83**, 1078 (1951).
(S 1) SMOLENSKII, G. A. and KOZHEVNIKOVA, N. V., *Doklady Akad. Nauk S.S.S.R.* **76**, 519 (1951).
(S 2) SMOLENSKII, G. A., *Izvest. Akad. Nauk S.S.S.R.*, *Ser. Fiz.* **20**, 166 (1956).
(S 3) SMOLENSKII, G. A., *Izvest. Akad. Nauk S.S.S.R.*, *Ser. Fiz.* **21**, 233 (1957).
(S 4) SHIRANE, G., DANNER, H., PAVLOVIC, A. and PEPINSKY, R., *Phys. Rev.* **93**, 672 (1954).
(S 5) SHIRANE, G., NEWNHAM, R. and PEPINSKY, R., *Phys. Rev.* **96**, 581 (1954).
(S 6) SMOLENSKII, G. A., ISUPOV, V. A., AGRAVNOVSKAYA, A. I. and SHOLOKHOVA, E. D., *Zhur. Tekh. Fiz.* **27**, 2528 (1957).
(S 7) SMOLENSKII, G. A., *Doklady Akad. Nauk S.S.S.R.* **70**, 405 (1950).
(S 8) SMOLENSKII, G. A., *Zhur. Tekh. Fiz.* **20**, 137 (1950).
(S 9) SHIRANE, G., HOSHINO, S. and SUZUKI, K., *Phys. Rev.* **80**, 1105 (1950).
(S 10) SHIRANE, G. and HOSHINO, S., *J. Phys. Soc. Japan* **6**, 265 (1951).
(S 11) SHIRANE, G., PEPINSKY, R. and FRAZER, B. C., *Acta Cryst.* **9**, 131 (1956).
(S 12) SHOLOKOVICH, M. L., FESENKO, E. G., KRAMAROV, O. P. and KHODAKOV, A. L., *Doklady Akad. Nauk S.S.S.R.* **111**, 1025 (1956).
(S 13) SHAFER, M. W. and ROY, R., *J. Am. Ceram. Soc.* **42**, 482 (1959).
(S 14) SHIRANE, G. and SUZUKI, K., *J. Phys. Soc. Japan* **6**, 274 (1951).
(S 15) SHIRANE, G. and TAKEDA, A., *J. Phys. Soc. Japan* **6**, 329 (1951).
(S 16) SUZUKI, K., *J. Phys. Soc. Japan* **6**, 340 (1941).
(S 17) SAUER, H. A., FLASCHEN, S. S. and HOERSTEREY, D. C., *J. Am. Ceram. Soc.* **42**, 363 (1959).
(S 18) SAWAGUCHI, E., MITSUMA, T. and ISHII, Z., *J. Phys. Soc. Japan* **11**, 1298 (1956).
(S 19) SAWAGUCHI, E. and CHARTERS, M. L., *J. Am. Ceram. Soc.* **42**, 157 (1959).
(S 20) SHIRANE, G. and TAKEDA, A., *J. Phys. Soc. Japan* **7**, 5 (1952).
(S 21) SHIRANE, G., SUZUKI, K. and TAKEDA, A., *J. Phys. Soc. Japan* **7**, 12 (1952).
(S 22) SHIRANE, G. and SUKUZI, K., *J. Phys. Soc. Japan* **7**, 333 (1952).
(S 23) SAWAGUCHI, E., *J. Phys. Soc. Japan* **8**, 615 (1953).
(S 24) SMOLENSKII, G. A., AGRANOVSKAYA, A. I. and KALINIMA, A. M., *Zhur. Tekh. Fiz.* **25**, 2134 (1955).
(S 25) SHIRANE, G., SAWAGUCHI, E. and TAKEDA, A., *Phys. Rev.* **80**, 485 (1950).
(S 26) SMOLENSKII, G. A., *Zhur. Tekh. Fiz.* **21**, 1045 (1951).
(S 27) SAWAGUCHI, E., MANIWA, H. and HOSHINO, S., *Phys. Rev.* **83**, 1078 (1951).
(S 28) SAWAGUCHI, E., *J. Phys. Soc. Japan* **7**, 110 (1952).
(S 29) SAWAGUCHI, E., SHIRANE, G. and TAKAGI, Y., *J. Phys. Soc. Japan* **6**, 333 (1951).
(S 30) SHIRANE, G., SAWAGUCHI, E. and TAKAGI, Y., *Phys. Rev.* **84**, 476 (1951).
(S 31) SAWAGUCHI, E. and KITTAKA, K., *J. Phys. Soc. Japan* **7**, 336 (1952).
(S 32) SMOLENSKII, G. A., *Zhur. Tekh. Fiz.* **27**, 1778 (1957).
(S 33) SHIRANE, G., *Phys. Rev.* **86**, 219 (1952).
(S 34) SHIRANE, G. and HOSHINO, S., *Acta Cryst.* **7**, 203 (1954).
(S 35) SMOLENSKII, G. A., KARAMESNEV, M. A. and ROZGACHEV, K. N. *Doklady Akad. Nauk S.S.S.R.* **79**, 53 (1951).

(S 36) SMOLENSKII, G. A., TARUTIN, N. A. and GRUDTSIN, N. P., *Zhur. Tekh. Fiz.* **24**, 1584 (1954).

(S 37) SMOLENSKII, G. A. and ISUPOV, V. A., *Zhur. Tekh. Fiz.* **24**, 1375 (1954).

(S 38) SMOLENSKII, G. A. and ROZGACHEV, K. N., *Zhur. Tekh. Fiz.* **24**, 1751 (1954).

(S 39) SMOLENSKII, G. A., AGRANOVSKAYA, A. I. and KRAINIK, N. N., *Doklady Akad. Nauk S.S.S.R.* **91**, 55 (1953).

(S 40) SHIRANE, G. and PEPINSKY, R., *Phys. Rev.* **91**, 812 (1953).

(S 41) SMOLENSKII, G. A., *Doklady Akad. Nauk S.S.S.R.* **85**, 985 (1952).

(S 42) SMOLENSKII, G. A. and AGRANOVSKAYA, A. I., *Zhur. Tekh. Fiz.* **28**, 1491 (1958).

(S 43) SMOLENSKII, G. A., AGRANOVSKAYA, A. I., POPOV, S. N. and ISUPOV, V. A., *Zhur. Tekh. Fiz.* **28**, 2152 (1958).

(S 44) SMOLENSKII, G. A., AGRANOVSKAYA, A. I. and POPOV, S. N., *Fiz. Tverdogo Tela* **1**, 167 (1959).

(S 45) SMOLENSKII, G. A., ISUPOV, V. A. and AGRANOVSKAYA, A. I., *Fiz. Tverdogo Tela* **1**, 170 (1959).

(S 46) SAWADA, S. and DANIELSON, G. C., *Phys. Rev.* **113**, 803, 1005, 1008 (1959).

(S 47) SMOLENSKII, G. A., ISUPOV, V. A. and AGRANOVSKAYA, A. I., *Doklady Akad. Nauk S.S.S.R.* **113**, 1053 (1957).

(S 48) SUBBARAO, E. C. and SHIRANE, G., *J. Am. Ceram. Soc.* **42**, 279 (1959).

(S 49) SAKUDO, T., *J. Phys. Soc. Japan* **12**, 1050 (1957).

(S 50) SUBBARAO, E. C., *J. Am. Ceram.. Soc.* **42**, 448 (1959).

(S 51) SABURI, O., *J. Phys. Soc. Japan* **14**, 1159 (1959).

(S 52) SAUER, H. A. and FLASCHEN, S. S., *Proc. Elect. Components Symposium*, 1956, p. 41; *Ceram. Ind.* **66**, 95 (1956).

(S 53) SMOLENSKII, G. A., AGRANOVSKAYA, A. I. and ISUPOV, V. A., *Fiz. Tverdogo Tela* **1**, 990 (1959).

(T 1) TRIEBWASSER, S., *Phys. Rev.* **101**, 993 (1956).

(T 2) TRIEBWASSER, S. and HALPERN, J., *Phys. Rev.* **98**, 1562 (1955).

(T 3) TRIEBWASSER, S., *Phys. Rev.* **114**, 63 (1959).

(T 4) TESSMAN, J. R., KAHN, A. H. and SHOCKLEY, W., *Phys. Rev.* **92**, 890 (1953).

(V 1) VOUSDEN, P., *Acta Cryst.* **4**, 545 (1951).

(V 2) VOUSDEN, P., *Acta Cryst.* **4**, 373 (1951).

(V 3) VENEVTSEV, YU. N. and ZHDANOV, G. S., *Izvest. Akad. Nauk S.S.S.R.*, SER. FIZ. **20**, 178 (1956).

(V 4) VON HIPPEL, A., BRECKENRIDGE, R. G., CHESLEY, F. G. and TISZA, L., *Ind. Eng. Chem.* **38**, 1097 (1946).

(V 5) VENEVTSEV, YU. N., KAPYSHEV, A. G. and SHUMOV, YU. V., *Kristallografiya* **2**, 233 (.1957)

(V 6) VENEVTSEV, YU. N., ZHDANOV, G. S., SOLOVEV, S. P. and ZUBOV, YU. A., *Kristallografiya* **3**, 473 (1958).

(V 7) VENEVTSEV, YU. N., ZHDANOV, G. S., SOLOVEV, S. P. and IBANOVA, V. V., *Kristallografiya* **4**, 255 (1959).

(W 1) WOOD, E. A., *Acta Cryst.* **4**, 353 (1951).

(W 2) WAINER, E. and WENTWORTH, C., *J. Am. Ceram. Soc.* **35**, 207 (1952).

(W 3) WEAVER, H. E., *J. Phys. Chem. Solids* **11**, 274 (1959).

(Z 1) ZHDANOV, G. S., SOLOVEV, S. P. and VENEVTSEV, YU. N., *Kristallografiya* **2**, 639 (1957).

# CHAPTER VI

# MISCELLANEOUS COMPLEX OXIDES

## 1. Introduction

Ferroelectric properties are also encountered in a rather large number of double and complex oxides that do not crystallize with a perovskite-type lattice. Typical examples of ferroelectrics of this type are cadmium (pyro) niobate, $Cd_2Nb_2O_7$, and lead (meta) niobate, $PbNb_2O_6$. It appeared, at first sight, that no relation exists between compounds of this sort and perovskite-like crystals, other than the fact that they are all mixed oxides. Closer examination, however, revealed that all these compounds are closely related to one another, as well as to the perovskite family, in their crystal–chemical and structural characteristics.

The common feature is provided by the fact that an essential element of the structural framework is the octahedron of oxygen ions surrounding the transition-metal ion. The role of the $BO_6$ octahedra has already been pointed out in the case of perovskite compounds of the type $ABO_3$, where the arrangement of the octahedra is such as to provide infinite linear chains $B-O-B-O-B$ along the $\langle 100 \rangle$ directions. This arrangement has been shown to be quite favorable for the action of long-range forces such as dipole interaction and thus for the occurrence of the phenomenon of ferroelectricity. Most of the compounds with which we are going to be concerned in the present chapter have chemical formulas of the type $A_2B_2O_7$ or $AB_2O_6$ and contain $BO_6$ octahedra exhibiting characteristics similar to but more complicated than those of the perovskite-type compounds. For this reason, all these ferroelectrics, including the perovskites, have often been referred to, in the literature, as "oxygen-octahedra ferroelectrics".

Of course, the existence of $BO_6$ octahedra as essential building blocks of the structural framework is not peculiar to these ferroelectric crystals but is a fairly common feature of complex oxides of transition metals. This point was particularly well illustrated by Hägg and Magneli (H 1) with their work on tungsten bronzes of the type $A_xWO_{3-x}$, where $A$ is an alkali metal such as K, Na or Li. It was shown, in fact, that the structure of these compounds is characterized by periodic arrangements of $WO_6$ octahedra within the (001)·plane, such as to realize linear $W-O-W$ chains along the [001] direction. More recent work by Magneli (M 1) and by Wadsley (W 1) has provided additional examples of similar structures in complex oxides containing V, Ti, Nb or Ta. It may be noted that these are the same metals that seem to be essential (when associated with large and highly polarizable atoms such as Pb, Cd, Ba or Sr) for the occurrence of ferroelectricity in the compounds to be dealt with below.

## 2. Pyrochlore-type Compounds

A number of complex oxides with the general formula $A_2B_2O_7$ have structures similar to that of the mineral pyrochlore, $CaNaNb_2O_6F$, and are consequently classified as pyrochlore (or pyrochlorite) types. According to the valence of the $A$ and $B$ ions involved, these compounds can be represented by the formulae $A_2^{2+}B_2^{5+}O_7$ and $A_2^{3+}B_2^{4+}O_7$. Within the former group, $Cd_2Ta_2O_7$, $Ca_2Ta_2O_7$ and $Pb_2Sb_2O_7$ have been found to crystallize with a cubic lattice (B 1), while $Pb_2Nb_2O_7$ and $Pb_2Ta_2O_7$ exhibit a rhombohedral distortion from the parent structure (J 1). Dielectric measurements failed to reveal any ferroelectric transition in these compounds. A flat peak in the dielectric constant was indeed found in lead (pyro) niobate, $Pb_2Nb_2O_7$, at 14 °K, but no evidence of ferroelectric activity was obtained (H 2), (S 1).

The second group of pyrochlore-type compounds $(A_{3+}^2B_2^{4+}O_7)$ was studied by Roth (R 1) who found that the following $A$ and $B$ ions combine to form a cubic pyrochlore structure, respectively:

| $A^{3+}$ | $B^{4+}$ |
|---|---|
| Sm, Gd, Dy, Y, Yb | Ti |
| La, Nd | Sn |
| La, Nd, Y | Zr |

No dielectric data for these compounds, however, are available at present.

The only ferroelectric representatives of the pyrochlore family are cadmium (pyro) niobate, $Cd_2Nb_2O_7$, and a few solid solutions in which Cd is replaced by Pb or Ca and Nb by Ta. It should be pointed out, however, that the pyrochlore structure can also be encountered in compounds with the formula $A_{1+x}A'_{1-x}B_2O_{6+x}C_{1-x}$. In fact, the mineral pyrochlore itself ($CaNaNb_2O_6F$) has a chemical formula of this type, which emphasizes the fact that the set of "seventh" oxygens of the $A_2B_2O_7$ compounds is not essential to the structural framework. It can indeed be left out, provided, of course, that charge neutrality is preserved by omitting some of the $A$ ions. An example of this case is the compound $Pb_{1.5}Nb_2O_{6.5}$ for which Cook and Jaffe (C 1) reported a cubic pyrochlore-type structure. It was later shown by Roth (R 2) that the formula of this compound can be written as $3PbO \cdot 2Nb_2O_5$, with reference to the phase diagram of the system $PbO - Nb_2O_5$. However, the compound exists only over a fairly narrow range, and this indicates that compounds of the type $A_{2-x}B_2O_{7-x}$ do not necessarily exist for any arbitrary value of $x$ between 0 and 1 after all. The point that we would like to make, however, is that further dielectric studies of this group of pyrochlore-type compounds may eventually permit the addition of new members to the family of ferroelectrics.

### Cadmium (pyro) Niobate, $Cd_2Nb_2O_7$

The ferroelectric properties of $Cd_2Nb_2O_7$ were discovered by Cook and Jaffe (C 1) in 1952. The Curie point lies at 185 °K and the behavior of the dielectric constant as a function of temperature is typical of that of the ferroelectrics discussed so far. The Curie–Weiss law $\varepsilon = C/(T - T_0)$ is obeyed, above the Curie point, with a Curie constant $C \cong 7 \times 10^4$ degrees. As shown in Fig. VI-1, a second, smaller, anomaly of the dielectric constant occurs at 85 °K, indicating a further phase transition (S 1), (H 2). Ferroelectric hysteresis loops can be observed down to 4 °K.

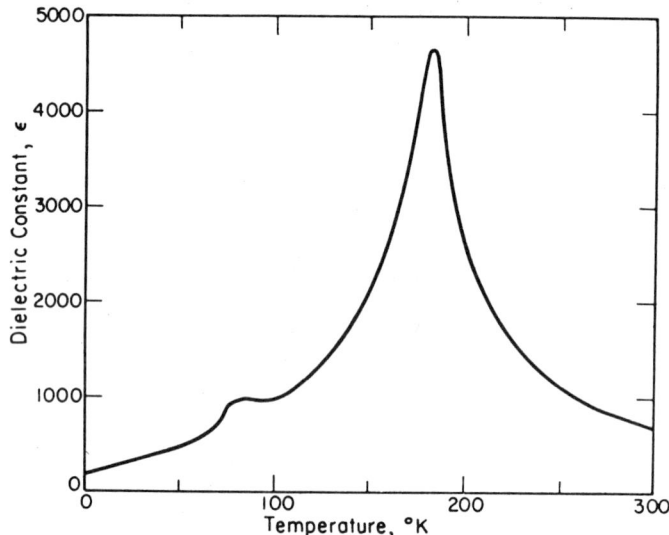

Fig. VI-1. Dielectric constant of $Cd_2Nb_2O_7$ ceramic as a function of temperature (according to Hulm (H 1)).

The lattice constant of the cubic phase at room temperature is:

$$a = 10.372 \text{ Å}; \quad \text{space group } Fd3m,$$

with eight formula units per unit cell. The structure was first studied by Bystroem (B 1) together with that of similar compounds. Figure VI-2 depicts a schematic projection of the structure on the (001) plane, showing the three-dimensional array of $NbO_6$ octahedra linked to one another by the corners. Thus, the structural framework has the composition $(Nb_2O_6)_\infty$. One parameter, $x$, determines the shape of the octahedra; the condition for regular octahedra is $x = 5/16 = 0.312$. In single crystals of $Cd_2Nb_2O_7$, this parameter was found to be $x = 0.305$ by Jona et al. (J 1), indicating that the $NbO_6$ octahedra are slightly distorted. A comparison of this structure with the perovskite structure shows that the main difference lies in the relative orientation of the oxygen octahedra. In $Cd_2Nb_2O_7$, the arrangement is such that the $O-Nb-O$ chains lie on zig-zag lines approximately along the $\langle 110 \rangle$ directions. The Cd ions and the seventh set of oxygens occupy open spaces in the structural framework.

Dielectric, optical and X-ray studies have been carried out on single crystals of $Ca_2Nb_2O_7$ by Jona et al. (J 1). The habit of the crystals grown by these authors was octahedral, so that tabular specimens had the major surfaces oriented perpendicular to the $\langle 111 \rangle$ directions. Thus, the dielectric measurements could be carried out only with dielectric fields applied along [111]. The value of the spontaneous polarization along this direction was found to be about $6 \times 10^{-6}$ C/cm². It is, however, uncertain if this the correct value of $P_s$ because the symmetry of the polar phase could not be determined despite the X-ray and optical investigations. Weissenberg photographs taken below the Curie point revealed a

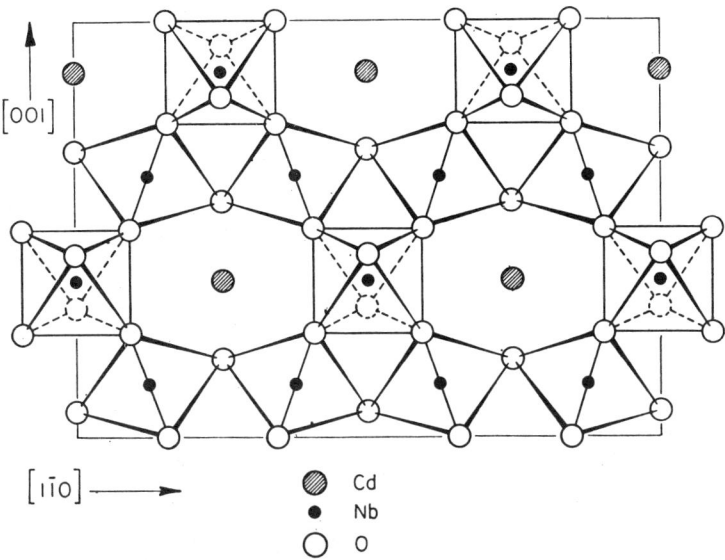

FIG. VI-2. Schematic projection of the structure of $Cd_2Nb_2O_7$ on (110) plane. The Cd ions, which overlap with the Nb ions at (1/8, 1/8) etc., and the seventh set of oxygens, which are located above and below the Cd ions, are not shown (according to Jona et al. (J 1)).

very small distortion from the cubic lattice which could be tentatively accounted for by assuming a tetragonal structure with:

$$a = 10.364 \text{ Å}; \quad c/a = 1.0005.$$

Calorimetric measurements on $Cd_2Nb_2O_7$ were carried out by Danner and Pepinsky (D 1) through the Curie point, where the specific heat exhibits an anomaly. The heat of transition was calculated to be 18 cal/mole. Table IV-2 shows that this value agrees satisfactorily with that calculated from the dielectric data according to Eq. (IV-7).

### Solid Solutions Containing $Cd_2Nb_2O_7$

Systems of solid solutions of $Cd_2Nb_2O_7$ and $Pb_2Nb_2O_7$, $Ca_2Nb_2O_7$ or $Cd_2Ta_2O_7$, respectively, have been studied by Jona et al. (J 1) with the results depicted in Fig. VI-3. The Curie temperature of pure $Cd_2Nb_2O_7$ is invariably decreased by these substitutions. The results for the solid solutions $Cd_2(Nb, Ta)_2O_7$ are not shown in Fig. VI-3(a) because they very nearly coincide with the curve for $(Cd, Pb)_2Nb_2O_7$. It seems that the temperature of the lower transition is also lowered by the substitutions.

The dependence of the lattice constant of the room-temperature phase upon composition is shown in Fig. VI-3 (b). In the $(Cd, Pb)_2Nb_2O_7$ system, solid solutions with the cubic pyrochlore structure are formed only with exclusion of a very narrow range near to pure $Pb_2Nb_2O_7$, which has a rhombohedrally distorted lattice. In the $(Cd, Ca)_2Nb_2O_7$ system, on the other hand, pyrochlore-type solid solutions are formed up to about 20% addition of $Ca_2Nb_2O_7$, which is, itself, not of the pyrochlore type.

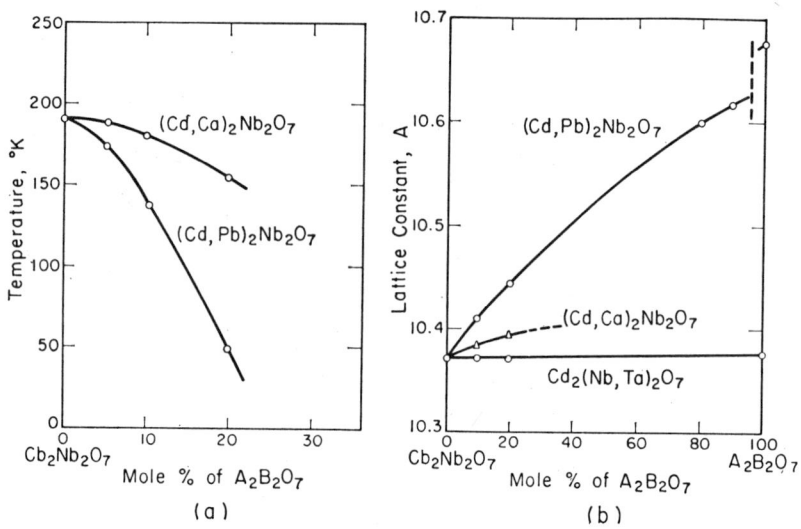

FIG. VI-3. Solid solutions of $Cd_2Nb_2O_7$ with other $A_2B_2O_7$ compounds.

(a) Curie temperature vs. composition.
(b) Lattice parameters vs. composition.
(According to Jona et al. (J 1)).

Isupov and Komutetskii (I 1) have also attempted the substitution of Cd by Sr, Zn, Mg and the "mixed" ion $(Na_{0.5}Bi_{0.5})$, respectively. In all cases, again, the solid solutions obtained have Curie temperatures lower than that of pure $Cd_2Nb_2O_7$.

### 3. Strontium (pyro) Tantalate, $Sr_2Ta_2O_7$

Ferroelectricity in $Sr_2Ta_2O_7$ has been reported by Smolenskii et al. (S 4) and by Isupov (I 2). The Curie temperature lies at $-80$ °C, at which point the dielectric constant exhibits an anomaly. Unfortunately, no detailed structural information concerning the polar and the non-polar phases of this compound is available at the present time of writing. It appears that the structure of $Sr_2Ta_2O_7$ is *not* of the pyrochlore type but the details are not known. Tetragonal symmetry has been reported, at room temperature, by Ismailzade (I 4), with the lattice parameters:

$$a = 10.63 \text{ Å}, \quad c = 10.91 \text{ Å}.$$

A number of solid solutions containing $Sr_2Ta_2O_7$ have been investigated by Smolenskii *et al.* (S 4), with particular attention being given to the dependence of the Curie temperature upon composition. A very striking effect was found for the substitution of Ta by Nb, which causes a drastic increase of the Curie temperature. This occurs at 400 °C for the solid solution $Sr_2(Ta_{0.8}Nb_{0.2})_2O_7$ which appears to be the solubility limit. Substitution of Sr by Ca also raises the Curie temperature, but in a less dramatic fashion. Replacement of Sr by Ba, on the other hand, causes a lowering of the Curie point. In all these solid solutions, the dielectric constant exhibits an anomaly at the ferroelectric transition but in no case is the peak very sharp.

### 4. Lead (meta) Niobate, $PbNb_2O_6$, Lead (meta) Tantalate, $PbTa_2O_6$ and Solid Solutions

#### Lead (meta) Niobate, $PbNb_2O_6$

The ferroelectric properties of $PbNb_2O_6$ were first discovered by Goodman (G1) in 1953. The Curie temperature lies at 570 °C and is the highest ferroelectric transition temperature known at the present time.* The dielectric anomaly is typical of the oxide ferroelectrics (see Fig. VI-4a) and the Curie–Weiss law $\varepsilon = C/(T - T_0)$ is obeyed with $C = 3 \times 10^5$ degrees. No reliable data are available for the value of the spontaneous polarization.

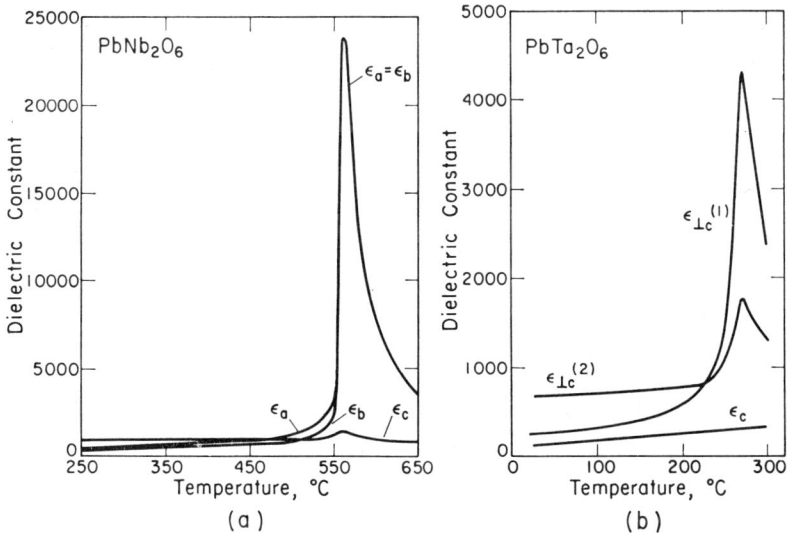

FIG. VI-4. Temperature dependence of the dielectric constant of single crystals of: (a) $PbNb_2O_6$ (according to Francombe and Lewis (F 1)). (b) $PbTa_2O_6$: superscripts (1) and (2) refer to different directions of the same sample (according to Subbarao *et al.* (S 3)).

* This statement is no longer true after the discovery, due Subbarao (*Phys. Rev.* **122**, 804 (1961)), of the ferroelectric $Bi_4Ti_3O_{12}$ with Curie point at 675 °C (see Table VI-2)

The ferroelectric phase has orthorhombic symmetry; the lattice parameters at room temperature are (F 1):

$$a = 17.51 \text{ Å}, \quad b = 17.81 \text{ Å}, \quad c = 2 \times 3.86 \text{ Å}$$

and the unit cell contains twenty formula units. The direction of the spontaneous polarization is not known with certainty but there is evidence that it lies within the (001) plane; most probably it is parallel to the $b$ axis, so that the axial ratio

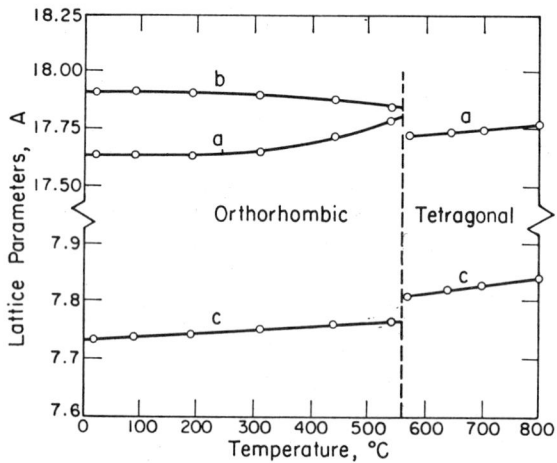

FIG. VI-5. Temperature dependence of the lattice parameters of $PbNb_2O_6$ (according to Francombe and Lewis (F 1)). Recent results of Subbarao (S 11) indicate, however, that the anomaly of the $a$ axis at the Curie temperature is an expansion when going from the orthorhombic into the tetragonal state.

$b/a = 1.017$ can be considered a measure of the spontaneous strain. Figure VI-5 depicts the temperature dependence of the lattice parameters: the symmetry of the non-polar phase (above the Curie point) is tetragonal, and the unit cell, which contains five formula units, has the lattice parameters:

$$a = 12.46 \text{ Å}, \quad c = 3.907 \text{ Å}.$$

The structure of this tetragonal phase closely resembles that of certain tetragonal tungsten bronzes, such as $K_{0.57}WO_3$, a fact that was recognized by Francombe and Lewis (F 1). The latter structure, in turn, can be obtained from that of $BaTiO_3$ by means of certain atomic rearrangements, so that, in the final analysis, there is a fairly marked structural similarity between $BaTiO_3$ and $PbNb_2O_6$. A schematic projection of this structure on (001) is depicted in Fig. VI-6. Not only is the $c$ dimension of the unit cell very close to that of the perovskite lattice, but we can recognize "perovskite units" within the network of octahedra depicted in Fig. IV-6. This type of structure was described by Wadsley (W 1) in terms of "tunnel"-type lattices, which are common to a number of compounds. The configuration of the oxygen octahedra is such as to give rise to tunnels that have a definite orientation in the crystal and are parallel to one another. In Wadsley's description, a tunnel may then be regarded as a continuous string of interstitial sites some of which are available to guest atoms. In tetragonal $PbNb_2O_6$, these tunnels are formed by five-membered rings of $NbO_6$ octahedra which may be occupied by ten co-ordinated Pb ions. Other Pb ions are eight co-ordinated at the center of four-membered rings of $NbO_6$ octahedra which are called "cages" and represent the perovskite units referred to above. It may be noted that only five out of six of the available lead sites are occupied in the $PbNb_2O_6$ structure.

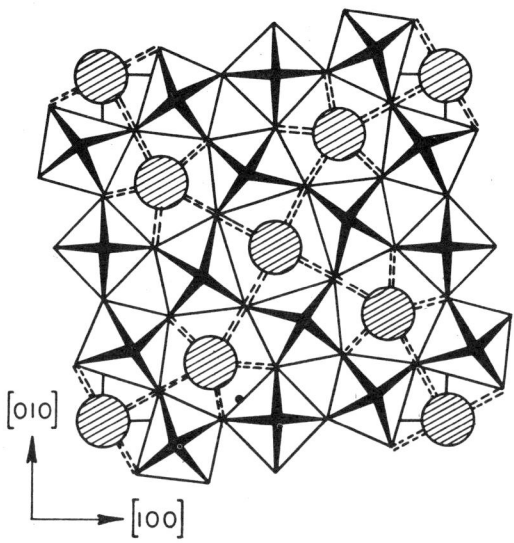

FIG. VI-6. Schematic projection of the structure of $PbNb_2O_6$-type compounds on the (001) plane. Tetragonal axes are used ($a = 12.46$ Å). $Pb^{2+}$ ions are shown as circles ($z = 0$) located in "tunnels" or "cages" formed by the $NbO_6$ octahedra ($z = \pm 1/2$) (according to Francombe and Lewis (F 1)).

In spite of the relation by way of the tetragonal tungsten bronzes, however, $PbNb_2O_6$ and $BaTiO_3$ differ substantially from one another in their ferroelectric properties. While the polar axis of tetragonal $BaTiO_3$ can assume any one of six equivalent directions, owing to the cubic symmetry of the non-polar phase, that of orthorhombic $PbNb_2O_6$ can only lie within the (001) plane of the non-polar tetragonal phase.

A peculiarity of $PbNb_2O_6$ is that the ferroelectric phase can only be obtained by quenching the material from temperatures above 1250 °C. The quenching process is not critical, and the natural cooling rate of ordinary furnaces is sufficient for the purpose, but heating above 1250 °C is a condition *sine qua non* for the achievement of the ferroelectric phase. When prepared below this temperature, $PbNb_2O_6$ is rhombohedral and not ferroelectric. The ferroelectric orthorhombic phase changes into the rhombohedral one when heated to 700 °C (R 3). The reader may recall the existence of a hexagonal phase in $BaTiO_3$ which may also be obtained by quenching from elevated temperatures (in this case, above 1460 °C).

In $BaTiO_3$, however, the stable phase at lower temperatures is the ferroelectric one. The polymorphism of $PbNb_2O_6$ was first pointed out by Francombe (F 2) and more thoroughly investigated by Roth (R 2) in a study of the phase diagram of the system $PbO-Nb_2O_5$.

### Lead (meta) Tantalate, $PbTa_2O_6$

$PbTa_2O_6$ was reported to be ferroelectric by Smolenskii and Agranovskaya (S 2) and subsequently studied by Isupov (I 5) and by Ismailzade (I 6) by means of ceramic specimens. Investigations with single crystals were carried out more recently by Subbarao et al. (S 3) The Curie point lies at 260 °C and the dielectric properties are very similar to those of $PbNb_2O_6$, discussed above (see Fig. VI-4). The Curie constant has the value $1.5 \times 10^5$ degrees and the spontaneous polarization of single crystals equals $10 \times 10^{-6}$ C/cm² at room temperature (S 3). The similarity between $PbTa_2O_6$ and $PbNb_2O_6$ extends also to their polymorphism: the ferroelectric phase of $PbTa_2O_6$ can be obtained only by firing the samples above 1150 °C. When prepared below this temperature the compound is rhombohedral and non-ferroelectric.

The ferroelectric phase is orthorhombic, and the lattice parameters, as measured with single crystals, are (S 3):

$$a = 17.68 \text{ Å}, \quad b = 17.72 \text{ Å}, \quad c = 2 \times 3.877 \text{ Å},$$

indicating a small orthorhombic distortion $b/a = 1.002$. Previous studies of polycrystalline $PbTa_2O_6$ had led Francombe and Lewis (F 1) to the assignment of tetragonal symmetry, and Ismailzade (I 6) to that of orthorhombic symmetry (with $b/a = 1.006$). One could therefore conclude, on the basis of these dielectric and crystallographic data, that $PbTa_2O_6$ and $PbNb_2O_6$ are isomorphous. A single-crystal study by Subbarao et al. (S 3), however, has shown that these compounds are not isostructural. They are rather to be considered, in their ferroelectric phase, as two different modifications of the same basic structure. This fact was confirmed by the study of solid solutions $Pb(Ta, Nb)_2O_6$ (to be discussed below) and by an optical investigation of single crystals of $PbTa_2O_6$.

The latter investigation has revealed that this compound approaches tetragonality with increasing temperature but does not quite attain it: above the Curie temperature, the structure is still orthorhombic although different, of course, from that of the ferroelectric phase. This behavior, it may be recalled, differs from that of $PbNb_2O_6$, and emphasizes an interesting difference in the ferroelectric characteristics of the two crystals. While in $PbNb_2O_6$ the polar axis can lie along any one of four equivalent directions within the (001) plane, in $PbTa_2O_6$ it can have only two antiparallel orientations, in accordance with the biaxial nature of the non-ferroelectric phase. Thus, the [100] and [010] axes are not interchangeable under the action of external electric fields and polarization reversal occurs only through 180° motions. A comparison between the properties of $PbNb_2O_6$ and $PbTa_2O_6$ is summarized in Table VI-1.

TABLE VI-1. SUMMARY OF CRYSTALLOGRAPHIC AND DIELECTRIC DATA OF $PbNb_2O_6$
AND $PbTa_2O_6$ (REFERENCES ARE GIVEN IN THE TEXT)

|  | $PbNb_2O_6$ | $PbTa_2O_6$ |
|---|---|---|
| Polymorphism | Below 1250 °C, the stable phase is rhombohedral. Above 1250 °C, the stable phase is the orthorhombic, ferroelectric phase that can be quenched to room temperature | When prepared below 1150 °C the compound is rhombohedral. Once heated above 1150 °C, the compound transforms into the orthorhombic ferroelectric phase |
| Crystallographic data of the ferroelectric phase at room temperature | Orthorhombic $a = 17.51$ Å $b = 17.81$ Å $c = 2 \times 3.86$ Å $b/a = 1.017$; 20 molecules/cell | Orthorhombic $a = 17.68$ Å $b = 17.72$ Å $c = 2 \times 3.877$ Å $b/a = 1.002$ |
| Curie temperature (°C) | 570 | 260 |
| Curie constant (°C) $C = \varepsilon(T - T_0)$ | $3 \times 10^5$ | $1.5 \times 10^5$ |
| Spontaneous polarization at room temperature ($C/cm^2$) | ? | $10 \times 10^{-6}$ |
| Crystallographic data of the non-polar phase | Tetragonal $a = 12.46$ Å $c = 3.907$ Å 5 molecules/cell | Orthorhombic |

## Solid Solutions of $PbNb_2O_6$ and $PbTa_2O_6$

The phase diagram of the solid solution system $Pb(Nb, Ta)_2O_6$ has been studied by Subbarao et al. (S 3). The dependence of the lattice parameters upon composition is depicted in Fig. VI-7. A definite discontinuity of the lattice parameters at the composition $Pb(Nb_{0.55}Ta_{0.45})_2O_6$ confirms the fact, discussed above, that the two end-members are not isostructural. The same effect is observed in the dependence of the Curie temperature upon composition, which is linear between 570 °C and 260 °C but changes slope at the composition mentioned above, thus indicating the existence of a phase boundary in the system.

Various other solid solutions involving $PbNb_2O_6$ and $PbTa_2O_6$ have been investigated by a number of authors. Substitution of Pb by Ba, Sr or Ca lowers the Curie temperature. The system $(Pb, Ba)Nb_2O_6$, in particular, exhibits interesting behavior (I 3), (I 6), (S 3), (S 12), (F 6). Figure VI-8 shows that the Curie temperature goes through a minimum at the composition with 35% Ba. For higher Ba concentrations, namely from 35% to 60%, the corresponding solid solutions exhibit a ferroelectric phase whose c spacing is approximately one-half of that found in $PbTa_2O_6$ and $PbNb_2O_6$.

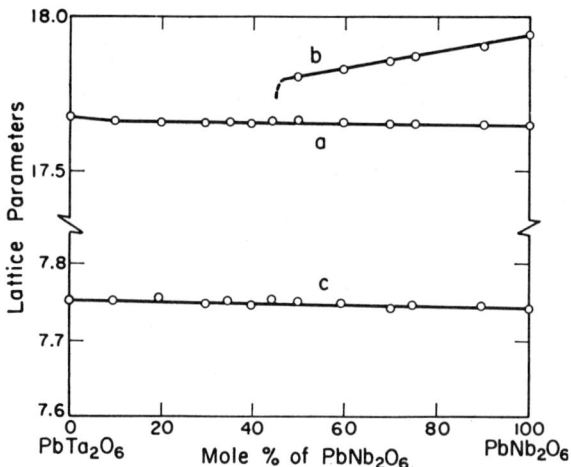

FIG. VI-7. Lattice parameters of solid solutions Pb(Ta,Nb)$_2$O$_6$ (according to Subbarao *et al.* (S 3)).

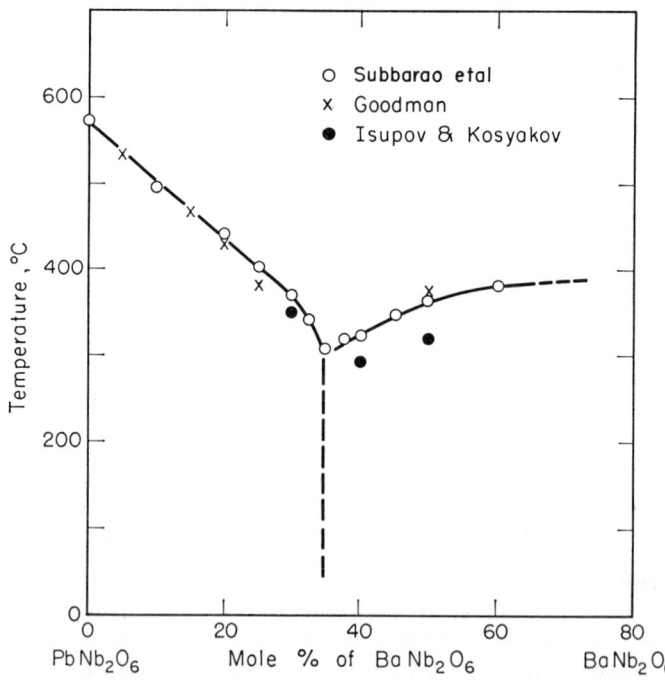

FIG. VI-8. Curie temperature of solid solutions (Pb,Ba)Nb$_2$O$_6$ (according to Subbarao *et al.* (S 3)).

It may be mentioned, before closing our discussion of these crystals, that general surveys of various compounds of the type $AB_2O_6$ have been published by Coates and Kay (C 2) and by Isupov (I 2). Ferroelectricity was reported for $BaNb_2O_6$, $MnTa_2O_6$ and $AlNb_3O_9$ (C 2), but it appears that further dielectric and structural studies of these compounds are needed.

## 5. Non-stoichiometric Solid Solutions of $AB_2O_6$-type Compounds

Owing to the existence of vacant sites in the "tunnels" and "cages" of the $PbNb_2O_6$ structure which are available for guest atoms (see above), non-stoichiometric solid solutions can be prepared more readily with this compound than with any of the other complex oxides that we have discussed so far. These non-stoichiometric solid solutions can be written thus:

$$Pb_{1+x}(B_x, Nb_{1-x})_2O_6 \quad \text{where} \quad B = \text{Ti, Zr or Sn};$$

$$(Pb_{1-x}, A_{2x})Nb_2O_6 \quad \text{where} \quad A = \text{Na, Li, Rb, K or Cs};$$

$$(Pb_{1-x}, A_{(2/3)x})Nb_2O_6 \quad \text{where} \quad A = \text{Y, La or Sm}.$$

A study of these systems has been made by Subbarao and Shirane (S 9), (S 10), and the system $(Pb, Na)Nb_2O_6$ was studied by Francombe and Lewis (F 3). The solubility of the various additions in $PbNb_2O_6$ is, of course, limited but it was noticed that the lead-substituted solid solutions can be described by the general formula $A_{1+x}B_2O_6$, where the variable $x$ varies between $-0.08$ and $0.15$. This is approximately the composition range in which potassium tungsten bronzes possess a lattice similar to that of $PbNb_2O_6$. With one exception, all the solid solutions considered above have a lower Curie temperature than that of pure $PbNb_2O_6$. The exception is the system $Pb(Ti, Nb)_2O_6$, in which the Curie temperature increases with increasing Ti content and reaches 590 °C at the solubility limit ($x = 0.1$).

The ferroelectric compound reported recently by Goodman (G 2), namely $BaNb_{1.5}Zr_{0.25}O_{5.25}$, belongs to the same family of non-stoichiometric solid solutions discussed here. An X-ray study of single crystals of this compound revealed that the symmetry is orthorhombic (pseudo-tetragonal) with the lattice parameters:

$$a = b = 12.67 \text{ Å}, \quad c = 4.017 \text{ Å}.$$

Ferroelectric hysteresis loops were observed in the temperature range between 20 °C and 300 °C, the value of the spontaneous polarization being $15 \times 10^{-6}$ C/cm$^2$ at 25 °C. The most interesting feature of this compound is that no definite Curie point in the usual sense can be observed: with increasing temperature, the spontaneous polarization approaches zero almost asymptotically, much in the way in which a second-order ferroelectric transition occurs under the action of biasing electric fields. The dielectric constant was found to be markedly dependent upon frequency.

Further investigations of this type of compound have been carried out by Roth (R 4) and by Fang et al. (F 4), (F 5). It was observed that solid solutions occur continuously within the range from $Ba(Nb_{1.6}, Zr_{0.2})O_{5.4}$ to $Ba(Nb_{1.34}, Zr_{0.33})O_{5.0}$ and that the X-ray powder diffraction patterns of these compounds exhibit a strong similarity with that of tetragonal $PbNb_2O_6$. We may point out, in fact, that the $a(= b)$ spacing of Goodman's ferroelectric is very close to the spacing along the [110] direction of orthorhombic $PbNb_2O_6$ and, of course, to the $a$ spacing of tetragonal $PbNb_2O_6$. In view of this similarity, it may be more consistent to rewrite the formulae of the solid solutions mentioned immediately above as:

$$Ba_{1.11}(Nb_{1.78}, Zr_{0.22})O_6; \quad Ba_{1.20}(Nb_{1.61}, Zr_{0.39})O_6,$$

and that of Goodman's ferroelectric as:

$$Ba_{1.14}(Nb_{1.71} Zr_{0.29})O_6.$$

Written in this form, the analogy between these compounds and non-stoichiometric solid solutions of $AB_2O_6$ compounds is quite evident. More work is needed, however, to elucidate the crystal–chemical interrelations among these types of ferroelectrics.

## 6. $PbBi_2Nb_2O_9$ and Related Compounds

Ferroelectricity in the complex oxide $PbBi_2Nb_2O_9$ has been reported by Smolenskii *et al.* (S 5). A marked anomaly of the dielectric constant of ceramic samples was observed at 526 °C (see Fig. VI-9) but no ferroelectric hysteresis

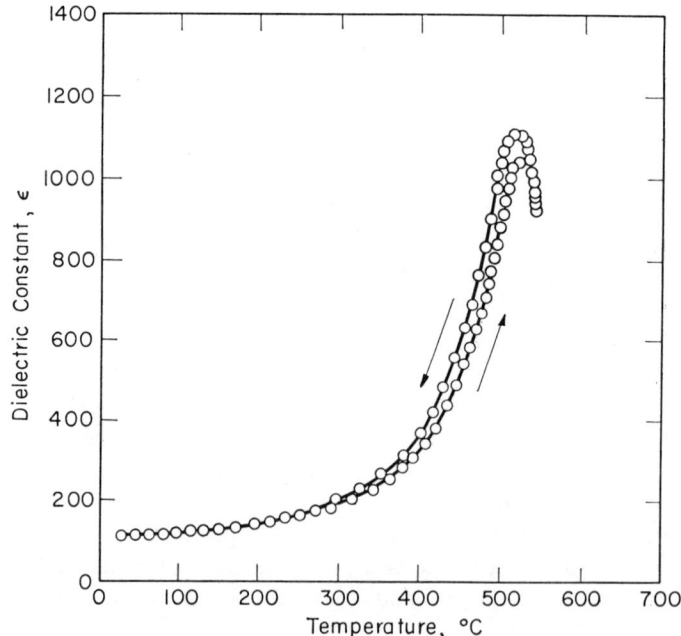

FIG. VI-9. Dielectric constant of $PbBi_2Nb_2O_9$ as a function of temperature (according to Smolenskii *et al.* (S 4)).

loops could be obtained from room temperature up to 200 °C even with applied fields as high as 32 kV/cm. It was assumed that this lack of direct evidence for ferroelectric properties could be attributed to very large values of the coercive field in the temperature range investigated, much as in the case of $PbTiO_3$.

From a crystal–chemical viewpoint, $PbBi_2Nb_2O_9$ may, more properly, be written as $(Bi_2O_2)^{2+}(PbNb_2O_7)^{2-}$ and belongs to a large family of so-called multi-layer interstitial compounds with the general formula:

$$(Bi_2O_2)^{2+}(A_{x-1}B_xO_{3x+1})^{2-}$$

where the metal $A$ is in twelve-coordination (Ca, Sr, Ba, etc.) and $B$ is six-co-ordinated (Ti, Nb, Ta). The value of $x$ may be 2, 3, 4 or 5. The structure of these compounds has been investigated by Aurivillius (A 1). It turns out that, in spite of the complicated chemical formula, the structural scheme is quite simple: for $x = 2$, it consists of an alternation of a layer of $(Bi_2O_2)^{2+}$ with two perovskite-like layers of oxygen octahedra. The structure of $PbBi_2Nb_2O_9$ is depicted schemati-

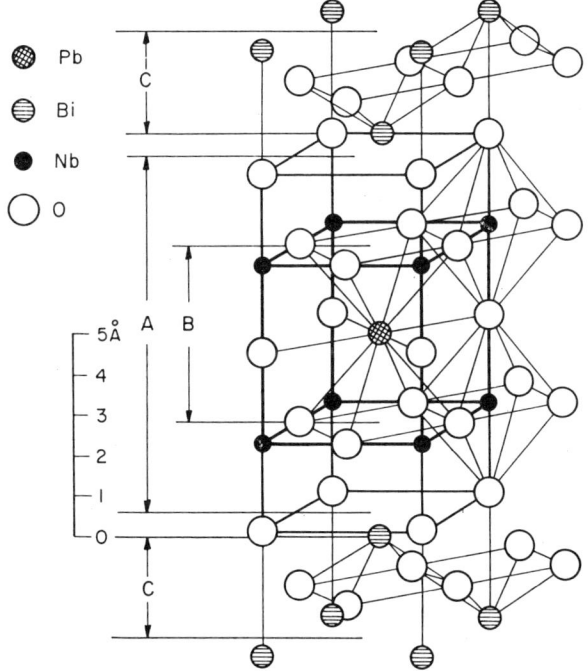

FIG. VI-10. Structure of $PbBi_2Nb_2O_9$ (according to Aurivillius (A 1)).

cally in Fig. VI-10 and suggests that the direction of the spontaneous polarization ought to be parallel to the plane of the layers, as this plane contains the $O-Nb-O$ chains that are known to be effective in the perovskite ferroelectrics. This sug-gestion is further supported by the fact that the crystal exhibits an orthorhombic distortion with the lattice parameters:

$$a = 5.492 \text{ Å}, \quad b = 5.503 \text{ Å}, \quad c = 25.53 \text{ Å}$$

whereas, e.g. $CaBi_2Nb_2O_9$ has tetragonal symmetry. An excellent survey of this type of multilayer substances can be found in Wadsley's review article on the crystal chemistry of non-stoichiometric compounds (W 1).

Recently, several new ferroelectrics have been discovered by Subbarao (S 13) in this family of compounds. A summarizing list is given in Table VI-2.

TABLE VI-2. FERROELECTRICS OF THE $PbBi_2Nb_2O_9$-TYPE *

These compounds belong to the family of multilayer interstitial compounds with the general formula $(Bi_2O_2)^{2+}(A_{x-1}B_xO_{3x+1})^{2-}$. The basic structure is tetragonal, the plane of the layers being perpendicular to the tetragonal c axis.

| x | Compound | c Spacing (Å) | Curie temperature (°C) | Reference |
|---|---|---|---|---|
| 2 | $PbBi_2Nb_2O_9$ | 25.5 | 550 | S 5, S 13 |
| 2 | $PbBi_2Ta_2O_9$ | 25.4 | 430 | S 13, a |
| 2 | $SrBi_2Nb_2O_9$ | 25.1 | 440 | a, b |
| 2 | $SrBi_2Ta_2O_9$ | 25.0 | 335 | a, b |
| 2 | $BaBi_2Nb_2O_9$ | 25.6 | 200 | a, b |
| 2 | $BaBi_2Ta_2O_9$ | 25.5 | 110 | b |
| 3 | $Bi_4Ti_3O_{12}$ | 32.8 | 675 | c, d, e |
| 4 | $BaBi_4Ti_4O_{15}$ | 41.9 | 395 | S 13, a, b, d |
| 4 | $SrBi_4Ti_4O_{15}$ | 41.0 | 530 | a |
| 4 | $PbBi_4Ti_4O_{15}$ | 41.4 | 570 | S 13, b |
| 4 | $Na_{0.5}Bi_{4.5}Ti_4O_{15}$ | 40.7 | 655 | b |
| 4 | $K_{0.5}Bi_{4.5}Ti_4O_{15}$ | 41.2 | 515 | b |
| 5 | $Pb_2Bi_4Ti_5O_{18}$ | 49.7 | 310 | b |
| 5 | $Sr_2Bi_4Ti_5O_{18}$ | 48.8 | 285 | b |

References:

a. SMOLENSKII, G. A., ISUPOW, V. A. and AGRANOVSKAYA, A. I., *Fiz. Tverdogo Tela* **3**, 895 (1961).
b. SUBBARAO, E. C., *J. Phys. Chem. Solids* (to be published).
c. SUBBARAO, E. C., *Phys. Rev.* **122**, 804 (1961).
d. FANG, P. H., ROBBINS, C. and FORRAT, F., *Compt. rend.* **252**, 683 (1961).
e. VITERT, L. G. VAN and EGÉRTON, L., *J. Appl. Phys.* **32**, 959 (1961).

* This table was revised in August, 1961 in order to include several new members to this family of ferroelectrics.

## 7. Lithium Niobate, $LiNbO_3$, and Lithium Tantalate, $LiTaO_3$

Both $LiNbO_3$ and $LiTaO_3$ were reported to be ferroelectric by Matthias and Remeika (M 2) in 1949. A few more details were given later by Matthias (M 3) in his review article on ferroelectricity. Hysteresis loops can reportedly be observed only in single crystals but not in ceramic materials, and, moreover, only above 200 °C. Some of the dielectric data pertinent to $LiTaO_3$ are depicted in Fig. VI-11 and similar behavior has been reported for $LiNbO_3$ at higher temperatures. It is quite evident that the dielectric properties are radically different

from those of any other ferroelectric known to date, as both the coercive field
strength and the spontaneous polarization increase with increasing temperature
and no Curie point is detected.

Structurally, $LiNbO_3$ and $LiTaO_3$ are isomorphous: the symmetry is rhombohedral and the compounds are closely related to ilmenite ($FeTiO_3$). It was ori-

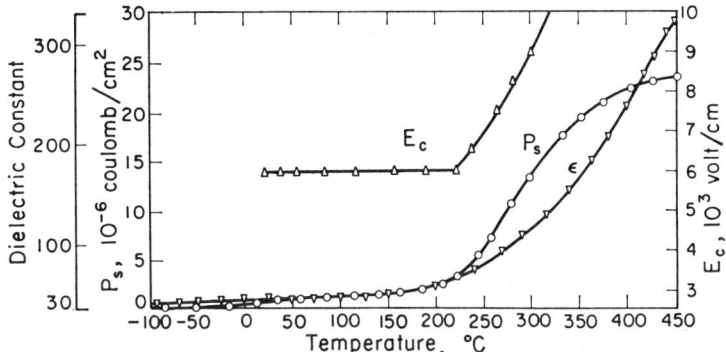

FIG. VI-11. Dielectric constant $\varepsilon$, spontaneous polarization $P_s$ and coercive field
strength of $LiTaO_3$ as functions of temperature (according to Matthias (M 3)).

ginally believed, in fact, that these lithium double oxides were of the ilmenite-
type, and a theory of the ferroelectric phenomenon in the ilmenite structure was
developed by Schweinler (S 6) along the lines of Slater's theory of $BaTiO_3$ (see
Section IV-11). It was later shown by Bailey (B 2), however, that the structure
of $LiNbO_3$ is significantly different from that of ilmenite. Both structures are
indeed based upon hexagonal close-packing of oxygen ions and contain the metal
ions aligned along the rhombohedral [111] direction, but whereas in ilmenite the
sequence is Ti−Fe−Fe−Ti, in $LiNbO_3$ it is Li−Nb–Li−Nb. The oxygen octahedra share faces and edges rather than corners only, as in the case of perovskite-
type compounds.

Megaw (M 4) advanced the suggestion that $LiNbO_3$ may be considered as a
distorted perovskite compound. The difficulty of this picture is that, in this case,
polarization reversal would require displacements of Li and O as large as 2.0 Å
and 0.7 Å respectively. It is preferable, however, to await the publication of
more experimental data before discussing   the details of possible theoretical
models.

## 8. Sodium Vanadate, $NaVO_3$, and Silver Vanadate, $AgVO_3$

Sawada and Nomura (S 7) reported ferroelectric properties for $NaVO_3$ below
380 °C. The dielectric constant does not undergo a pronounced anomaly at this
temperature, and no data concerning the spontaneous polarization were published.

The structure of $NaVO_3$ has often been classified as one of the perovskite-
type but it is actually quite different as was shown by Sorum (S 8). The lattice

contains $VO_4$ tetrahedra which share two corners, and can be regarded as being constructed of infinite linear chains of $VO_3^-$ ions. The importance of this compound in the ferroelectric family is that it seems to be the only example amongst the ferroelectric oxides that does *not* crystallize with a framework of oxygen octahedra.

Recently, Hellicar and White (H 3) have observed ferroelectric properties in silver vanadate, $AgVO_3$, the Curie temperature being 170–180 °C. The compound, however, is probably not isomorphous with $NaVO_3$.

## 9. $BaAl_2O_4$-type Ferroelectrics

Recently, Dunne and Stemple (D 2) have reported ferroelectric activity in single crystals of complex oxyfluorides with the formulas:

$$Ba(Al_{1.4}Li_{0.6})(O_{2.8}F_{1.2}) \quad \text{and} \quad Ba(Al_{1.7}Li_{0.3})(O_{3.4}F_{0.6}).$$

The former has a Curie point at 150 °C, the latter at 140 °C. The crystals are hexagonal, with the lattice parameters, at room temperature:

$$a = 10.44 \text{ Å}, \quad c = 8.77 \text{ Å}$$

i.e. very close to those of barium aluminate, $BaAl_2O_4$. The general formula of these compounds can in fact be written as

$$Ba(Al_{2-x}Li_x)(O_{4-2x}F_{2x}),$$

but it is not known, at the present stage, just how large the range of $x$ values is, within which the crystals are ferroelectric.

The dielectric constant along the polar hexagonal axis is about 10 at room temperature, and only exhibits a small, broad anomaly at the Curie point. At room temperature, $P_s = 0.1 \times 10^{-6}$ C/cm$^2$, and the 60 c/s coercive field $E_c = 5$ kV/cm. Hysteresis loops cannot be observed below approximately $-20$ °C, where the coercive field becomes very large.

## 10. Concluding Remarks

If we review the content of the preceding and the present chapters, we see that the family of double and complex oxide ferroelectrics is already very large. From the crystal-chemical viewpoint, a characteristic feature common to most of these ferroelectric compound is the role of centered oxygen octahedra as essential building elements of the structural scheme. From the dielectric point of view, it appears that the compounds containing Pb and Nb usually exhibit the highest Curie points within one and the same subgroup.

The structural and the dielectric characteristics of these ferroelectrics, however, vary over a rather wide range, a tendency that is very noticeable in the group of water-soluble ferroelectrics as well (see Chapters VIII and IX). It is quite likely that this tendency will be even more pronounced in the near future; as a result, a unified treatment of the ferroelectric phenomenon may become more and more difficult.

## BIBLIOGRAPHY

(A 1)  AURIVILLIUS, B., *Arkiv Kemi* **1**, 463 (1949); **2**, 519 (1950).

(B 1)  BYSTROEM, A., *Arkiv Kemi Mineral. Geol.* **A 18**, No. 21 (1944).

(B 2)  BAILEY, P., Thesis, Bristol (1952), Quoted by MEGAW, H. D., *Acta Cryst.* **7**, 187 (1954).

(C 1)  COOK, JR., W. R. and JAFFE, H., *Phys. Rev.* **88**, 1426 (1952); **89**, 1297 (1953).

(C 2)  COATES, R. V. and KAY, H. F., *Phil. Mag.*, Ser. 8, **3**, 1449 (1958).

(D 1)  DANNER, H. and PEPINSKY, R., *Phys. Rev.* **99**, 1215 (1955).

(D 2)  DUNNE, T. G. and STEMPLE, N. R., *Phys. Rev.* **120**, 1949 (1960).

(F 1)  FRANCOMBE, M. H. and LEWIS, B., *Acta Cryst.* **11**, 696 (1958).

(F 2)  FRANCOMBE, M. H., *Acta Cryst.* **9**, 683 (1956).

(F 3)  FRANCOMBE, M. H. and LEWIS, B., *J. Electronics* **2**, 387 (1957).

(F 4)  FANG, H. P., BROWER, W. S., ROTH, R. S. and MARZULLO, S., *Bull. Am. Phys. Soc.* Ser. II, **4**, 64 (1959).

(F 5)  FANG, P. H., ROTH, R. S. and BROWER, W. S., *Am. Ceram. Soc. Bull.* **38**, 183 (1959).

(F 6)  FRANCOMBE, M. H., *Acta Cryst.* **13**, 131 (1960).

(G 1)  GOODMAN, G., *J. Am. Ceram. Soc.* **36**, 368 (1953).

(G 2)  GOODMAN, G., *J. Am. Ceram. Soc.* **43**, 105 (1960).

(H 1)  HÄGG, G. and MAGNELI, A., *Rev. Pure Appl. Chem.* **4**, 235 (1954).

(H 2)  HULM, J. K., *Phys. Rev.* **92**, 504 (1953).

(H 3)  HELLICAR, N. J. and WHITE, E. A. D., Private communication (May 1960).

(I 1)  ISUPOV, V. A. and KHOMUTETSKII, O. K., *Zhur. Tekh. Fiz.* **27**, 2704 (1956).

(I 2)  ISUPOV, V. A., *Izvest. Akad. Nauk S.S.S.R.*, Ser. *Fiz.* **21**, 402 (1957).

(I 3)  ISUPOV, V. A. and KOSYAKOV, V. I., *Zhur. Tekh. Fiz.* **28**, 2175 (1958).

(I 4)  ISMAILZADE, I. G., *Izvest. Akad. Nauk S.S.S.R.*, Ser. *Fiz.* **22**, 1485 (1958).

(I 5)  ISUPOV, V. A., *Fiz. Tverdogo Tela* **1**, 242 (1959).

(I 6)  ISMAILZADE, I. G., *Kristallografiya* **4**, 658 (1959).

(J 1)  JONA, F., SHIRANE, G. and PEPINSKY, R., *Phys. Rev.* **98**, 903 (1955).

(M 1)  MAGNELI, A., *J. Inorg. & Nuclear Chem.* **2**, 330 (1956).

(M 2)  MATTHIAS, B. T. and REMEIKA, J. P., *Phys. Rev.* **76**, 1886 (1949).

(M 3)  MATTHIAS, B. T., *Science* **113**, 591 (1951).

(M 4)  MEGAW, H. D., *Acta Cryst.* **7**, 187 (1954).

(R 1)  ROTH, R. S., *J. Research. Natl. Bur. Standards* **56**, 17 (1956).

(R 2)  ROTH, R. S., *J. Research. Natl. Bur. Standards* **62**, 27 (1959).

(R 3)  ROTH, R. S., *Acta Cryst.* **10**, 437 (1957).

(R 4)  ROTH, R. S., *Bull. Am. Phys. Soc.* II, **4**, 63 (1959).

(S 1)  SHIRANE, G. and PEPINSKY, R., *Phys. Rev.* **92**, 504 (1953).

(S 2)  SMOLENSKII, G. A. and AGRANOVSKAYA, A. I., *Doklady Akad. Nauk S.S.S.R.* **97**, 237 (1954).

(S 3)  SUBBARAO, E. C., SHIRANE, G. and JONA, F., *Acta Cryst.* **13**, 226 (1960).

(S 4)  SMOLENSKII, G. A., ISUPOV, V. A. and AGRANOVSKAYA, A. I., *Doklady Akad. Nauk S.S.S.R.* **108**, 232 (1956); **113**, 803 (1957).

(S 5)  SMOLENSKII, G. A., ISUPOV, V. A. and AGRANOVSKAYA, A. I., *Fiz. Tverdogo Tela* **1**, 169 (1959).

(S 6)  SCHWEINLER, H. C., *Phys. Rev.* **87**, 5 (1952).

(S 7)  SAWADA, S. and NOMURA, S., *J. Phys. Soc. Japan* **6**, 192 (1951).

(S 8)  SORUM, H., *Kgl. Norske Videnskab. Selskabs. Forh.* **16**, 39 (1943); see also reference (W 2).

(S 9)  SUBBARAO, E. C., *J. Am. Ceram. Soc.* **42**, 448 (1959).

(S 10)  SUBBARAO, E. C. and SHIRANE, G., *J. Chem. Phys.* **32**, 1846 (1960).

(S 11)  SUBBARAO, E. C., *J. Am. Ceram. Soc.* **43**, 439 (1960).

(S 12)  SMOLENSKII, G. A., ISUPOV, V. A. and AGRANOVSKAYA, A. I., *Fiz. Tverdogo Tela* **1**, 442 (1959).

(S 13)  SUBBARAO, E. C., *J. Chem. Phys.* **34**, 695 (1961).

(W 1)  WADSLEY, A. D., *Rev. Pure Appl. Chem.* **5**, 165 (1955).

(W 2)  WELLS, A. F., *Structural Inorganic Chemistry*, p. 499, Clarendon Press, Oxford (1950).

# CHAPTER VII

# ROCHELLE SALT AND ISOMORPHOUS CRYSTALS

## 1. Introduction

Rochelle salt is the common name given to the double tartrate of sodium and potassium crystallizing with four molecules of water: $NaKC_4H_4O_6 \cdot 4H_2O$. First prepared around 1655 by Seignette in La Rochelle (France) (V 4), it is also known under the name of "Seignette salt", particularly in the German and early Russian literature. It is ferroelectric below approximately 24 °C; it is the oldest, and was for a long time the only, ferroelectric crystal known. Its anomalously large piezoelectric effect had been noticed in 1880 by the Curie brothers in their pioneer investigation of piezoelectricity, and quantitative measurements of its piezoelectric and electro-optic effects were carried out in 1894 by Pockels (P 1), but its anomalous dielectric properties were discovered only around 1921 by Valasek (V 1). It was he, in fact, who recognized the ferromagnetic analogy of this dielectric phenomenon and introduced the term "Curie point" for the first time in the physics of dielectrics. Owing to the parenthood of Rochelle salt, in fact, the phenomenon known today under the name of ferroelectricity, was more often called, in the early literature, "Seignette-electricity" or "Rochelle-electricity", names which, unfortunately, have now become obsolete.

The most outstanding property of Rochelle salt is that it exhibits *two* Curie points. The ferroelectric state is confined to the region between approximately $-18$ °C and $+24$ °C. The lower critical temperature (of $-18$ °C) is called the "lower Curie point". It is characterized by a large anomaly of the dielectric constant, just as the otherwise normal Curie point at $+24$ °C which, by antonymy, is called the "upper Curie point". The symmetry of the non-polar phase above $+24$ °C is orthorhombic, point group 222, thus piezoelectric, and is the same as that of the non-polar phase below $-18$ °C. The dimensions of the orthorhombic unit cell are, at 35 °C (U 1):

$$a = 11.878 \text{ Å}, \quad b = 14.246 \text{ Å}, \quad c = 6.218 \text{ Å}.$$

The ferroelectric phase belongs to the monoclinic crystal class 2, the polar axis being parallel to the direction of the original orthorhombic [100] axis.

The volume of literature pertinent to the physical properties of Rochelle salt is enormous, but unfortunately most of the early publications are now obsolete and confusing because they were written at a time when the behavior of this crystal was little understood. The first clarifying contributions were given in a series of papers by Mueller (M 1–M 5). A critical and detailed review of the work up to 1946 can be found in the book of Cady (C 1), to which the reader is referred for a number of experimental and theoretical details. A concise review of the properties and the uses of Rochelle salt crystals can also be found in Mason's book (M 6). Despite its rather poor mechanical strength and low temperature of disintegration (55 °C), Rochelle salt was and is used for a number of important applications, especially in the field of acoustics.

Chemically, the crystal is stable only under limited conditions of temperature and humidity. At 25 °C, for example, relative humidity of less than about 40% causes the crystal to lose water of crystallization (efflorescence), whereas above 85% humidity the crystal absorbs water from the surrounding atmosphere on its surface and slowly dissolves (deliquescence) (M 6). At 55.6 °C, the crystal decomposes into a mixture of sodium and potassium tartrates and their saturated solution, and at 58 °C, these salts are completely dissolved in the water (C 1). This irreversible process has commonly been referred to as the "melting" of Rochelle salt.

The solubility of the crystal in water is very large and strongly temperature dependent: 1.50 moles (430 g) at 0 °C and 4.90 moles (1390 g) at 30 °C. The latter property makes the growth of large single crystals of Rochelle salt from water solutions a particularly easy task. Supersaturation of the solution can be achieved either by cooling or by evaporation. Unless special precautions are taken, however, the growth of crystals from solution must take place at temperatures below 40 °C, since above this temperature sodium tartrate is deposited. For a detailed description of the various methods employed in industry and in the laboratory the reader is referred to Cady's book (C 1). Crystals weighing as much as 2 kg can be grown rather easily, although the crystallizing process mostly requires several days up to a couple of weeks.

A method aiming at the rapid production of Rochelle salt crystals has been recently described by Aliavdin et al. (A 1): these authors found that the dissociation of the compound into the separate tartrates (starting at 40 °C and completed around 55 °C) is due not only to the increase in temperature but also to the presence of nuclei of the individual tartrates. If, however, a solution of Rochelle salt is heated for a long time at about 100 °C, these nuclei are destroyed and the crystallization process can then be carried out at temperatures between 56 °C and 40 °C. This technique reportedly allows the growth of crystals weighing about 1 kg in only 60 hr. With a gradual decrease in temperature from 52.5 °C to 51 °C, the linear rate of growth along the [001] axis is of the order of 3 mm/hr.

The kinetics of crystallization of Rochelle salt at constant temperature and supersaturation has been studied by Kozlovskii (K 1), who also described a growing process involving continuous enrichment of the mother solution. It was found that the rate of crystallization depends on the size of the seed crystal used, and that the linear velocity of growth of the individual faces is constant.

Although the crystal class of orthorhombic Rochelle salt is enantiomorphous, the crystals are usually of the right-handed form, with axial ratio $a : b : c = 0.8325 : 1 : 0.4334$ (C 1). The most prominent and typical forms are the $c$ faces $\{001\}$ and the prismatic $m$ faces $\{110\}$; less developed are the prismatic $n$ $\{120\}$, $l$ $\{210\}$ and $b$ $\{010\}$ faces, while the $a$ faces $\{100\}$ are mostly very small or completely absent. The crystals can be easily oriented by measuring the interfacial angles with a simple circle spectrometer; a schematic drawing of a (001) cross-section for this purpose is depicted in Fig. VII-1. The angles indicated are only accurate to within 30′ but this accuracy is usually sufficient for the identification of the various faces.

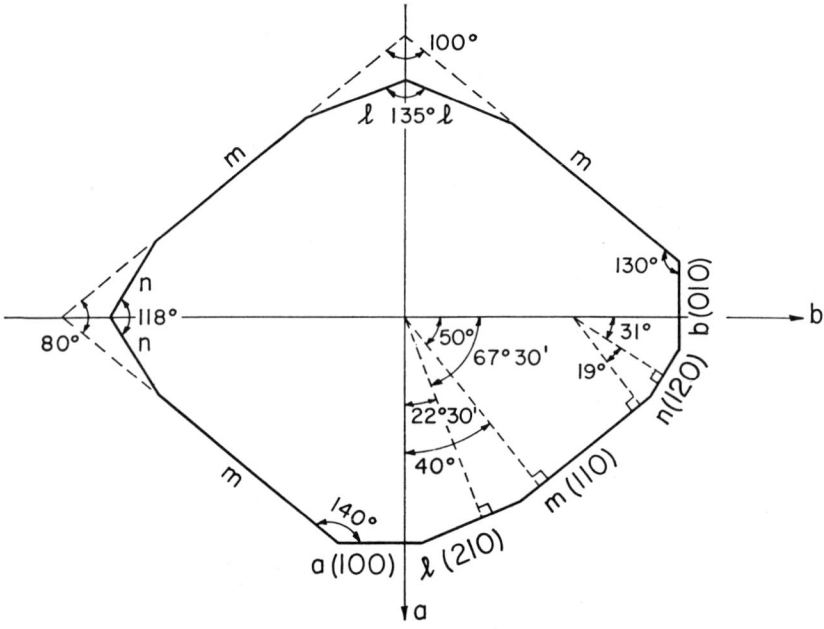

FIG. VII-1. Morphology of Rochelle salt. The drawing represents the cross-section perpendicular to [001] of an ideal crystal (schematic). The angles indicated are only accurate within approximately 30′.

The crystals are commonly cut with a wet string technique described by Cady (C 1) and a number of other investigators. Details of the procedures to be followed for the preparation of plates and bars are also given in Cady's book. The evaporation of noble metal electrodes upon the crystal surfaces under vacuum requires proper care, as the crystal loses water rather rapidly in a vacuum. A method for improving the surface of Rochelle salt samples through recrystallization by contact with water vapor has been described by Asselmeyer and Bienert (A 2). The treatment improves the stability of the surface in vacuum and dry air.

## 2. Dielectric and Optical Properties

### The Dielectric Constant

The measurement of the small-signal dielectric constant along the orthorhombic $a$ axis ($\varepsilon_a$) reveals very pronounced anomalies at the upper and lower Curie points, i.e. at $+24$ °C and $-18$ °C, respectively. The dielectric constants $\varepsilon_b$

and $\varepsilon_c$ (measured along [010] and [001], respectively) have a smooth and normal temperature dependence. Figure VII-2 shows the result of Hablützel's measurement of $\varepsilon_a$ as a function of temperature (H 1). The character of the anomalies and the fact that no thermal hysteresis is observed are indicative of transitions of the second order. Above the upper Curie point, more precisely from 25 °C to 32 °C, the Curie–Weiss law $\varepsilon_a = C/(T - T_0)$ is obeyed (M 1) with $C = 2.24 \times 10^3$

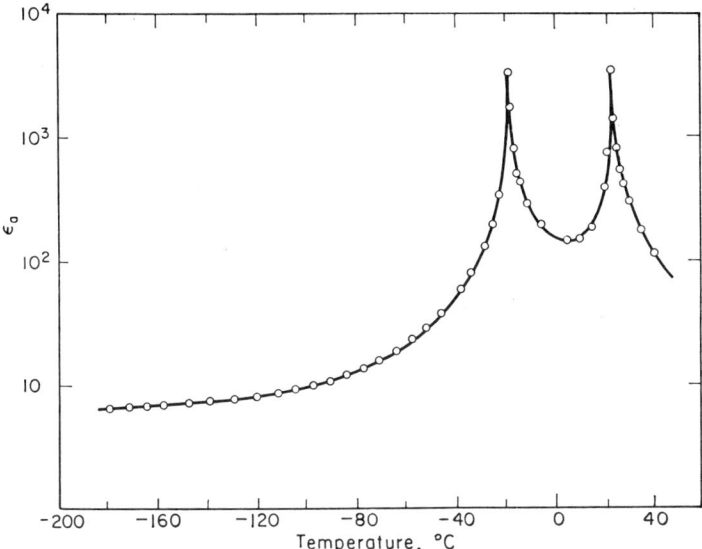

FIG. VII-2. Temperature dependence of the small-signal dielectric constant $\varepsilon_a$ of Rochelle salt. The measuring field was 4 V/cm above $-40$ °C, 50 V/cm below $-40$ °C. Frequency 1000 c/s (according to Hablützel (H 1)).

degrees and $T_0 = 23.0 \pm 0.5$ °C. From 34 °C to 50 °C, the law still holds very well, but the constants become $C = 1.71 \times 10^3$ degrees and $T_0 = 25.3 \pm 0.05$ °C. In a narrow temperature range below the lower Curie point, namely from $-18$ °C to $-28$ °C, the dielectric constant $\varepsilon_a$ follows again a law of the Curie–Weiss type, $\varepsilon_a = C'/(T'_0 - T)$, with $C' = 1.18 \times 10^3$ degrees and $T'_0 = -17.9$ °C. At lower temperatures, the dielectric constant keeps decreasing and assumes a constant value of about 7 below $-160$ °C. The decrease of the dielectric constant in the ferroelectric region is to be ascribed to a spontaneous saturation effect, as discussed in Chapter III for the case of $KH_2PO_4$.

High-frequency measurements of the dielectric constant have been carried out by a number of investigators (B 1), (B 2), (M 3), (A 3), (A 4). In general, the results concur in establishing that the Curie–Weiss law is still obeyed at frequencies above piezoelectric resonance. As discussed in Section III-2, at these frequencies the crystal is clamped by inertia, but only partially, because the spontaneous polarization cannot be suppressed in this way. The results, however, indicate that the difference between the reciprocal susceptibilities of the free and the clamped crystal is essentially temperature independent (see Eq. III-1). The dielectric dispersion above piezoelectric resonance was investigated by Akao and co-workers (A 3), (A 4) with particular attention to the frequency dependence of the dielectric-constant maxima at the Curie points. These authors reported a drastic drop of *both* these maxima from about 250 to 60 between $3 \times 10^8$ c/s and $3 \times 10^9$ c/s. At frequencies as high as $2.5 \times 10^{10}$ c/s the temperature depen-

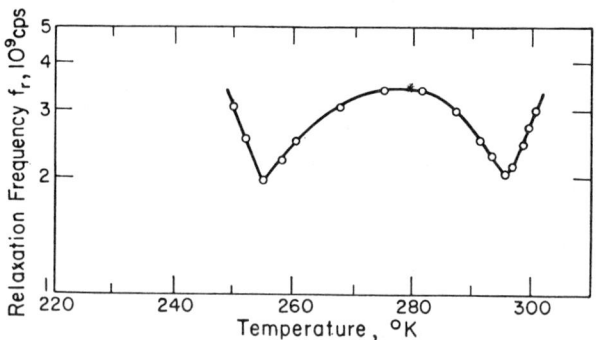

FIG. VII-3. Temperature dependence of the dielectric relaxation frequency of Rochelle salt (according to Akao and Sasaki (A 4)).

dence of the dielectric constant is practically smooth and free of anomalies. Figure VII-3 shows that the relaxation frequency exhibits minima at the Curie points (A 4), in a way somewhat recalling the behavior of the reciprocal dielectric constant as a function of temperature.

### The Spontaneous Polarization and the Coercive Field

The temperature dependence of the spontaneous polarization $P_s$ of Rochelle salt is depicted in Fig. VII-4. The onset at the Curie points is steep but continuous,

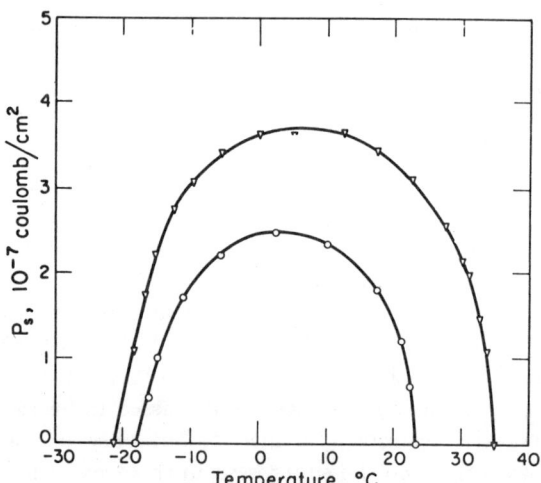

FIG. VII-4. Spontaneous polarization of Rochelle salt (circles) and deuterated Rochelle salt (triangles) as functions of temperature (according to Hablützel (H 1)).

in accordance with the character of a second-order transition. The maximum value of $P_s$ occurs in the vicinity of 5 °C and equals $0.25 \times 10^{-6}$ C/cm².

A somewhat similar behavior is exhibited by the coercive field strength $E_c$, with a maximum of about 200 V/cm occurring between 5 and 15 °C. A dependence of $E_c$ upon the thickness of the samples has been reported by Mueller (M 1), Nakamura (N 3), and Wieder (W 4) and upon the strength of the applied field by David (D 1). Ballistic measurements by Takahashi and Hara (T 1) confirmed the dependence of the shape of the hysteresis loops and of the coercive field upon the frequency of the applied field.

## Optical Properties

Orthorhombic Rochelle salt is biaxial positive with the refractive indices $n_a = 1.4954$, $n_b = 1.4920$, $n_c = 1.4900$ (C 1). The plane of the optic axes is the (010) plane, the $a$ axis being the acute bisectrix and the optical axial angle 69°40′. The temperature dependence of the refractive indices was measured by Valasek (V 1) from about $-70$ °C to 40 °C and, more recently, by Baranskii et al. (B 3) from 16 °C to 28 °C. Neither of these authors found great changes of the refractive indices in these temperature ranges; the thermal behavior is reportedly linear, with no departure from linearity at the Curie points. This is in contradiction with the small anomaly of the birefringence reported by Mueller (M 1), but is probably due to the lower experimental accuracy of the direct determination of the $n$'s as compared to that of the $\Delta n$'s. The birefringence values determined by Vergnoux et al. (V 2), in fact (0.00119; 0.00536; 0.00648), differ markedly from those computed from the refractive indices. An interesting dispersion effect of the birefringence in the near infrared was reported by the latter authors: at 20 °C, the birefringence in the $XZ$ plane becomes equal to that in the $XY$ plane at the wavelength $1.39\mu$, where the crystal becomes uniaxial. Above the upper Curie point (at 33 °C) the wavelength dependence of the birefringence is essentially the same as at 20 °C and the crystal becomes uniaxial at $1.42\mu$.

A general theory of the additional scattering of light in the vicinity of second-order transitions has been developed and applied to Rochelle salt by Krivoglaz and Rybak (K 2).

## The Electro-optic Effect

As the general theory of the electro-optic effect has been discussed in Section III-5 as an introduction to the case of $KH_2PO_4$, we consider here directly the specialization for the case of Rochelle salt. Let us first study the effect in the clamped crystal, i.e. a crystal in which all strains $x_{ik}$ are kept equal to zero. From Eq. (III-17), we obtain that the change in the polarization constants $a_{ik}$ due to a dielectric polarization $P$ is:

$$a_{ik} - \delta_{ik}\frac{1}{n_i^2} = \sum_{j=1}^{3} r_{ikj} P_j \tag{VII-1}$$

where $\delta_{ik}$ is the Kronecker delta. For crystals of the orthorhombic class 222, the electro-optic coefficients $r_{ikj}$ are such that above equations are reduced to:

$$a_{11} = 1/n_1^2 \, ; \quad a_{22} = 1/n_2^2 \, ; \quad a_{33} = 1/n_3^2 \, , \\ a_{23} = r_{41} P_1 \, ; \quad a_{31} = r_{52} P_2 \, ; \quad a_{12} = r_{63} P_3 \, . \tag{VII-2}$$

In the present discussion, we are going to limit ourselves to the case in which the electric field is applied along the [100] axis, so that $P_2 = P_3 = 0$. The equation of the deformed indicatrix (see Eq. III-16) becomes then:

$$\frac{x^2}{n_1^2} + \frac{y^2}{n_2^2} + \frac{z^2}{n_3^2} + 2r_{41} P_1 yz = 1 . \tag{VII-3}$$

If we send the light in the [010] or [001] direction ($y = 0$, or $z = 0$, respectively), the linear electro-optic effect cannot be observed (transverse effect). If we send the light in the [100] direction (longitudinal effect) we see that the effect consists in a rotation of the pertinent ellipse within the $xy$ plane by an angle $\varphi$ that can be easily computed from (VII-3):

$$\tan 2\varphi = \frac{2r_{41} P_1}{1/n_2^2 - 1/n_3^2} . \tag{VII-4}$$

The change in the principal refractive indices is proportional to $(r_{41} P_1)^2$, thus small of the second order, and the change in the value of birefringence $n_b - n_c$ can be computed to be (the formulas are similar to those for the elasto-optic effect given, e.g. by Jona (J 1)):

$$\delta(n_2 - n_3) \cong \frac{n_2^2 \, n_3^2 \, (n_2^3 + n_3^3) \, r_{41}^2 \, P_1^2}{2 \, (n_3^2 - n_2^2)} . \tag{VII-5}$$

The linear electro-optic effect can be observed, however, if we send the light along the [011] direction, since in such a case the principal refractive indices are given by the ellipse (obtained from (VII-3) for $z = -y$):

$$\frac{x^2}{n_1^2} + \left( \frac{1}{n_2^2} + \frac{1}{n_3^2} - 2r_{41} P_1 \right) y^2 = 1 . \tag{VII-6}$$

Measurements of the coefficient $r_{41}$ were carried out by Pockels long before the ferroelectric properties of Rochelle salt were known, and later re-evaluated by Cady (C 1). Obviously, the experiments could not be carried out on clamped specimens, and thus the influence of piezoelectric deformations had to be taken into account for the computation of the "true" electro-optical coefficients $r_{ik}$ from the experimental data. This makes the reported value of $r_{41}$ considerably uncertain because the measurements were carried out around room temperature where both the dielectric susceptibility and the piezoelectric modulus are very large and very strongly temperature dependent (see Section 4). The reported values are (C 1):

$$r_{41} \cong -0.14 \times 10^{-8} \, ; \quad r_{52} = -5.9 \times 10^{-8} \, ; \quad r_{63} = +2.3 \times 10^{-8} \, ;$$

in c.g.s. units, and are of the same order of magnitude as, for example, the electro-optic coefficients of quartz. The values of $r_{52}$ and $r_{63}$ were obtained, of course, with fields applied along [010] and [001], and the light along [101] and [110], respectively. It is important to keep in mind that these values of the coefficients $r_{ik}$ represent the "true" electro-optic effect. The overall effect is so much larger that the corresponding constant (which we may call $r_{41}^*$) is probably orders of magnitude larger than the $r_{41}$ reported above.

In the actual experiments, carried out on free rather than clamped crystals, the indirect effect caused by the piezoelectric deformations is so large that a change in birefringence can be observed even in the case where the field is applied along [100] and the light sent along [010] or [001]. This is a second-order effect and has been investigated extensively by Mueller (M 1), (M 4). Application of an electric field along [100] causes the crystal to become monoclinic with the [100] direction as the polar axis. For crystals in this class the equations for the linear electro-optic effect become

$$
\begin{aligned}
a_{11} - 1/n_1^2 &= r_{11}P_1; \quad a_{22} - 1/n_2^2 = r_{21}P_1; \\
a_{33} - 1/n_3^2 &= r_{31}P_1; \quad a_{23} = r_{41}P_1; \\
a_{31} &= a_{12} = 0.
\end{aligned}
\qquad\text{(VII-7)}
$$

Since the appearance of the three new coefficients $r_{11}$, $r_{21}$ and $r_{31}$ is the result of piezoelectric deformations induced by the field, these parameters are proportional to the strain and, therefore, also the polarization $P_1$ (see also Section III-4). Thus, the change in the polarization constants $a_{11}$, $a_{22}$ and $a_{33}$ is proportional to $P_1^2$ and the changes in birefringence for light traveling along [010] or [001] can be computed to be (M 4) (transverse effect):

$$
\begin{aligned}
\delta(n_3 - n_1)/\lambda &= N_2 P_1^2 \\
\delta(n_1 - n_2)/\lambda &= N_3 P_1^2,
\end{aligned}
\qquad\text{(VII-8)}
$$

for light traveling along [100] (longitudinal effect):

$$
\delta(n_2 - n_3)/\lambda = N_1 P_1^2
\qquad\text{(VII-9)}
$$

where $N_1$, $N_2$, and $N_3$ are functions of the refractive indices that can be determined experimentally. Figure VII-5 shows the experimental results of the transverse effect for temperatures above the upper Curie point. It may be noted that the effect appears to be far from quadratic; this is due to the fact when the change in birefringence is expressed in terms of the field $E$, rather than the polarization $P$ as in (VII-8), the non-linearity of the dielectric behavior causes a substantial deviation from a square law. It may also be pointed out that while the effect is about one million times greater than the Kerr effect in most liquids (typical values of the Kerr constant are of the order of $10^{-15}$), it is still remarkably small as compared to the linear electro-optic effect of the phosphate ferroelectrics (see Section III-5). A field of $10\,\text{kV/cm}$ causes a change of approximately 0.4 in the quantity $\Delta n/\lambda$ at the upper Curie point of Rochelle salt (M 4), thus about 200 times smaller than the linear effect in $KD_2PO_4$. It has been suggested by Zwicker and Scherrer (Z 1) that this quadratic effect of Rochelle salt may receive a substantial contribution from the electrostrictive effect (see Eq. III-20).

In the temperature region between the Curie points, the spontaneous polarization $P_s$ along [100] creates a *spontaneous* Kerr effect that can be described by equations of the type (VII-8) and (VII-9). This gives rise to an anomalous

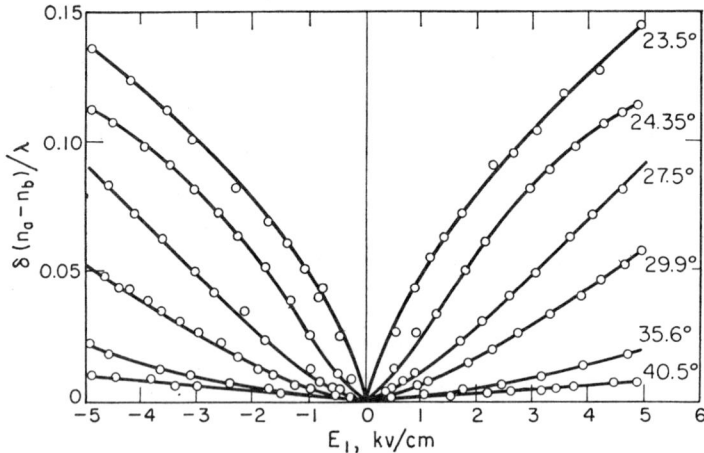

FIG. VII-5. Second-order transversal electro-optic effect in Rochelle salt above the upper Curie point. Field parallel to [100], light parallel to [001] (according to Mueller (M 1)).

temperature variation of the natural birefringence, as measured by Mueller (M 1), (M 4). In the ferroelectric region, the crystal is truly monoclinic, so that the electro-optic effect measured under the conditions described above is really the linear effect of the monoclinic phase. This can be shown, as indicated by Mueller (M 4), by considering that electric fields larger than the coercive field cause the polarization to increase from $P_s$ to $P_s + P_{ind}$, where the induced polarization $P_{ind}$ is much smaller than $P_s$. Thus, the change in birefringence in the monoclinic phase is given by:

$$\delta(\varDelta n) = \mathrm{N}(P_s + P_{ind})^2 - \varDelta n_0 = 2\mathrm{N}\,P_s\,\varepsilon\,E_1,$$

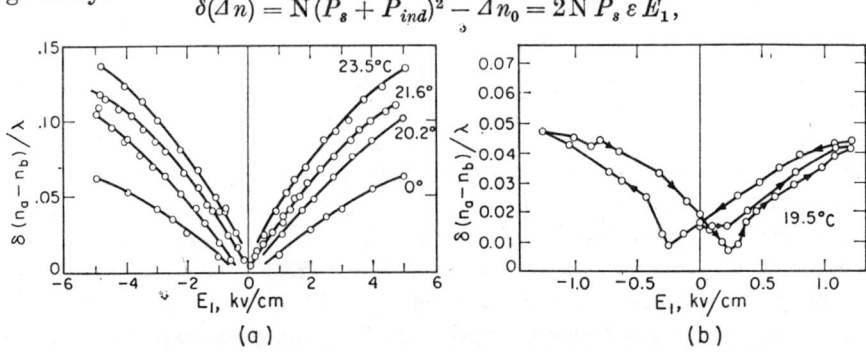

FIG. VII-6. Electro-optic effect of Rochelle salt in the ferroelectric phase.
(a) Almost linear effect for fields larger than the coercive field, applied along [100]. Light traveling along [001]. For fields smaller than ±1 kV/cm, this figure should be completed with (b).
(b) Quadratic ("butterfly") hysteresis loop for electric fields of the order of the coercive field. (According to Mueller (M 1).)

where $\varepsilon$ is the dielectric constant and $E_1$ the field applied along [100]. This explains why the observed change in birefringence is almost linear with the field at temperatures between but not too close to the Curie points (Fig. VII-6a). The sign of this effect, however, cannot be reversed by reversing the field direction because this reverses the direction of the spontaneous polarization (the product $P_s E_1$ is always positive when $E_1$ is larger than the coercive field). For smaller fields, the dependence of $\delta(\varDelta n)$ upon field is such as to give a quadratic hysteresis curve ("butterfly loop", Fig. VII-6b) that is similar to and can be understood on the same basis as the quadratic hysteresis of the piezoelectric effect of $BaTiO_3$ (see Fig. IV-28).

## 3. Temperature and Pressure Effects

### Thermal Expansion

Measurements of the lattice expansion have been made by a number of authors (V 1), (V 3), (H 2), (U 1), (M 7), but the results are not quite in agreement with one another. Early investigations found a normal behavior of the expansion coefficients as functions of temperature, except for a slight change in slope at the upper Curie point (V 1), (V 3), but the rate of temperature change was probably too rapid for reliable results. Hablützel took care of stabilizing the temperature before each measurement and found anomalies in the expansion coefficients along [100] and [011] at the Curie points (H 2). In all cases, however, the samples investigated were most probably multi-domain in the ferroelectric phase, so that the latter results may be only partially representative of the true thermal behavior or Rochelle salt. These results can be qualitatively understood on the basis of the inverse quadratic piezoelectric effect in the polar phase. The X-ray data of Ubbelohde and Woodward (U 1), in principle, ought to be more reliable, but unfortunately only average values of the expansion coefficients over wide temperature ranges were reported, so that no reliable conclusion can be reached about the dependence of volume upon temperature. Careful measurements of Mason and Armstrong (M 7) of the expansion coefficients along [010], [001] and [011] seem to indicate the occurrence of a volume contraction in the ferroelectric phase.

### The Lattice Distortion

As mentioned in Section I, the polar phase of Rochelle salt has monoclinic symmetry. In this connection, it may be of historical interest to point out that our present knowledge of ferroelectrics recognizes the lowering of symmetry in the polar phase as obvious, but it took until 1937 for such a recognition to be made clearly in what was then the only ferroelectric known. Full credit for this must be given to Jaffe (J 2).

Crystallographic conventions for monoclinic crystals would require one to designate the unique polar axis of the ferroelectric phase as $b$ and the angle between the two other axes perpendicular to $b$ by $\beta$. However, it has become customary for Rochelle salt, ever since Jaffe's suggestion (J 2), to maintain the original notations of the orthorhombic axes also in the polar phase, and thus to call $a$ the polar twofold screw axis of this phase. Accordingly, the plane of the two other monoclinic axes, labeled $b$ and $c$, respectively, is non-polar, and the monoclinic angle should be called $\alpha$. Although the latter angle has often been labeled $\beta$, in the literature, the orthorhombic notation has been adopted consistently and will also be adhered to in the following in order to avoid confusion.

The spontaneous lattice deformation consists therefore in a shear $y_z$ within the (100) plane that causes the angle between the $b$ and $c$ axes to differ slightly from 90°. The magnitude of this spontaneous shear $y_z$ can be estimated (C 1) from Hinz's investigation (H 3) of the converse piezoelectric effect, and was measured directly by Mueller (M 2) from the variation with temperature of the angle between the $b$ and $c$ faces of a Rochelle salt block. Mueller's results indicate that this angle differs from 90° by 3' ($y_z = 8.7 \times 10^{-4}$) at 0 °C. Despite the fact that Mueller's samples were definitely not single-domain the above values appear to be too large when compared to the X-ray data of Ubbelohde and Woodward (U 1) of 1'48'' ($y_z = 5.2 \times 10^{-4}$) at 0 °C. The latter data are also in better agreement with more recent determinations of the spontaneous strain by means of optical methods by Indenbom and Chernysheva (I 1) (see Section 6). The temperature dependence of the spontaneous strain reflects that of the spontaneous polarization, as is to be expected from the direct proportionality between piezoelectric strain and induced polarization characteristic of the non-polar phase.

### The Specific Heat

Measurements of the specific heat of Rochelle salt have been carried out by a number of authors (K 3), (R 1), (H 4), (B 4), (W 1), (W 2), (W 3), but again the various results are not in agreement with one another. What is certain is that the heat of transition is very small and barely detectable at both Curie points. The orders of magnitude for the entropy change are 1.0 cal mole$^{-1}$ °C$^{-1}$ (negative) at the lower Curie point and probably even less (but positive) at the upper Curie point (W 1). The negative sign of the anomaly at the lower Curie temperature is in accordance with the expectation from Ehrenfest relation (F 1), but is in contradiction with the expectation from the thermodynamic theory of Mueller (H 5).

### The Effect of Pressure

Early investigations by Eremeev (E 1) of the effect of hydrostatic pressure upon the dielectric constant of Rochelle salt revealed that the upper Curie point increases with increasing pressure. Systematic measurements were later carried out by Bancroft (B 5) for pressures ranging up to 10,000 atm and temperatures from $-20$ °C to $+60$ °C. Figure VII-7 (a) shows the dependence upon pressure of the reciprocal susceptibility measured at 40 °C. At atmospheric pressure, this temperature is higher than the upper Curie point, and the diagram shows that the reciprocal susceptibility along [100] becomes almost zero, i.e. a transition occurs, at approximately 1400 atm. The trend of the curve suggests also that a further transition will occur at sufficiently high pressures. Thus, both Curie points are raised under increasing pressure and the interval between them, the ferroelectric range, becomes greater, as is seen in Fig. VII-7(b). The rates of increase with pressure are $10.4 \times 10^{-3}$ °C/atm for the upper and $3.65 \times 10^{-3}$ °C/atm for the lower Curie point. Thus, at a pressure of about 12,000 atm a lower Curie point should be found at the same temperature as that of the upper Curie point at atmospheric pressure. Above 24 °C, one can in principle always obtain two transitions by keeping the temperature constant and varying the pressure. One can

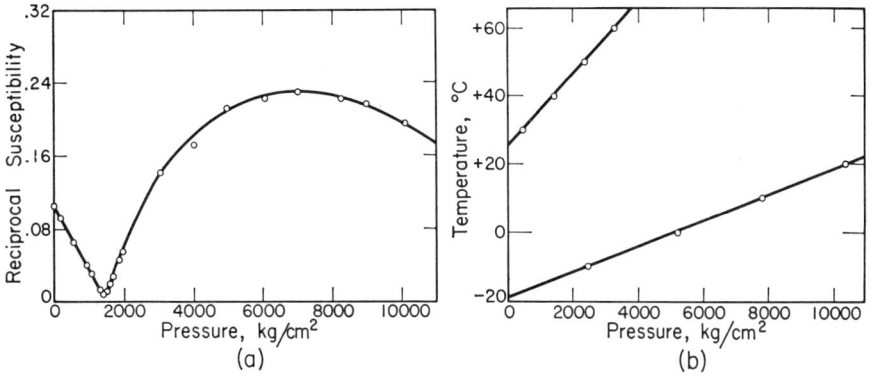

FIG. VII-7. Effect of hydrostatic pressure on Rochelle salt.
(a) Reciprocal susceptibility along [100] as a function of pressure at 40 °C.
(b) Dependence of the upper and lower Curie points upon pressure.
(According to Bancroft (B 5).)

also obtain two transitions by varying the temperature at constant volume instead of constant pressure. This feature is particularly important when pursuing an atomic model of Rochelle salt (see Section 9).

As discussed in Section II-6, the displacement of a second-order transition temperature $T_c$ with pressure is governed by Ehrenfest relation (II-14). This displacement can be computed from the knowledge of the volume expansion coefficients and the specific heat at constant pressure, both above and below $T_c$. Although the latter quantities are not known with great accuracy, one finds order-of-magnitude agreement between calculated and observed values.

A study of the variation with pressure of the electromechanical $Q$ factor of Rochelle salt was carried out by Mayer (M 8) in the range from $10^{-3}$ mm Hg to 12 atm. This variation was found to be similar to that of quartz.

The effects of mechanical and electrical bias upon the shape of the ferroelectric hysteresis loops was also investigated extensively (V 1), (M 1), (S 1), (S 3), (D 1). Application of equal and opposite pressures along the [011] direction causes the loops to appear distorted and unsymmetrical. In this case, a piezoelectric polarization is superposed upon that due to the electric field and thus shifts the position of the origin in the hysteresis loop. If two pairs of equal and opposite pressures, one pair along [011], the other along [01$\bar{1}$], are applied, they produce equal and opposite contributions to the polarization. The bias is thus reduced to zero. This restores the symmetry of the hysteresis loops but the polarization appears diminished, as the crystal is no longer free. Experiments on the equivalence of mechanical and electrical bias were carried out by David (D 1), who proved that the asymmetry of mechanical origin could be completely removed by application of a suitable field opposing the piezoelectric polarization. A discussion of these biased hysteresis loops can be found in Cady's book (C 1).

## 4. Piezoelectric and Elastic Properties

The piezoelectric and elastic anomalies of Rochelle salt are very similar to those of $KH_2PO_4$. For a general discussion and the definition of terms, the reader is therefore referred to Section III-4. The matrix of the piezoelectric moduli for orthorhombic Rochelle salt is:

$$\begin{matrix} 0 & 0 & 0 & d_{14} & 0 & 0 \\ 0 & 0 & 0 & 0 & d_{25} & 0 \\ 0 & 0 & 0 & 0 & 0 & d_{36}; \end{matrix}$$

that of the elastic compliances $s_{ik}$ is:

$$\begin{matrix} s_{11} & s_{12} & s_{13} & 0 & 0 & 0 \\ s_{12} & s_{22} & s_{23} & 0 & 0 & 0 \\ s_{13} & s_{23} & s_{33} & 0 & 0 & 0 \\ 0 & 0 & 0 & s_{44} & 0 & 0 \\ 0 & 0 & 0 & 0 & s_{55} & 0 \\ 0 & 0 & 0 & 0 & 0 & s_{66} \end{matrix}$$

and that of the elastic constants $c_{ik}$ is identical to the latter.

As the anomalous component of the polarization is $P_1$, the set of equations involving anomalous coefficients will be:

$$\left. \begin{aligned} P_1 &= -d_{14}X_4 + k_1^X E_1, \\ x_4 &= -s_{44}^E X_4 + d_{14}E_1. \end{aligned} \right\} \qquad \text{(VII-10)}$$

We recall that $X_4$ is the shear stress $Y_z$ in the $yz$ plane, $k_1^X$ the component along [100] of the dielectric susceptibility of the free crystal, and $x_4$ the shear strain $y_z$.

Numerous measurements of the direct (V 1), (S 1), (S 2), (K 5) and the converse (S 3), (B 6), (V 3), (N 1), (H 3), (M 9) effect were reported in the literature, and a detailed analysis of the older investigations can be found in Cady's book (C 1). In most of these, the experimental conditions were not sufficiently specified; the stresses applied for the measurement of the direct effect were often too large, so that domain-wall motions were induced in the ferroelectric phase. An example of the latter case is the result of Valasek's investigations (V 1) depicted in Fig. VII-8. The curve for $d_{14}$ fails to show maxima at the Curie points, as expected from the behavior of the dielectric constant $\varepsilon_a$ (Fig. VII-2), because the applied stress was too large. Maxima were observed, however, by other investigators (S 1), (S 2) and characteristic curves of $d_{14}$ showing a decrease in the ferroelectric region were published by Mason (M 9) and by Kawai (K 5). The peak value of $d_{14}$ has been measured as large as $32,500 \times 10^{-8}$ c.g.s. (C 1), i.e. more than fourteen times larger than that appearing in Fig. VII-8. The analogy between the piezoelectric

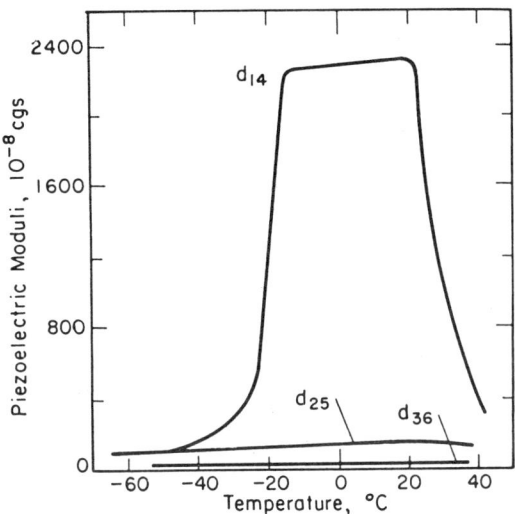

FIG. VII-8. Temperature dependence of the piezoelectric moduli of Rochelle salt. The curve for $d_{14}$ does not look like that of $\varepsilon_a$ (Fig. VII-2) because the measuring stress was too large (according to Valasek (V 1)).

and the dielectric anomalies has indeed been established by a number of experiments; the piezoelectric strain induced by an applied field shows both saturation and hysteresis; the modulus $d_{14}$ shows the same dispersion as the dielectric constant in the ferroelectric region, and the ratio $d_{14}/\varepsilon_a$ is independent of frequency (K 6); the piezoelectric Curie–Weiss law is obeyed. Between 25 °C and 34 °C, the modulus $d_{14}$ is given by the formula:

$$d_{14} = \frac{B}{T - T_0} \tag{VII-11}$$

with $B = 8.67 \times 10^{-5}$ c.g.s., while above 34 °C the piezoelectric Curie constant becomes smaller, $B = 5.17 \times 10^{-5}$ c.g.s., exactly as in the dielectric case. The relation between polarization and strain is given by an equation similar to (III-9), thus:

$$P_1 = \frac{k_1^X}{d_{14}} x_4 \tag{VII-12}$$

and again the ratio between polarization and strain is found to be almost temperature independent (K 5) and equal to that between spontaneous polarization and spontaneous strain. It is remarkable that the value of this ratio, computed from the dielectric and piezoelectric Curie constants, viz.:

$$P_1/x_4 = k_1^X/d_{14} = C/4\pi B = 2.05 \times 10^6, \qquad \text{(VII-13)}$$

is almost the same as for $KH_2PO_4$ (see Section III-4). The relationship between piezoelectric and dielectric behaviors is further evidenced by the fact that the piezoelectric moduli $d_{25}$ and $d_{36}$ exhibit no anomalies at the Curie point, just as the dielectric constant $\varepsilon_b$ and $\varepsilon_c$.

A quadratic piezoelectric effect also occurs in Rochelle salt, owing to lowering of the symmetry that occurs spontaneously in the ferroelectric phase and is induced by an electric field in the non-polar phases. In the monoclinic modification, new piezoelectric moduli come into being: those related to a polarization in the [100] direction are $d_{11}$, $d_{12}$ and $d_{13}$. As these moduli are proportional to $P$, it follows that the corresponding strains $x_1$, $x_2$ and $x_3$ are proportional to $P_1^2$. Measurements of these spontaneous strains in the ferroelectric phase were carried out by Mason (M 7) in connection with an investigation of the electrostrictive effect. The quadratic piezoelectric effect was studied by Mueller (M 4) at temperatures slightly above the upper Curie point. In the ferroelectric region, this effect ought to be analogous to the quadratic electro-optic effect discussed in Section 2, thus exhibiting the same characteristics of irreversibility and "butterfly" loops (see Fig. VII-6). Also, the quadratic piezoelectric effect should be responsible for the anomalous thermal contraction measured in the ferroelectric phase by Hablützel (H 2).

### The Elastic Properties

The investigations of the elastic behavior of Rochelle salt are also very numerous. Both static (M 10), (H 3) and dynamic methods (M 2), (M 3), (L 1), (M 11), (H 6), (M 6), (M 9), (P 2), (J 1) were employed; the agreement among the published values of the elastic compliances $s_{ik}$ (or elastic constants $c_{ik}$) can be considered satisfactory. The temperature dependence of all the elastic constants $c_{ik}$ (except $c_{44}$) between $-50$ °C and $+40$ °C was determined by Jona (J 1) by the method of Schäfer–Bergmann, discussed in Section III-4. This temperature dependence was found to be small, and no anomalies were detected at the Curie points. As for the compliance $s_{44} = 1/c_{44}$, the various experimental results indicate that the constant-polarization compliance $s_{44}^P$ is nearly independent of stress and temperature, whereas the constant-field compliance $s_{44}^E$ varies greatly with both temperature and stress. Above the upper Curie point, the difference $(s_{44}^E - s_{44}^P)$ obeys an elastic Curie–Weiss law (M 2):

$$s_{44}^E - s_{44}^P = \frac{D}{T - T_0} \qquad \text{(VII-14)}$$

where $D = 6.7 \times 10^{-11}$ cm² °C/dyn, which is again of the same order of magnitude as the corresponding constant for $KH_2PO_4$.

Application of a biasing electric field along the [100] direction causes the elastic constant $c_{44}^E$ to increase (M 3), (P 2). The effect is analogous to the so-called $\Delta E$ effect of ferromagnetics and was investigated in detail by Matthias (M 11). A concise discussion of the phenomena involved can be found in Känzig's review article (K 7). The influence of electric fields upon the absorption of ultrasonic pulse waves traveling along [100] was investigated by Yakovlev et al. (Y 1).

## 5. Effects of Radiation and Impurities

Extensive investigations of the effects of prolonged exposure of Rochelle salt to X-rays and other types of radiation were carried out by the Russian workers. Eisner (E 2) found that the dielectric constant decreases with increasing exposure to 30 kV, 10 mA Fe radiation. The hysteresis loops become biased and assume the appearance of the double loops already discussed in connection with the analogous effect in tri-glycine sulfate; after 15 hr exposure, the value of the spontaneous polarization was found to decrease about 30% with respect to unexposed specimens. Zheludev, Yurin and co-workers (Z 2), (Z 3), (Y 2), (Y 3) reported analogous effects of $\gamma$-radiation from $Co^{60}$ (c.f. also Toyoda et al. (T 4)). A dosage rate of 10 r/hr causes the distortion of the normal to a double hysteresis loop after only 2 hr exposure and the complete disappearance of the loop after 11 hr. The $P-E$ curves are then strictly linear. Simultaneous observations of the double loops and the domain structure showed that reorientation of the domains begins in fact only at field strengths equal to the critical value on the double hysteresis loops (see also Okada (O 1)). Radiation also causes decrease of the dielectric constant $\varepsilon_a$ with progressive flattening and final disappearance of the peaks at the Curie points. It is remarkable, however, that the piezoelectric activity does not seem to be affected by exposure to radiation, as the piezoelectric constant $b_{14}$ has the same value for irradiated as for non-irradiated crystals. Qualitatively similar results for small dosages were recently reported by Krueger (K 11) in relation to the elastic constants and the $Q$-factor.

A possible explanation for these effects has been proposed by Yurin (Y 3) in terms of decomposition processes. Exposure to radiation reportedly causes the evolution of gases ($H_2$, CO, $CO_2$, $CH_4$) in amounts of about five molecules per 100 eV, and thus a yellow coloration of the crystals. Such decomposition products may be trapped in interstitial positions of the crystal lattice and produce stresses that clamp the ferroelectric domains (see also Abe (A 6)). Thus, appreciably greater fields are required for the reorientation of domains in irradiated samples; when irradiation is prolonged, the polarization reversal is completely suppressed.

Somewhat similar phenomena have been observed also in Rochelle salt crystals containing various non-isomorphous impurities (K 8), (E 3). In general, small additions of Mo, B or Cu alter the dielectric properties of Rochelle salt in that the Curie points are shifted, the hysteresis loops distorted and the loss tangent reduced. Most pronounced are the anomalies exhibited by crystals grown from solutions containing 2% $CuCO_3$. The crystals appear of sky-blue color and the variation of the color intensity is evidence for inhomogeneous distribution of the impurities. The dielectric constant decreases and the ferroelectric range becomes narrower with increasing impurities.* The hysteresis loops are biased (double loops) as in the case of mechanically or electrically biased crystals (see Section 3). This anomalous behavior is probably again caused by the stresses due to the impurities in the lattice. It is interesting that the double hysteresis loops can be made to disappear temporarily by proper heat treatments. Annealing of the impure crystals at 40 °C for 5–6 hr followed by slow cooling to room temperature causes normal hysteresis loops to appear, but storage at room temperature for two days causes the reappearance of the double loops. A complete explanation of all these phenomena is not available at the present stage.

It may be worth pointing out, in this connection, that distorted hysteresis loops and anomalous non-linear effects may occur also in reputedly pure Rochelle salt crystals (S 4), (M 12). The phenomena have been ascribed to non-uniform stress distributions within the crystals investigated and can be eliminated by the application of a.c. fields (M 12).

## 6. Properties of the Domain Structure

### The Static Properties of Rochelle Salt Domains

Direct and indirect experimental evidence for the existence of domains with opposite polarization in Rochelle salt crystals, between the Curie points, have been available for a long time. Most of the early publications are now only of qualitative and historical interest, but the experiments on the electric Barkhausen effect (K 9), (M 1) and the pyroelectric tests with sulphur–lead powders (M 5), (C 1) may be worth mentioning. Miyake carried out extensive investigations of the effects of temperature and stress upon the intensities of X-ray reflections from Rochelle salt crystals (M 13). Application of suitable mechanical stresses and electric fields causes elimination of domain walls, and thus the lattice becomes more perfect; the extinction effects are thereby increased and the X-ray reflection intensities and line breadth are decreased. These effects do not occur for reflections of type $(h00)$, because the inhomogeneous strain resulting from the domain structure has only the $y_z$ component, so that the regularity of the $(h00)$ lattice planes remains undisturbed. X-ray investigations of the domain structure were also carried out by Ubbelohde and Woodward (U 1).

More recently, studies of the ferroelectric domains have been done on crystal surfaces etched with water (S 5), (S 6) and also by bombardment with neon ions (S 7).

The most fruitful method of investigation, however, is again the observation with the polarizing microscope. It may be interesting to point out that direct

---

*Similar effects are also observed in the case of isomorphous substitutions, i.e. when the K of Rochelle salt is replaced by $NH_4$, Rb or Tl. The effects of these substitutions, in particular that of K by $NH_4$, are discussed in Section 7.

observation of ferroelectric domains in Rochelle salt was accomplished a few years later than in $BaTiO_3$, despite the fact that the discoveries of these two ferroelectrics are spaced 25 years apart. The reason is that the question as to whether or not Rochelle salt domains can be observed directly under the polarizing microscope does not have as obvious an answer as that concerning tetragonal $BaTiO_3$. The strong anisotropy of the non-polar phase of Rochelle salt leads us to expect only antiparallel domain orientations and 180° walls. As discussed in Section 2, the onset of a spontaneous polarization $P_s$ along [100] causes a spontaneous Kerr effect. If we send the light along [100] and consider any given domain, we know that the corresponding cross-section of the indicatrix (i.e. the extinction positions of the domain) appears to be rotated, with respect to the original orthorhombic position, by an angle $\varphi$ that is given by Eq. (VII-4):

$$\tan 2\varphi = \frac{2 r_{41}^* P_s}{(1/n_2^2 - 1/n_3^2)} \qquad \text{(VII-15)}$$

where $r_{41}^*$ represents the coefficient related to the overall electro-optic effect of the free crystal. If we consider an oppositely polarized domain immediately adjacent to the one picked first, we expect that the corresponding angle of rotation $\varphi$ will be equal and opposite to that above, since $P_s$ has reversed its sign. Thus, the extinction positions of two adjacent domains will differ from one another by an angle equal to $2\varphi$. The formula given immediately above tells us that, all other quantities being given, the magnitude of $r_{41}^*$ determines whether or not the angle $2\varphi$ will be large enough to be observable. The value of $r_{41}^*$ is not known, but, as pointed out in Section 2, it is not too unreasonable to expect that it will be of the order of $2$–$3 \times 10^{-8}$ c.g.s. in which case $2\varphi$ would be of the order of $1$–$2$ ° and thus easily detectable. The experimental measurements indicate that these are indeed the orders of magnitude involved (see also more recent experiments on deuterated Rochelle salt by Wieder and Collins (W 5)).

Detailed studies of the geometry of Rochelle salt domains have been carried out in Russia (K 10), (C 2), (C 3), (I 1), (I 2) and, independently, in Japan (F 2), (M 14), (N 2). The domain patterns can be represented schematically in the same way as for $KH_2PO_4$. Figure III-22 therefore applies also to the case of Rochelle salt, if the axes, of course, are labeled $c$ and $b$ and the shear distortion $y_z$, respectively. Mitsui and Furuichi (M 14) have established that the domains are slabs parallel either to the (001) or the (010) plane, called $c$ and $b$ domains and corresponding to the configurations I and II of Fig. III-22, respectively. These two configurations are, of course, not equivalent, owing to the anisotropy of the (100) plane of the non-polar phase. Figure VII-9 shows photographs of an $a$-cut plate between crossed nicols. The angle $2\varphi$ between the two extinction positions of antiparallel domains was found to be 2.5° at 17 °C (M 14). Indenbom and Chernysheva (I 1) reported, however, that this angle, at any given temperature in the ferroelectric region, is larger after the first than after any subsequent cooling cycle: the stationary value at 7 °C, after nine cycles, is only 1.5°. The temperature dependence of $2\varphi$ is similar to that of the spontaneous polarization (N 2), (I 1), (A 5), as is to be expected from Eq. (VII-15). The relation between the sense of the polarization

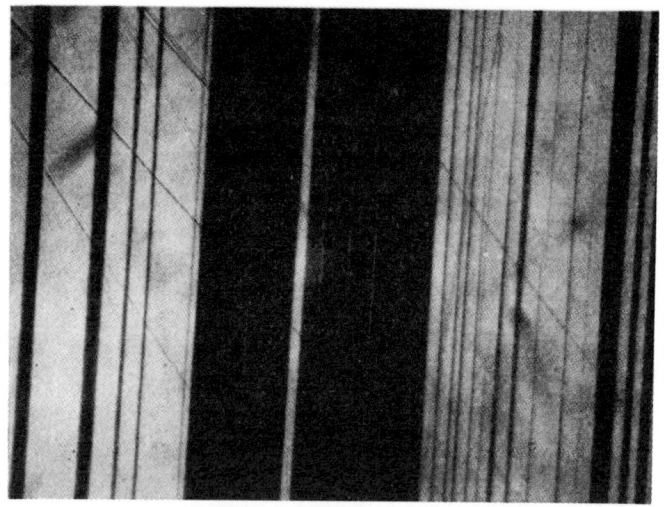

*a)*

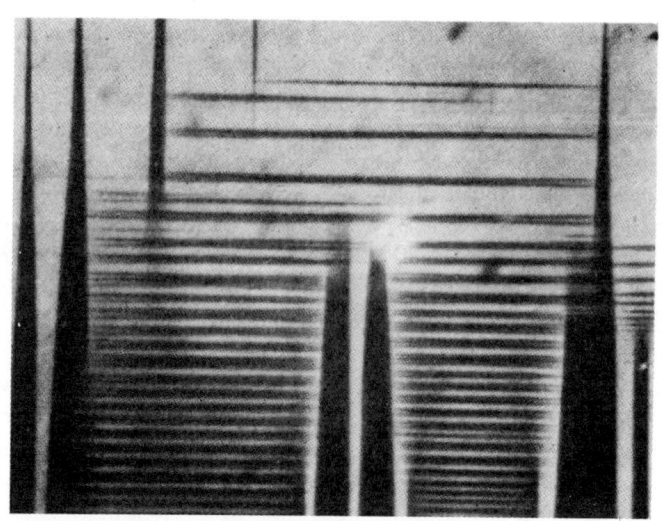

*b)*

FIG. VII-9. Ferroelectric domains in Rochelle salt.
(a) *c* domains. The photograph was taken at the extinction position of one of the two sets of antiparallel domains.
(b) Coexistence of wedge-shaped *b* and *c* domains.
(According to Mitsui and Furuichi (M 14).)

and the arrangement of the optical axes in a given domain is indicated schematically in Fig. VII-10. Nakamura (N 2) found that in crystals grown at temperatures higher than the upper Curie point and then cooled into the ferroelectric phase, $c$ domain configurations are stable, while in crystals grown at temperatures within the ferroelectric range, $b$ domains are stable. More generally, however, mixtures of $c$ and $b$ domains are observed (Fig. VII-9b). A domain terminating within the crystal becomes generally wedge-shaped. Marutake (M 15) found that the domains can also be wedge-shaped in the $a$ direction, so that $c$ and $b$ domains can occur in layers (see also Chernysheva (C 3)). The relation between the con-

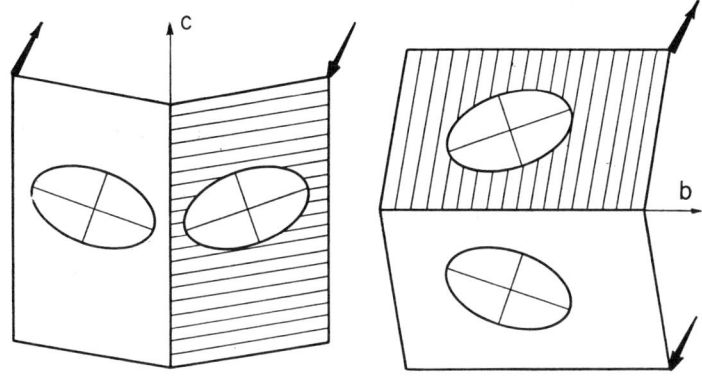

FIG. VII-10. Relation among the direction of the spontaneous polarization, the spontaneous shear and the rotation of the optical indicatrix in $c$ domains (left) and $b$ domains (right) of Rochelle salt (schematic). Arrows indicate the direction of spontaneous polarization into or out of the plane of the drawing (according to Indenbom and Chernysheva (I 1)).

ditions of crystal growth and the observed domain patterns were investigated by Müser and Flunkert (M 16), who reported that $b$ domain regions have a Curie point slightly higher than $c$ domain regions.

The effect of temperature and annealing upon the domain structure has been studied in detail by Mitsui and Furuichi (M 14) and by Nakamura (N 2). The former authors found that if a crystal is heated above the upper Curie point for a short time and then cooled again, the domain structure is similar but not quite identical to the original one. If the crystals are kept at temperatures above the upper Curie point for a long time and cooled, the domain structure changes considerably. These effects may be attributed to the electrostatic energy of polarization surface charges in the former case, and to changes in the non-uniform strain distribution in the latter case, where efflorescence, among others, may play an important role. Nakamura (N 2), in fact, has proved that if the crystals are annealed in an atmosphere in which Rochelle salt is stable, no changes in the domain structure can be observed. This may indicate that the elastic energy is playing the most important role in establishing a given domain pattern, while the electrostatic energy plays only a minor role. In fact, if the latter were the major agent, it would be reasonable to expect that a crystal grown at temperatures within the ferroelectric region in a conducting medium such as a Rochelle-salt solution ought to be a single domain, but this is not what one observes in practice (N 2). The experiments of Mitsui and Furuichi (M 14), at any rate, are in accordance with the theoretical expectations (K 7) in establishing a proportionality between the domain width $W$ and the square root of the thickness $d$:

$$W = 1.1 \times 10^{-3} \sqrt{d}. \tag{VII-16}$$

### The Thickness and the Energy of the Domain Walls

If we assume that the crystal is well annealed, it is possible to compute the domain structure that corresponds to a minimum of the internal energy by minimizing the sum of the electrostatic self-energy (or "depolarizing" energy) $E_{el}$ of the crystal in its depolarizing field and the total wall energy $E_{wall}$ (M 14), (K 7). Assuming that the crystal is a flat plate perpendicular to the ferroelectric axis with thickness $d$ and with a number of antiparallel domains of width $W$, the electrostatic potential can be calculated by solving the Laplace equations outside and inside the plate with proper consideration of the boundary conditions. The solution of these equations yields the following expression for the electrostatic energy of the specimen (M 14), (K 7):

$$E_{el} = \frac{3.4}{1 + \sqrt{(\varepsilon_a \varepsilon_b)}} W P^2, \tag{VII-17}$$

where $P$ is the saturation polarization. The total wall energy, on the other hand, can be written as:

$$E_{wall} = \sigma \frac{d}{W}, \tag{VII-18}$$

where $\sigma$ is the surface energy per unit area of the wall. Therefore, the domain width which makes the sum $(E_{el} + E_{wall})$ a minimum is given by the formula:

$$W = \left[ \frac{\{1 + \sqrt{(\varepsilon_a \varepsilon_b)}\} d \sigma}{3.4} \right]^{1/2} \frac{1}{P}, \tag{VII-19}$$

which is in agreement with the experimental result expressed by (VII-16). The importance of Eq. (VII-19) is that it allows the estimate of an upper limit for the wall-energy density $\sigma$ from the measurement of the domain width $W$ on well-annealed samples. This upper limit is computed to be 0.12 erg/cm². If compensation of the surface charges through conductivity is taken into account, the calculation yields (K 7):

$$\sigma = 0.08 \text{ erg/cm}^2.$$

The same order of magnitude of wall energy was reached by Zhirnov (Z 4) by minimizing the free energy of the domain wall (see also Section IV-7). The expression of the free energy was written in terms of anisotropic energy, elastic energy, piezoelectric effect, electrostriction and non-uniform distribution of the polarization within the wall. The numerical results of Zhirnov's calculations for the wall energy $\sigma$ and the wall thickness $t$ are:

$$\text{at } T = 0 \text{ °C}, \quad \sigma \cong 0.06 \text{ erg/cm}^2, \quad t = 12 \text{ Å};$$

$$\text{at } T = 20 \text{ °C}, \quad \sigma \cong 0.012 \text{ erg/cm}^2, \quad t = 220 \text{ Å}.$$

Mitsui and Furuichi (M 14) have developed a continuum theory for the domain wall of Rochelle salt by taking into account the electrostatic as well as the elastic energy, but disregarding the electrostrictive contribution considered by Zhirnov (Z 4) and that of the piezoelectric strain $y_z$ (the latter assumption has been criticized by Känzig (K 7)). The results are expressed in terms of the saturation polarization $P$ thus:

$$\sigma = (1.4 \times 10^{-10}) P^3 \text{ erg/cm}^2 \, ; \; t = \frac{1.8 \times 10^{-4}}{P} \qquad \text{(VII-20)}$$

and therefore:

$$\text{at } T = 0 \text{ °C}, \quad \sigma = 0{,}057 \text{ erg/cm}^2, \quad t = 24 \text{ Å};$$

$$\text{at } T = 20 \text{ °C}, \quad \sigma = 0.010 \text{ erg/cm}^2, \quad t = 43 \text{ Å}.$$

The agreement between the results of different theoretical treatments is satisfactory, as far as orders of magnitude are concerned. They all concur in indicating that the domain walls are very thin, of the order of a few lattice constants. This is in agreement with the results obtained for $BaTiO_3$ and is consistent with the general character of the boundaries between ferroelectric domains.

### Dynamic Properties of the Domains

Extensive investigations have been carried out by a number of authors about the effect of mechanical stresses (C 2), (M 13), (M 14), (N 2) and electric fields (C 2), (M 14), (M 17), (M 15), (N 3), (Z 5), (A 5), (W 4) upon the domain structure of Rochelle salt. After due consideration of the differences in lattice distortion, in a broad sense the cases of Rochelle salt and barium titanate appear to be very similar to one another in the processes of polarization reversal. Most of the observations on Rochelle salt were concerned with the motions of the domain walls under the action of static fields. No significant difference was detected, in this respect, between $b$ and $c$ domains. The domain wall motions show hysteresis and relaxation characteristics, each single domain wall showing its own (M 15). It was established that some of the domains can actually never be reversed. Observations of the changes in domain structures, extended up to 10 hr after the application of static fields, revealed that no wall motion occurs for fields smaller than 40 V/cm, but saturated square loops can be obtained for fields slightly larger than 60 V/cm (N 3). Under these conditions, the coercive field strength equals approximately 62 V/cm and the fraction $f$ of the total volume of a crystal that is reversed by the applied field depends exponentially upon the time:

$$f(t) = 1 - e^{-t/\tau}, \qquad \text{(VII-21)}$$

where the time constant $\tau$ is of the order of 1 hr at 10 °C. These observations represent another indication of the marked frequency dependence of the coercive field in general and the pronounced relaxation effects of Rochelle salt in particular: it may be recalled that, at 50 c/s the relation between $P$ and $E$ is almost perfectly linear for fields smaller than about 150 V/cm, while well saturated hysteresis loops can be observed with fields of about 400 V/cm.

The process of polarization reversal occurs by way of nucleation and growth of wedge-shaped domains in the direction of the polar axis, and also by sideway motion of the domain walls. The nucleation rate* depends exponentially upon the applied field (M 14):

$$\partial n/\partial t = \text{constant} \times e^{-\alpha/E}, \qquad\qquad \text{(VII-22)}$$

while the sidewise velocity of the domain walls depends linearly upon the applied field:

$$v = \mu(E - E_0) \qquad\qquad \text{(VII-23)}$$

where the mobility $\mu$ and the critical field $E_0$ are structure sensitive and equal $1.36 \times 10^{-3}$ cm$^2$ V$^{-1}$ sec$^{-1}$ and 80 V/cm, respectively, at 11 °C (M 14). Thus, at this temperature and for a field of 200 V/cm the wall velocity is approximately 0.14 cm/sec.

These numerical results are in striking contrast with those obtained by Wieder (W 4) from pulsing experiments similar to those carried out by Merz on BaTiO$_3$ (see Section IV-8). Similarly as in the latter case, the dependence of the maximum switching current $i_{\max}$ was found to be exponential with the field at lower field amplitudes (smaller than about 200 V/cm):

$$i_{\max} \propto \exp(-\alpha/E), \qquad\qquad \text{(VII-24)}$$

and linear at larger amplitudes:

$$i_{\max} \propto \mu(E - E_0). \qquad\qquad \text{(VII-25)}$$

The wall mobility calculated from these results is $\mu = 110$ cm$^2$ V$^{-1}$ sec$^{-1}$ and the wall velocity $v = 10^3$ cm/sec at 11 °C. Moreover, no threshold field could be found down to fields of the order of 10 V/cm. The large differences between these and the results reported above cannot be explained on the basis of different experimental methods alone and must be ascribed, according to Wieder (W 4), to differences between the fast reversal process characteristic of nucleation and wall propagation in the direction of the ferroelectric axis and the slow sideway motion of $b$ and $c$ domain walls. Qualitatively, the situation is analogous to that encountered in the early stages of the same problem in BaTiO$_3$, and there is reason to believe that, *mutatis mutandis*, the solution of the two problems will be qualitatively similar. It should be kept in mind, however, that the changes in strain involved in 180° switching are very different in the two crystals.

* An interesting study of the conditions under which domain nuclei can survive has recently been initiated by Takahashi *et al.* (T 3). The study was carried out on an idealized two-dimensional model but bearing in mind the case of Rochelle salt crystals. The nuclei were assumed to be cylindrical domains anchored to dislocation lines, the calculation showing that the domain shape is plate-like in the absence of an external field and wedge-like in the presence of a field.

## 7. Isomorphous Crystals

### The Isotope Effect

As in the case of $KH_2PO_4$ and isomorphous crystals, the suspicion that hydrogen bonds (and water molecules) may play an important role in the ferro-electric behavior of Rochelle salt led the researchers to investigate the properties of the deuterated compound (H 1), (H 7). Crystals of deuterated (or "heavy") Rochelle salt were prepared by dissolving in highly concentrated $D_2O$ the ordinary salt that had been thoroughly desiccated at higher temperatures. Not all the hydrogens can be replaced by deuterium in this way, but only the water of crystallization and the hydrogen atoms in the OH groups. The chemical formula of the deuterated salt is thus:

$$NaKC_4H_2D_2O_6 \cdot 4D_2O.$$

The Curie points of this compound are at $-22\ °C$ and $+35\ °C$, respectively, indicating that the hydrogen substitution widens the region of stability of the ferro-electric phase. The spontaneous polarization is increased, with respect to the ordinary salt (Fig. VII-4), and reaches a maximum of about $0.35 \times 10^{-6}\ C/cm^2$ at $6\ °C$. The coercive field strength exhibits a similar relative behavior, the maximum value being of the order of $400\ V/cm$ at approximately $18\ °C$.

This isotope effect is clearly not as spectacular as in the $KH_2PO_4$-type ferro-electrics, but it does involve an increase of the ferroelectric range by about $40\%$. Unfortunately it is not possible to decide, on this basis only, exactly which kind of hydrogen atom is responsible for the effect.

### Isomorphous Replacements

Extensive investigations have been carried out on a series of solid solutions of Rochelle salt and other isomorphous double tartrates such as $NaNH_4C_4H_4O_6 \cdot 4H_2O$ (occasionally also called "ammonium Rochelle salt"), $NaRbC_4H_4O_6 \cdot 4H_2O$ and $NaTlC_4H_4O_6 \cdot 4H_2O$ (K 4), (B 6), (E 4). The former system, in particular, has been the object of more recent re-examinations (M 18), (M 19), and will be discussed in the following.

The interesting dielectric properties of this system are best visualized on the basis of the phase diagram $Na(K, NH_4)C_4H_4O_6 \cdot 4H_2O$ depicted in Fig. VII-11. The mixed crystals were grown from solutions containing mixtures of the $NaNH_4$ with the NaK salt. The composition of the crystals depends upon the composition of the mother solution in a definite but not quite linear fashion, so that only small crystals grown from large volumes of solution can be considered sufficiently uniform in composition. The addition of only 1 mole $\%$ of the $NH_4$ salt to pure Rochelle salt reduces the temperature range of the ferroelectric region by about one-half. Addition of about 3 mole $\%$ of the $NH_4$ salt completely eliminates the ferroelectric properties, leaving only a maximum in the curve of dielectric constant vs. temperature. Thus, region I in Fig. VII-11 represents the ferroelectric phase discussed so far in the present chapter. In crystals containing from 3 to about 18 mole % of the $NH_4$ salt, no ferroelectric activity can be detected, although the dielectric constant $\varepsilon_a$ exhibits broad and flat maxima at temperatures decreasing with increasing $NH_4$ content. This trend is represented by the phase line II in Fig. VII-11: this may indicate that the upper and the lower non-polar phases of Rochelle salt are possibly not quite isostructural.

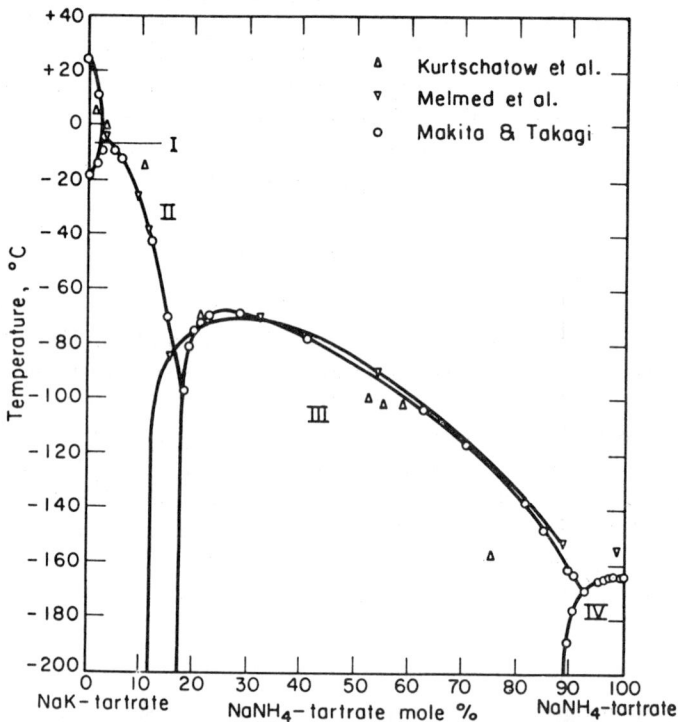

FIG. VII-11. Phase diagram of the system $Na(K, NH_4)C_4H_4O_6 \cdot 4H_2O$. Regions I and III denote ferroelectric phases. Line II may separate two similar but not quite isostructural phases. Region IV represents a polar phase whose spontaneous polarization cannot be reversed by electric fields but can indeed be reversed by mechanical stresses (from results of Kurchatov (K 4), Melmed *et al.* (M 18) and Makita and Takagi (M 19)).

The most interesting feature, however, is the reappearance of ferroelectric properties, at lower temperatures, in crystals containing more than about 18 mole % of the $NH_4$ salt. The new ferroelectric phase, which is probably different from that of pure Rochelle salt, is characterized by a single Curie point (region III in Fig. VII-11), and there is reason to believe that the transitions leading to this phase are of the first order. In this range, both the spontaneous polarization and the coercive field increase with decreasing temperature, as normally observed in ferroelectrics. A crystal containing 55 mole % $NaNH_4$ salt, for example, exhibits at $-190\,°C$ a spontaneous polarization of about $1 \times 10^{-6}\,C/cm^2$ and a coercive field strength of $10^4\,V/cm$ (K 4). The new ferroelectric phase extends up to a $NH_4$ content of about 85 mole %. Beyond this concentration, a new phase appears (region IV in Fig. VII-11) which extends up to pure $NaNH_4$ tartrate. The physical properties of this phase are exceedingly interesting and deserve some consideration.

The dielectric properties of pure $NaNH_4C_4H_4O_6 \cdot 4H_2O$ were investigated by Jona and Pepinsky (J 3). The dielectric constant measured along all three orthorhombic axes exhibits a small anomaly at 109 °K. As it appears from Fig. VII-12, this anomaly consists in a drop of about 17% and is therefore quite different from that exhibited by Rochelle salt at the Curie points. No ferroelectric hysteresis loops were observed in the low temperature phase. Nevertheless, it was established by Takagi and Makita (T 2) that the crystal is pyroelectric below 109 °K, the polar axis being parallel to the [010] axis, while the [100] and the [001] axes are not polar. This phase, then, is different from that of ferroelectric Rochelle salt, whose polar axis is [100]. Optically, there are already differences between the non-polar phases of the two crystals, $NaNH_4$-tartrate being optically negative, with (100) as the plane of the optic axes and [001] as the acute bisectrix. Below 109 °K, domain patterns were observed in plates cut perpendicular to [010], which are strongly reminiscent of those observed in ferroelectric Rochelle salt. The angle between the extinction positions of neighboring domains is about 3.5° and independent of temperature. This is indicative of the existence of a spontaneous strain $z_x$ and monoclinic symmetry. The spontaneous polarization cannot be reversed by electric fields as high as 20 kV/cm, as no dielectric loops can be observed, but *can indeed be reversed by a mechanical stress $Z_x$* of the order of 100 kg/cm². The relationship between polarization and stress is thus represented by an electromechanical hysteresis loop. The "coercive stress" in terms of pressure along [101] is about 100 kg/cm² at 92 °K, and the spontaneous polarization is $0.21 \times 10^{-6}$ C/cm², independent of temperature. Takagi and Makita find it justified, therefore, to call the crystal ferroelectric below 109 °K, although reversal of the polarization can only be brought about by mechanical stresses, so that the term "ferroelectric" assumes a broader meaning than heretofore understood.

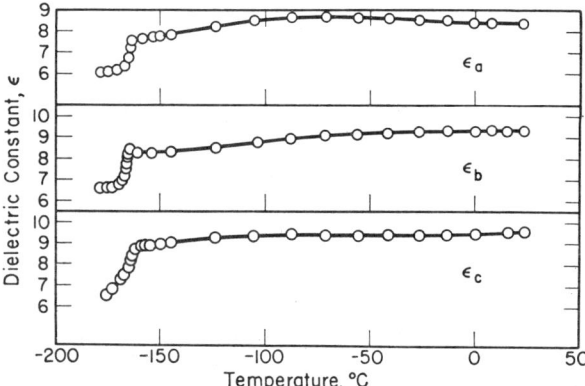

FIG. VII-12. Dielectric constants of $NaNH_4C_4H_4O_6 \cdot 4H_2O$ vs. temperature (according to Jona and Pepinsky (J 3)).

It may be reasonable to ask why a mechanical stress can achieve something that an electric field cannot. It is well known that the polarization is related to the stress and to the field by the equation: $P_2 = -d_{25} X_5 + k_2 E_2$ (see Eq. I-5b), but the conditions of the stress experiment are such (insulated crystal) that: $E_2 + 4\pi P_2 = 0$, so that it follows:

$$E_2 = \frac{4\pi d_{25} X_5}{\varepsilon_2}.$$

Using Mason's values (M 6) of $d_{25} = 95 \times 10^{-8}$ c.g.s., $\varepsilon_2 = 8.9*$, and putting $X_5 = (1/2) \times 100$ kg/cm$^2$, one would compute that the field corresponding to a pressure of 100 kg/cm$^2$ along [101] is about 19.7 kV/cm. This is of the same order of magnitude as the maximum field of 20 kV/cm that was applied to the crystal while attempting to obtain hysteresis loops, and yet this field was unable to reverse the polarization. The fallacy of this argument has been instructively pointed out by Takagi and Makita (T 2) and is worth mentioning. One should actually compare the field and the stress which give rise to the same *strain*, not to the same polarization. From Eq. (I-4a), we write:

$$\left.\begin{array}{l} X_5 = -c_{55} x_5 + a_{25} P_2, \\ E_2 = -a_{25} x_5 + \chi_2 P_2, \end{array}\right\} \tag{VII-26}$$

where, as usual, $x_5 = z_x$. Putting the stress equal to zero, we have:

$$X_5 = 0, \quad x_5 = \frac{a_{25}}{c_{55}} P_1, \tag{VII-27}$$

and thus:

$$E_2 = \left(\chi_2 - \frac{a_{25}^2}{c_{55}}\right) P_2, \tag{VII-28}$$

which is representative of the experiment involving the application of an electric field. On the other hand, putting the field equal to zero:

$$E_2 = 0, \quad x_5 = \frac{\chi_2}{a_{25}} P_2, \tag{VII-29}$$

we have:

$$X_5 = \left(a_{25} - \frac{c_{55} \chi_2}{a_{25}}\right) P_2, \tag{VII-30}$$

which is the case of the experiment involving the application of a mechanical stress. If we were to plot (VII-28) and (VII-30), we would see (using Mason's value (M 6) $s_{55} = 1/c_{55} = 33 \times 10^{-2}$ c.g.s.) that a pressure of 100 kg/cm$^2$ along [101] is equivalent to a field of about 500 kV/cm, which is far above the breakdown field of NaNH$_4$ tartrate. This is the reason why coercivity is attainable with a mechanical stress but not with an electric field.

---

* For estimates of orders of magnitude we may use these values, although they were determined at room temperature, since the temperature coefficients of both $d_{25}$ and $\varepsilon_2$ are small.

The properties of $NaNH_4$ tartrate described above are characteristic of all the compositions encompassed by region IV in the phase diagram of Fig. VII-11. Of particular interest is the crystal containing 90.5 mole % of the $NaNH_4$ salt, as, according to Fig. VII-11, it undergoes a transition into phase III at $-163$ °C and subsequently one into phase IV at $-176$ °C. Makita and Takagi (M 19) have established that the upper transition is evidenced by a large peak in $\varepsilon_a$, thus characteristic of phase III, while the lower transition is accompanied by small anomalies of all three dielectric constants $\varepsilon_a$, $\varepsilon_b$ and $\varepsilon_c$, characteristic of phase IV. Normal dielectric hysteresis loops can be observed in the intermediate temperature range (phase III) for fields applied along [100], and pyroelectric activity can accordingly only be detected along this direction. Below $-176$ °C, elastomechanical hysteresis loops can be observed for stresses $Z_x$, and pyroelectric activity detected only along [010]. Accordingly, (100) plates of this crystal exhibit domain patterns within phase III, but the patterns disappear when the temperature is lowered below $-176$ °C; (010) plates, on the other hand, exhibit domain structures only when they are cooled below this temperature. All these observations indicate that a field-reversible spontaneous polarization along [100] (connected with a spontaneous strain $y_z$) sets in at the upper transition point and increases with decreasing temperature until the lower transition point is reached. At this point the direction of the spontaneous polarization rotates suddenly by 90° into the [010] axis (involving a spontaneous strain $z_x$) and is only stress-reversible. It may be objectionable, from a certain viewpoint, to include into the ferroelectric family crystals whose polarization can be reversed by stresses only, as such a step may eventually lead to the inclusion of a much larger number of pyroelectric crystals than heretofore suspected, but at the present stage this extension of the definition does not appear to be too confusing.

## 8. Structural Characteristics

As pointed out in Section 1, the structure of Rochelle salt is orthorhombic above the upper Curie point. Accurate data of the lattice parameters in all three phases were reported by Ubbelohde and Woodward (U 1); at $+35$ °C:

$$a = 11.878 \text{ Å}, \quad b = 14.246 \text{ Å}, \quad c = 6.218 \text{ Å}.$$

The structure was first investigated by Beevers and Hughes (B 7), with X-ray diffraction techniques in the vicinity of room temperature. The space group of the upper non-polar phase was reported to be $P2_12_12$, with four formula units per cell. Although the exact temperature at which this X-ray study was carried out is not known, the results are generally assumed to be valid for the upper non-polar phase, the reason being that X-rays do not seem to be capable of distinguishing between the polar and the non-polar phases. The orthorhombic symmetry requires, namely, that reflections of the type $(h00)$ and $(0k0)$ should not be observed for $h$ and $k$ odd, respectively, and this is the case with X-rays at any temperature outside as well as within the ferroelectric region. One could infer from this, knowing that X-rays are nearly blind with respect to the hydrogen atoms, and also that the symmetry of the polar phase is undoubtedly monoclinic, that the deviation from orthorhombic symmetry is caused essentially by the hydrogens in the structure. It was, in fact, proved by Frazer *et al.* (F 4) that, using neutrons rather than X-rays, reflections $(0k0)$ with $k$ odd can indeed be observed within the polar region, and disappear at the Curie points, whereas the odd $(h00)$ reflections are missing in all phases. With the axial choice discussed in Section 3, this means that the polar phase is indeed monoclinic, with the unique axis parallel to [100]. The structure evolved by the X-ray study of Beevers and Hughes (B 7) is shown in projection in Fig. VII-13. A refinement of this structure was later carried out by Mazzi *et al.* (M 21), who also investigated the structure of the non-polar low-temperature phase at $-64\,°C$. In addition to improving the $x$ and $y$ co-ordinates of Beevers and Hughes for all the atoms except the hydrogens, the work of Mazzi *et al.* disclosed strongly anisotropic thermal vibrations of the potassium atoms, while confirming the fact that the differences among the three phases of Rochelle salt, as detectable by X-rays, are extremely small. The hydrogen-bond system was originally *assumed*, on the basis of the X-ray results, by Beevers and Hughes (B 7), and particular attention was devoted, by these and a few other investigators (U 1), (M 6), to the hydrogen bond between the oxygen atom labeled $O_1$ and the water molecule labeled $(H_2O)_{10}$ in Fig. VII-13. This hydrogen bond is directed approximately along the polar $a$ axis, and was assumed to be responsible for the anomalous dielectric properties

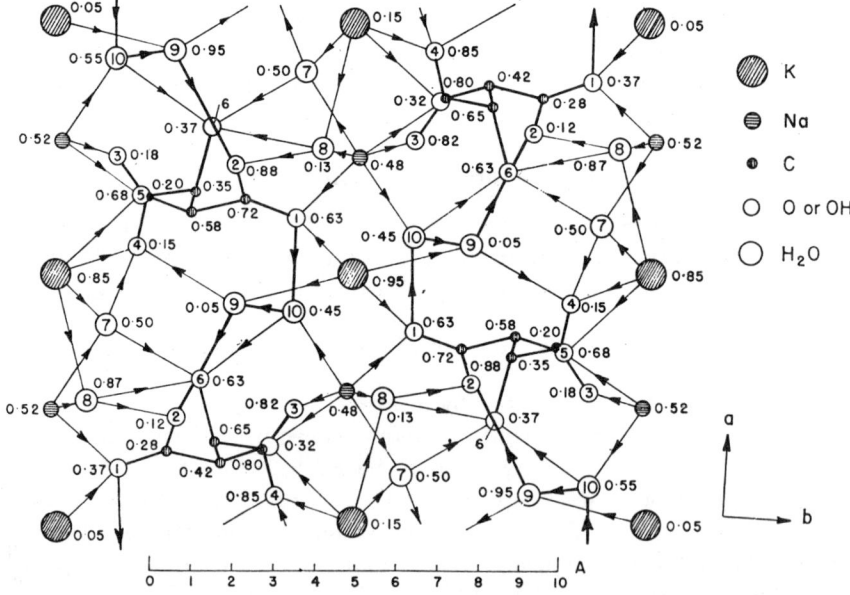

FIG. VII-13. Projection on (001) plane of the structure of Rochelle salt. The oxygen atoms are numbered from 1 to 10. Numbers near the atoms indicate $z$ co-ordinates (according to Beevers and Hughes (B 7)).

through the motion of H atoms between two equilibrium positions along the bond axis (see Section 9).

The neutron diffraction study of Frazer *et al.* (F 4), however, concluded that this bond plays no part in the phenomenon of spontaneous polarization. The neutron study, on the other hand, revealed the importance of the orientation of the hydroxyl group $(O-H)_5$ of the tartrate molecule. In one of its orientations, this group is directed toward the carboxyl oxygen $O_4$ (in the next cell) in what might be called weak hydrogen bonding. In the other position, the $(O-H)_5$ group is not directed toward any particular oxygen, but is just in a generally electronegative region (F 5). The result of this orientational change is a net proton displacement along the polar $a$ axis, which is probably the major contribution to the polarization process. There are probably also a number of other minor atomic displacements in the structure, but, unfortunately, there is a considerable degree of interaction between the enormous number of positional parameters and the anisotropic temperature factors, which makes an accurate solution of this structural problem extremely difficult. It may be interesting to point out, however, that comparison between the neutron data of deuterated and hydrogenous crystals shows that the deviation from orthorhombic symmetry cannot be explained in terms of hydrogen displacements alone. If this were true, one ought to be able to match the odd $(0k0)$ reflections of the deuterated crystal to those of the hydrogenated crystal by a simple scale factor, which is not the case even in a qualitative sense (F 5). It is not possible, at the present stage, to pick out all the atomic displacements occurring in Rochelle salt at the Curie points, so that one can only conclude at present that the most relevant contribution to the phase change arises from the $(O-H)_5$ groups. An intermediate result of the neutron diffraction analysis is depicted in Fig. VII-14.

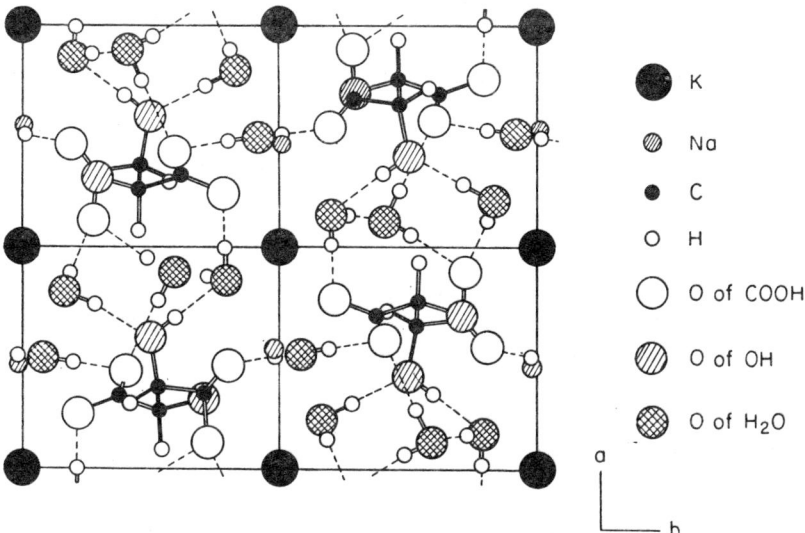

FIG. VII-14. The hydrogen bond system in Rochelle salt. Projection of the structure on (001) plane (according to Frazer *et al.* (F 4)).

Information about the role played by the hydrogen atoms in the Rochelle salt structure can also be inferred from the interpretation of Raman spectra (C 5), (C 6), (C 4), (L 2) (additional references are given in reference (B 9)), and polarized infrared absorption spectra (B 8), (B 9). Strong Raman lines are observed at 2938 cm$^{-1}$ and 2983 cm$^{-1}$, which are attributed to C$-$H vibrations, and whose frequency is independent of temperature. Other Raman bands in the region between 3130 cm$^{-1}$ and 3560 cm$^{-1}$ are assigned to vibrations of the water molecules and O$-$H groups. Similarly, the three strong bands observed in the 5000 cm$^{-1}$ region of the polarized infrared spectra are assigned to vibrations of the water molecules. These assignments, together with the knowledge of the rough features of the structure of Rochelle salt as obtained from the X-ray study of Beevers and Hughes, allowed Baker and Webber (B 9) to calculate two sets of the possible positions of all but the hydroxyl hydrogens, which are in reasonable agreement with the results of the optical studies. It is encouraging to see that one of these sets, although derived for the upper non-polar phase and independently of the results of the neutron analysis, is indeed in good agreement with the temporary results of the latter study, as depicted in Fig. VII-14, with the exception of the location of one of the hydrogen atoms belonging to water molecule 8. It cannot be said, at this stage, whether this difference is real.

## 9. Theoretical Treatments

The first formulation of a theory of Rochelle salt was published by Kobeko and Kurchatov in 1930 and later re-elaborated by Kurchatov (K 4). Based on Debye's theory of electric dipoles and similar to the Langevin–Weiss theory of ferromagnetism, this molecular theory assumes rotatable dipoles and attributes the occurrence of a spontaneous polarization to the dipolar interaction. In order to account for the lower Curie point, it postulates the formation of antiparallel chains, so that the number of dipoles effective in producing a spontaneous polarization decreases with temperature. In the low-temperature phase, the crystal would therefore be antipolar.

A similar conclusion was arrived at by Fowler (F 3) with a different approach, although the polarized state was again interpreted, as in Kurchatov's theory, as intermediate between the state of complete disorder at high temperatures and that of complete order at low temperatures. Both theories identify the rotatable dipoles with the water molecules in the crystal lattice. Although we know now that the structural studies failed to reveal free rotation of these molecules in the upper non-polar phase, the most serious objection to both Kurchatov and Fowler theories is that raised by Cady (C 1) on the neglect of the piezoelectric effect. For these reasons, these theories are now considered obsolete, but they still represent important contributions to the understanding of polarization in solids.

The most successful attempts at theoretical treatments of ferroelectricity in Rochelle salt are those of Mason, Devonshire and Mitsui (to be discussed below). These are essentially statistical local-field theories. A truly atomistic theory of

Rochelle salt will have to await completion of the structural investigations in all three phases. The thermodynamic formalism, on the other hand, is fairly well understood and will briefly summarized presently.

*The phenomenological treatment* of Rochelle salt developed by Mueller (M 1–5) and by Cady (C 1) is, in fact, parent to the general theory of ferroelectricity later expanded by Devonshire (D 2) and already outlined in Section I-5. A complete expression for the expansion of the free energy $A(P, x)$ including terms in the fourth power of $P$ has been given recently by Mitsui (M 20). The properties of the clamped crystal polarized along [100] are given by the free energy:

$$A(P; x = 0) = \frac{1}{2} \chi_1^x P_1^2 + \frac{1}{4} \xi^x P_1^4, \tag{VII-31}$$

while those of the free crystal are described by.

$$A(P; X = 0) = \frac{1}{2} \chi_1^X P_1^2 + \frac{1}{4} \xi^X P_1^4. \tag{VII-32}$$

The relation between the reciprocal susceptibilities of the clamped and free crystal is given by (D 2), (M 20):

$$\chi_1^x - \chi_1^X = b_{14}^2 c_{44}^P, \tag{VII-33}$$

where $b_{14}$ is the piezoelectric constant that gives the ratio between shear strain $x_4 = y_z$ and polarization $P_1$ (see Eq. I-4b), and $c_{44}^P$ is the elastic constant relating the shear stress $X_4$ to the shear strain $x_4$ at constant polarization (see Eq. I-4a). Both $b_{14}$ and $c_{44}^P$ are "true" constants and have no anomalies at the transitions, being almost temperature independent (see Section 4). Their measurement therefore allows the calculation of their product at all temperatures. It turns out that (M 2), (M 6):

$$\chi_1^x - \chi_1^X = 0.04. \tag{VII-34}$$

When plotted as a function of temperature, $\chi_1^x$ shows a vanishing value in the vicinity of 0 °C (see Fig. VII-16b). This very peculiar result indicates that the temperature range of the spontaneously polarized state is just about reduced to zero (at atmospheric pressure) when the strains are suppressed. Thus, while the free crystal exhibits a spontaneous polarization in a narrow temperature range, delimited by second-order transitions, the clamped crystal never quite reaches the polar state and only exhibits an anomaly in the dielectric constant. It may be recalled that, in $BaTiO_3$, the effect of clamping is to change the character of the ferroelectric transition from first to second order, while in $KH_2PO_4$ the same effect would reduce the Curie temperature by about 4 °C. In deuterated Rochelle salt, on the other hand, clamping would reduce the range of ferroelectric activity to about one-third of that for the free crystal.

As for the coefficients of $P^4$ in Eqs. (VII-31) and (VII-32), their difference can be expressed by:

$$\xi^x - \xi^X = F, \tag{VII-35}$$

where $F$ is a function of the elastic, piezoelectric and electrostrictive constants (M 20). $\xi^X$ has been measured by Mueller (M 3) as $(5.8 \pm 0.7) \times 10^{-8}$ c.g.s., and $\xi^x$ has been estimated by Mitsui (M 20) to have an order of magnitude ranging between $10^{-8}$ and $10^{-6}$ c.g.s.

*The theory of Mason* (M 6) was the first successful treatment of the Rochelle salt problem from a microscopic point of view. It was based upon the hydrogen-bond system that was inferred from the early X-ray analysis of Beevers and Hughes (B 7) and the suggestions of Ubbelohde and Woodward (U 1), according to which the hydrogen bond between the oxygen $O_1$ and the water molecule $(H_2O)_{10}$ (see Fig. VII-13) is responsible for the dielectric behavior of Rochelle salt. We have already pointed out in the preceding section that the neutron-diffraction studies have de-emphasized the role played by this bond in favor of the role played the $(OH)_5$ hydroxyl group of the tartrate molecule. These results, however, only affect that part of Mason's theory which is concerned with cal-culations of the dipole moment and the energy barrier opposed to reversal, but do not alter the essential feature of the theoretical model. This essential feature consists in recognizing that certain hydrogen atoms have two possible positions of equilibrium within the lattice so that in each position the system can be con-sidered as a dipole with reversible sign. The dipole, however, has different energies in the two positions, and can be described in terms of an asymmetric potential well such as depicted schematically in Fig. VII-15(a). The symmetry of the polar phase requires that such dipoles occur in sets of two related to one another by the twofold screw axis along [100]. One set of dipoles has minimum energy when pointing, say, in the positive [100] direction, while the other set has mini-mum energy when pointing in the negative [100] direction. As the projection of the structure on the (001) plane has a center of symmetry in the non-polar phases, it follows that the dipoles are arranged in an antiparallel way as indicated in Fig. VII-15(b). It is possible that such an antiparallel arrangement obtains at

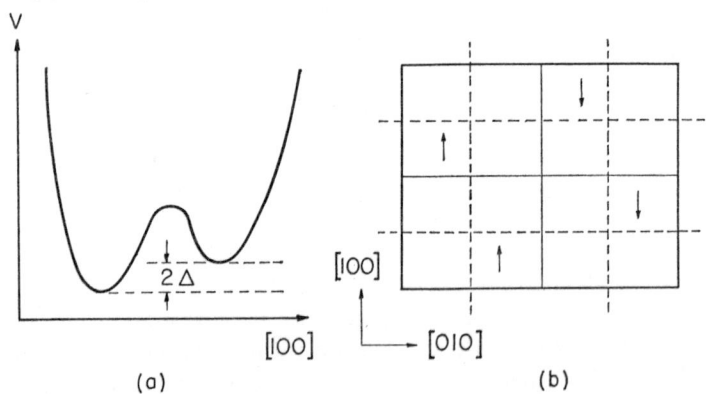

(a)                                        (b)

FIG. VII-15. Theoretical model of Rochelle salt.

(a) Asymmetric potential well of the hydrogen atom.

(b) Symmetry of the non-polarphase: projection on (001) plane (schematic). Arrows represent the dipoles, dashed lines show the positions of the twofold screw axes. Note the change in the direction of [100] between (a) and (b).

low temperature, complete disorder exists at high temperatures and a parallel arrangement occurs in an intermediate temperature range.

On the basis of this model, the theory was developed by writing an expression for the Helmholtz free energy $A = U - TS$. Let $2N$ be the total number of dipoles in the unit volume; assume that, at any time, a number equal to $(1/2)\,N(1 + n_1)$ of dipoles of the first set, and a number equal to $(1/2)\,N(1 + n_2)$ of the second set are pointing in the positive [100] direction. Then the polarization $P$ can be written as:

$$P = N\mu(n_1 + n_2), \tag{VII-36}$$

where $\mu$ is the dipole moment. Calling $2\varDelta$ the difference in energy between the two dipole positions and assuming a single coefficient $\gamma$ for the description of the internal field (dipole interaction energy equal to $(-1/2)\,\gamma P^2$), the internal energy $U$ can be written thus:

$$U = -\tfrac{1}{2}\,\gamma N^2\mu^2(n_1 + n_2)^2 + \varDelta(2 + n_2 - n_1)N. \tag{VII-37}$$

The entropy function $S$ was assumed to be only configurational, and with this the free energy $A$ could be written in terms of the quantity $(n_1 + n_2)$ and two constants $\alpha$ and $\beta$ defined as:

$$\alpha = \frac{\gamma N \mu^2}{kT}, \quad \beta = \frac{\varDelta}{kT}. \tag{VII-38}$$

The procedure of minimization of the free energy allowed Mason to find a non-vanishing solution for the quantity $(n_1 + n_2)$, provided that the constants $\alpha$ and $\beta$ satisfy certain conditions. This means that, if these conditions were satisfied, the crystal would exhibit a spontaneous polarization. However, in order to account for two Curie points, Mason was forced to assume that the dipole moment $\mu$ is dependent upon temperature. This assumption is difficult to accept in the case of a clamped crystal. On the other hand, the dielectric investigations under hydrostatic pressure (B 5) have shown that at high pressures and constant volume the crystal should still exhibit two Curie points. We can therefore conclude that Mason's treatment may possible explain the existence of two Curie points at atmospheric pressure but cannot account for the same effect at constant volume.

*The theory of Devonshire* (D 3) takes care of this incompleteness of Mason's treatment by introducing the assumption that the two states of the hydrogen atom (Fig. VII-15a) have different weights. This corresponds to considering that the function $S$ consists not only of configurational but also of vibrational entropy, as, in a classical picture, the hydrogen atom would have different vibration frequencies in its two equilibrium positions. With this slight modification of Mason's theory, Devonshire was able to account for the occurrence of two transition temperatures at constant volume. This modified theory, however, is yet not completely satisfactory, in that it has difficulty in explaining why the range of the polarized state increases with decreasing volume, as the Bancroft results indicate that it does. The theory shows that the experimental results can be explained by assuming a pressure dependence of either the internal-field coefficient $\gamma$, or the total number of dipoles $N$, or the dipole moment $\mu$. As both $\gamma$ and $N$ vary only little with volume, it should be concluded that the main change with volume lies in $\mu$. This change, however, is in the wrong direction, because, in order to match the experimental results, the theory ought to require that the dipole moment increases with decreasing volume, which is obviously not what one would expect. Thus, the treatments of Mason and Devonshire cannot be considered entirely satisfactory.

*The theory of Mitsui* (M 20) avoids these difficulties with a somewhat more rigorous treatment of the internal field, while keeping the same general model as discussed above. The dipole interaction is described by *two* parameters ($b$ and $c$) rather than by a single one ($\gamma$) as in Mason–Devonshire's theory. This implies considerable complication of the mathematical formalism, for which the reader is referred to the original paper. It may suffice to point out here that the relative values of the two parameters $b$ and $c$ determine the state of the crystal. This is indicated in Fig. VII-16(a), which is a plot of $b$ vs. $c$. If the value of these parameters is such that the representative point of the crystal falls into region II of the figure, then the crystal will be ferroelectric in a temperature range limited by two transition temperatures. Rochelle salt is a special case insofar as its representative point must fall on the lower boundary of region II (the reciprocal dielectric susceptibility of the clamped crystal has a single zero value in the vicinity of 0 °C). This allows the calculation of $\chi^x$ from the parameters of the theory: the results seem to agree satisfactorily with the curve obtained indirectly by Mueller using Eq. (VII-34), as is indicated in Fig. VII-16(b). Mitsui's theory

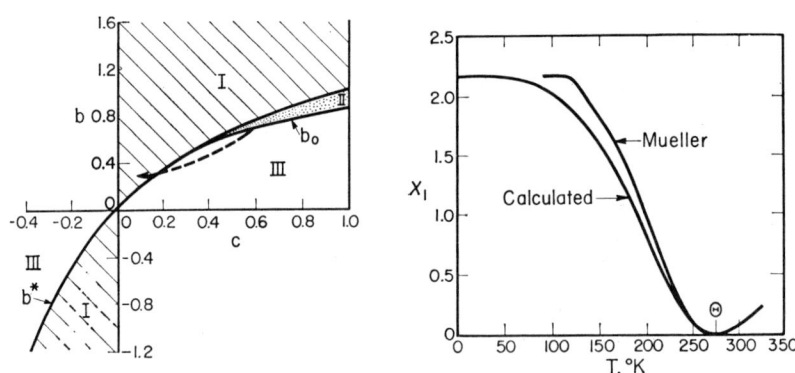

FIG. VII-16. Theory of Rochelle salt.
(a) Conditions for ferroelectricity in the plane of the parameters $b$ and $c$, which are functions of internal field coefficients. Region I: ferroelectricity below some Curie temperature. Region II: ferroelectricity between two Curie temperatures. Region III: Non-polar properties at all temperatures. The dashed arrow shows displacement of the representative point of mixed crystals of NaK and $NaNH_4$ tartrates with increasing $NH_4$ content.
(b) Reciprocal dielectric susceptibility of the clamped crystal vs. temperature (according to Mitsui (M 20)).

can also explain, qualitatively, the behavior of the mixed crystals of Rochelle salt and $NaNH_4$ tartrate discussed in Section 7, as well as the hydrostatic pressure experiments of Bancroft. The former is understood by assuming that the representative point of the mixed crystals migrates, with increasing $NaNH_4$ content, along the dashed line of Fig. VII-16(a), the compounds becoming successively non-ferroelectric and then ferroelectric below a single Curie point with a first-order transition. The effects of hydrostatic pressure, on the other hand, may be explained without difficulty by the fact that both $b$ and $c$ could be very sensitively dependent upon pressure.

The theory of Mitsui can be considered as the most satisfactory treatment of Rochelle salt-type crystals available to date. Unfortunately, it has to be only qualitative, owing to the incomplete knowledge of the coefficients of the free energy expansion. It shows that a given crystal, satisfying the conditions imposed by the model and having the proper values of the parameters $b$ and $c$ which are functions of the internal field coefficients, would exhibit a dielectric behavior similar to that of Rochelle salt. All one can conclude from this is that Rochelle salt happens to fulfill these conditions. What the theory cannot say, however, is just why and how Rochelle salt does have such very special values of the internal field parameters.

## BIBLIOGRAPHY

(A 1)  ALIAVDIN, N. V., SHEFTAL, N. N. and FROLOVA, Z. I., *Kristallografiya* **2**, 193 (1957).
(A 2)  ASSELMEYER, F. and BIENERT, W., *Z. angew. Phys.* **9**, 253 (1957).
(A 3)  AKAO, H., TAKAKURA, T. and SASAKI, T., *J. Phys. Soc. Japan* **7**, 361 (1952).
(A 4)  AKAO, H. and SASAKI, T., *J. Chem. Phys.* **23**, 2210 (1955).
(A 5)  ABE, R., *J. Phys. Soc. Japan* **13**, 244 (1958).
(A 6)  ABE, R., *J. Phys. Soc. Japan* **15**, 795 (1960).
(B 1)  BUSCH, G. and SCHERRER, P., *Helv. Phys. Acta* **6**, 234 (1933).
(B 2)  BANTLE, W. and BUSCH, G., *Helv. Phys. Acta* **10**, 262 (1937).
(B 3)  BARANSKII, K. N., GRIBOV, L. A. and PRIKHOD'KO, V. P., *Kristallografiya* **1**, 368 (1956).
(B 4)  BANTLE, W., *Helv. Phys. Acta* **15**, 373 (1942).
(B 5)  BANCROFT, D., *Phys. Rev.* **53**, 587 (1938).
(B 6)  BLOOMENTHAL, S., *Physics* **4**, 172 (1933).
(B 7)  BEEVERS, L. A. and HUGHES, W., *Proc. Roy. Soc. (London)* **A 177**, 251 (1941).
(B 8)  BERNARD, M. P., *Compt. rend.* **242**, 1012 (1956).
(B 9)  BAKER, A. N. and WEBBER, D. S., *J. Chem. Phys.* **27**, 689 (1957).
(C 1)  CADY, W. G., *Piezoelectricity*, McGraw-Hill, New York (1946).
(C 2)  CHERNYSHEVA, M. A., *Doklady Akad. Nauk S.S.S.R.* **74**, 247 (1950); **81**, 1065 (1951); **91**, 87 (1953).
(C 3)  CHERNYSHEVA, M. A., *Izvest. Akad. Nauk S.S.S.R., Ser. Fiz.* **21**, 289 (1957).
(C 4)  CHAPELLE, J., CHAMPIER, G. and DELAIN, C., *J. chim. phys.* **50**, No. 9 (1953).
(C 5)  CHAPELLE, J., *Compt. rend.* **223**, 993 (1946); **226**, 401 (1948); *J. chim. phys.* **45**, 79 (1948); *Bull. soc. franç. minéral. et crist.* **73**, 511 (1950).
(C 6)  CHAPELLE, J. and GALY, A., *J. phys. radium* **13**, 111 (1952).
(D 1)  DAVID, R., *Helv. Phys. Acta* **8**, 431 (1935).
(D 2)  DEVONSHIRE, A. F., Theory of Ferroelectrics, *Phil. Mag. Suppl.* **3**, 85 (1954).
(D 3)  DEVONSHIRE, A. F., *Phil. Mag.* **2**, 1027 (1957).

(E 1) EREMEEV, M., cited by KURCHATOV, I. V., *Le champ moléculaire dans les diélectriques* (*le sel de seignette*), Hermann, Paris (1936).

(E 2) EISNER, I. YA., *Kristallografiya* **2**, 296 (1957).

(E 3) EISNER, I. YA., *Izvest. Akad. Nauk S.S.S.R.*, *Ser. Fiz.* **21**, 334 (1957).

(E 4) EVANS, R. C., *Phil. Mag.* **24**, 70 (1937).

(F 1) FORSBERGH, P. W., JR., *Piezoelectricity, Electrostriction and Ferroelectricity, Handbuch der Physik.* vol. 17, p. 389, Springer-Verlag, Berlin (1956).

(F 2) FURUICHI, J. and MITSUI, T., *Phys. Rev.* **80**, 93 (1950).

(F 3) FOWLER, R. H., *Proc. Roy. Soc. (London)* **149**, 1 (1935).

(F 4) FRAZER, B. C., MCKEOWN, M. and PEPINSKY, R., *Phys. Rev.* **94**, 1435 (1954).

(F 5) FRAZER, B. C., Private communication (March 1960).

(H 1) HABLÜTZEL, J., *Helv. Phys. Acta* **12**, 489 (1939).

(H 2) HABLÜTZEL, J., *Helv. Phys. Acta* **8**, 499 (1935).

(H 3) HINZ, H., *Z. Physik* **111**, 617 (1939).

(H 4) HICKS, J. F. G. and HOOLEY, J. G., *J. Am. Chem. Soc.* **60**, 2994 (1938).

(H 5) HOSHINO, S., MITSUI, T., JONA, F. and PEPINSKY, R., *Phys. Rev.* **107**, 1255 (1957).

(H 6) HUNTINGTON, H. B., *Phys. Rev.* **72**, 321 (1947).

(H 7) HOLDEN, A. N. and MASON, W. P., *Phys. Rev.* **57**, 54 (1940).

(I 1) INDENBOM, V. L. and CHERNYSHEVA, M. A., *Kristallografiya* **2**, 526 (1957).

(I 2) INDENBOM, V. L. and CHERNYSHEVA, M. A., *Zh. Eksptl. i Teoret. Fiz.* **32**, 697 (1957).

(J 1) JONA, F., *Helv. Phys. Acta* **23**, 795 (1950).

(J 2) JAFFE, H., *Phys. Rev.* **51**, 43 (1937).

(J 3) JONA, F. and PEPINSKY, R., *Phys. Rev.* **92**, 1577 (1953).

(K 1) KOZLOVSKII, M. I., *Kristallografiya* **2**, 760 (1957).

(K 2) KRIVOGLAZ, M. A. and RYBAK, S. A., *Zh. Eksptl. i Teoret. Fiz.* **33**, 139 (1957).

(K 3) KOBEKO, P. and NELIDOW, J. G., *Physik, Z. Sowjetunion* **1**, 382 (1932).

(K 4) KURCHATOV, I. V., *Le champ moléculaire dans les diélectriques* (*le sel de seignette*), Hermann, Paris (1936).

(K 5) KAWAI, H., *J. Phys. Soc. Japan* **2**, 113 (1947); **3**, 111 (1948).

(K 6) KAWAI, H. and MARUTAKE, M., *J. Phys. Soc. Japan* **3**, 8 (1948); **4**, 91 (1949).

(K 7) KÄNZIG, W., *Ferroelectrics and Antiferroelectrics, Solid State Physics*, vol. 4, pp. 1−197. Academic Press, New York (1957).

(K 8) KONSTANTINOVA, V. P. and YURIN, V. A., *Kristallografiya* **2**, 294 (1957).

(K 9) KLUGE, M. and SCHÖNFIELD, H., *Naturwiss.* **21**, 194 (1933).

(K 10) KLASSEN-NEKLIUDOVA, M. V., CHERNYSHEVA, M. A. and SHTERNBERG, A. A., *Doklady Akad. Nauk S.S.S.R.* **63**, 527 (1948).

(K 11) KRUEGER, H. H. A., *Bull. Am. Phys. Soc.* Ser. II, **4**, 425 (1959).

(L 1) LÜDY, W., *Helv. Phys. Acta.* **15**, 527 (1942).

(L 2) LAROCHE, A. M. and CHAPELLE, J., *Compt. rend.* **244**, 876 (1957).

(M 1) MUELLER, H., *Phys. Rev.* **47**, 175 (1935).

(M 2) MUELLER, H., *Phys. Rev.* **57**, 829 (1940).

(M 3) MUELLER, H., *Phys. Rev.* **58**, 565 (1940).

(M 4) MUELLER, H., *Phys. Rev.* **58**, 805 (1940).

(M 5) MUELLER, H., *Ann. N. Y. Acad. Sci.* **40**, 321 (1940).

(M 6) MASON, W. P., *Piezoelectric Crystals and Their Application to Ultrasonics*, Van Nostrand Co., New York (1950).

(M 7) MASON, W. P. and JAFFE, *Bell System Tech. J.* **26**, 80 (1947).

(M 8) MAYER, H., *Compt. rend.* **243**, 246 (1956).

(M 9) MASON, W. P., *Phys. Rev.* **55**, 775 (1939).

(M 10) MANDELL, W., *Proc. Roy. Soc. (London)* **116**, 623 (1927).

(M 11) MATTHIAS, B., *Helv. Phys. Acta* **16**, 99 (1943).

(M 12) MÜSER, H. E., *Z. angew. Phys.* **10**, 249 (1958).

(M 13) MIYAKE, S., *J. Phys. Soc. Japan* **2**, 98 (1947); *Proc. Phys. Math. Soc. Japan* **23**, 371 (1941); **23**, 810 (1941); *Acta Cryst.* **2**, 192 (1949).

(M 14) MITSUI, T. and FURUICHI, J., *Phys. Rev.* **90**, 193 (1953).

(M 15) MARUTAKE, M., *J. Phys. Soc. Japan* **7**, 25 (1925).
(M 16) MÜSER, H. E. and FLUNKERT, H., *Z. Physik* **150**, 21 (1958).
(M 17) MITSUI, T. and FURUICHI, J., *Phys. Rev.* **95**, 558 (1954).
(M 18) MELMED, A., JONA, F. and PEPINSKY, R., Tech. Rep. No. 20, Wright Air Development Center, Contract AF 33 (616)−2133 (1954). Unpublished.
(M 19) MAKITA, Y. and TAKAGI, Y., *J. Phys. Soc. Japan* **13**, 367 (1958).
(M 20) MITSUI, T., *Phys. Rev.* **111**, 1259 (1958).
(M 21) MAZZI, F., JONA, F. and PEPINSKY, R., *Z. Krist.* **108**, 359 (1957).
(N 1) NORGORDEN, O., *Phys. Rev.* **49**, 820 (1936); **50**, 782 (1936).
(N 2) NAKAMURA, T., *J. Phys. Soc. Japan* **11**, 624 (1956).
(N 3) NAKAMURA, T., *J. Phys. Soc. Japan* **12**, 477 (1957).
(O 1) OKADA, K., *J. Phys. Soc. Japan* **15**, 363 (1960); **16**, 1647 (1961).
(P 1) POCKELS, F., *Lehrbuch der Kristalloptik*, Teubner, Leipzig (1906).
(P 2) PRICE, W. J., *Phys. Rev.* **75**, 946 (1949).
(R 1) RUSTERHOLZ, A. A., *Helv. Phys. Acta* **7**, 643 (1934); **8**, 39 (1935).
(S 1) SCHWARTZ, E., *Elek. Nachr. Tech.* **9**, 481 (1932).
(S 2) SHULVAS-SOROKINA, R. D., *Z. Physik* **73**, 700 (1932); **77**, 541 (1932).
(S 3) SAWYER, C. B. and TOWER, C. H., *Phys. Rev.* **35**, 269 (1930).
(S 4) STANKOWSKI, J., *Bull. acad. polon. sci., ser. sci. math. astron. phys.* **6**, 599 (1958).
(S 5) STRAUBEL-FISCHER, E., *Naturwiss.* **44**, 230 (1957).
(S 6) SPIVAK, G. V., IGRAS, E. and ZHELUDEV, I. S., *Doklady Akad. Nauk S.S.S.R.* **122**, 54 (1958).
(S 7) SPIVAK, G. V., KROKHINA, A. I., YAVORSKAYA, T. V. and DURASOVA, YU. A., *Doklady Akad. Nauk S.S.S.R.* **114**, 1001 (1957).
(T 1) TAKAHASHI, H. and HARA, S., *J. Phys. Soc. Japan* **4**, 257, 261 (1949).
(T 2) TAKAGI, Y. and MAKITA, Y., *J. Phys. Soc. Japan* **13**, 272 (1958).
(T 3) TAKAHASHI, H., NAKAMURA, T. and ISHIBASHI, Y., *J. Phys. Soc. Japan* **15**, 853 (1960).
(T 4) TOYODA, K., SHIMADA, A. and TANAKA, T., *J. Phys. Soc. Japan* **15**, 536 (1960).
(U 1) UBBELOHDE, A. R. and WOODWARD, I., *Proc. Roy. Soc. (London)* **A 185**, 448 (1946).
(V 1) VALASEK, J., *Phys. Rev.* **17**, 475 (1921); **19**, 478 (1922); **20**, 639 (1922); **24**, 560 (1924); *Science* **65**, 235 (1927).
(V 2) VERGNOUX, A. M., BLANC, J. and VIERNE, R., *Compt. rend.* **244**, 580 (1957).
(V 3) VIGNESS, I., *Phys. Rev.* **46**, 255 (1934); **48**, 198 (1935).
(V 4) VAN KLOOSTER, H. S., *J. Chem. Educ.* **36**, 314 (1959).
(W 1) WILSON, A. J. C., *Phys. Rev.* **54**, 1103 (1938).
(W 2) WEHRLE, W., Thesis, E. T. H. Zürich (1937); cited by BANTLE, W., *Helv. Phys. Acta* **15**, 373 (1942).
(W 3) WILDBERGER, A., Thesis, E. T. H. Zürich (1938); cited by BANTLE, W., *Helv. Phys. Acta* **15**, 373 (1942).
(W 4) WIEDER, H. H., *Phys. Rev.* **110**, 29 (1958).
(W 5) WIEDER, H. H. and COLLIUS, D. A., *Phys. Rev.* **120**, 725 (1960).
(Y 1) YAKOVLEV, I. A., VELICHINA, T. S. and BARANSKII, K. N., *Zhur. Eksptl. i Teoret. Fiz.* **33**, 1075 (1957).
(Y 2) YURIN, V. A., *Kristallografiya* **1**, 734 (1956).
(Y 3) YURIN, V. A., *Izvest. Akad. Nauk S.S.S.R., Ser. Fiz.* **21**, 329 (1957).
(Z 1) ZWICKER, B. and SCHERRER, P., *Helv. Phys. Acta* **27**, 346 (1944).
(Z 2) ZHELUDEV, I. S., PROSKURNIN, M. A., YURIN, V. A. and BABERKIN, A. S., *Doklady Akad. Nauk S.S.S.R.* **103**, 207 (1955).
(Z 3) ZHELUDEV, I. S. and YURIN, V. A., *Izvest. Akad. Nauk S.S.S.R., Ser. Fiz.* **20**, 211 (1956).
(Z 4) ZHIRNOV, V. A., *Zhur. Eksptl. i Teoret. Fiz.* **35**, 1175 (1958).
(Z 5) ZHELUDEV, I. S. and SIT'KO, R. IA., *Izvest. Akad. Nauk S.S.S.R., Ser. Fiz.* **21**, 286 (1957).

# CHAPTER VIII

# FERROELECTRIC SULFATES AND RELATED COMPOUNDS

## 1. Introduction

Up to 1955, ferroelectricity was believed to be a rather rare phenomenon, peculiar to a limited number of compounds. It was known to occur only in certain tartrates, orthophosphates and arsenates, and in a number of double oxides. Dielectrically, the phenomenon was characterized by the existence of (at least) one Curie point, causing a very large anomaly of the dielectric constant, and by general adherence to the Curie–Weiss law. The discovery of ferroelectricity in guanidinium aluminum sulfate hexahydrate, in 1955, marked the beginning of a very rapid expansion of the ferroelectric family, expansion which eventually led to a revision of the older concepts. The first group of ferroelectric sulfates was enlarged almost immediately by the discovery of ferroelectric properties in some alums, where the existence of dielectric anomalies had been known for a long time. This, in turn, revealed not only the importance of tetrahedral ions such as $(NH_4)^{1+}$ and $(SO_4)^{2-}$, but also the necessity of re-examining all the numerous compounds that were known to undergo polymorphic transitions and exhibit dielectrics anomalies. The first success of this re-examination was the discovery of ferroelectricity in one of the simplest sulfates, namely $(NH_4)_2SO_4$. The research for new materials led subsequently to tri-glycine sulfate and its isomorphs, already discussed in Chapter II, and to a variety of hydrogen-bonded crystals, of which the most important is probably $LiH_3(SeO_3)_2$.

All of these new materials have considerably modified previous ideas on the dielectric aspects of ferroelectricity. It was found that one need not have a large dielectric anomaly nor even a Curie–Weiss law; in fact, one need not have a Curie point at all in some cases. These aspects will be made evident by the description of various ferroelectrics in the present and in the following chapter. The structural characteristics of most of these ferroelectrics are not sufficiently known, at the present time, to permit discussion of possible models. Even the classical phenomenological theory fails, in many cases, so that we must limit ourselves for the moment to little more than a description of the collected data. The present chapter is concerned with the hydrogen-bonded ferroelectrics containing tetrahedral ions such as $(SO_4)^{2-}$, $(SeO_4)^{2-}$, $(BeF_4)^{2-}$ and pyramidal ions such as $(SeO_3)^{2-}$.

The coefficients of thermal expansion are markedly anisotropic, the coefficient along the trigonal axis being about nine times larger than that at right angle to this axis (E 1), (H 4). Preparation of thin plates for dielectric measurements is facilitated by the existence of cleavage parallel to (0001) (perfect in GASH, not quite as distinct in GVSH).

The dielectric properties of these guanidinium ferroelectrics were investigated by Holden and co-workers (H 1), (M 1), (H 3). At room temperature, the small-signal dielectric constant along the (polar) trigonal axis is remarkably small ($\varepsilon_c = 6$), while that measured at right angles to this axis is $\varepsilon_a = 5$. Practically no change in these values can be observed up to 100 °C, above which temperature dehydration causes unreproducibility in the results. The spontaneous polarization of GASH is $0.35 \times 10^{-6}$ C/cm² at room temperature and, as indicated in Fig. VIII-1 (a), it decreases almost linearly with increasing temperature. Extrapolation of the line toward high temperatures seems to indicate that the spontaneous polarization would vanish in the vicinity of 300 °C, if the crystal did not decompose before this temperature is reached. The crystals isomorphous with GASH show a very similar behavior. The sulfates (Ga, Cr and V) have $P_s$ values varying between 0.35 and $0.38 \times 10^{-6}$ C/cm², while the selenates (Al, Ga and Cr) have slightly larger values, ranging between 0.45 and $0.48 \times 10^{-6}$ C/cm² (Fig. VIII-1 a). In general, the hysteresis loops exhibited by these salts are biased, i.e, asymmetrical with respect to the polarization axis. This effect is particularly pronounced in GASH (much less in GVSH), where Holden *et al.* (H 3) found that the bias and the shape of the hysteresis loop vary considerably among specimens cut from different regions of the parent crystal. A satisfactory ex-

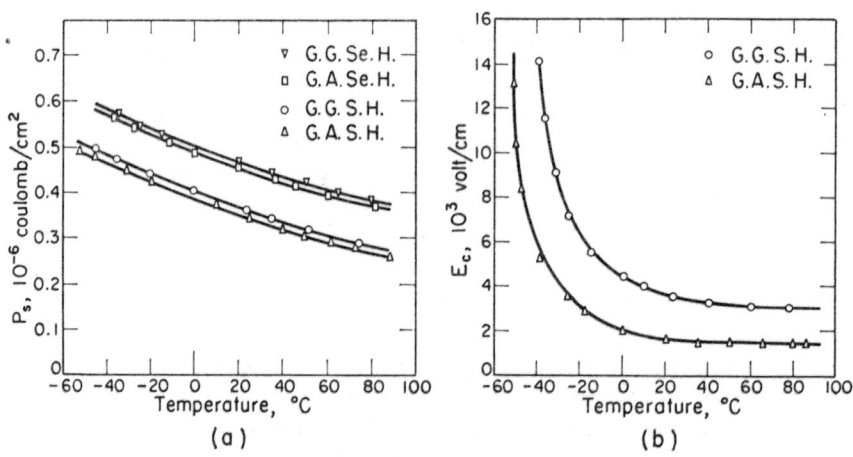

FIG. VIII-1. Dielectric properties of the guanidinium ferroelectrics.

(a) Spontaneous polarization $P_s$ vs. temperature for the aluminum and gallium sulfates and selenates.

(b) Coercive field strength $E_c$ vs. temperature for the aluminum and gallium sulfates.

(According to Holden *et al.* (H 2).)

planation for these effects was not given, but it may conceivably be linked with the presence of dislocations (see below), which limit the motion of domain walls.

Owing to the frequent distortion of the hysteresis loops, reliable values of the coercive field strength can hardly be obtained. The order of magnitude of $E_c$ measured at 60 c/s is 1000–3000 V/cm. The temperature dependence of $E_c$ is depicted in Fig. VIII-1 (b) for GASH and GGSH. The steep increase of $E_c$ around $-40\ °C$ makes it difficult to saturate the crystal at 60 c/s and to obtain hysteresis loops at low temperatures. This increase, however, is not due to a transition of any sort, but rather to the fact that the relaxation time associated with the switching process increases rapidly with decreasing temperature. Chynoweth (C 2) found that the polarization of the crystal can indeed be reversed, even at $-180\ °C$, by applying a d.c. field in the appropiate direction, because in so doing, the direction of the pyroelectric signal can be reversed. The rate at which this reversal takes place depends on the field strength. For example, when a d.c. field was applied at $-180\ °C$, with a magnitude equal to that of the 60 c/s coercive force at room temperature, reversal took about 30 sec to be completed. This large temperature dependence of the switching time accounts for the apparent increase in the coercive field strength and its frequency dependence (H 2) when studied with an alternating field. The pyroelectric measurements of Chynoweth (C 2) have established, in fact, that the spontaneous polarization of GASH keeps increasing linearly as the temperature is lowered, reaching the value of $0.60 \times 10^{-6}\ C/cm^2$ at $-175\ °C$. This linear increase in $P_s$ with decreasing temperature is a peculiarity of the guanidinium ferroelectrics which is not shared with any other material.

Just as peculiar is the dependence of the spontaneous polarization upon hydrostatic pressure. Merz (M 2) has found that the value of $P_s$ in GASH increases linearly with pressure, at room temperature. Under a pressure of 5000 atm, the value of $P_s$ is $0.5 \times 10^{-6}\ C/cm^2$, which is the value that the crystal exhibits, at atmospheric pressure, at the temperature of about $-110\ °C$. Merz, in fact, was able to prove that, at least as far as orders of magnitude are concerned, the change in $P_s$ is proportional to the change in volume, regardless of whether the latter change is obtained by hydrostatic pressure or by lowering of the temperature. The factor of proportionality, however, is about twenty to twenty-five times larger than one would expect from the sole change in the number of dipoles per unit volume, so that one is led to conclude that application of hydrostatic pressures must cause the dipole moments to increase. This is indeed very unusual behavior. A possible explanation is that the spontaneous polarization may be principally due to those hydrogen bonds that have large components along the ferroelectric axis (M 2). This seems to be substantiated by the fact that the piezoelectric effect in GASH crystals is rather small (H 2), so that piezoelectric coupling cannot be held responsible for the large polarization increase. Also, the elastic anisotropy of GASH is among the largest ever observed on any crystal (H 2), so that the crystal is most easily deformed in the direction of the polar axis.

Extensive investigations of the *elastic properties* of GASH, GGSH, GASeH and GGSeH, carried out by Haussühl (H 2), have in fact established that the abnormally large elastic anisotropy (represented by the ratio of elastic constants $c_{33}/c_{11}$) is common to all these materials. An interesting comparison between the elastic properties of these salts and those of the alums (which are crystal–chemically related to the guanidinium salts discussed here), seems to indicate that this large elastic anisotropy is a consequence of the much larger axial ratio $c/a$ in GASH with respect to that of the alums (referred to a trigonal cell). The origin of the marked anisotropy of GASH is therefore not a different character of the bonds, but rather the fact that the cell appears to be stretched along the threefold axis and compressed in the perpendicular directions, with respect to the corresponding cell of the alums. Accordingly, of the six independent elastic constants of GASH and isomorphs:

$$c_{11} = c_{22}; \ c_{33}; \ c_{44} = c_{55}; \ c_{12}; \ c_{13} = c_{23}; \ c_{14} = -c_{24},$$

only those which are affected by this lattice distortion differ markedly from those of the (isotropic) alums, while those which are less influenced by the deformation ($c_{13}$ and $c_{44}$) have values similar to the corresponding quantities of the alums (H 2).

From this viewpoint, it is reasonable to expect that the large lattice distortion (together with the fact that GASH-type ferroelectrics can only be grown in their polar phase) causes non-uniform strain and eventually a number of lattice imperfections such as dislocations. The existence of dislocations in GASH crystals was, in fact, proved by Nakamura (N 2) with a careful investigation of the etch pits created on cleaved (0001) surfaces by a mixture of alcohol and water. These etch pits are generally symmetrical or cocked triangles with edges parallel to three edges of the (0001) face. Their average density is approximately $10^4$ per cm$^2$ and their number can be increased by mechanical shocks. The correspondence between the distributions of etch pits obtained on the matching faces of the two halves of a cleaved crystal is indicative of dislocations that were crossing the plane before cleavage. The symmetrical pits represent dislocations crossing the cleavage plane at 90°, the cocked pits indicate dislocations crossing this plane at an angle which was estimated to vary between 15° and 30°. It is probable, as mentioned above, that the dislocations interact with the motions of domain walls during polarization reversal, thus providing an explanation for the biased hysteresis loops observed by Holden *et al.* (H 3) and for the pronounced field and frequency dependence of the coercive force.

The etching technique does not seem to reveal much of the ferroelectric domain structure in GASH crystals. Neither could the domains be observed under the polarizing microscope, unless a straining field was applied. But the powder-pattern technique involving colloidal suspensions of sulfur and red lead oxide (P 1) (see also Section II-7) was capable of revealing domain patterns that reflect the distribution of polarity on the (0001) faces. Pearson and Feldmann (P 1) were able to establish a direct correlation between the biases of the hysteresis loops and the domain patterns revealed by the powder technique.

*The dynamic properties of the domains* can be inferred from the studies of the process of polarization reversal (Wieder (W 2) and Prutton (P 2), (P 3)). The results of such studies indicate that this process is analogous to that occurring in tetragonal BaTiO$_3$ (see Section IV-8). Under pulsing conditions and at low pulse amplitudes, the switching time $t_s$ depends exponentially upon the field:

$$\frac{1}{t_s} = \frac{1}{t_\infty} e^{-\alpha/E}.$$

The activation field $\alpha$ is equal to $4.1 \times 10^3$ V/cm, which is of the same order as that of tetragonal BaTiO$_3$. Both the activation field $\alpha$ and the coercive field depend upon the thickness of the sample investigated, so that one is led to assume the existence of either a space charge or a dehydrated layer at the surface. No threshold field exists, below which reversal would not take place in any length of time. At higher field strengths, the reciprocal switching time increases linearly with the applied field, so that one can define a domain wall mobility:

$$\mu = \frac{d}{t_s (E - E_0)},$$

where $d$ is the crystal thickness and $E_0$ the extrapolated value of $E$ at $t_s = \infty$. The domain wall mobility is of the order of $2-8 \times 10^{-2}$ cm$^2$/V sec, thus considerably smaller than the corresponding value for BaTiO$_3$ ($\mu \cong 2.5$ cm$^2$/V sec). Recent investigations of Fatuzzo (F 3) confirm that at high electric fields the domain wall motion time is longer than the nucleation time and hence controls the switching process (see Section II-7). The hysteresis losses are also much smaller in GASH than in BaTiO$_3$ (by about a factor of 10), but the upper frequency limit for observing ferroelectric phenomena is about the same (25–50 kc/s) in both crystals (W 2).

### Structural Characteristics

An X-ray analysis of the structure of guanidinium gallium sulfate hexahydrate was carried out by Geller and Booth (G 1). The essential features of this structure, which are probably common to all the guanidinium ferroelectrics known to date, are the following. The metal ions (Ga$^{3+}$) lie on the threefold axes, and are surrounded by somewhat distorted octahedra of water molecules. The latter, in turn, are connected to sulfate tetrahedra by hydrogen bonds, while these tetrahedra are slightly tilted with respect to the (0001) plane in amounts varying between 5° and 16°. The guanidinium ions appear to be rather loosely bound to the rest of the structure, and it is possible that they may be in a state of disorder or rotation. Despite the fact that thermal motion of the atoms was reported to appear highly anisotropic, only an isotropic temperature factor was applied to the whole structure. This may account, in part, for the comparatively high values of the reliability factor (in GGSH: 18% for the (0$kil$) data and 21% for the ($h0il$) data; in GASH: 23% for the ($h0il$) data), but several large individual discrepancies between calculated and observed structure factors (including some of the non-observed reflections) may indicate significant errors in atomic positions. For some of the reflections, consideration of the hydrogen atoms may improve the agreement, but as a whole, the atomic co-ordinates do not appear to be sufficiently refined to allow a clear picture of the atomistic mechanism of polarization reversal. Three-dimensional X-ray analysis and neutron diffraction studies are likely to provide the refinement necessary for a model of the ferroelectric process. At the present stage, however, any such model would, at best, be only speculative.

It should be mentioned that a different structure of the guanidinium ferroelectrics has been suggested by Varfolomeeva et al. (V 1). The differences between this structure and that of Geller and Booth (G 1) involve the parameters of the metal ions which lie on the threefold axes, the distances from the metal ions to the water molecules, and the positions of the guanidiniums ions. The information regarding this structure, however, is rather incomplete. Geller (G 2) has pointed out that, although the trial structure proposed by Varfolomeeva et al. looks very plausible, particularly from the viewpoint of efficient packing, this trial structure leads to poorer overall agreement between calculated and observed structure factors.

A number of resonance experiments have been carried out on GASH-type compounds, but, unfortunately, the results do not contribute much to the understanding of the ferroelectric phenomenon in these crystals. The nuclear quadrupole coupling of Al in GASH appears to be strongly temperature dependent, and is indicative of an axially symmetric field gradient for the two non-equivalent Al sites (B 2), in agreement with the X-ray results. The paramagnetic spectrum of $Cr^{3+}$ in GASH shows no changes consequent of the dielectric polarization (B 1), (D 1). Proton resonance experiments seem to provide some evidence for the fact that the guanidinium groups may be rotating in the (0001) plane (M 3), (S 1). There is no reliable information, however, on the question as to whether the guanidinium ions are planar or not, although the infrared data of guanidinium halides seem to indicate that they are indeed planar in these latter crystals (A 1). The Raman and infrared spectra of GASH and the partially deuterated isomorph have been studied by Lafon (L 5).

## 3. The Alums

.Alum is the name common to a very large family of double salts with the general formula:

$$M^{1+} M^{3+} (R O_4)_2 \cdot 12 H_2O,$$

where $M^{1+}$ is a monovalent ion, such as K, Rb, Cs, Tl, $NH_4$, $CH_3NH_3$, etc; $M^{3+}$ is a trivalent metal, such as Al, Fe, Cr, V, In, Ga, etc; and $R$ is S, Se or Te. All these crystals have cubic symmetry at room temperature. They can be very easily prepared in the form of large single crystals from saturated aqueous solutions. A large number of them undergo transitions at low temperatures. Because they are paramagnetic, the physical properties of chromium and iron alums have been studied extensively by both experimental and theoretical physicists. An excellent review of the properties of various chromium alums was published by Eisenstein (E 2). The low-temperature transitions occurring in these and other types of alums were detected by a number of investigators in the course of a variety of experiments: optical absorption (K 1), (C 3), specific heat measurements (B 3), (J 3), paramagnetic resonance (B 4), (W 3), (B 5), (G 3) and nuclear quadrupole coupling (B 7). Measurements of the elastic constants of various alums were recently carried out by Haussühl (H 6), (H 2). As far as the dielectric properties are concerned, extensive work was done mainly by the French investigators: anomalies of the dielectric constant were reported by Guillien (G 4) and by Granier (G 5) while the anomalies of the dielectric loss at microwave frequencies were studied by Freymann et al. (F 1) and also by Griffiths and Powell (G 6).

It was only in 1956, however, that Pepinsky et al. (P 4) identified the transitions in methylammonium aluminum alum and urea chromium alum as ferroelectric. Ferroelectricity was subsequently discovered in a large number of ammonium and ammonium-substituted alums (P 5), although not all the compounds that were found to exhibit dielectric anomalies turned out to be ferroelectric. The properties of the alums which were more extensively investigated are summarized in Table VIII-1. Hydroxylamine and hydrazine alums were found to shatter when cooled through the corresponding transition points. Ferroelectric hysteresis loops could *not* be observed in ammonium aluminum, gallium and chromium alums, while the remaining ammonium alums and all the methylammonium alums investigated were found to be ferroelectric at low temperatures. Deuteration of ammonium iron alum does not affect the lattice dimensions or the Curie point. Similar results were obtained in the case of methylammonium aluminum alum. The selenate salts generally exhibit higher transition temperatures than the sulfates. Mixed crystals of members of the same subgroup (ammonium or methylammonium salts) can be grown easily. Within the systems $NH_4(Fe, In)(SO_4)_2 \cdot 12H_2O$ and $CH_3NH_3(Al, Cr)(SO_4)_2 \cdot 12H_2O$, the transition temperatures vary linearly with composition, the low-temperature phases being always ferroelectric (P 5). Earlier solid solution studies of the system $(Tl, NH_4)Al$, $(Tl, K)Al$ and $(K, NH_4)Cr$ sulfate alums, by Klug and Alexander (K 2), had already revealed the linear dependence of the lattice constant upon composition. Finally, it may be mentioned that the attempts at growing substituted higher amine alums have been always unsuccessful (see also Kraus and Nutting (K 1)).

As compared with most of the ferroelectrics encountered so far in the present review, the alums stand out for two characteristic features. One is the relatively small value attained by the dielectric constant at the Curie point, the other is the extremely rapid increase of the 60 c/s coercive field with decreasing temperature. This increase is so pronounced, in fact, that saturated hysteresis loops can be observed only in temperature ranges of about 13–15° in methylammonium alums and 2–3° in ammonium alums. The dielectric properties of these two subgroups have been investigated in detail by Jona et al. for the representative crystals methylammonium aluminum sulfate dodecahydrate (J 1) (abbreviated

TABLE VIII-1. AMMONIUM AND AMMONIUM-SUBSTITUTED ALUMS, $M^{1+} M^{3+}(RO_4)_2 \cdot 12 H_2O$

The compounds are listed in increasing order of ionic radius of $M^{3+}$ (Ahren's radii, in Å)
$Al^{3+} = 0.51$; $Ga^{3+} = 0.62$; $Cr^{3+} = 0.63$; $Fe^{3+} = 0.64$; $V^{3+} = 0.74$; $In^{3+} = 0.81$.

$a$ = lattice parameter in Å at the temperature given in parenthesis; $\varrho$ = density in g/cm³; $T_c$ = transition temperature in °K, $(F)$ in this column indicates ferroelectricity below $T_c$; $P_s$ = spontaneous polarization in $10^{-6}$ C/cm² about 2° below $T_c$; $E_c$ = 60 c/s coercive field in kV/cm about 2° below $T_c$

| Formula | $a$ | | $\varrho$ | $T_c$ | $P_s$ | $E_c$ | Reference |
|---|---|---|---|---|---|---|---|
| $NH_4Al(SO_4)_2 \cdot 12 H_2O$ | 12.240 | (15 °C) | 1.642 | 71 | — | — | P 5 |
| Ga | 12.267 | (25 °C) | 1.784 | 110 | — | — | K 4, D 2 |
| Cr | 12.274 | (25 °C) | 1.718 | 85 | — | — | P 5 |
| Fe | 12.318 | (25° C) | 1.713 | 88 $(F)$ | 0.4 | 33 | J 2, P 5 |
| V | 12.355 | (12 °C) | 1.680 | 116 $(F)$ | 1.0 | 12 | P 5 |
| In | 12.427 | (10 °C) | 1.872 | 127 $(F)$ | 1.2 | 10 | P 5 |
| $ND_4Fe(SO_4)_2 \cdot 12 D_2O$ | 12.317 | (22 °C) | 1.812 | 88 $(F)$ | 0.4 | 30 | P 5, J 2 |
| $NH_4Cr(SeO_4)_2 \cdot 12 H_2O$ | — | — | — | 108 | — | — | P 5 |
| $CH_3NH_3Al(SO_4)_2 \cdot 12 H_2O$ | 12.502 | (23 °C) | 1.585 | 177 $(F)$ | 1.0 | 6 | J 1 |
| Ga | — | — | — | 171 $(F)$ | — | — | D 2 |
| Cr | 12.559 | (20 °C) | 1.650 | 164 $(F)$ | 1.0 | 6 | P 5 |
| Fe | 12.60 | (12 °C) | 1.647 | 169 $(F)$ | 1.3 | 6 | P 5 |
| V | 12.594 | (10 °C) | 1.633 | 157 $(F)$ | 0.9 | 6 | P 5 |
| In | 12.688 | (0 °C) | 1.805 | 164 $(F)$ | 1.2 | 6 | P 5 |
| $CH_3ND_3Al(SO_4)_2 \cdot 12 D_2O$ | — | — | — | 177 $(F)$ | 1.0 | 6 | P 5 |
| $CH_3NH_3Al(SeO_4)_2 \cdot 12 H_2O$ | 12.698 | (22 °C) | 1.820 | 216 $(F)$ | 1.2 | 9 | P 5 |
| $NH_3OHAl(SO_4)_2 \cdot 12 H_2O$ | 12.314 | (21 °C) | 1.668 | 261 | — | — | P 5 |
| $(NH_2)_2HAl(SO_4)_2 \cdot 12 H_2O$ | — | — | — | 155 | — | — | P 5 |
| Cr | — | — | — | 225 | — | — | P 5 |
| $CO(NH_2)_2HCr(SO_4) \cdot 12 H_2O$ | — | — | — | 160 $(F)$ | 0.2 | 10 | P 4 |

MASD, for convenience), and ferric ammonium sulfate dodecahydrate (J 2). The results of these studies are reported below.

The temperature dependence of the small-signal dielectric constant of MASD, measured along [100], is depicted in Fig. VIII-2. A small temperature hysteresis of about 1.5 °C is found between the Curie points determined with increasing and decreasing temperature. The sharp discontinuity and the finite value of the dielectric constant at the Curie point are indicative of a first-order transition. Above 177 °K, the Curie–Weiss law $[\varepsilon = C/(T - T_0)]$ is obeyed, with $C = 500$ degrees and $T_0 = 168.5$ °K, only in a narrow temperature interval of about 6°; from 183 °K to 203 °K, $C = 700$ degrees, $T_0 = 160.0$ °K; and from 203 °K to 243 °K, $C = 1200$ degrees, $T_0 = 120.0$ °K, a behavior that is reminiscent of that of Rochelle salt (see Section VII-2). The effect of a d.c. biasing field upon the small-signal dielectric constant is depicted in Fig. VIII-3 as measured on crystals of ferric ammonium alum (Curie point 88 °K). It is seen that the effect is different when the d.c. bias is applied along [100] or along [111] In both cases, the Curie point is raised at the rate of approximately $10^{-5}$ °C/V cm⁻¹, but in the former case, the value of the dielectric constant is slightly increased above, and decreased below the Curie point of the unbiased crystal, while in the case of fields applied along [111], the dielectric constant is decreased both above and below the Curie point. This behavior will be made understandable in the following.

Figure VIII-4 depicts the spontaneous polarization $P_s$ of MASD, measured along the pseudo-cubic [100], [110] and [111] directions, as a function of temperature. It is evident, from this figure, that the polar axis of the ferroelectric phase lies along the original cubic [100] direction, as the values of $P_s$ measured

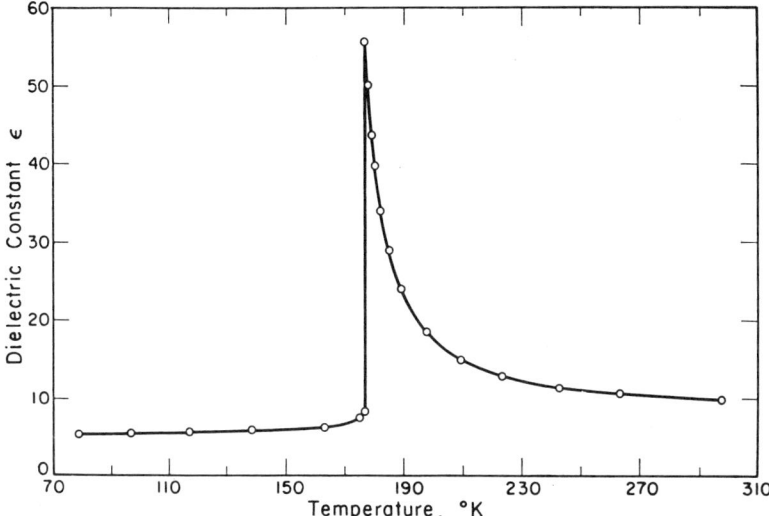

FIG. VIII-2. Temperature dependence of the small-signal dielectric constant of methylammonium aluminum sulfate alum, measured along [100] (according to Jona *et al.* (J 1)).

along [110] and [111] are $\sqrt{2}$ and $\sqrt{3}$ times smaller, respectively, than that measured along [100]. The 60 c/s coercive field, as mentioned above, increases very rapidly with decreasing temperature: with applied fields of the order of 20 kV/cm, in MASD, saturated hysteresis loops can be observed only within a range of about 15° below the Curie point.

The X-ray study of MASD (J 1) indicates that the lattice parameter of the non-polar cubic phase decreases linearly with temperature down to the Curie point. The symmetry of the polar phase is monoclinic, space group $P2_1$, the polar axis lying, of course, along the twofold screw axis, which is parallel to one of the original cubic $\langle 100 \rangle$ directions. The monoclinic $a$ and $c$ axes are approximately parallel to the two other $\langle 100 \rangle$ directions. The spontaneous strain consists therefore in a shear within a $\{001\}$ plane. The values of the lattice parameters of MASD at three different temperatures are given in Table VIII-2, which seems to indicate that the crystal exhibits a volume contraction at the transition from the cubic to the monoclinic phase.

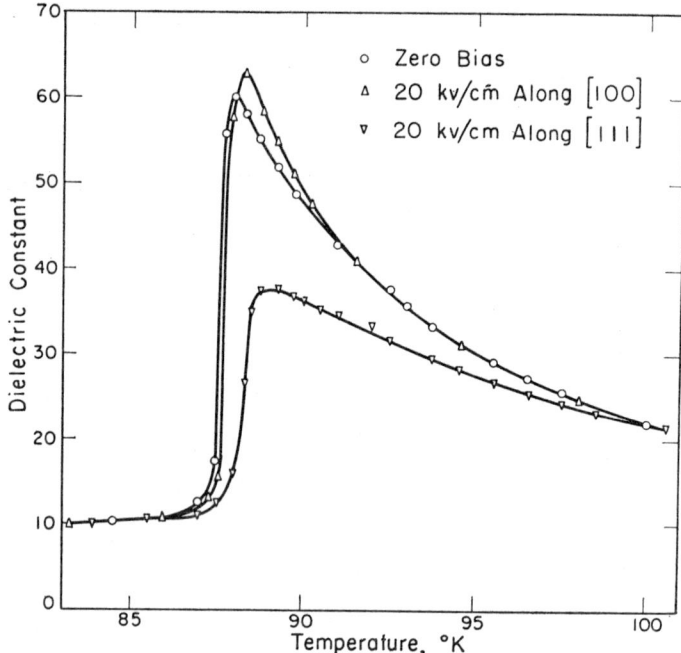

FIG. VIII-3. Effect of d.c. biasing fields applied along [100] and [111] upon the dielectric constant of ferric ammonium sulfate alum (according to Jona *et al.* (J 2)).

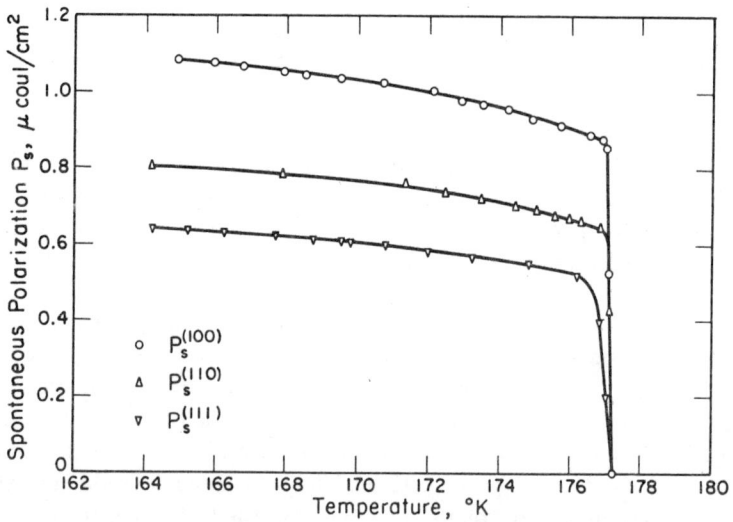

FIG. VIII-4. Spontaneous polarization $P_s$ of methylammonium aluminum sulfate alum along [100], [110] and [111] as a function of temperature (according to Jona *et al.* (J 1)).

## 2. Guanidinium Aluminum Sulfate Hexahydrate and Isomorphous Crystals

In 1955, a new class of ferroelectric crystals was discovered by Holden *et al.* (H 1) and later supplemented by Remeika and Merz (R 1). These crystals, all isomorphous to one another, are hydrated double sulfates of guanidinium and a trivalent metal, and can be described by the general formula:

$$[C(NH_2)_3][M(H_2O)_6] \cdot [AO_4]_2$$

where $M = Al^{3+}$, $Ga^{3+}$, $Cr^{3+}$ or $V^{3+}$, $A = S$ or $Se$, and the water of crystallization may be replaced by $D_2O$. This formula is more meaningful, from a structural point of view, than that more commonly used: $C(NH_2)_3 M(AO_4)_2 \cdot 6H_2O$. As both the formulas and the chemical names are somewhat cumbersome, abbreviations have been introduced for each one of these compounds. The aluminum sulfate member, which was the object of most of the dielectric investigations, is briefly referred to as GASH (for guanidinium aluminum sulfate hexahydrate), the gallium sulfate member, which was the object of structural studies, as GGSH, the selenate as GASeH and GGSeH, respectively, and so on. The symmetry of all these crystals is trigonal.

Large specimens of these compounds can be grown from saturated water solutions. Details on the best growing procedure for GASH have been published by Chapelle and Chollot (C 1) and a number of other investigators (R 2), (N 1), (H 2). The solubility of GASH in water is large (750 g/l. at room temperature), but varies very little up to 60 °C, so that growth in this temperature range is best achieved through slow evaporation of the saturated solution. Between 60 °C and 100 °C, on the other hand, the solubility increases steeply. In this range, it is advantageous to work at as high a temperature as possible. Below 85 °C the rate of growth along the trigonal axis is almost zero, so that only rather thin plates can be grown. Between 85 °C and 95 °C, this growth rate is considerable, and thick crystals are obtainable. It was also found that the addition to the GASH solution of $Cr_2(SO_4)_3$ in concentrations of the order of 1/100 mole/l. of water greatly facilitates the growth of large flawless crystals.

From a chemical viewpoint, the discovery of ferroelectricity in these salts was important insofar as these were the first ferroelectric sulfates (or selenates) to be discovered. Also, they occupy a somewhat special position in the ferroelectric family because they exhibit no Curie point. This, however, is only accidental. These guanidinium salts decompose at temperatures of the order of 200 °C, whereas the Curie points are expected to lie in the neighborhood of 300 °C or higher (see below). We therefore know only the polar phase of these ferroelectrics, which, as mentioned above, is trigonal. Crystallographic data have been reported by Wood (W 1), Ezhkova *et al.* (E 1) and by Haussühl (H 2). The lattice constants of GASH are at room temperature (W 1):

$$a = 11.77 \pm 0.04 \text{ Å}, \quad c = 8.98 \pm 0.03 \text{ Å}$$

with three formula units per unit cell, space group $C_{3v}^2 - P31m$. The crystals are unaxial negative, with birefringences of the order of 0.09 for $Na_D$ light (H 2).

TABLE VIII-2. TEMPERATURE DEPENDENCE OF THE LATTICE PARA-
METERS OF METHYLAMMONIUM ALUMINUM SULFATE ALUM
(ACCORDING TO JONA et al. (J 1))

| | $T = 23\,°C$ (cubic) | $T = -82\,°C$ (cubic) | $T = -165\,°C$ (monoclinic) |
|---|---|---|---|
| Lattice parameters (Å) | 12.502 | 12.462 | $a = 12.47$ $b = 12.33$ $c = 12.28$ $\beta = 90°40'$ $V^{1/3} \doteq 12.36$ |

MASD represents, therefore, the first example of a ferroelectric that trans-forms directly from cubic into monoclinic symmetry. It is not difficult to prove, both with geometrical as well as analytical considerations, that in this case 90° domain walls are prohibited in the stress-free crystal (J 1). Thus, an unstrained plate of MASD, in the ferroelectric phase, will consist of antiparallel domains separated by 180° walls, a configuration that cannot be recognized under the polarizing microscope (unless, of course, straining fields were applied as was done in the case of BaTiO$_3$, but this observation may present considerable experimental difficulties at the low temperatures required by MASD). If, however, the crystal plate is strained non-uniformly, for example by cementing it to a microscope slide before cooling it, then 90° domains can be observed, which are obviously formed in order to relieve the mechanical constraint. The domain patterns that can be observed under such conditions are usually very complicated. Two exam-ples of such patterns of a strained plate are show in the photomicrographs of Fig. VIII-5. Careful analysis of these domain structures has revealed that they consist of three regions with different values of birefringence. This is exactly what one would expect when looking down a pseudo-cubic ⟨100⟩ direction and allowing all possible orientations of the monoclinic axes with respect to the ori-ginal cube edges. These three regions are recognizable in Fig. VIII-5, namely (i) the spikes oriented at 45° to the pseudo-cubic axes (upper portions of the photo-micrographs of Fig. VIII-5); (ii) the bright horizontal stripes with diffused boundaries, running approximately parallel to the pseudo-cubic edges; and (iii) the dark regions surrounding both the spikes and the stripes. All three regions extinguish parallel to the pseudo-cubic ⟨100⟩ directions. Within the spikes and the stripes the polar axis lies in the plane of the plate, horizontally and vertically, respectively, whereas in the dark regions the polar axis is perpendicular to the surface of the plate. These latter regions, in fact, appear to be almost perpendicular to an optic axis, so that we may conclude that the polar axis is the acute bisectrix of the monoclinic phase and the optic axial angle is very small. The domain boundaries in such strained plates are probably {110} and {111} planes of the pseudo-cubic lattice: they involve considerable mismatch of the crystalline structure and probably also neutralized head-to-head (or tail-to-tail) arrange-ments of the polar axes of adjacent domains. This causes the boundaries to be locked in place, so that they cannot be moved with electric fields smaller than the breakdown field. Thus, only the 180° walls are mobile and contribute to the process of polarization reversal evidenced by the hysteresis loops. Reliable measurements of the spontaneous polarization require therefore that care be

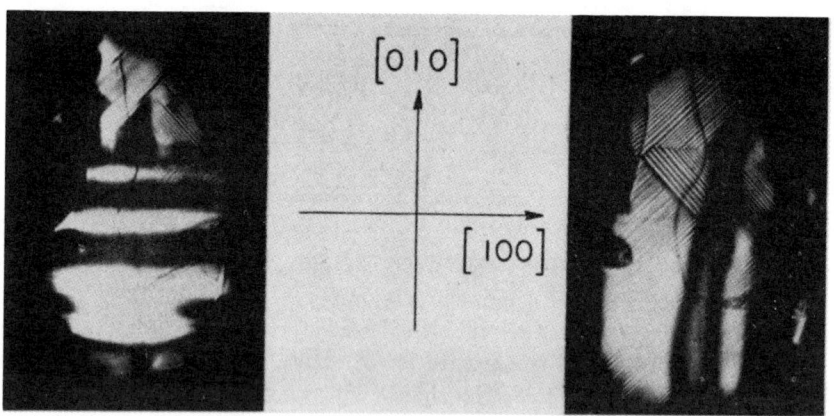

FIG. VIII-5. Domain patterns of a strained (100) plate of methylammonium aluminum sulfate alum in the ferroelectric phase. The plates were photographed at 45° from the extinction position (according to Jona *et al.* (J 1)).

exerted not to clamp the crystals with the electrodes of the holder. Occasionally, it can happen that clamping is such as to prevent observation of hysteresis loops altogether.

### Thermodynamic Treatment

In trying to develop a thermodynamic treatment of the ferroelectric transitions in the alums, it is necessary to know the point group symmetry of the crystals in the non-polar phase. As we are going to see later in this section, there are some ambiguities in the point-group assignments for the ammonium and methylammonium alums, the choice being between the centrosymmetric point group $m3$ and the non-centric point group 23. In the latter case, the crystals should be piezoelectric. Unfortunately, not much effort on proving or disproving piezoelectricity in these alums has been made. The negative results of Giebe–Scheibe tests are not sufficient reason to eliminate the possibility of piezoelectric activity. If the crystals do belong to point group 23, then a piezoelectric polarization could be generated only by shear stresses, and this polarization would be directed along a $\langle 100 \rangle$ direction. The converse effect would imply that a polarization induced along, say, [100] would cause a shear strain $y_z$. We know, on the other hand, that the direct proportionality between polarization and strain remains unaffected by the onset of ferroelectricity (see Section III-4), and this may allow us to go a step further. Since the spontaneous strain in the ferroelectric phase is indeed a shear, and the polar axis lies along the pseudo-cubic [100], it is more probable that the point group of the non-polar phase is 23, rather than $m3$. In the latter case, in fact, the strains connected with a polarization along the cube edge are electrostrictive, and the predicted symmetry for the low-temperature phase would be orthorhombic. Of course, these arguments are only qualitative, and it is better to keep in mind that both point groups are possible for the non-polar phases of ammonium and methylammonium alums.

Since we are concerned with the dielectric properties only, we limit ourselves to considering the expansion of the free energy of the stress-free crystal in terms of polarization. For the point group 23 (including terms of the sixth order), this expansion is:

$$A(P) = (1/2)\chi(P_1^2 + P_2^2 + P_3^2) + \omega_{123}\,P_1 P_2 P_3 +$$
$$(1/4)\xi_{11}(P_1^4 + P_2^4 + P_3^4) + (1/2)\xi_{12}(P_1^2 P_2^2 + P_1^2 P_3^2 + P_2^2 P_3^2) +$$
$$\psi_{11123}(P_1^3 P_2 P_3 + P_2^3 P_3 P_1 + P_3^3 P_1 P_2) +$$
$$(1/6)\zeta_{111}(P_1^6 + P_2^6 + P_3^6) + \qquad\qquad\text{(VIII-1)}$$
$$(1/2)\zeta_{112}(P_1^4 P_2^2 + P_1^4 P_3^2 + P_2^4 P_1^2 + P_2^4 P_3^2 + P_3^4 P_1^2 + P_3^4 P_2^2) +$$
$$(1/2)\zeta_{123} P_1^2 P_2^2 P_3^2 + \ldots\ldots,$$

where, as usual, $P_1$, $P_2$ and $P_3$ are the components of the polarization vector along the cubic axes. For a crystal in the symmetry class $m3$, the free energy expression is equal to (VIII-1) but with $\omega_{123} = \psi_{11123} = 0$. Putting $P_1 = P$, $P_2 = P_3 = 0$, we obtain the familiar formula:

$$A(P) = (1/2)\chi P^2 + (1/4)\xi_{11} P^4 + (1/6)\zeta_{111} P^6. \qquad\text{(VIII-2)}$$

If we know the temperature dependence of the reciprocal dielectric susceptibility $\chi$ and the value of the spontaneous polarization at the Curie point, we can compute the values of the coefficients $\xi_{11}$ and $\zeta_{111}$ from the formulas given by Devonshire (D 4) and reported in Eqs. (I-13) and (I-14). The results obtained with MASD and ferric ammonium alum are listed in Table VIII-3, where they are compared with the values of other ferroelectrics.

These coefficients $\xi_{11}$ and $\zeta_{111}$ can of course also be determined from the double hysteresis loops that characterize the relationship between polarization $P$ and field $E$ at temperatures slightly above the Curie point of a first-order transition (see Section IV-4). In the case of the alums, we can go a step further, because, in contrast to $BaTiO_3$, we can grow large single crystals from which we can cut plates of arbitrary orientations, and thus apply fields in different crystallographic directions. This has been done with ferric ammonium sulfate alum (Curie point 88 °K) by Jona et al. (J 2). Figure VIII-6 depicts a series of oscillograms of $P$–$E$ curves obtained with fields applied along [100], [111] and [110] a few degrees above the Curie point. In the former case, the $P$–$E$ curve is characterized by a pronounced upward curvature and finally by the familiar double loops. In the case of a field along [111], the curvature is downward, and similarly in the case of a field along [110]. These differences can be understood on the basis of Eq. (VIII-1) as follows.

TABLE VIII-3. COMPARISON BETWEEN THE VALUES OF THE HIGH-ORDER COEFFICIENTS OF
THE FREE-ENERGY EXPANSION OF THE STRESS-FREE CRYSTAL:

$$A = (1/2)\,\chi P^2 + (1/4)\,\xi_{11}P^4 + (1/6)\,\zeta_{11}P^6,$$

IN DIFFERENT FERROELECTRICS

(The values reported here are those measured in the vicinity of the respective Curie points.
Methods of measurement: $TR$ = from the thermodynamic relations (I-13) and (I-14);
$DL$ = double hysteresis loop experiment; $PD$ = polarization dependence of dielectric constant; $PT$ = from the dependence of $P^2$ on temperature.)

| Crystal | $\xi_{11}$ (c.g.s. units) | $\zeta_{11}$ (c.g.s. units) | Method | Reference |
|---|---|---|---|---|
| BaTiO$_3$ | $-10.8 \times 10^{-13}$ | $28.2 \times 10^{-23}$ | $TR$ | M 4 |
| | $-\ 6.8 \times 10^{-13}$ | $22.8 \times 10^{-23}$ | $DL$ | M 4 |
| | $-10.0 \times 10^{-13}$ | $-$ | $PD$ | D 3 |
| Tri-glycine sulfate | $8.0 \times 10^{-10}$ | $5.04 \times 10^{-18}$ | $PT$ | T 1 |
| | $7.7 \times 10^{-10}$ | $-$ | $PD$ | T 1 |
| Tri-glycine fluoberyllate | $13.5 \times 10^{-10}$ | $-$ | $PT$ | H 5 |
| Tri-glycine selenate | $4.7 \times 10^{-10}$ | $-$ | $PT$ | J 4 |
| Methylammonium aluminum sulfate alum | $-\ 8.9 \times 10^{-8}$ | $7.4 \times 10^{-15}$ | $TR$ | J 1 |
| Ferric ammonium sulfate alum | $\{\ \ \begin{matrix}-\ 7.8 \times 10^{-7}\\ -\ 3.3 \times 10^{-7}\end{matrix}$ | $\begin{matrix}5.3 \times 10^{-13}\\ 2.6 \times 10^{-13}\end{matrix}$ | $\begin{matrix}TR\\ DL\end{matrix}$ | $\begin{matrix}\text{J 2}\\ \text{J 2}\end{matrix}$ |
| Colemanite | $9.4 \times 10^{-8}$ | $12.6 \times 10^{-15}$ | $PT$ | W 4 |

(i) When the field is applied along [100], the $P$–$E$ curve is described analytically by the derivative of Eq. (VIII-2), i.e.

$$E = \chi P + \xi_{11}P^3 + \zeta_{111}P^5. \qquad\qquad \text{(VIII-3)}$$

The values of $\xi_{11}$ and $\zeta_{111}$ were determined as those which give the best fit between the theoretical expression (VIII-3) and the experimental curves. The results, reported in Table VIII-3, are only fairly in agreement with those obtained from Eqs. (I-13) and (I-14), the difference probably being due mainly to a temperature dependence of the coefficients involved rather than to experimental inaccuracy.

(ii) When the field is applied along [111], the theoretical $P$–$E$ curve can be obtained from Eq. (VIII-1) by putting $P_1 = P_2 = P_3 = P/\sqrt{3}$ and differentiating with respect to $P$. For point group $m3$, we obtain:

$$E = \chi P + \left(\frac{\xi_{11} + 2\xi_{12}}{3}\right)P^3 + \left(\frac{\zeta_{111} + 6\zeta_{112} + \zeta_{123}}{9}\right)P^5. \qquad \text{(VIII-4)}$$

It is interesting to see, incidentally, that in point group 23 the existence of the coefficients $\omega_{123}$ and $\psi_{11123}$ would cause the $P$–$E$ curves to be non-symmetrical with respect to the origin. Unfortunately, a choice between the two point groups cannot be done on this basis, because the experimental accuracy is insufficient.

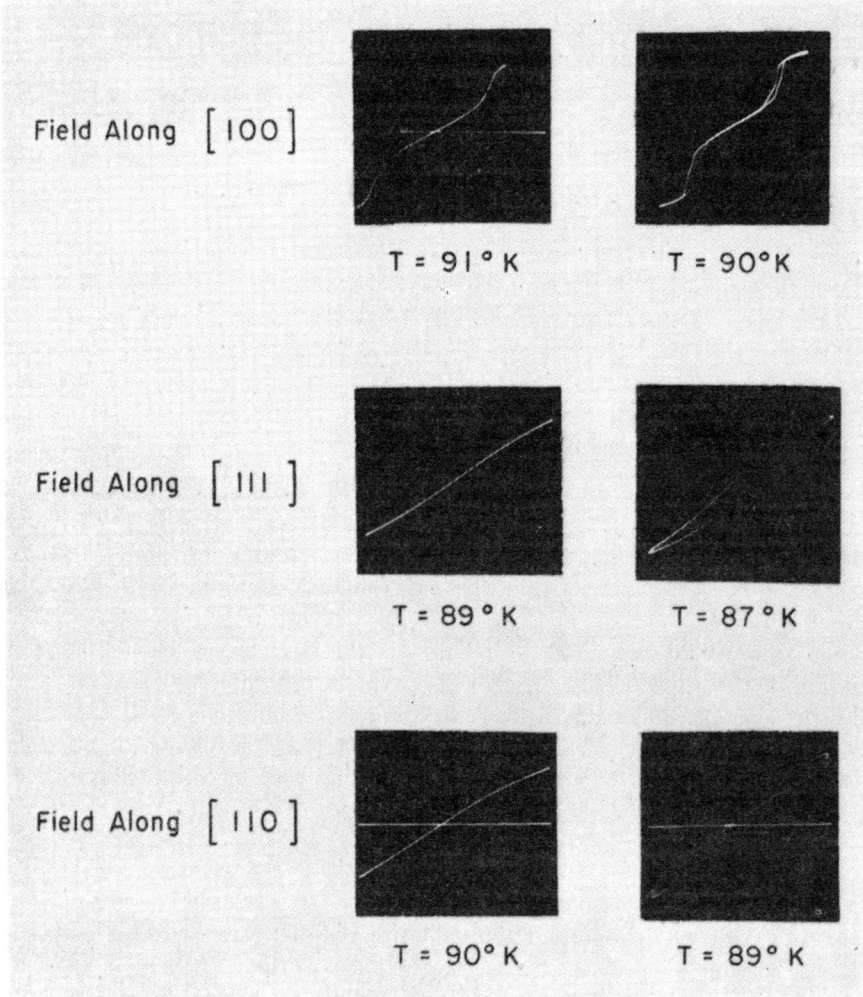

Field Along $\begin{bmatrix} 100 \end{bmatrix}$

T = 91° K          T = 90° K

Field Along $\begin{bmatrix} 111 \end{bmatrix}$

T = 89° K          T = 87° K

Field Along $\begin{bmatrix} 110 \end{bmatrix}$

T = 90° K          T = 89° K

FIG. VIII-6. Oscillograms depicting the relationship between polarization $P$ and field $E$ a few degrees above the Curie point of ferric ammonium sulfate alum. Top: 33 kV/cm applied along [100]. Middle: 24 kV/cm applied along [111]. Bottom: 36 kV/cm applied along [110] (according to JONA et al. (J 2)).

We therefore limit ourselves to considering the case expressed by Eq. (VIII-4). Putting:

$$\xi^* = \frac{\xi_{11} + 2\xi_{12}}{3}, \tag{VIII-5}$$

we see that the downward curvature of the experimental $P$–$E$ curves must be due to a positive value of $\xi^*$, just as the upward curvature in the previous case was due to the negative value of $\xi_{11}$. Since we know the latter coefficient from the latter case, we are now in the position to determine $\xi_{12}$ by fitting the experimental curves with Eq. (VIII-4), and then using Eq. (VIII-5). The result is: $\xi_{12} = 5.3 \times 10^{-7}$ c.g.s. units.

(iii) When the field is applied along [110], $P_1 = P_2 = P/\sqrt{2}$, and $P_3 = 0$. The $P$–$E$ curves are given by:

$$E = \chi P + \left(\frac{\xi_{11} + \xi_{12}}{2}\right) P^3 + \left(\frac{\zeta_{111} + 4\zeta_{112}}{4}\right) P^5. \qquad \text{(VIII-6)}$$

Since both $\xi_{11}$ and $\xi_{12}$ are known from the two previous cases, we can now predict what the shape of the $P$–$E$ curves will be in the present case. Putting:

$$\xi^{**} = \frac{\xi_{11} + \xi_{12}}{2}, \qquad \text{(VIII-7)}$$

we obtain from the preceding results that $\xi^{**} = 1.0 \times 10^{-7}$ c.g.s. units. Thus, we expect the $P$–$E$ curve to exhibit a downward curvature, which is what we observe experimentally (Fig. VIII-6). The direct determination of $\xi^{**}$ from the experimental curves yields $\xi^{**} = 0.93 \times 10^{-7}$ c.g.s. units, which is considered in satisfactory agreement, within the experimental accuracy, with the value computed above.

Just as we could explain the differences in the $P$–$E$ curves depending on the direction of the applied field, we can also explain the different behavior of the dielectric constant under d.c. fields with different orientations (as it was shown in Fig. VII-4). We have shown in Section IV-4 that the change in dielectric constant consequent to the application of a d.c. bias is given, in the first approximation (Eq. IV-19), by the formula:

$$\varepsilon_E(T) - \varepsilon_0(T) = -\frac{3\varepsilon_0^4 \xi}{(4\pi)^3} E^2, \qquad \text{(VIII-8)}$$

where $\varepsilon_E(T)$ and $\varepsilon_0(T)$ represent the values of the dielectric constant measured with and without a d.c. bias $E$, respectively, and $\xi$ is the coefficient in $P^4$ of the free-energy expansion. We know from Eq. (VIII-3) that when $E$ is parallel to [100], $\xi = \xi_{11}$, and this value is negative. Thus the dielectric constant is increased by the d.c. bias. When the field is applied along [111], $\xi = \xi^*$ (see Eqs. VIII-4 and VIII-5) and this value is positive, implying that the dielectric constant is decreased by such a d.c. bias. For fields applied along [110], $\xi = \xi^{**} > 0$ and the effect should be similar, although smaller, than in the latter case. The experimental results, not shown in Fig. VIII-4, confirm this expectation.

All the above experiments demonstrate once more the internal consistency of the thermodynamic treatment of ferroelectric transitions. A glance at Table VIII-3 shows that the values of the coefficients of the free-energy expansion vary rather widely from compound to compound, although the orders of magnitude are roughly similar within a given family of ferroelectrics. A meaningful interpretation of these orders of magnitude cannot be given before detailed model theories become available. It may be recalled, in this respect, that according to Slater's theory of $BaTiO_3$ the coefficients $\xi_{11}$ and $\xi_{12}$ are measures of the coefficients $b_1$ and $b_2$ of the anharmonic terms of the potential energy of a displaced ion (see Eq. IV-76 and IV-79).

The thermodynamic theory, finally, allows an estimate of the entropy change $\Delta S$ and the heat of transition $\Delta Q$ from the relation:

$$\Delta S = (2\pi/C)P_0^2,$$

where $P_0$ is the value of the spontaneous polarization at the Curie point. We obtain, for MASD: $\Delta S = 0.79$ cal/mole °C; $\Delta Q = 140$ cal/mole, and for ferric ammonium sulfate alum $(C = 400$ degrees): $\Delta S = 0.15$ cal/mole.

### Structural Characteristics

The general features of the alum structures were first studied by Lipson and Beevers (L 1), as early as 1935, by means of X-ray diffraction analyses of the potassium salts. The space group was reported as the cubic $P2_1/a3$, and the unit cell as containing four formula units. A subsequent investigation of the methylammonium aluminum sulfate alum, by Lipson (L 2), first raised the question of the space group assignment of ammonium-substituted alums in general, but finally confirmed the centrosymmetrical space group cited above. The problem is concerned with the fact that both the monovalent and the trivalent ions must be located on centers of symmetry on the threefold axes. When the monovalent ion is $NH_4$, it is quite evident that the hydrogen atoms cannot conform in any reasonable manner with the requirements of symmetry. The situation is even more puzzling when $NH_4$ is substituted with $CH_3NH_3$ or similar ionic groups. In trying to give an answer to this question, Lipson actually found that the structure of MASD, for example, is not quite the same as that of potassium alum, but found no evidence for noncentric symmetry. These structural differences in the alums deserve some attention. Lipson (L 3) reported existence of three different structure types, which he denoted as $\alpha$, $\beta$ and $\gamma$, and later defined in some more detail (L 4). The $\alpha$ structure is typical of "medium-sized" ions, the $\beta$, of the larger ones, and the $\gamma$, of the small Na ion. In all structures, each of the trivalent anions is surrounded by six water molecules in a nearly octahedral arrangement; the remaining six water molecules link together the $SO_4$ tetrahedron, the $M^{3+}(H_2O)_6$ octahedra and the monovalent ion. In particular, each of these waters is in contact with two oxygens, one water at one corner of the octahedron, and one monovalent ion, in such a way that the arrangement of the bonds is approximately tetrahedral. Now, if we were to start from such an $\alpha$-structure and increase the radius of the monovalent ion, the water molecules in contact with it would be pushed toward the other groups, and this would tend to alter the orientation of the latter in order to preserve correct interatomic distances. In particular, the equilibrium position of the $SO_4$ groups will vary with respect to that assumed in $\alpha$-structures, and eventually the whole distribution of bonds will be altered, leading to the $\beta$-structure (Fig. VIII-7). If, on the other hand, the radius of the monovalent ion is made as small as that of Na, the water molecules in contact with it will move closer and cause a readjustment of the $SO_4$ groups, which end up being oppositely orientated along the triad axes. This, essentially, is the $\gamma$-structure (see Fig. VIII-8).

These differences among alum structures are not sufficiently well defined to allow confident predictions of the structure type that a given alum will assume; it is also somewhat difficult to understand why the transition from the $\alpha$- to the

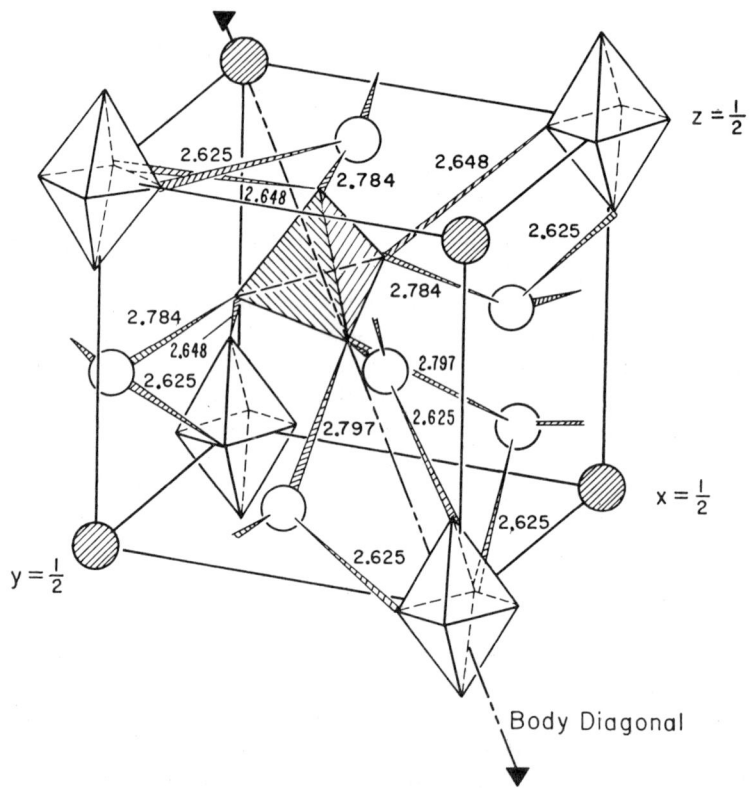

Fig. VIII-7. Room-temperature structure of methylammonium aluminum sulfate alum. The figure represents one quarter of the unit cell, and shows schematically the relative orientation of the $Al(H_2O)_6$ octahedra with respect to the $SO_4$ tetrahedron and the six remaining water molecules. Interatomic distances are given in Å (according to Okaya *et al.* (O 1)).

$\beta$-structure does not take place continuously when certain ions are replaced by others. The rubidium and the thallium aluminum alums, for example, have the $\alpha$-structure, while the rubidium and the thallium chromium alum have the $\beta$-structure, but rubidium gallium alum has again the $\alpha$-structure (K 3), (K 4). As far as the ferroelectric alums are concerned, it may be only tentatively concluded that the $\beta$-structure seems to be more likely to exhibit ferroelectric properties than the $\alpha$-structure, in view of the fact that the methylammonium aluminum and chromium sulfate alums were found to belong to the $\beta$-type, while the ammonium aluminum and chromium sulfate alums were ascribed to the α-type. It would be interesting, in this respect, to investigate the dielectric properties of α-type MASD, which, according to Lipson (L 4), can be deposited from solution on rubidium alum.

The structure of MASD has been recently re-examined by Okaya *et al.* (O 1) by means of X-ray diffraction (Fig. VIII-7). The most important result of this

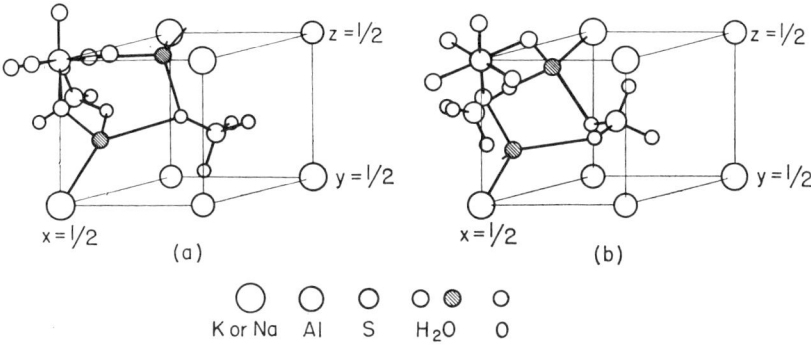

FIG. VIII-8. Comparison of the structures of α and γ alums: (a) potassium alum (α); (b) sodium alum (γ). The figures show the $Al(H_2O)_6$ octahedra at $(1/2, 0, 1/2)$ and the arrangement of the bonds to the remaining water molecules (according to Lipson (L 4)).

analysis is the fact that the $CH_3NH_3$ ions do not lie along the cubic body diagonal. but are arranged statistically at an angle to this diagonal. It is not clear whether this statistical arrangement is the consequence of thermal disorder or a true statistical arrangement over a number of unit cells, but there are indications that the latter case obtains. If this is true, then the structure is only statistically cubic, i.e. no single unit cell is cubic, and the choice between space groups $P2_1/a3$ and $P2_13$ is trivial. It is not immediately apparent from the room-temperature structure what the mechanism of the ferroelectric transition could be, particularly because the complex hydrogen-bond system can only approximately described on the basis of the X-ray data.

A neutron diffraction analysis of potassium chromium sulfate alum has been carried out by Bacon and Gardner (B 6), but the results concerning the hydrogen-bond system need not apply to the case of the ferroelectric alums, particularly because the potassium chromium salt reportedly has the α-structure. One feature, however, seems to be common to the latter crystal and to MASD: the sulfur atoms have normal values of the temperature factor, but the oxygen surrounding it in a regular tetrahedral arrangement have rather large values, compared with the rest of the structure. This instability of the sulfate group seems to be due to asymmetrical distribution of the bonding forces to this group (B 6). It is not unlikely that the large thermal vibrations of the sulfate oxygens play a role in the phase transition occurring at lower temperatures.

## 4. Di-ammonium Di-cadmium Sulfate, $(NH_4)_2Cd_2(SO_4)_3$

Ferroelectric activity in $(NH_4)_2Cd_2(SO_4)_3$ was reported by Jona and Pepinsky (J 5) below 95 °K. The crystal belongs to a rather large group of compounds with the general formula: $(M^{1+})_2(M^{2+})_2(SO_4)_3$, where $M^{1+} = $ K, Rb, Tl or $NH_4$, and $M^{2+} = $ Zn, Co, Mn, Mg, or Cd. All these crystals are commonly classed as langbeinite types, from the name of the mineral $K_2Mg_2(SO_4)_3$. The ammonium cadmium salt is the only known ferroelectric in this family, to date, but it is possible that the $NH_4Mn$ and the $NH_4Mg$ salts may behave similarly.

Chemical and crystallographic information about $(NH_4)_2Cd_2(SO_4)_3$ is sparse. Formation of the double salt $2\,CdSO_4 \cdot (NH_4)_2SO_4$ was first reported by Veres (V 2) and later specified in more detail by Benrath and Thiemann (B 8). Rather large crystals (up to 1 cm³) can be grown from acidic solutions of $CdSO_4$ and $(NH_4)_2SO_4$ in water at 85 °C. The habit is tetrahedral and cubic–octahedral–dodecahedral with $\{111\}$, $\{110\}$ and $\{100\}$ as the prominent forms. The structure is cubic, at room temperature, with:

$$a = 10.360 \text{ Å},$$

space group $P2_13$, density $\varrho = 3.28$ g/cm³, and four formula units per unit cell. The non-centric symmetry was confirmed by the existence of piezoelectric activity.

The dielectric properties were investigated by Eastman *et al.* (E 3). Figure VIII-9 depicts the temperature dependence of the small-signal dielectric constant measured along [100]. Both the peak value of the dielectric constant at the Curie temperature of 95 °K, and the shape of the anomaly are strongly dependent not only upon the quality and the degree of clamping of the crystals, but also upon the rate of temperature change during the measurements. The shape of the anomaly depicted in Fig. VIII-9 is obtained with very slow cooling rates. Such a behavior of the dielectric constant is quite unusual for ferroelectrics and was subsequently found to occur in $(NH_4)_2SO_4$ as well (see Section 5). Obviously, it makes no sense to speak of a Curie–Weiss law above the Curie point. The transition appears to be of the first order, although the thermal hysteresis is very small, if it exists at all. The onset of the spontaneous polarization $P_s$ is just as discontinuous and the temperature dependence of $P_s$ is very flat, so that the curve $P_s$ vs. $T$ has a very rectangular appearance, recalling that reported by Takagi and Makita for $NaNH_4$-tartrate tetrahydrate (see Section VII-7). At 93 °K, $P_s = 0.5 \times 10^{-6}$ C/cm² along the pseudo-cubic [100], and the 60 c/s coercive field equals 10 kV/cm. The values of $P_s$ along the pseudo-cubic [110] and [111] directions are found to be $\sqrt{2}$ and $\sqrt{3}$ times smaller, respectively, than that measured along [100], thus establishing that the polar axis of the ferroelectric phase is parallel to one of the original cubic [100] directions.

The situation appears, therefore, to be similar to that encountered in the alums, as far as the symmetry of the low-temperature phase is concerned. Optical studies reveal, again, that unstrained plates become uniformly birefringent below the Curie point and consist most probably of 180° domains undistinguishable with the polarizing microscope. Non-uniformly strained plates, on the other hand, exhibit very complicated domain patterns which are strongly reminiscent of those observed in ferroelectric MASD (see Fig. VIII-5). Thus, although no direct determination of the low-temperature symmetry was carried out by X-ray methods, it seems reasonable to assume that the structure of the polar phase of $(NH_4)_2Cd_2(SO_4)_3$ is monoclinic (space group $P2_1$). Since the room-temperature phase belongs to point group 23, the expected symmetry of the polar phase should indeed be monoclinic.

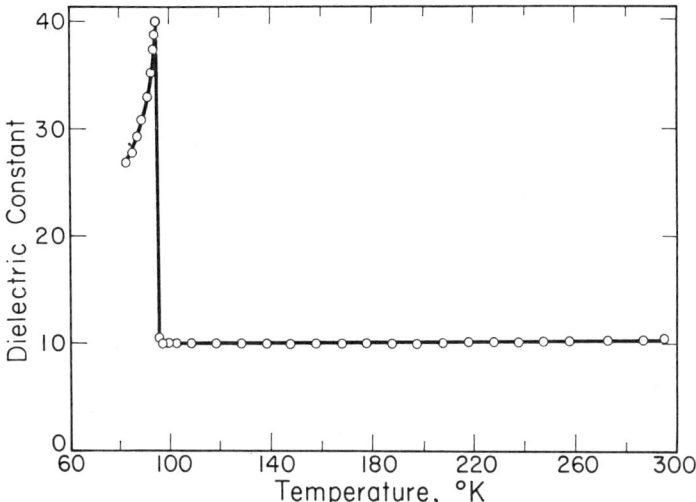

FIG. VIII-9. Temperature dependence of the small-signal dielectric constant of $(NH_4)_2Cd_2(SO_4)_3$ (according to Eastman *et al.* (E 3)).

Dilatometric measurements indicate a linear decrease of the lattice parameter with temperature, the coefficient of linear thermal contraction being $11.4 \times 10^{-6}\,°C^{-1}$. The anomaly at the Curie point consists of a contraction of $0.2\%$ of the linear dimensions for decreasing temperature. Differential thermal analysis reveals a rather large anomaly at the Curie point, but the heat of transition can be only very roughly estimated to be of the order of several hundred calories per mole.

The crystal structure of $(NH_4)_2Cd_2(SO_4)_3$ is not known from direct diffraction studies, but its general features can be inferred from an X-ray analysis of langbeinite itself, $K_2Mg_2(SO_4)_3$, carried out by A. Zemann and J. Zemann (Z 1). According to this analysis, each oxygen atom links a sulfur to a magnesium, in such a way that the magnesium atoms are surrounded by six oxygens in a slightly distorted octahedral arrangement. The structural framework is thus provided by $[Mg_2(SO_4)_3]^{2-}$ blocks, the K ions occupying the interstices between blocks. Zemann (Z 2) reported also the lattice parameters of a number of langbeinite-type compounds.

## 5. Ammonium Sulfate and Ammonium Fluoberyllate

### Ammonium Sulfate $(NH_4)_2SO_4$

The existence of a dielectric anomaly in ammonium sulfate had been known for a number of years (G 7), but ferroelectricity was first recognized in this crystal only in 1956. This discovery is due to Matthias and Remeika (M 5), who reported the Curie temperature at $-49.5\ °C$. Detailed dielectric and thermal investigations have been subsequently carried out by Hoshino et al. (H 7) and will be discussed presently.

At room temperature, $(NH_4)_2SO_4$ is orthorhombic, with the lattice parameters (W 5):

$$a = 7.729\ \text{Å}, \quad b = 10.560\ \text{Å}, \quad c = 5.951\ \text{Å},$$

and space group $Pnam$. This axial designation follows the crystallographic convention for orthorhombic crystals, $b > a > c$. The axial lengths reported above are referred to the so-called "basic cell" of $(NH_4)_2SO_4$, since it has been reported by Okaya et al. (O 2) that this salt may occasionally exhibit super-structure. The room-temperature phase belongs crystallographically to the group of $K_2SO_4$-type crystals, many of which are known to undergo polymorphic transitions at various temperatures.

Below the Curie point, the mirror symmetry disappears and the space group of $(NH_4)_2SO_4$ becomes $Pna2_1$. Thus the symmetry is still orthorhombic, but the $c$ axis has become polar, as was confirmed by the dielectric measurements.

The behavior of the dielectric constant $\varepsilon_c$ (H 7) shows the same characteristics as that of $(NH_4)_2Cd_2(SO_4)_3$ depicted in Fig. VIII-9, except for the fact that the peak at the Curie point is of the order of 160. The dielectric constants $\varepsilon_a$ and $\varepsilon_b$ have a normal temperature dependence, with no anomalies at the Curie point. The shape of the $\varepsilon_c$ anomaly was found to depend upon the rate of temperature change and upon frequency, and in no case was a law of the Curie–Weiss type obeyed. No effect of a d.c. biasing field upon the Curie temperature was detected by Hoshino et al. (H 7), in striking contrast to the very large effects reported by Kamiyoshi (K 5). No explanation for these differing results seems possible at present, but the results of Hoshino et al. are more reasonable in the light of the present experience in ferroelectric behavior.

The temperature dependence of the spontaneous polarization is depicted in Fig. VIII-10(a). This behavior is very similar to that encountered in $(NH_4)_2Cd_2(SO_4)_3$ and described in the preceding section. Below the Curie point, the spontaneous polarization is independent of temperature, the value being $0.45 \times 10^{-6}\ C/cm^2$. The behavior of the 60 c/s coercive field $E_c$ is also quite peculiar: down to about $-60\ °C$, $E_c$ varies very little from a value of about $4\ kV/cm$; then it increases linearly with decreasing temperature, reaching $12\ kV/cm$ at approximately $-73\ °C$.

Specific heat measurements on $(NH_4)_2SO_4$ were carried out by Nitta and Suenaga (N 3), by Shomate (S 2), and by Hoshino et al. (H 7). The anomaly found at the Curie point is of the $\lambda$-type, the heat of transition $\Delta Q$ and the entropy change $\Delta S$ being (H 7):

$$\Delta Q = 0.93 \text{ kcal/mole}; \quad \Delta S = 4.2 \text{ cal/mole } ^\circ\text{C}.$$

Thermal expansion data reveal anomalies of all three lattice parameters at the Curie point, the volume anomaly consisting of an expansion for decreasing temperatures.

It is interesting to note that no isotope effect was detected in ammonium sulfate. The deuterated salt, $(ND_4)_2SO_4$, exhibits the same dielectric and thermal properties as the common hydrogenated compound (H 7). This behavior is

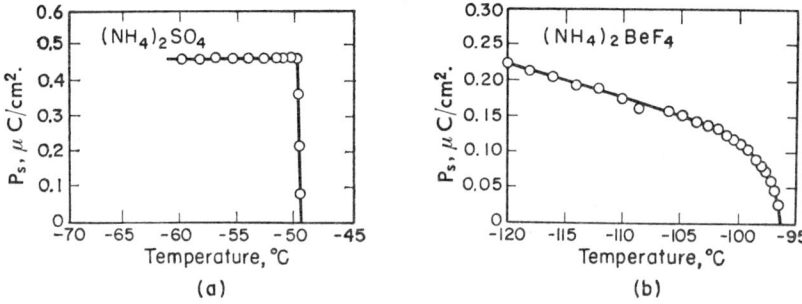

FIG. VIII-10. Temperature dependence of the spontaneous polarization of:
(a) $(NH_4)_2SO_4$; (b) $(NH_4)_2BeF_4$ (according to Hoshino et al. (H 7)).

analogous to that found in the guanidinium ferroelectrics and in the alums, but is quite different from the rather larger isotope effects observed in the "older" hydrogen-bonded ferroelectrics discussed in previous chapters.

### Ammonium Fluoberyllate $(NH_4)_2BeF_4$

It has been known for some time that, in several compounds, the $(SO_4)^{2-}$ groups can be partially or totally replaced by $(BeF_4)^{2-}$ or $(ZnCl_4)^{2-}$ groups. A case in which the former substitution leads to an isomorphous salt has already been encountered in the tri-glycine ferroelectrics (Chapter II). $(NH_4)_2BeF_4$ was originally reported to be isomorphous with $(NH_4)_2SO_4$, with the lattice parameters, at room temperature (M 6):

$$a = 7.49 \text{ Å}, \quad b = 10.39 \text{ Å}, \quad c = 5.89 \text{ Å},$$

and space group $Pnam$. This led Pepinsky and Jona (P 6) to the examination of the dielectric properties of this crystal and to the discovery of its ferroelectricity below $-97$ °C. Thus, in these ammonium salts the substitution of $(SO_4)^{2-}$ by $(BeF_4)^{2-}$ leads to a lowering of the Curie temperature, in contrast to the results found in the tri-glycine ferroelectrics. It was later discovered by Okaya et al. (O 2), however, that $(NH_4)_2SO_4$ and $(NH_4)_2BeF_4$ are *not* truly isomorphous. The lattice parameters reported above refer to the "basic" cell of the fluoberyllate salt, but accurate $X$-ray study revealed superstructuring, the true room-temperature cell having the dimensions:

$$a_{\text{true}} = a, \quad b_{\text{true}} = 2b, \quad c_{\text{true}} = 2c,$$

with space group $Acam$. The polar phase of the fluoberyllate salt is again different from that of the sulfate: while in the latter case, the polar axis lies along $c$, in the former case it lies along $b$. The "basic" ferroelectric cell of the fluoberyllate has the dimensions:

$$a^f \cong 2a, \quad b^f \cong b, \quad c^f \cong c,$$

and space group $Pn2_1a$, while the true ferroelectric cell is multiple, thus (O 2):

$$a^f_{\text{true}} \cong 2a, \quad b^f_{\text{true}} \cong 2b, \quad c^f_{\text{true}} \cong 2c.$$

The dielectric constant $\varepsilon_b$ of $(NH_4)_2BeF_4$ exhibits a sharp anomaly at the Curie temperature $-97$ °C, while the dielectric constants $\varepsilon_a$ and $\varepsilon_c$ are almost temperature independent over the whole range from liquid nitrogen to room temperature. The spontaneous polarization has the value $0.19 \times 10^{-6}$ C/cm² at $-110$ °C; its temperature dependence is depicted in Fig. VIII-10(b). It is probable that the onset of $P_s$ is actually discontinuous at the Curie point, although this does not appear from the figure (see Ref. (P 6)). Double hysteresis loops can be observed in a narrow temperature range above the Curie point, confirming the first order of this transition.

The anomaly of the specific heat, on the other hand, is markedly broad and flat (H 7), quite different from the $\lambda$-type exhibited by $(NH_4)_2SO_4$, and rather reminiscent of the anomalies found at the transitions of the glycine ferroelectrics. Transition energy $\Delta Q$ and entropy change $\Delta S$ are:

$$\Delta Q = 0.31 \text{ kcal/mole}; \quad \Delta S = 1.90 \text{ cal/mole °C}.$$

The volume anomaly is also an expansion, as in the sulfate, but it is interesting to note that the variations of the lattice constants of the two salts through their Curie points are the opposite of one another along any corresponding axis.

The peculiar dielectric and thermal behavior of $(NH_4)_2BeF_4$, and, in particular, the superstructure observed along the $b$ and $c$ axes of the room-temperature phase, led Hoshino et al. (H 7) to advance the hypothesis that this crystal may be antiferroelectric above its Curie point. Below this temperature, the $a$ axis would be antipolar, while the $b$ axis is polar. Attempts at finding a further transition from the antipolar to a non-polar state at higher temperatures (attempts which were carried out in order to test the hypothesis) were unsuccessful. The establishment of the character of the room-temperature phase will therefore have to await completion of the structural studies (H 7). It is also possible that the ferroelectric phase detected below $-97$ °C may be induced by the electric fields applied.

The deuterated compound, $(ND_4)_2BeF_4$, was found to have a Curie temperature only 3° higher than that of $(NH_4)_2BeF_4$, and slightly larger transition energy and entropy ($\Delta Q = 0.38$ kcal/mole; $\Delta S = 2.27$ cal/mole °C), but otherwise exhibits the same shape of dielectric and specific heat anomalies as the hydrogenated compounds. The values of the spontaneous polarizations are also the same in the two crystals (H 7).

## The system $(NH_4)_2SO_4$–$(NH_4)_2BeF_4$

The system of solid solutions of $(NH_4)_2SO_4$ and $(NH_4)_2BeF_4$ was investigated by Hoshino et al. (H 7), with the results depicted in Fig. VIII-11. The Curie points of both end-members are lowered by the additions of the opposite member. Solid solutions are apparently formed for all compositions, but in the range from about 20 to 70 mole % of $(NH_4)_2BeF_4$ a new phase appears (space group $P2_12_12_1$) which undergoes no transition at temperatures above that of liquid nitrogen. It may be interesting to point out that nuclear magnetic resonance data of Burns (B 9) indicate that the second moment of fluorine undergoes a sharp change at the same temperature for all solid solutions in this system irrespective of composition. This temperature lies at $-66$ °C and represents the limit above which hindered rotation of the $(BeF_4)^{2-}$ groups occurs. The fact that this temperature is practically independent of composition seems to suggest that freezing-in of the $(BeF_4)^{2-}$ groups is not directly involved in the ferroelectric transition. Deformation of these and the $NH_4$ groups in ammonium fluoberyllate (and of the $(SO_4)^{2-}$ groups in ammonium sulfate) below the Curie temperature has been established by Blinc and Levstek (B 10) on the basis of infrared data.

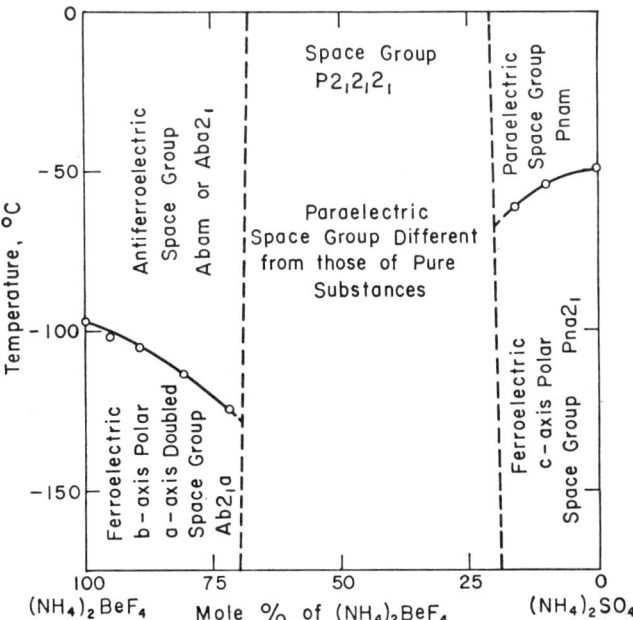

Fig. VIII-11. Phase diagram of the $(NH_4)_2BeF_4$–$(NH_4)_2SO_4$ system (according to Hoshino et al. (H 7)).

## 6. Ammonium and Rubidium Bisulfates

### Ammonium Bisulfate $(NH_4)HSO_4$

Ferroelectricity in $(NH_4)HSO_4$ was discovered by Pepinsky et al. (P 7). The dielectric behavior of this compound is reminiscent of that of Rochelle salt, insofar as spontaneous polarization occurs only in a temperature range delimited by two Curie points, at $-3$ °C and $-119$ °C, respectively. It differs from Rochelle salt, however, in that the symmetry of the phase above $-3$ °C is not equal to that below $-119$ °C. The room-temperature structure is centrosymmetric monoclinic, pseudo-orthorhombic; the ferroelectric phase is (polar) monoclinic, pseudo-orthorhombic; and the low-temperature phase, stable below $-119$ °C, is triclinic. Crystallographic data for all three phases are summarized in Table VIII-4.

Single crystals of $(NH_4)HSO_4$ can best be grown from the melt (melting point 146.9 °C), but good specimens can also be obtained by slow hydrolysis of chloroacetamide in dilute $H_2SO_4$ (molar ratio 1 : 1). The crystals show excellent cleavage in the (001) plane, which is perpendicular to the axis exhibiting reversible polarity (P 7).

The temperature dependence of the dielectric constant $\varepsilon_c$ is strongly reminiscent of that of the mixed crystals $Na(K, NH_4)C_4H_4O_6 \cdot 4H_2O$ with compositions such as to undergo transitions first into phase III and then into phase IV (see Fig. VII-11 and Section VII-7). It is not impossible, in fact, that the properties of $(NH_4)HSO_4$ may be very similar to those of the mixed tartrates mentioned. At $-3$ °C, the anomaly of $\varepsilon_c$ consists in a very pronounced peak which has all the characteristics of a transition of the second order. At $-119$ °C, the anomaly

TABLE VIII-4. CRYSTALLOGRAPHIC DATA FOR $(NH_4)HSO_4$ (ACCORDING TO PEPINSKY et al. (P 7))

(Axial lengths are given in Å units; $Z$ = number of formula units per cell.)

| Room-temperature (centrosymmetric) | $T = -30$ °C· (polar, ferroelectric) | $T = -140$ °C (polar, non-ferroelectric) |
|---|---|---|
| Monoclinic cell | Monoclinic cell | Triclinic cell |
| $a = 14.51$ | $a = 14.26$ | $a = 14.24$ |
| $b = 4.54$ | $b = 4.62$ | $b = 4.56$ |
| $c = 14.90$ | $c = 14.80$ | $c = 15.15$ |
| $\beta = 120°18'$ | $\beta = 121°18'$ | $\beta = 123°24'$ |
| space group $P\,2_1/c$ | space group $Pc$ | $\alpha \approx 90°, \gamma \approx 90°$ |
| $(Z = 8)$ | | space group $P\,1$ |
| Monoclinic, pseudo-orthorhombic cell: | Monoclinic, pseudo-orthorhombic cell: | Triclinic, pseudo-orthorhombic cell: |
| $a = 24.90$ | $a = 24.37$ | $a = 24.43$ |
| $b = 4.54$ | $b = 4.62$ | $b = 4.56$ |
| $c = 14.90$ | $c = 14.80$ | $c = 15.15$ |
| $\beta = 90°18'$ | $\beta \approx 90°$ | $\beta = 91°12'$ |
| space group $B\,2_1/a$ | space group $Ba$ | $\alpha \approx 90°, \gamma \approx 90°$ |
| $(Z = 16)$ | | space group $B\,1$ |

consists merely in a drop of $\varepsilon_c$ by approximately 50%. Ferroelectric hysteresis loops can be observed for fields along [001] in the intermediate temperature range. Figure VIII-12 depicts the thermal behavior of the spontaneous polarization $P_s$ along the $c$ direction: the maximum value attained by $P_s$, just above the lower Curie point, is approximately $0.8 \times 10^{-6}$ C/cm². The 60 c/s coercive field, at the same temperature, is of the order of 1.1 kV/cm. The lower transition exhibits a thermal hysteresis of about 7°, and is therefore probably of the first order.

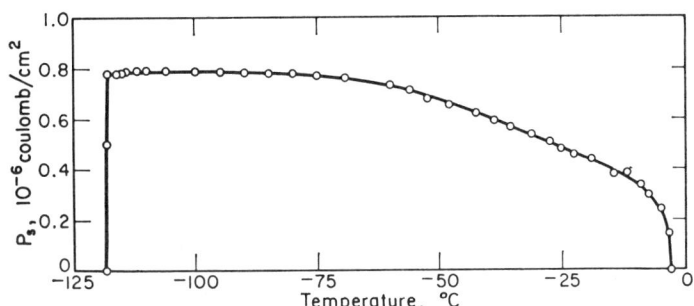

FIG. VIII-12. Temperature dependence of the spontaneous polarization along [001] in $(NH_4)HSO_4$ (according to Pepinsky *et al.* (P 7)).

The specific heat exhibits anomalies at both Curie points: the heat of transition $\Delta Q$ and the entropy change $\Delta S$ are, at $-3$ °C: $\Delta Q \cong 0.12$ kcal/mole, $\Delta S \cong 0.5$ cal/mole °C; at $-119$ °C: $\Delta Q \cong 0.34$ kcal/mole, $\Delta S \cong 2.1$ cal/mole °C.

It may be interesting to point out that tri-ammonium bisulfate, $(NH_4)_3H(SO_4)_2$, also shows dielectric anomalies at two temperatures; namely, $-26$ °C and $-130$ °C, but no evidence for ferroelectricity was found at any temperature (P 7).

### Rubidium Bisulfate, RbHSO$_4$

Recently, Pepinsky and Vedam (P 8) reported ferroelectric activity in RbHSO$_4$. The compound seems to be isomorphous with the ammonium salt discussed above, but, in contrast to this, exhibits only one Curie point at $-15$ °C. The upper (non-polar) phase is monoclinic pseudo-orthorhombic with cell dimensions:

$$a = 14.36 \text{ Å}, \quad b = 4.62 \text{ Å}, \quad c = 14.81 \text{ Å}, \quad \beta = 121.0°$$

space group $P2_1/c$ and $Z = 8$. The crystal remains monoclinic in the polar phase as the twofold screw axis disappears and the space group becomes $Pc$, the $c$ axis being the direction of reversible polarity.

The dielectric constant $\varepsilon_c$ shows a sharp anomaly at the Curie temperature, reminiscent of the behavior exhibited by lithium ammonium tartrate monohydrate (cf. Fig. IX-1). The onset of the spontaneous polarization is not discontinuous, showing that the transition is probably of the second order. At $-170$ °C, $P_s = 0.65 \times 10^{-6}$ C/cm² and the 60 c/s coercive field $E_c = 23$ kV/cm.

## 7. Lithium Hydrazinium Sulfate, $Li(N_2H_5)SO_4$

Pepinsky and co-workers (P 9) reported ferroelectric activity in $Li(N_2H_5)SO_4$ from approximately $-15\,°C$ to $80\,°C$. These temperatures do not indicate phase transitions, but merely the fact that, below $-15\,°C$, the 60 c/s coercive field increases very rapidly, so that good hysteresis loops can no longer be observed at this frequency; while above $80\,°C$ detection of the loops is hindered by the sharply rising electrical conductivity. The structure of $Li(N_2H_5)SO_4$ is ortho-rhombic, with cell dimensions:

$$a = 8.97\ \text{Å}, \quad b = 9.91\ \text{Å}, \quad c = 5.18\ \text{Å},$$

and space group $Pbn2_1$. The density is $\varrho = 1.966\ \text{g cm}^{-3}$, and thus there are four formula units per cell. The ferroelectric axis, as indicated by the space group assignment, lies along [001].

The dielectric constant $\varepsilon_c$ equals 14 at room temperature and is almost in-dependent of temperature. No anomalies can be observed, in fact, either in the dielectric constants or in the specific heat (between $-196\,°C$ and $140\,°C$). Decom-position starts approximately at the latter temperature and is completed, together with mild explosiveness, at about $285\,°C$. Thus, similarly as in the guanidinium ferroelectrics, no Curie point can be detected in this crystal. At room temperature, the value of spontaneous polarization is $0.30 \times 10^{-6}\ \text{C/cm}^2$ and that of the 60 c/s coercive field, $320\ \text{V/cm}$.

## 8. Lithium and Sodium Tri-hydrogen Selenites

### $LiH_3(SeO_3)_2$

It has been reported by Pepinsky and Vedam (P 10) that $LiH_3(SeO_3)_2$, also described as $LiHSeO_3 \cdot H_2SeO_3$, exhibits ferroelectric properties in the temperature range from $-196\,°C$ to $90\,°C$. At and above the latter temperature, dielectric measurements are impeded by the large value of the loss factor. Dielectric and specific-heat data reveal no transition up to the melting point ($110\,°C$). There is therefore no Curie point.

The compound crystallizes in the monoclinic system, with axial dimensions (P 10):

$$a = 6.26\ \text{Å}, \quad b = 7.89\ \text{Å}, \quad c = 5.43\ \text{Å}, \quad \beta = 105.2°,$$

space group $Pn$, and $Z = 4$. Large crystals can be grown easily from aqueous solutions of one mole equivalent of LiOH and two moles of $H_2SeO_3$. The ferro-electric direction is perpendicular to the (001) plane and approximately parallel to the $[40\bar{1}]$ direction (P 10). The crystals exhibit very pronounced cleavage parallel to the ferroelectric direction, and hence extreme care must be exercised in cutting and handling samples meant for dielectric investigations (F 2). The crystal structure has recently been determined by Vedam et al. (V 3) using X-rays, which give no information about the positions of the weakly scattering hydrogen and lithium ions. These ions hold together the framework of the pyramidal selenite ions. From the short oxygen–oxygen distances revealed by the X-ray analysis it appears that the hydrogen bonds are fairly strong, and oriented almost perpendicularly to the polar direction, a situation that is familiar from the cases of $KH_2PO_4$ (Section III-7) and tri-glycine sulfate (Section II-9).

At room temperature, the dielectric constant along the ferroelectric axis is equal to 30, while $\varepsilon_a = 29$ and $\varepsilon_b = 13$. The value of the spontaneous polarization is constant over the entire temperature range mentioned above and equals $15 \times 10^{-6}$ C/cm$^2$. This is the largest value observed, to date, in water-soluble ferro-electrics, and compares favorably even with the value of $26 \times 10^{-6}$ C/cm$^2$ observed in single crystals of tetragonal $BaTiO_3$. The 60 c/s coercive field has the value 1.4 kV/cm at room temperature and becomes as high as about 20 kV/cm at the temperature of liquid nitrogen.

The mechanism of polarization reversal in $LiH_3(SeO_3)_2$ has been studied by Fatuzzo (F 2) with a technique involving electrical pulses. The field dependence of the switching time is qualitatively similar to that found in $BaTiO_3$. At low field strengths, the switching time $t_s$ follows the exponential law:

$$t_s = t_\infty \exp(\alpha/E); \tag{VIII-9}$$

at higher fields (from 5 kV/cm up to 50 kV/cm), it follows a power law of the form:

$$t_s = \text{constant} \times E^{-5/2}. \tag{VIII-10}$$

This latter result is particularly interesting, in that it recalls the power law governing the switching mechanism of $BaTiO_3$ (see Section IV-8) and of thiourea (see Section IX-13) at high field strengths, although in these latter cases the exponent is 3/2 rather than 5/2. Thermal and electrical treatments of $LiH_3(SeO_3)_2$ samples led Fatuzzo to the conclusion that up to fields of 50 kV/cm the switching is determined by the nucleation time, which therefore determines both the exponential and the 5/2 power laws cited above. The experimental evidence for this can be summarized in the following three points (see also Section II-7):

(i) The shape of the switching pulse remains unaffected when Barkhausen noise is present. This seems to imply that the slow and hence the dominating process is that of nucleation.

(ii) Heating of the surface of the sample for a short time causes a decrease of the switching time by a factor of 5–6 in just a few seconds. This short heating cycle cannot affect the bulk of the sample, thus the prompt response of the switching pulse suggests again that nucleating of new domains plays the pre-dominant role. It was pointed out by Fatuzzo that these results are in contrast to those obtained with GASH crystals, where the switching time varies only by 20% after heating periods of the order of 5 min. This evidence indicates that, in GASH, the switching process is primarily determined by domain-wall motions.

(iii) Switching in $LiH_3(SeO_3)_2$ can be accomplished in steps by applying a few short voltage pulses. The single small current pulses observed in such a case add up to the switching transient that is usually obtained by applying a single very long voltage pulse. This confirms once more that the time required for nucleation is longer than that required by domain-wall motion. If this were not the case, in fact, the growing domains would collapse back each time that the field is removed.

Comparison of the above properties of $LiH_3(SeO_3)_2$ with other ferroelectrics led Fatuzzo to suggest that the field dependence of the switching time is probably similar in every ferroelectric material. According to Fatuzzo, at low field strengths the switching time is governed by nucleation: this covers both the range in which the exponential law $t_s = t_\infty \exp(\alpha/E)$ holds, as well as the range in which the switching time follows a power law, $t_s = \text{constant} \times E^{-n}$. At higher field strengths, the switching time is determined by the domain-wall motion, and $1/t_s$ depends linearly upon the applied field $E$.

It is interesting to note (F 2) that $LiH_3(SeO_3)_2$ is, in many respects, very similar to $BaTiO_3$ in its dielectric behavior: it has a large spontaneous polarization, similar switching characteristics, and exhibits a rather pronounced decay when it is subjected to severe pulsing.

## $NaH_3(SeO_3)_2$

Ferroelectric properties were reported in this compound by Pepinsky and Vedam (P 10). The crystal belongs to the monoclinic system and exhibits a Curie point at $-79$ °C. Ferroelectric activity was reported from this temperature down to that of liquid nitrogen. No more information is available about this compound at the time of the present writing.

## BIBLIOGRAPHY

(A 1)  ANGELL, C. L., SHEPPARD, N., YAMAGUCHI, A., SHIMANOUCHI, T., MIYAZAWA, T. and MIZUSHIMA, S., *Trans. Faraday Soc.* **53**, 589 (1957).

(B 1)  BOGLE, G. S., GABRIEL, J. R. and BOTTOMLEY, G. A., *Trans. Faraday. Soc.* **53**, 1058 (1957).

(B 2)  BURNS, G., *Bull. Am. Phys. Soc.* Ser. II, **3**, 371 (1958); Ser. II, **5**, 159 (1960); Ser. II, **5**, 253 (1960).

(B 3)  BENZIE, R. J. and COOKE, A. H., *Proc. Phys. Soc. (London)* **A 63**, 213 (1950).

(B 4)  BLEANY, B. and PENROSE, R. P., *Proc. Phys. Soc. (London)* **60**, 395 (1948).

(B 5)  BOWERS, K. D. and OWEN, J., *Repts. Progr. in Phys.* **18**, 305 (1955).

(B 6)  BACON, G. E. and GARDNER, W. E., *Proc. Roy. Soc. (London)* **A 246**, 78 (1958).

(B 7)  BURNS, G., *J. Chem. Phys.* **32**, 1585 (1960).

(B 8)  BENRATH, A. and THIEMANN, W., *Z. anorg. Chem.* **208**, 177 (1932).

(B 9)  BURNS, G., Private communication (1960).

(B 10) BLINC, R. and LEVSTEK, I., *J. Phys. Chem. Solids* **12**, 295 (1960).

(C 1)  CHAPELLE, J. and CHOLLOT, B., *Compt. rend.* **244**, 1185 (1957).

(C 2)  CHYNOWETH, A. G., *Phys. Rev.* **102**, 1021 (1956).

(C 3)  COUTURE, L., JACQUINOT, P. and TSUJIKAWA, I., Conference on low-temperature crystallography, Oxford, April 1956, *Brit. J. Appl. Phys.* **7**, 434 (1956).

(D 1)  DANIELS, J. M. and WESEMEYER, H., *Can. J. Phys.* **36**, 144 (1958).

(D 2)  DUNNE, T., Unpublished work (1959).

(D 3)  DROUGARD, M. E., LANDAUER, R. and YOUNG, D. R., *Phys. Rev.* **98**, 1010 (1955).

(D 4)  DEVONSHIRE, A., Theory of Ferroelectrics, *Phil. Mag.* Suppl. **3**, 85 (1954).
(E 1)  EZHKOVA, Z. I., ZHDANOV, G. S. and UMANSKII, M. M., *Kristallografiya* **3**, 231 (1958).
(E 2)  EISENSTEIN, J., *Revs. Modern Phys.* **24**, 74 (1952).
(E 3)  EASTMAN, D., JONA, F. and PEPINSKY, R., Unpublished work (1957).
(F 1)  FREYMANN, M., ROLLAND, M. T. and FREYMANN, R., *Compt. rend.* **232**, 2312 (1951).
(F 2)  FATUZZO, E., *Helv. Phys. Acta* **32**, 302 (1959); **33**, 21 (1960).
(F 3)  FATUZZO, E., *Helv. Phys. Acta* **33**, 429 (1960).
(G 1)  GELLER, S. and BOOTH, D. P., *Z. Krist.* **111**, 117 (1959).
(G 2)  GELLER, S., *Z. Krist.* **114**, 148 (1960).
(G 3)  GARIF'IANOV, N. S., *Zhur. Eksptl. i Teoret. Fiz.* **35**, 612 (1958).
(G 4)  GUILLIEN, R., *Compt. rend.* **209**, 21 (1939); **213**, 991 (1941); *Cahiers phys.* **11**, 17 (1942).
(G 5)  GRANIER, J., *Les Diélectriques*, Dunod, Paris (1948).
(G 6)  GRIFFITHS, J. H. E. and POWELL, J. A., *Proc. Phys. Soc.* (*London*) **A 65**, 289 (1952).
(G 7)  GUILLIEN, R., *Compt. rend.* **208**, 980 (1939); *Ann. Phys.* **17**, 334 (1942).
(H 1)  HOLDEN, A. N., MATTHIAS, B. T., MERZ, W. J. and REMEIKA, J. P., *Phys. Rev.* **98**, 546 (1955).
(H 2)  HAUSSÜHL, S., *Z. Krist.* **111**, 321 (1959).
(H 3)  HOLDEN, A. N., MERZ, W. J., REMEIKA, J. P. and MATTHIAS, B. T., *Phys. Rev.* **101**, 962 (1956).
(H 4)  HAUSSÜHL, S. and TROST, F., *Z. Naturforsch.* **14a**, 437 (1959).
(H 5)  HOSHINO, S., MITSUI, T., JONA, F. and PEPINSKY, R., *Phys. Rev.* **107**, 1255 (1957).
(H 6)  HAUSSÜHL, S., *Fortschr. Mineral.* **36**, 75 (1958).
(H 7)  HOSHINO, S., VEDAM, K., OKAYA, Y. and PEPINSKY, R., *Phys. Rev.* **112**, 405 (1958).
(J 1)  JONA, F., MITSUI, T. and PEPINSKY, R., Unpublished work (1957).
(J 2)  JONA, F., VEDAM, K., MITSUI, T. and PEPINSKY, R., Unpublished work (1957).
(J 3)  JOHNSTON, H. J., HU, J. H. and HORTON, W. S., *J. Am. Chem. Soc.* **75**, 3922 (1953).
(J 4)  JONA, F. and SHIRANE, G., *Phys. Rev.* **117**, 139 (1960).
(J 5)  JONA, F. and PEPINSKY, R., *Phys. Rev.* **103**, 1126 (1956).
(K 1)  KRAUS, D. L. and NUTTING, G. C., *J. Chem. Phys.* **9**, 133 (1941).
(K 2)  KLUG, H. P. and ALEXANDER, L., *J. Am. Chem. Soc.* **62**, 2993 (1940); **64**, 1819 (1942).
(K 3)  KLUG, H. P., *J. Am. Chem. Soc.* **62**, 1492 (1940); **62**, 2992 (1940).
(K 4)  KLUG, H. P. and KIEFFER, G. L., *J. Am. Chem. Soc.* **65**, 2071 (1943).
(K 5)  KAMIYOSHI, K., *J. Chem. Phys.* **26**, 218 (1957).
(L 1)  LIPSON, H. and BEEVERS, C. A., *Proc. Roy. Soc.* (*London*) **A 148**, 664 (1935).
(L 2)  LIPSON, H., *Phil. Mag.* (7) **19**, 887 (1935).
(L 3)  LIPSON, H., *Nature* **135**, 912 (1935).
(L 4)  LIPSON, H., *Proc. Roy. Soc.* (*London*) **A 151**, 347 (1935).
(L 5)  LAFON. J., *Compt. rend.* **247**, 2120 (1958).
(M 1)  MERZ, W. J., REMEIKA, J. P., HOLDEN, A. N. and MATTHIAS, B. T., *Phys. Rev.* **99** 626 (1955).
(M 2)  MERZ, W. J., *Phys. Rev.* **103**, 565 (1956).
(M 3)  McCALL, D. W., *J. Chem. Phys.* **26**, 706 (1957).
(M 4)  MERZ, W. J., *Phys. Rev.* **91**, 513 (1953).
(M 5)  MATTHIAS, B. T. and REMEIKA, J. P., *Phys. Rev.* **103**, 262 (1956).
(M 6)  MUKHERJEE, P. L., *Indian J. Phys.* **18**, 148 (1944).
(N 1)  NISHIOKA, A. and SEKIKAWA, K., *J. Phys. Soc. Japan* **13**, 1237 (1958).
(N 2)  NAKAMURA, T., *J. Phys. Soc. Japan* **14**, 1022 (1959).
(N 3)  NITTA, I. and SUENAGA, K., *Bull. Chem. Soc. Japan* **13**, 36 (1938).
(O 1)  OKAYA, Y., AHMED, M. S., PEPINSKY, R. and VAND, V., *Z. Krist.* **109**, 367 (1957).
(O 2)  OKAYA, Y., VEDAM, K. and PEPINSKY, R., *Acta Cryst.* **11**, 307 (1958).
(P 1)  PEARSON, G. L. and FELDMANN, W. L., *J. Phys. Chem. Solids* **9**, 28 (1959).
(P 2)  PRUTTON, M., *Proc. Phys. Soc.* (*London*) **B 70**, 702 (1957).
(P 3)  PRUTTON, M., *Proc. Phys. Soc.* (*London*) **B 70**, 1064 (1957).
(P 4)  PEPINSKY, R., JONA, F. and SHIRANE, G., *Phys. Rev.* **102**, 1181 (1956).
(P 5)  PEPINSKY, R., VEDAM, K. and JONA, F., Unpublished work (1957).

(P 6)   PEPINSKY, R. and JONA, F., *Phys. Rev.* **105**, 344 (1957).
(P 7)·  PEPINSKY, R., VEDAM, K., HOSHINO, S. and OKAYA, Y., *Phys. Rev.* **111**, 1508 (1958).
(P 8)   PEPINSKY, R. and VEDAM, K., *Phys. Rev.* **117**, 1502 (1960).
(P 9)   PEPINSKY, R., VEDAM, K., OKAYA, Y. and HOSHINO, S., *Phys. Rev.* **111**, 1467 (1958).
(P 10)  PEPINSKY, R. and VEDAM, K., *Phys. Rev.* **114**, 1217 (1959).
(R 1)   REMEIKA, J. P. and MERZ, W. J., *Phys. Rev.* **102**, 295 (1956).
(R 2)   REZ, I. S., BELIAKOVA, IU. A. and VARFOLOMEEVA, L. A., cited by Z. I. EZHKOVA,
        G. S. ZHDANOV and M. M. UMANSKII, *Kristallografiya* **3**, 231 (1958).
(S 1)   SPENCE, R. D. and MULLER, J., *J. Chem. Phys.* **26**, 706 (1957).
(S 2)   SHOMATE, C. H., *J. Am. Chem. Soc.* **67**, 1096 (1945).
(T 1)   TRIEBWASSER, S., *I.B.M. J. Research Developm.* **2**, 212 (1958).
(V 1)   VARFOLOMEEVA, L. A., ZHADNOV, G. S. and UMANSKII, M. M., *Kristallografiya* **3**,
        368 (1959).
(V 2)   VERES, M., *Compt. rend.* **158**, 39 (1914).
(V 3)   VEDAM, K., OKAYA, Y. and PEPINSKY, R., *Phys. Rev.* **119**, 1252 (1960).
(W 1)   WOOD, E., *Acta Cryst.* **9**, 618 (1956).
(W 2)   WIEDER, H. H., *Proc. I.R.E.* **45**, 1094 (1957).
(W 3)   WHITMER, C. A. and WEIDNER, R. T., *Phys. Rev.* **84**, 159 (1951).
(W 4)   WIEDER, H. H., *J. Appl. Phys.* **30**, 1010 (1959).
(W 5)   WYCKOFF, R. W. G., *Crystal Structures*, vol. 2, ch. 8, Table page 47, Interscience, New
        York (1951).
(Z 1)   ZEMANN, A. and ZEMANN, J., *Acta Cryst.* **10**, 409 (1957).
(Z 2)   ZEMANN, J., *Fortschr. Mineral.* **35**, 155 (1958).

CHAPTER IX

# MISCELLANEOUS FERROELECTRICS

## 1. Introduction

The present chapter is devoted to the description of the ferroelectric compounds that have not been considered previously in this book. The ferroelectric tartrates other than Rochelle salt are discussed here because they are only chemically and structurally related to, but not isomorphous with Rochelle salt. A chemical relationship exists also with the monocarboxylic acid derivatives (P 2) that have numerous representatives in the ferroelectric family, such as the glycine compounds, and generally the acetates or substituted acetates, such as ammonium chloro-acetate and dicalcium strontium propionate. In this and in almost all other chemical species that we are going to encounter in the present chapter, the number of ferroelectrics is probably going to increase in the future. From a crystal–chemical point of view, the most interesting case is that of the molecular crystal of thiourea, whose ferroelectricity may represent the closest example to a realization of the classical picture of rigid, semirotatable macrodipoles in a three-dimensional lattice. From a structural point of view, the most interesting among the ferro-electrics discussed in the present chapter is sodium nitrite, whose crystal structure is very simple and thus leads to a very reasonable model of the ferroelectric phenomenon.

## 2. Ferroelectric Tartrates

### Lithium Ammonium Tartrate Monohydrate

In a search for ferroelectric tartrates other than Rochelle salt, Matthias and Hulm (M 1) and, independently, Merz (M 2) discovered ferroelectric activity at low temperatures in $LiNH_4C_4H_4O_6 \cdot H_2O$ (abbreviated LAT). The possibility of ferroelectricity in this compound had been previously pointed out by Mason (M 3) on the basis of piezoelectric investigations carried out by Scholz (S 1). The Curie point of LAT was reported to be at 98.5 °K by Matthias and Hulm (M 1), at 106 °K by Merz (M 2), and later confirmed to be at 98 °K by Jona and Pepinsky (J 1). The symmetry of the room-temperature phase is orthorhombic, as is that of non-polar Rochelle salt, but the two crystals are not isomorphous. Table IX-1 summarizes the crystallographic and dielectric data of LAT, Rochelle salt and $LiRbC_4H_4O_6 \cdot H_2O$ (LRT), the latter crystal not being ferroelectric down to liquid helium temperatures. The symmetry of the polar phase of LAT is probably

TABLE IX-1. COMPARISON BETWEEN LITHIUM AMMONIUM TARTRATE MONOHYDRATE (LAT), LITHIUM RUBIDIUM TARTRATE MONOHYDRATE (LRT), AND ROCHELLE SALT (RS)

| Crys. | Lattice constants (Å) | | | Space group | Properties |
|---|---|---|---|---|---|
| | $a$ | $b$ | $c$ | | |
| LAT | 7.86 | 14.60 | 6.47 | $P2_12_12$ ($Z = 4$) | Ferroelectric below 98 °K, $b$ axis polar |
| LRT | 7.87 | 14.68 | 6.35 | $P2_12_12$ ($Z = 4$) | Not ferroelectric |
| RS | 11.93 | 14.30 | 6.17 | $P2_12_12$ ($Z = 4$) | Ferroelectric between $-18$ °C and $+24$ °C, $a$ axis polar |

monoclinic, since neutron-diffraction experiments have shown (P 1) that the two-fold screw axis along $a$ (and hence also the twofold axis along $c$) disappears below the Curie temperature. The dielectric studies confirm that the polar axis of the ferroelectric phase is parallel to the orthorhombic $b$ axis.

The dielectric constants $\varepsilon_a$, $\varepsilon_b$ and $\varepsilon_c$ are approximately equal to one another at room temperature ($\sim 8$), but their temperature dependences are quite different. Fig. IX-1 depicts $\varepsilon_b$ as a function of temperature (M 2). $\varepsilon_a$ shows a smaller anomaly at the Curie point; it slowly increases, with decreasing temperature, up to a value of approximately 12 at the transition point, then drops very rapidly (J 1). $\varepsilon_c$, on the other hand, is practically independent of temperature. Fig. IX-1 shows that one can hardly speak of a Curie–Weiss behavior of $\varepsilon_b$. Attempts to fit the curve with a law such as:

$$\varepsilon = \varepsilon_0 + \frac{C}{T - T_0} \tag{IX-1}$$

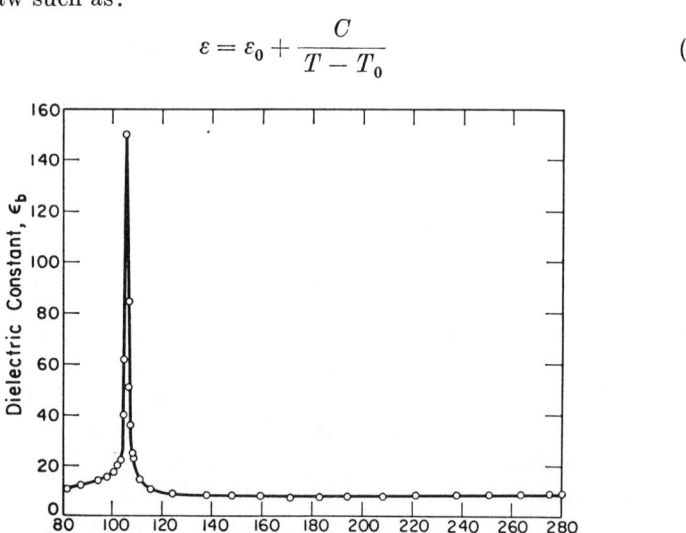

FIG. IX-1. Temperature dependence of the dielectric constant $\varepsilon_b$ of lithium ammonium tartrate monohydrate (according to Merz (M 2)).

yield the values $\varepsilon_0 = 8.7$, $C = 37°$, $T_0 = 93.8\ °\mathrm{K}$, but agreement between the experimental curve and Eq. (IX-1) is limited to a range of about 15° above the Curie point (J 1).

Deuteration of LAT was attempted by recrystallizing the compound in highly concentrated $D_2O$. Since only eight out of the ten hydrogens contained in LAT can be replaced by deuterium, the density of the crystals grown from $D_2O$ indicated the formula $LiND_4C_4H_4(D_{1.25}H_{0.75})O_6 \cdot (D_{1.25}H_{0.75})O$. For these crystals, the Curie temperature was found to lie at 95 °K, thus approximately 3° lower than that of non-deuterated LAT (J 1).

The spontaneous polarization of LAT is equal to $0.22 \times 10^{-6}\ \mathrm{C/cm^2}$ at 78 °K (M 1), thus almost the same as the maximum value of the same quantity measured in Rochelle salt. Specific heat measurements of LAT through the Curie point are indicative of an extremely small anomaly at the critical temperature (D 1). Dilatometric measurements show that the volume anomaly is an expansion for decreasing temperature (J 1).

The structure of LAT at room temperature has been determined by Vernon and Pepinsky (V 1) with X-ray diffraction techniques. The structure was not sufficiently refined for the elucidation of the ferroelectric mechanism, but the co-ordinates of all atoms (with exception of lithium and the hydrogens) were determined with an accuracy that allows interesting comparisons with non-isomorphous Rochelle salt. It turns out that the steric arrangement of the tartrate molecules in the latter crystal and in LAT are essentially identical. The bond lengths between C and O, and C and C atoms are very similar in the two compounds. The most important difference is the fact that Rochelle salt has four molecules of water of crystallization, while LAT has only one. Thus, in LAT, the absence of three water molecules causes a shrinkage of the distances between neighboring tartrate groups along the $a$ axis, as compared to Rochelle salt (see Table IX-1), but the axial lengths along $b$ and $c$ remain very similar in the two crystals. A more detailed discussion of the LAT structure can be found in the review article by Shirane et al. (S 2). Although the hydrogen positions are unknown, the X-ray results seem to indicate a chain of hydrogen bonds connecting the two crystallographically non-equivalent $NH_4$ ions by way of oxygen atoms approximately along the $b$ direction. This chain may conceivably account for a net polarization in the direction of the $b$ axis and perhaps also for a very small polarization in the direction of the $a$ axis (V 1), in accordance with the small anomaly exhibited by $\varepsilon_a$ at the Curie point.

### Lithium Thallium Tartrate Monohydrate

$LiTlC_4H_4O_6 \cdot H_2O$ (abbreviated LTT) is reportedly isomorphous with LAT, as is the LiRb salt. In contrast to the latter, however, the LiTl salt was found to be ferroelectric, namely by Matthias and Hulm (M 1). The Curie temperature is the lowest known in any ferroelectric, at present, namely 10 °K. The surprising point, however, is not so much the fact that replacement of $NH_4$ by the nearly isosteric Tl ion causes such a drastic decrease of the Curie temperature, but rather the fact that in LTT the ferroelectric axis is parallel to the orthorhombic $a$ axis, whereas in LAT it is parallel to the $b$ direction. This situation is somewhat reminiscent of that occurring in Rochelle salt and ammonium-Rochelle salt (see Section VII-7), and also in ammonium sulfate and ammonium fluoberyllate (see Section VIII-5), but in the former case the reversibility of the spontaneous polarization has different character, and in the latter case the two crystals involved are not completely isomorphous in their non-polar phases.

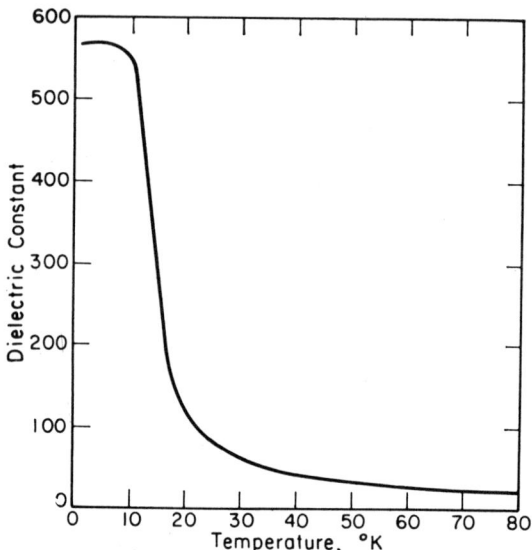

FIG. IX-2. Temperature dependence of the dielectric constant $\varepsilon_a$ of lithium thallium tartrate monohydrate (according to Matthias and Hulm (M 1)).

The temperature dependence of the dielectric constant $\varepsilon_a$ of LTT is depicted in Fig. IX-2. The maximum at the Curie point is very flat; in the ferroelectric phase, $\varepsilon_a$ seems to tend to a steady value as the absolute zero is approached. This behavior seems to be characteristic of phase transitions occurring at very low temperatures (the other example is the perovskite-type $KTaO_3$, discussed in Section V-2), as is the fact that deviations from the Curie–Weiss law can be found within a certain range above the Curie point. As compared to LAT, however, LTT exhibits a very nice Curie–Weiss behavior in the quantity $\varepsilon_a$. The spontaneous polarization equals $0.14 \times 10^{-6}$ C/cm² at 1.3 °K, and the 60 c/s coercive field is only 120 V/cm (M 1). No information is available about the symmetry of the low-temperature phase. A study of the system of solid solutions of LTT and LAT would probably prove to be quite interesting.

### 3. Di-glycine Nitrate

Pepinsky et al. (P 2) reported a Curie point at $-67$ °C in di-glycine nitrate, $(NH_2CH_2COOH)_2 \cdot HNO_3$. Good crystals of this compound can be obtained from aqueous solutions of stoichiometric amounts of the components. Crystallographic data were reported as follows (P 2): monoclinic symmetry, with lattice parameters

$$a = 9.50 \text{ Å}, \quad b = 5.11 \text{ Å}, \quad c = 9.35 \text{ Å}, \quad \beta = 98.8°,$$

space group $P2_1/a$, density 1.581 g cm$^{-3}$, and two formula units per cell. Below the Curie temperature, the twofold screw axis disappears, and the space group is $Pa$. The direction of the reversible polarization lies along the [101] direction.

It is, in fact, only the dielectric constant along [101] that exhibits an anomaly at the Curie point, while the dielectric constants along [010] and [10$\bar{1}$] are practically independent of temperature. The spontaneous polarization along [101] shows a thermal behavior typical of a transition of the second order. At $-77$ °C, $P_s = 0.60 \times 10^{-6}$ C/cm$^2$, and at $-190$ °C, $P_s = 1.5 \times 10^{-6}$ C/cm$^2$. The 60 c/s coercive field rises accordingly with decreasing temperature, being 400 V/cm at $-77$ °C and about 1.5 kV/cm at $-120$ °C. The specific heat exhibits a flat anomaly extending from $-65$ °C to $-80$ °C and very similar to that found in tri-glycine sulfate and fluoberyllate. The heat of transition was estimated to be $\Delta Q = 0.15$ kcal/mole and the entropy change $\Delta S = 0.74$ cal/mole °C.

### 4. Glycine Silver Nitrate

Ferroelectricity in $NH_2CH_2COOH \cdot AgNO_3$ was reported by Pepinsky et al. (P 3). The Curie point lies at $-55$ °C. The room temperature structure is monoclinic with cell constants:

$$a = 5.53 \text{ Å}, \quad b = 19.58 \text{ Å}, \quad c = 5.51 \text{ Å}, \quad \beta = 100°,$$

space group $P2_1/a$, density 2.83 g cm$^{-3}$, and four formula units per cell. The crystals (which can be grown from aqueous solutions of stoichiometric amounts of the components in the dark), exhibit perfect cleavage perpendicular to the $a$ axis and good cleavage perpendicular to $c$. The ferroelectric direction is that of the $b$ axis, the dielectric constant $\varepsilon_b$ being anomalous at the Curie point. The onset of the spontaneous polarization has the character of a second-order transition; at $-195$ °C, $P_s = 0.55 \times 10^{-6}$ C/cm$^2$, and the 60 c/s coercive field $E_c = 740$ V/cm. Differential thermal analysis indicates that the specific heat anomaly at the Curie point is probably extremely small.

Partial substitution of Ag with Tl or Li results in crystals that are still ferroelectric, but the Curie temperature of these mixed crystals is raised with respect to that of the pure silver nitrate salt (P 2), (P 4). The dielectric data of two such mixed compounds are listed in Table IX-2 together with those of related ferroelectrics.

### 5. Di-glycine Manganous Chloride Dihydrate

Ferroelectric properties in $(NH_2CH_2COOH)_2 \cdot MnCl_2 \cdot 2H_2O$ were discovered by Pepinsky and co-workers (P 4) in the temperature range from $-195$ °C to $+55$ °C. The latter temperature is not a Curie point, but rather marks the beginning of a dehydration process that destroys the compound irreversibly. The struc-

TABLE IX-2. FERROELECTRIC CRYSTALS OF THE SUBSTITUTED-ACETATE CLASS
$T_c$ = Curie point ([$hkl$] = direction of the reversible spontaneous polarization with respect to the axial choice given in the text; $P_s$ = value of spontaneous polarization at the temperature given. The last compound listed is tetragonal, all others are monoclinic.)

| Substance | $T_c$ (°C) | [$hkl$] | $P_s$ ($10^{-6}$ C/cm²) | Reference |
|---|---|---|---|---|
| Tri-glycine sulfate $(NH_2CH_2COOH)_3 \cdot H_2SO_4$ | 49 | [010] | 2.8 at 20 °C | see Chap. II |
| Tri-glycine selenate $(NH_2CH_2COOH)_3 \cdot H_2SeO_4$ | 22 | [010] | 3.2 at 0 °C | see Chap. II |
| Tri-glycine fluoberyllate $(NH_2CH_2COOH)_3 \cdot H_2BeF_4$ | 70 | [010] | 3.2 at 20 °C | see Chap. II |
| Di-glycine nitrate $(NH_2CH_2COOH)_2 \cdot HNO_3$ | $-67$ | [101] | 1.5 at $-190$ °C | P 2 |
| Glycine silver nitrate $NH_2CH_2COOH \cdot AgNO_3$ | $-55$ | [010] | 0.55 at $-195$ °C | P 3 |
| $NH_2CH_2COOH \cdot (Ag_{0.82}Tl_{0.18})NO_3$ | $-38$ | [010] | 0.17 at $-48$ °C | P 2 |
| $NH_2CH_2COOH \cdot (Ag_{0.82}Li_{0.18})NO_3$ | $-38$ | [010] | 0.20 at $-48$ °C | P 2 |
| Di-glycine $\cdot$ MnCl$_2$ $\cdot$ dihydrate $(NH_2CH_2COOH)_2 \cdot MnCl_2 \cdot 2 H_2O$ | – | [010] | 1.3 at 25 °C | P 4 |
| Ammonium monochloroacetate $ClCH_2COONH_4$ | $-150$ | [$10\bar{1}$] | 0.12 at $-170$ °C | P 5, P 4 |
| Di-calcium strontium propionate $Ca_2Sr(CH_3CH_2COO)_6$ | 8.5 | [001] | 0.3 at $-45$ °C | M 4 |

ture is monoclinic, with the following values of lattice parameters, at room temperature:
$$a = 9.96 \text{ Å}, \quad b = 8.53 \text{ Å}, \quad c = 6.86 \text{ Å}, \quad \beta = 107°,$$

space group $P2_1$, density 1.875 g cm$^{-3}$, and two formula units per cell. The ferroelectric direction is therefore that of the monoclinic $b$ axis. The dielectric constants $\varepsilon_a = 6.6$, $\varepsilon_b = 8.1$, $\varepsilon_c = 7.4$ are practically independent of temperature, and so is the spontaneous polarization $P_s = 1.3 \times 10^{-6}$ C/cm². The 60 c/s coercive field increases from 6 kV/cm at room temperature to about 20 kV/cm at $-100$ °C. It has not been possible to obtain isomorphous compounds by substituting Br for Cl or any suitable metal for Mn (P 4).

## 6. Ammonium Monochloroacetate

It has been reported by Pepinsky et al. (P 5) that the compound $ClCH_2COONH_4$ crystallizes in two crystallographically different forms at room temperature. One form is orthorhombic and non-ferroelectric, the other is monoclinic, with axial lengths:
$$a = 8.42 \text{ Å}, \quad b = 11.63 \text{ Å}, \quad c = 9.82 \text{ Å}, \quad \beta = 110°,$$

space group $C2/c$, density 1.558 g cm$^{-3}$ and eight formula units per cell. This monoclinic modification undergoes a ferroelectric transition at $-150$ °C, the space group of the polar phase being probably $Cc$ ($P$ 4). The direction of the reversible spontaneous polarization is [10$\bar{1}$] and its value is $0.12 \times 10^{-6}$ C/cm$^2$ at $-170$ °C, where the 60 c/s coercive field $E_c = 10$ kV/cm.

## 7. Di-calcium Strontium Propionate

The ferroelectric properties of $Ca_2Sr(CH_3CH_2COO)_6$ have been reported by Matthias and Remeika (M 4). The Curie temperature is at 8.5 °C. Crystals of this compound can be obtained by slow evaporation of a solution of the two metal oxides in propionic acid. The room-temperature phase is tetragonal, with lattice parameters (O 1):

$$a = 12.48 \text{ Å}, \quad c = 17.28 \text{ Å},$$

space group $P4_12_12$ or $P4_32_12$, density $\varrho = 1.48$ g cm$^{-3}$, and $Z = 4$. Ferroelectric activity is observed, below the Curie temperature, in the direction of the tetragonal $c$ axis, but no crystallographic information is available about the polar phase. Measurements of the dielectric constant have not been published, but the onset of the spontaneous polarization at the Curie point (M 4) seems to indicate that the transition is of the second order. At $-45$ °C, $P_s = 0.3 \times 10^{-6}$ C/cm$^2$ and the 60 c/s coercive field $E_c = 13$ kV/cm.

The crystal structure of the room-temperature phase has been studied with $X$-ray diffraction techniques by Orioli and Pieroni (O 1), and is very closely related to that of the isomorphous di-calcium lead propionate, $Ca_2Pb(CH_3CH_2COO)_6$ (F 1). The Sr (or Pb) atoms appear to be distributed on a slightly distorted diamond lattice; the Ca atoms are arranged almost tetrahedrally around the Sr (or Pb) atoms, but each Ca is not situated exactly midway between two neighboring metal atoms.

Despite the very close structural similarity between the Sr and the Pb salts, it appears that the latter does not become ferroelectric (M 4). It may also be interesting to point out that the Ba salt of this series, $Ca_2Ba(CH_3CH_2COO)_6$, is cubic at room temperature ($a = 18.20$ Å) (N 1) and undergoes a transition at $-6$ °C, but does not become ferroelectric in the low-temperature phase (R 1). Extensive studies of the specific heat and volume anomaly of this Ba salt have been carried out by Seki et al. (S 3), (S 4). Cubic mixed crystals of $Ca_2(Ba, Pb)(CH_3CH_2COO)_6$ can be grown up to a composition of about 14.5 mole % of the Pb salt. Beyond 44.4 mole % of the latter salt, tetragonal mixed crystals are obtained, while at any concentration between these two limits no solid solutions are formed (S 3). No dielectric data of this series of mixed crystals are available at the present time.

## 8. Tetramethylammonium Trichloromercurate

Ferroelectric activity has recently been reported in $[N(CH_3)_4] \cdot HgCl_3$ by Fatuzzo and Nitsche (F 2). No Curie point can be detected, as the compound decomposes above 200 °C. Ferroelectricity is found between 200 °C and $-80$ °C where the 60 c/s coercive field becomes so large as to hinder observation of hysteresis loops. The compound is reportedly orthorhombic ($a = 8.68$ Å, $b = 15.75$ Å, $c = 7.69$ Å), but it is not known which one of the crystallographic axes is polar. At room temperature, the dielectric constant along the polar direction is equal to 25, the spontaneous polarization $P_s = 1.0 \times 10^{-6}$ C/cm² and the 60 c/s coercive field $E_c = 3$ kV/cm. Replacement of Cl by Br results in crystals which are also ferroelectric.

### 9. Potassium Ferrocyanide Trihydrate and Isomorphous Crystals

Most peculiar dielectric and optical behavior is exhibited by crystals of $K_4Fe(CN)_6 \cdot 3H_2O$ (W 1), (W 7). At room temperature, crystals can be grown that belong to either of two main modifications, tetragonal or monoclinic. The monoclinic modification is pseudo-tetragonal, with the monoclinic $b$ axis as the pseudo-tetragonal axis. Complicated lamellar twinning can therefore occur not only among two or more monoclinic individuals but also among monoclinic and tetragonal ones, giving rise to a large variety of optical anomalies (T 2). Both the monoclinic and the tetragonal modification show perfect cleavage in the (010) and (001) plane, respectively.

The monoclinic modification, with lattice parameters (T 2), (K 1)

$$a = 9.38 \text{ Å}, \quad b = 16.84 \text{ Å}, \quad c = 9.40 \text{ Å}, \quad \beta = 90° \pm 3'$$

and space group $C2/c$, has a Curie point at $-24.5$ °C, where it looses its center of symmetry, the space group becoming $Cc$ (cf. ammonium monochloroacetate, Section IX-6). The dielectric constant along $[10\bar{1}]$ exhibits a pronounced anomaly at the Curie point, while the dielectric constants measured along [101] and [010] show only minor anomalies. The spontaneous polarization lies along the $[10\bar{1}]$ direction, being about $1.4 \times 10^{-6}$ C/cm² at $-40$ °C; the coercive field increases rapidly with decreasing temperature so that saturated hysteresis loops can only be obtained in a temperature range of about 20° below the Curie point.

The tetragonal modification, with lattice parameters (T 2):

$$a = 9.41 \text{ Å}, \quad c = 33.67 \text{ Å}$$

and space group $I4_1/a$, is apparently only metastable at room temperature. When cooled below $-55$ °C, it transforms irreversibly into the monoclinic ferroelectric phase that has the Curie point at $-24.5$ °C.

Twinned crystals of both modifications exhibit very complicated dielectric behavior. An example of what one can observe in such crystals is illustrated in Fig. IX-3 from earlier work of Waku et al. (W 1). Pronounced dielectric anomalies occur along [101] and $[10\bar{1}]$, and minor anomalies along [010]. The dielectric constant along [010] exhibits a very small peak at about $-22$ °C, the constants along [101] and $[10\bar{1}]$ show marked peaks at $-22$ °C, $-26$ °C, $-33$ °C, $-43$ °C and broad maxima at $-80$ °C. Ferroelectric hysteresis loops can be observed below $-22$ °C for fields applied along the [101] and $[10\bar{1}]$ directions. The spontaneous polarization of these twinned crystals is generally smaller than that of a single monoclinic crystal reported above, and often undergoes an abrupt change at about $-33\,°C$, corresponding to the third peak in the dielectric constant along [101] (see Fig. IX-3). The explanation for these various anomalies in twinned crystals is probably to be sought in the fact that the various twins are

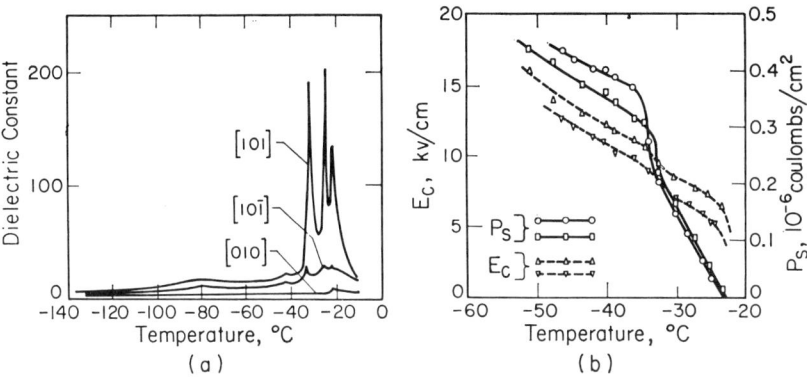

FIG. IX-3. Dielectric properties of potassium ferrocyanide trihydrate. (a) Dielectric constants along [010], [101] and [10$\bar{1}$] as functions of temperature; (b) Spontaneous polarization and coercive field along [101] and [10$\bar{1}$] as functions of temperature. (according to Waku *et al.* (W 1)).

subjected to different internal stresses which in turn may affect their stability and transformation temperature. There is indeed some experimental evidence for the fact that the stability of each phase and the transition temperatures of potassium ferrocyanide crystals are very sensitive to mechanical stresses (T 2).

Waku and co-workers (W 7) (W 8) have recently discovered ferroelectric properties in a number of crystals isomorphous with $K_4Fe(CN)_6 \cdot 3H_2O$, where Fe is replaced by Ru, Os or Mn. The isotope effect (replacement of H by D) consists in a small shift of the Curie point toward higher temperatures. Table IX-3 summarizes the data available at the present time.

TABLE IX-3.  CURIE TEMPERATURES $T_c$ AND SPONTANEOUS POLARI-
ZATION $P_s$ OF FERROELECTRICS OF THE POTASSIUM FERROCYANIDE
GROUP WITH GENERAL FORMULA $K_4M(CN)_6 \cdot 3T_2O$
(From data reported by Waku and co-workers (W 1), (W 7), (W 8), (T 2).)

| $M$ | $T_c$ (°C) | | $P_s$ ($10^{-6}$ Cb/cm²) | |
|---|---|---|---|---|
| | $T = H$ | $T = D$ | $T = H$ | $T = D$ |
| Fe | $-24.5$ | $-18.0$ | 1.4 at $-40\,°C$ | 1.5 at $-40\,°C$ |
| Ru | $-14.5$ | $-7.3$ | 1.4 at $-65\,°C$ | 1.5 at $-35\,°C$ |
| Os | $-2.4$ | $+1.8$ | 3.5 at $-45\,°C$ | 1.3 at $-20\,°C$ |
| Mn | $-40.0$ | $-$ | $-$ | $-$ |

## 10. Potassium Nitrate

The existence of polymorphic transitions in $KNO_3$ has been known for a number of years from the results of dilatometric, calorimetric and X-ray measurements (C 2), (K 2), (B 1), (M 5), (K 3), (F 3). The occurrence of three phases may be summarized in the following diagram:

$$
\begin{array}{ccc}
 & 130\ °C & \\
\alpha & \xrightarrow{\hspace{2cm}} & \beta \\
110\ °C & \diagdown \quad \diagup & 124\ °C \\
 & \gamma &
\end{array}
$$

At room temperature, the stable phase ($\alpha$) is orthorhombic, space group $Pmnb$, with the lattice parameters (at 26 °C) (W 4):

$$a = 6.431 \text{ Å}, \quad b = 9.164 \text{ Å}, \quad c = 5.414 \text{ Å}$$

and $Z = 4$. A transition occurs, upon heating, at approximately 130 °C, into a phase $\beta$, which is rhombohedral (space group $R\bar{3}c$). Upon cooling, however, the phase $\beta$ transforms first into a new phase ($\gamma$) at approximately 124 °C and then into $\alpha$ at about 110 °C. There are indications that the appearance of the $\gamma$-phase is somewhat dependent upon the rate of cooling, but the literature data are in contradiction with one another on this point (C 2), (M 5). The $\gamma$-phase reportedly can also be induced from the $\alpha$-modification by applying hydrostatic pressures (B 2), (B 3). The structure of the $\gamma$-phase is also rhombohedral, but polar, with lattice parameters (at 120 °C) (B 3):

$$a = 4.365 \text{ Å}, \quad \alpha = 76°56'.$$

and space group $R3m$.

It has recently been reported by Sawada et al. (S 5) that the $\gamma$-phase is ferroelectric. The dielectric measurements were carried out only on samples obtained directly from solidification of melted $KNO_3$ (melting point 334 °C) between two metal electrodes. As the orientation of these samples with respect to the crystallographic axes was not known, it is possible that the numerical data reported will have to be modified. The dielectric constant obeys the Curie–Weiss law, in the $\beta$-phase:

$$\varepsilon \cong \frac{4300}{T - 293\ °K}.$$

At 121 °C, the spontaneous polarization is $6.3 \times 10^{-6}$ C/cm$^2$ and the 50 c/s coercive field 4.5 kV/cm.

No direct information about the structure of the $\gamma$-phase is available in the literature. It is known (E 1) that the $\alpha$-phase, which is pseudo-hexagonal, has the aragonite structure, and is approximately described by hexagonal close-packing of layers of $K^+$ separated by double layers of $NO_3^-$. The $\beta$-phase has the calcite structure (this, however, was disputed by Barth (B 3)), approaching cubic close-packing, with the double layer of nitrate ions coalesced into a single layer. It is also known that the planar $NO_3^-$ ion is geometrically similar to the planar $CO_3^{2-}$ ion. It may be interesting to point out, in this connection, that the mineral

CaCO$_3$ does not only occur in the two modifications mentioned above (aragonite and calcite), but rarely also in a hexagonal form called vaterite. This latter modification exhibits a strong positive birefringence, while calcite and aragonite are both optically negative.

## 11. Sodium Nitrite

NaNO$_2$ represents another example of crystals that were found to be ferroelectric a number of years after the discovery of their phase transitions. A change from a non-centrosymmetric to a centric structure was known to occur in NaNO$_2$ at approximately 160 °C (S 9). This transition was discoverd to be a ferroelectric one by Sawada *et al.* (S 6) in 1958. Above the Curie point, the structure is orthorhombic with lattice constants (at 205 °C) (W 5):

$$a = 5.33 \text{ Å}, \quad b = 5.68 \text{ Å}, \quad c = 3.69 \text{ Å}$$

and space group *Immm*. In the polar phase, the mirror symmetry perpendicular to [010] disappears, the space group becoming *Im2m*, and the lattice constants (at 26 °C) (W 5):

$$a = 5.390 \text{ Å}, \quad b = 5.578 \text{ Å}, \quad c = 3.570 \text{ Å}.$$

The ferroelectric axis is therefore the $b$ axis.

The dielectric constant $\varepsilon_b$ exhibits a sharp anomaly at the Curie point (see Fig. IX-4), and obeys the Curie–Weiss law, above this temperature, in the form (S 6):

$$\varepsilon_b \cong \frac{5000 \text{ °K}}{T - 433 \text{ °K}}.$$

The transition is obviously one of the first order. The spontaneous polarization is rather large, namely of the same order as that of KNO$_3$. At 143 °C, $P_s = 6.4 \times 10^{-6}$ C/cm$^2$, and the 50 c/s coercive field $E_c = 2.3$ kV/cm. The coercive field is strongly field and temperature dependent, and is so large, at room temperature, that hysteresis loops cannot be obtained with fields of the order of 25 kV/cm (S 6).

X-ray structural investigations of NaNO$_2$ were carried out by Ziegler (Z 1), Carpenter (C 3) and by Truter (T 1). The structure strongly suggests that the ferroelectric activity is probably the result of the favorable atomic configuration in the NO$_2^-$ groups. The O$-$N$-$O bond is not linear, but rather forms an angle of approximately 115°, the N$-$O distances being equal to 1.24 Å (see Fig. IX-5). Above the Curie point, these ions may oscillate along the [010] axis about positions which can be described in terms of the centrosymmetrical space group *Immm* (W 5), but there are also indications that this centric symmetry may be the result of random disorder (K 5). A refinement of the polar structure has been recently carried out via neutron diffraction by Kay and Frazer (K 5), with particular attention to the parameters of thermal vibrations. The usefulness of a neutron diffraction study, in this case, is provided by the fact that the neutron scattering lengths are independent of the scattering angle, and nitrogen appears

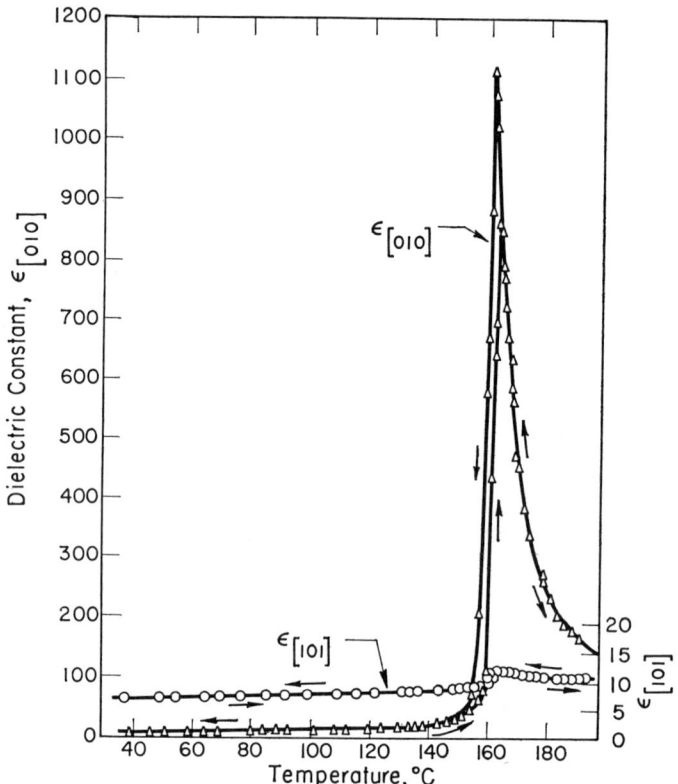

Fig. IX-4. Dielectric constant of sodium nitrite measured along [010] and [101] as a function of temperature (according to Sawada *et al.* (S 6)).

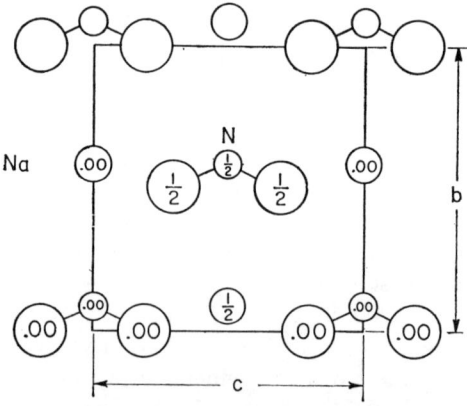

Fig. IX-5. Schematic projection on (001) of the orthorhombic structure of sodium nitrite. Atoms of the bent $NO_2^-$ ions are joined by lines; the $Na^+$ ions are shown by circles of intermediate size (according to Wyckoff (W 5)). Following the axial choice made in the text, the *c* spacing of this figure should be relabeled *a*.

as the "heaviest" atom in the structure, whereas in the case of X-rays (due to uncertainty in the electronic distribution of the $NO_2^-$ groups) there is some question as to the proper X-ray scattering factors for nitrogen and oxygen, and, moreover, nitrogen is the "lightest" atom. The results of the neutron study are in excellent agreement with those obtained with X-rays, as far as positional parameters are concerned, but indicate a measurable degree of thermal anisotropy only in the oxygen atoms. The spontaneous polarization calculated on the basis of a strictly ionic structure ($Na^{1+}$, $N^{3+}$, $O^{2-}$) turns out to be $74 \times 10^{-6}$ C/cm², which is about ten times larger than the experimental value. This, of course, is a confirmation of the highly covalent character of the nitrite group. Kay and Frazer (K 5) computed the effective charges $n_N$ and $n_O$ of nitrogen and oxygen, respectively, that would provide agreement between the calculated and observed values of the spontaneous polarization. The results: $n_N = -0.36$ and $n_O = -0.32$, (assuming $Na^{1+}$), are qualitatively significant in showing that the effective charge of nitrogen is just as strongly negative as that of oxygen. This may suggest that the role played by the $Na^{1+}$ ions is to exert a strong counter-polarizing influence upon the nitrite groups (K 5).*

### 12. Colemanite

The ferroelectric activity of colemanite was first discovered by Goldsmith in 1956 (G 1). Colemanite is a natural mineral, chemically a hydrated calcium borate with the formula $Ca_2B_6O_{11} \cdot 5H_2O$ (also written $2CaO \cdot 3B_2O_3 \cdot 5H_2O$). At room temperature, the substance is monoclinic, space group $P2_1/a$, with the lattice parameters (C 4):

$$a = 8.743 \text{ Å}, \quad b = 11.264 \text{ Å}, \quad c = 6.102 \text{ Å}, \quad \beta = 110°7'$$

and two formula units per cell. The Curie point was reported by Goldsmith (G 1) to lie at $-2.5$ °C, in agreement with Davisson's pyroelectric tests (D 2), while Chynoweth (C 1) found it at $-6$ °C, also on the basis of pyroelectric investigations, and Wieder (W 2) at $-7$ °C. The structure of the polar phase is still monoclinic, with space group $P2_1$ (B 4), (H 1), and the ferroelectric axis is, of course, along the twofold $b$ axis. The prismatic natural crystals exhibit perfect cleavage perpendicular to [010], which is very fortunate for the preparation of samples for dielectric measurements.

A detailed study of the dielectric properties of colemanite has been carried out by Wieder (W 2). While the dielectric constant $\varepsilon_a$ ($\cong 20$) is practically independent of temperature between $-20$ °C and $+20$ °C, the constant $\varepsilon_b$ (of the order of 50 at room temperature) exhibits a very pronounced anomaly at the Curie point, reminiscent of that shown, e.g. by tri-glycine sulfate. The Curie–Weiss law is very nicely obeyed, with a Curie constant $C$ of the order of 500 degrees. The spontaneous polarization shows the typical temperature dependence of a

---

* Studies of the domain structure and thermal properties of $NaNO_2$ have recently been published by Nomura et al. (*J. Phys. Soc. Japan* **16**, 917 (1961); **16**, 1352 (1961)).

second-order transition. At $-20$ °C, $P_s = 0.45 \times 10^{-6}$ C/cm², and $E_c = 2$ kV/cm; at $-70$ °C, $P_s = 0.65 \times 10^{-6}$ C/cm². By fitting the experimental curve of $P_s$ vs. temperature with the thermodynamic equation (II-2), Wieder was able to compute the values of the coefficients $\xi$ and $\zeta$ of the free-energy expansion which are reported in Table VIII-3. A detailed study of the electrocaloric effect showed that this effect, although not as large as that observed in $KH_2PO_4$, is considerable in colemanite: sudden application of a field of 10 kV/cm causes a temperature change of approximately 0.2°.

It has been proved first by Wieder (W 2) and more recently by Fatuzzo (F 4) that most colemanite crystals are intrinsically biased by space charge fields of varying magnitude. Such an intrinsic bias cannot affect the Curie point but can indeed affect the temperature at which the dielectric constant exhibits its peak (see also Sections III-2 and IV-4). Since the temperature of the dielectric-constant peak is often identified with the Curie point, the existence of an intrinsic bias can therefore explain some of the discrepancies among the Curie temperatures reported by different authors. It can be shown (W 2) that the temperature of the dielectric-constant peak in an unbiased crystal can be shifted by applying a d.c. field along the polar [010] direction, and that this temperature shift $\Delta T_{\text{peak}}$ is proportional to the 2/3 power of the applied field $E_b$: $\Delta T_{\text{peak}} \propto (E_b)^{2/3}$. Accordingly, if an internal bias (space charge field) $E_x$ exists, this will have to be added to the external bias $E_b$. The exact relationship becomes then (W 2):

$$\Delta T_{\text{peak}} = 0.14(E_b + E_x)^{2/3}. \tag{IX-2}$$

Figure IX-6 shows the effect of an external biasing field $E_b$ upon the reciprocal dielectric susceptibility $\chi$ of a crystal which exhibits a peak of the dielectric constant at 0 °C. A field $E_b = -47$ kV/cm shifts the minimum of $\chi$ to $-6$ °C. If we assume, following Wieder (W 2), that the Curie temperature of a completely unbiased crystal is $-7$ °C, we may conclude in this case that $\Delta T_{\text{peak}} = 1$. Eq. (IX-2) yields then the value $E_x = 49.9$ kV/cm for the internal bias. Therefore, a positive external field $E_b = +47$ kV/cm ought to produce a shift $\Delta T_{\text{peak}} = 11$°, and the minimum of $\chi$ should occur at $+4$ °C, in good agreement with the experimental result of $+4.5$ °C. For an applied field $E_b = -29.4$ kV/cm, a minimum is predicted at $-4$ °C, which is again in excellent agreement with the experimental data (Fig. IX-6). The space charge fields in the natural samples could be due, according to Wieder, to the substitution of trivalent boron with tetravalent silicon, which was found to be present in amounts as large as 0.6% in the most biased specimens. It is also possible that the different Curie temperatures of different samples could be due in some cases to substitution of Sr for Ca (which was also found to be present in concentrations ranging between 0.2 and 0.5%) without any space charge effects (W 4). Fatuzzo (F 4), however, has confirmed the existence of an intrinsic bias above the Curie point (for which he reported values ranging from 0.2 to 7 kV/cm) and proved that such a bias is responsible for the pyroelectric effect experimentally observed in the non-polar phase (D 2). Fatuzzo's analysis shows that the maximum of the pyroelectric coefficient $dP_s/dT$ occurs always at the Curie point: the effect of the intrinsic bias is only to decrease the peak value of the pyroelectric coefficient at the Curie temperature. There are therefore two ways to determine the Curie point of a second-order transition independently of the existence of an internal bias. One is to plot the square of the spontaneous polarization ($P_s^2$) vs. temperature and extrapolate to $P_s^2 = 0$, the other is to measure the temperature dependence of the pyroelectric coefficient and to locate its peak.

The switching characteristics of colemanite crystals were also studied in detail by Wieder (W 3) using the technique of electrical pulses. Qualitatively, the

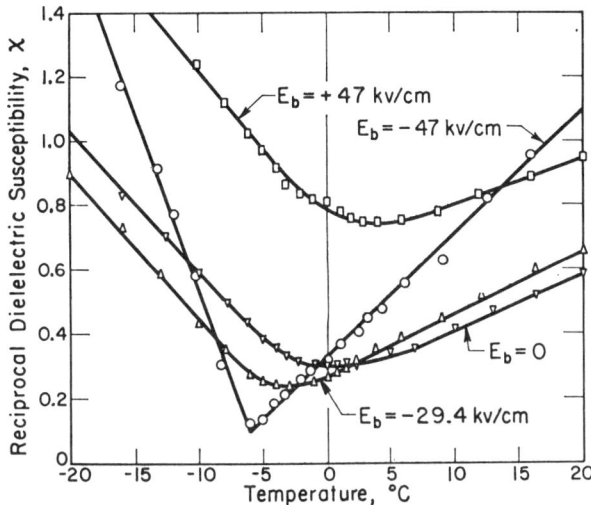

FIG. IX-6. Temperature dependence of the reciprocal dielectric susceptibility $\chi$ of a strongly biased crystal of colemanite. The minimum of $\chi$ measured with no external bias ($E_b = 0$) was 0 °C. Note the difference in shift of the minimum of $\chi$ as a function of polarity and magnitude of $E_b$, illustrating the predominance of one type of charge carriers (according to Wieder (W 2)).

characteristics of the polarization reversal transients were found to be the same as in other ferroelectrics. At low field strengths, the maximum switching current $i_{max}$, and the reciprocal switching time $1/t_s$, depend exponentially upon the applied field in the form $\exp(-\alpha/E)$, where the activation field $\alpha$ is of the order of 40 kV/cm about 100° below the Curie temperature. At high electric fields $i_{max}$ rises linearly with field $E$: no evidence for a 3/2 or 5/2 power law, as found in BaTiO$_3$ and in LiH$_3$(SeO$_3$)$_2$, respectively, was found in colemanite.

Wieder's contribution to the problem of polarization reversal consists in showing that consistent results can be obtained upon assuming random nucleation in the direction of the ferroelectric axis: a large electric field is assumed to cause the nucleation of 180° domains in the form of thin spikes which thereafter grow in the forward direction at a faster rate than either the nucleation or the sub-sequent two-dimensional sidewise growth of the generated nuclei. This model leads to a direct proportionality between $i_{max}$ and the probability of nucleation $n_p$

(see Section IV-8). In colemanite, this relationship assumes the form:

$$i_{\max} = 0.765\, n_p\, P_s \qquad\qquad (IX\text{-}3)$$

and is found to be in excellent agreement with the experimental data. In addition, for low field amplitudes, the field dependence of the nucleation probability $(\partial n_p/\partial E)_T$ is found to be proportional to $\exp(-\alpha/E)$, where $\alpha$ is, within the experimental error, the same activation field appearing in the dependence of $i_{\max}$ upon field. The time dependence of the switching transient in colemanite is apparently quite different from that of $BaTiO_3$, but the essential features of the polarization reversal process are, once more, qualitatively similar to those of all other ferroelectrics investigated.

The crystal structure of colemanite was studied with X-ray diffraction techniques by Christ *et al.* (C 4). The results of this study indicate that the structure contains infinite boron–oxygen chains running parallel to the $a$ axis, the chain element consisting of two $BO_4$ tetrahedra and a planar $BO_3$ triangle linked at corners, to form a ring of composition $[B_3O_4(OH)_3]^{2-}$. The structurally correct formula of colemanite should, therefore, be written $CaB_3O_4(OH)_3 \cdot H_2O$, since the $B-O$ chains are joined to each other laterally by ionic bonds through the $Ca^{2+}$ ions to form sheets extending parallel to (010). The sheets are tied together by hydrogen bonds, which are supposedly responsible for the ferroelectic transition. Both the X-ray evidence (C 4) and a nuclear magnetic resonance study of the boron sites (H 1) speak in favor of a centrosymmetrical structure above the Curie point. Lack of a center of symmetry below this critical temperature, on the other hand, can reportedly be detected not only by the resonance experiments (H 1) but also by X-ray measurements (B 4), which may indicate that ordering of the hydrogen bonds is not solely responsible for the ferroelectric phenomenon.

## 13. Thiourea

The discovery of ferroelectricity in thiourea, $SC(NH_2)_2$, by Solomon (S 7), was particularly significant in revealing that the phenomenon of reversible spontaneous polarization could occur in molecular crystals as well. The ferroelectric behavior of this compound was later studied in detail by Goldsmith and White (G 2), and is interesting, despite its complexity, because it was proved to be related to relative displacements of entire molecules rather than to the motion of single atoms or ions within the lattice.

Single crystals of thiourea can best be grown from saturated methanol solutions by slow evaporation at 30 °C, as the samples grown from water solutions are mostly cloudy and imperfect. The crystals are orthorhombic, with the following axial lengths at room temperature (K 4):

$$a = 7.655\ \text{Å}, \quad b = 8.537\ \text{Å}, \quad c = 5.520\ \text{Å}$$

and exhibit pronounced cleavage in the plane perpendicular to [100] which, at low temperatures, becomes the ferroelectric direction. It should be pointed

out that the axial lengths reported above are consistent with the modern crystallographic convention for axial choice in orthorhombic crystals ($b > a > c$), and
do *not* correspond to the older setting (G 3), (D 3), (W 6) adhered to by Goldsmith and White (G 2). Accordingly, the latter authors' $a, b, c$ axes are relabeled
$c, a, b$, respectively, in the present discussion.

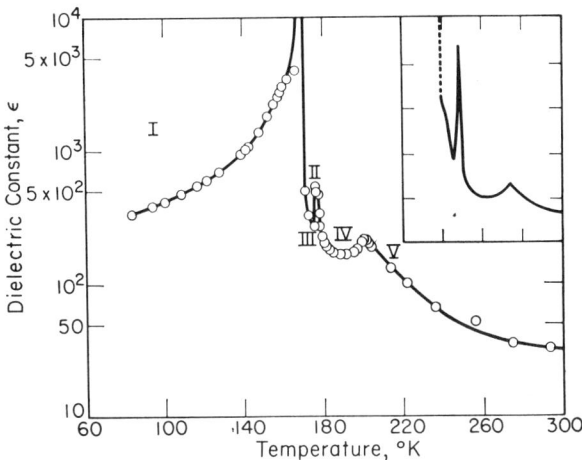

FIG. IX-7. Temperature dependence of the dielectric constant $\varepsilon_a$ of thiourea
(according to Goldsmith and White (G 2)).

The dielectric constants $\varepsilon_b$ and $\varepsilon_c$ are practically temperature independent,
but the dielectric constant $\varepsilon_a$ exhibits a number of anomalies at lower temperatures. Figure IX-7 shows that three prominent maxima occur at 202 °K, 177 °K
and 169 °K, respectively, while an inflection of the curve appears at 170.5 °K,
which probably corresponds to a fourth dielectric anomaly (G 2). For convenience, the various temperature ranges defined by the dielectric peaks are
called I, II, III, IV and V, respectively, as indicated in Fig. IX-7 and in Table IX-4.

TABLE IX-4.    SPECIFICATION OF THE VARIOUS PHASES OF THIOUREA AS INDICATED BY THE
DIELECTRIC ANOMALIES (Fig. IX-7)

(According to Goldsmith and White (G 2); temperature $T$ in °K.)

|  | $T < 169$ | $169 < T < 173$ | $173 < T < 179$ | $179 < T < 202$ | $T > 202$ |
|---|---|---|---|---|---|
| Phase | I<br><br>Ferroelectric | II<br>Non-or<br>anti-polar | III<br><br>Ferroelectric | IV<br>Non-or<br>anti-polar | V<br>Non-or<br>anti-polar |

The Curie–Weiss law, $\varepsilon_a = C/(T - T_0)$, is nicely adhered to in region V, with
$C = 3.7 \times 10^3$ degrees and $T_0 = 185$ °K (see Fig. IX-8). Ferroelectric hysteresis
loops can be observed only in regions I and III. The existence of double loops in
region II, at temperature close to the 169 °K transition, indicates merely that the

latter is of the first order, but otherwise the $P$–$E$ relationship is linear in the regions II, IV and V. The onset of the spontaneous polarization at 169 °K is in accord with the first-order character of this ferroelectric transition (see Fig. IX-9 a). At 140 °K, $P_s = 3 \times 10^{-6}$ C/cm² and the 60 c/s coercive field is $E_c = 165$ V/cm.

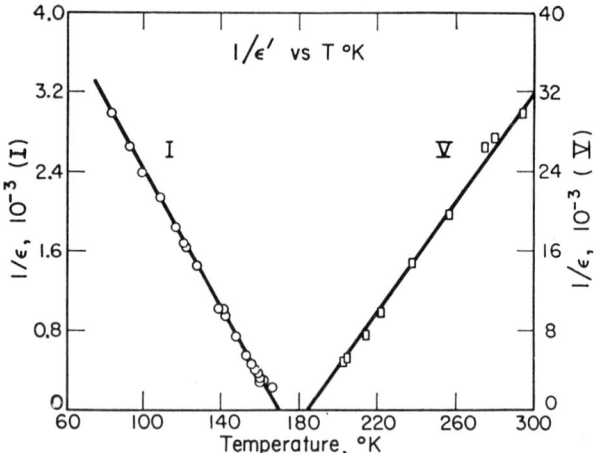

Fɪɢ. IX-8.  Reciprocal dielectric constant $\varepsilon_a$ of thiourea as a function of temperature (according to Goldsmith and White (G 2)).

Of particular interest is the occurrence of the ferroelectric phase III in a narrow temperature range between approximately 173 °K and 179 °K. Qualitatively, the behavior of the spontaneous polarization in this phase is reminiscent of that of Rochelle salt (see Chapter VII), but the maximum value of the spontaneous polarization, occurring approximately at 176 °K, is very much smaller, namely $2.5 \times 10^{-9}$ C/cm² (see Fig. IX-9 b). The anomaly of the dielectric constant in

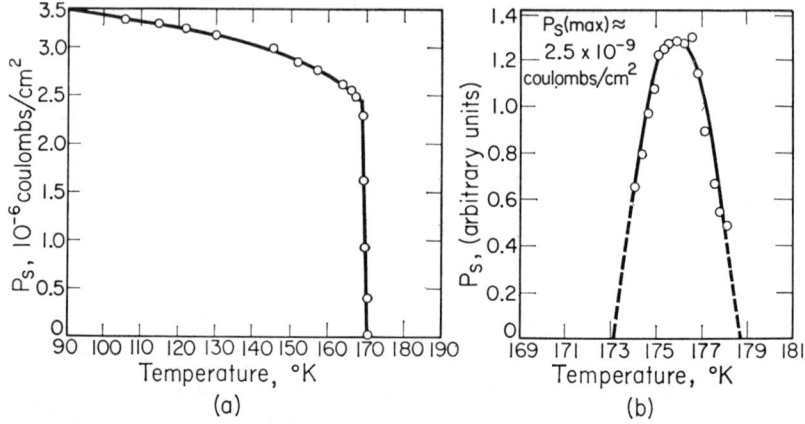

Fɪɢ. IX-9.  Spontaneous polarization of thiourea as a function of temperature: (a) in region I; (b) in region III (see Fig. IX-7). (According to Goldsmith and White (G 2).)

this region would be expected, of course, to consist in two peaks at the temperatures corresponding to the onset and the vanishing of the spontaneous polarization. However, the closeness of these two Curie points is probably the reason for the single dielectric peak actually observed.

Goldsmith and White (G 2) examined also the switching characteristics of thiourea in the phase I. The field-dependence of the switching time $t_s$ was found to be similar to that of other ferroelectrics. For fields smaller than about 100 V/cm, $t_s = t_\infty \exp(\alpha/E)$, with $t_\infty = 10^{-4}$ sec and $\alpha = 2.7 \times 10^2$ V/cm; whereas between field strengths of 100 V/cm and 3000 V/cm the switching time follows a 3/2 power law ($t_s =$ constant $\times E^{-3/2}$), as observed in $BaTiO_3$ (see Section IV-8) and, with a different exponent, in $LiH_3(SeO_3)_2$ (see Section VIII-8). It is thus reasonable to expect that polarization reversal in thiourea is governed by the same processes of nucleation and domain-wall motions that are active, e.g. in $BaTiO_3$.

It may be interesting to point out that, although hydrogen bonding is not expected to occur in thiourea, a small but not negligible isotope effect does occur upon substitution of the hydrogens with deuterium atoms. The spontaneous polarization and the coercive field in phases I and II show negligible changes, but the temperatures at which the three major dielectric anomalies of the deuterated crystal occur appear to be shifted upwards by 16°, 16° and 11°, respectively, with respect to those of the hydrogen salt. It has been suggested that the only effect of isotopic substitution is in the dipole moment of the molecule as a whole and that this change is relatively small. The increase in transition temperature with increasing mass of the molecule is in accordance with the corresponding decrease in vibrational energy at any given temperature (G 2). The infrared spectra show an anomalously large shift of the lower NH frequencies upon deuterium substitution (B 5).

X-ray data (K 4) and proton resonance experiments (F 2) are in agreement in revealing hindered rotation of the $SC(NH_2)_2$ molecule about the carbon–sulfur bond axis. The latter experiments indicate that this rotational vibration stops at about 198 °K (E 2), which may correspond to the dielectric peak occurring at 202 °K (see Fig. IX-7). The fact that the molecule, excluding hydrogen atoms, is very precisely planar is evident not only from the X-ray studies (D 3), (W 6), (K 4), but also from infrared absorption data (S 8) and proton magnetic resonance experiments on urea (A 1). As far as the orientation of the single molecules in the lattice is concerned, the X-ray evidence seems to indicate that the room-temperature phase may be anti-polar. The first X-ray analysis of thiourea was carried out by Demeny and Nitta (D 3) as early as 1928, later repeated by Wyckoff and Corey (W 6), and more recently refined by Kunchur and Truter (K 4). The results of the latter authors, in particular, differ substantially from those of Zvonkova and Tashpulatov (Z 2) as far as bond lengths and angles within the thiourea molecule are concerned, but the relative distribution of the molecular groups within the lattice is very closely the same in all pertinent investigations. The space group of the room-temperature phase is $Pnma$, with the lattice parameters given above, density 1.405 g cm$^{-3}$ and four molecules per unit cell. The carbon and sulfur atoms lie in the mirror plane and they still do so in the low-temperature phase. A detailed structural study of this phase at 120 °K, by Goldsmith and White (G 2), has revealed that the diagonal glide plane $n$ vanishes in the phase I, and the two crystallographically non-equivalent molecules in the unit cell have different tilts with respect to the $a$ axis, being slightly shifted along $c$ with respect to the room-temperature modification. It is interesting, in this respect, that a polymorphic transition in thiourea under pressure had been reported to occur around room temperature by Bridgman (B 6). The X-ray study at 120 °K yields the following values of the orthorhombic lattice parameters (G 2):

$$a = 7.516 \text{ Å}, \quad b = 8.519 \text{ Å}, \quad c = 5.494 \text{ Å},$$

with space group $P2_1ma$ and $Z = 4$. The dimensions of the thiourea molecule, excluding hydrogen atoms, are identical, within the experimental accuracy, to those found at room temperature, and the planarity of the molecule is not destroyed by the phase change. An illustration of the mechanism of polarization reversal in thiourea is given in Fig. IX-10, which is a schematic projection of the structure on the (010) plane. The figure shows the different tilts of the two non-equivalent pairs of molecules with respect to the polar $a$ axis. The net result of this configuration is that the dipole components in the [100] direction do not cancel, and this direction is therefore polar. This, as pointed out by Goldsmith and White (G 2), is the closest example of an electric analog to the phenomenon

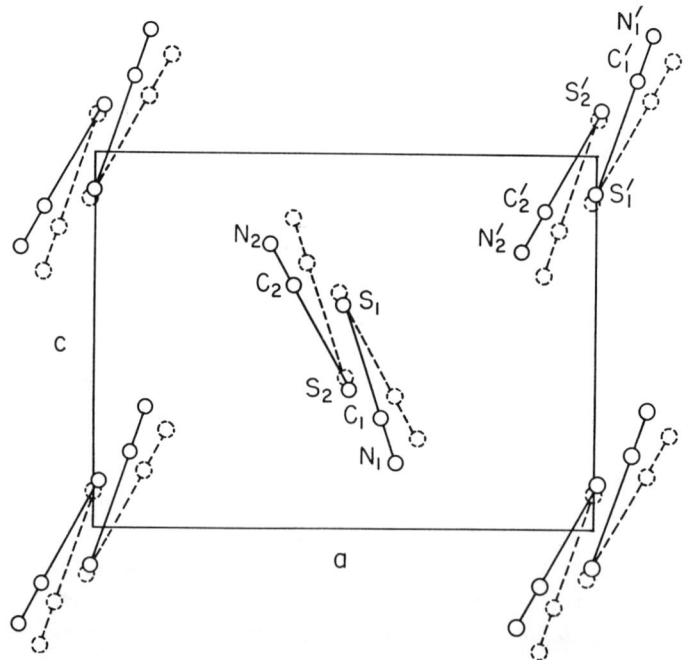

Fig. IX-10. Schematic illustration of the polarization reversal in thiourea: projection of the structure on (010). The full circles represent atomic positions for one direction of the applied field; the dotted circles represent the new atomic positions upon reversal of the field (according to Goldsmith and White (G 2)).

of ferrimagnetism, and it may therefore be justified to call it a case of *ferri*-electricity. The reversal of polarity implies, as indicated in Fig. IX-10, that the two non-equivalent pairs of molecules in each unit cell interchange their inclinations with respect to the *a* axis. The atomic displacements required by this mechanism are fairly large (approximately 0.5 Å for the nitrogen atoms), but appear to be reasonable in view of the fact that they occur in the direction of the largest thermal vibration. This may explain the rather small coercivity of the ferroelectric reversal.

No satisfactory explanation can be given, at the present stage, for the minor dielectric anomalies encountered above 169 °K, and, in particular, for the occurrence of phase III.

## 14. Questionable Ferroelectrics

There are a few compounds for which ferroelectricity has been claimed, in the literature, but not conclusively proved. They are worth mentioning because of the interest they may arouse from a crystal–chemical point of view.

One such compound is the mineral called boracite, chemically a magnesium chloroborate with the formula $Mg_3B_7O_{12}Cl$. This crystal is cubic above 265 °C, and orthorhombic, pseudo-cubic, below this temperature, where it shows a very strong pyroelectric effect and often exhibits twinning. All these facts seem to indicate that boracite might well be ferroelectric below its transition point, a possibility first suggested by Cady and Jaffe (C 5). A more detailed dielectric investigation of natural crystals, by Le Corre (L 1), gave some additional evidence substantiating this possibility. However, a subsequent study of polycrystalline artificial samples, by Jona (J 2), failed to confirm ferroelectricity in boracite as well as in a number of isomorphous compounds.

Another interesting substance is the mineral pyrolusite, which is a polycrystalline sintered form of $MnO_2$ with a structure of the rutile type. According to Das (D 4) and Bhide and Damle (B 8), the specimens of this mineral have a dielectric constant of the order of $10^5$, which exhibits anomalies in a temperature range varying between 20 °C and 50 °C, and furthermore, obeys a Curie–Weiss law. The relation between polarization and field was reported to be non-linear, but it is doubtful whether this is due to ferroelectric activity of conductivity of the samples. The piezoelectric effect seems to be large and the electromechanical coupling coefficient larger than those of quartz and barium titanate.

Ferroelectricity was also claimed in sintered specimens of $TiO_2$ by Nicolini (N 2). The reported effects are probably real, but there is considerable room for doubt as to the exact chemical composition of the samples investigated. It is well known that single crystals of rutile do not show ferroelectric properties. Nicolini's samples, on the other hand, exhibit dielectric anomalies at 350 °C, − 10 °C and − 60 °C, which strongly recalls the dielectric behavior of some mixed titanates containing $BaTiO_3$ (see Chapter V).

Finally, it should be mentioned that interesting dielectric properties have been measured in the yellow modification of lead oxide by Bergstein (B 7).

Yellow, orthorhombic lead oxide is known to be stable above 488 °C, but it can be quenched to room temperature, where the stable form of lead oxide is the common red, tetragonal modification. The dielectric constant of quenched samples of the yellow phase exhibits a small anomaly at about 60 °C, when measured at a frequency of 1 kc/s. The $P$–$E$ curves are non-linear and recall the shape of hysteresis loops. These investigations need to be confirmed by more detailed dielectric studies. The interest in this compound lies in the fact that the ferroelectric phenomenon is not known to occur, to date, in any simple oxide.

## BIBLIOGRAPHY

(A 1)  ANDREW, E. R. and HYNDMAN, D., *Discussions Faraday Soc.* **19**, 195 (1955).
(B 1)  BORCHERT, W., *Krist. Z.*, **95**, 28 (1936).
(B 2)  BRIDGMAN, P. W., *Proc. Am. Acad. Arts Sci.* **51**, 581 (1916).
(B 3)  BARTH, T. F. W., *Z. physik. Chemie. (Leipzig)* **B 43**, 448 (1939).
(B 4)  BLOCK, S. and BURLEY, G., cited by C. L. CHRIST, J. R. CLARK and H. T. EVANS, JR., *Acta Cryst.* **11**, 761 (1958).
(B 5)  BADGER, R. M. and WALDRON, R. D., *J. Chem. Phys.* **26**, 255 (1957).
(B 6)  BRIDGMAN, P. W., *Proc. Am. Acad. Arts Sci.* **72**, 227 (1938).
(B 7)  BERGSTEIN, A., *Czechoslov. J. Phys.* **6**, 164 (1956).
(B 8)  BHIDE, V. G. and DAMLE, R. V., *Physica* **26**, 33 (1960); **26**, 513 (1960).
(C 1)  CHYNOWETH, A. G., *Acta Cryst.* **10**, 511 (1957).
(C 2)  COHEN, E. and BREDÉE, H. L., *Z. physik. Chem. (Leipzig)* **A 140**, 391 (1929).
(C 3)  CARPENTER, G. B., *Acta Cryst.* **5**, 132 (1952).
(C 4)  CHRIST, C. L., CLARK, J. R. and EVANS, H. T., JR., *Acta Cryst.* **11**, 761 (1958).
(C 5)  CADY, W. G., *Piezoelectricity*, p. 230, McGraw-Hill, New York (1946).
(D 1)  DANNER, H. and PEPINSKY, R., Unpublished work (1953).
(D 2)  DAVISSON, J. W., *Acta Cryst.* **9**, 9 (1956); *Bull. Am. Phys. Soc.* Ser. II **1**, 177 (1956).
(D 3)  DEMENY, L. and NITTA, I., *Bull. Chem. Soc. Japan* **3**, 128 (1928).
(D 4)  DAS, J. N., *Z. Physik* **155**, 465 (1959).
(E 1)  EDWARDS, D. A., *Z. Krist.* **80**, 154 (1931).
(E 2)  EMSLEY, J. W. and SMITH, J. A. S., *Proc. Chem. Soc. (London)* **53** (1958); *Arch. sci. (Geneva)* **12**, 122 (1959).
(F 1)  FERRONI, E. and ORIOLI, P., *Z. Krist.* **111**, 362 (1959).
(F 2)  FATUZZO, E. and NITSCHE, R., *Phys. Rev.* **117**, 936 (1960).
(F 3)  FINBAK, C. and HASSEL, O., *Z. physik. Chem. (Leipzig)* **B 37**, 75 (1937).
(F 4)  FATUZZO, E., *J. Appl. Phys.* **31**, 1029 (1960).
(G 1)  GOLDSMITH, G. J., *Bull. Am. Phys. Soc.* Ser. II **1**, 322 (1956).
(G 2)  GOLDSMITH, G. J. and WHITE, J. G., *J. Chem. Phys.* **31**, 1175 (1959).
(G 3)  GROTH, P., *Chemische Krystallographie*, vol. 3, Engelmann, Leipzig (1910).
(H 1)  HOLUJ, F. and PETCH, H. E., *Can. J. Phys.* **36**, 145 (1958).
(J 1)  JONA, F. and PEPINSKY, R., *Phys. Rev.* **92**, 845 (1953).
(J 2)  JONA, F., *J. Phys. Chem.* **63**, 1750 (1959).
(K 1)  KIRIYAMA, R. and TABATA, K., *Symposium of Structural Chemistry*, The Chemical Society of Japan, October (1956); cited in reference (W 1).
(K 2)  KRACEK, F. C., *J. Phys. Chem.* **34**, 225 (1930).
(K 3)  KENNEDY, S. W., UBBELOHDE, A. R. and WOODWARD, I., *Proc. Roy. Soc. (London)* **A 219**, 303 (1953).
(K 4)  KUNCHUR, N. R. and TRUTER, M. R., *J. Chem. Soc.* 2551 (1958).
(K 5)  KAY, M. I. and FRAZER, B. C., *Acta Cryst.* **14**, 56 (1961):
(L 1)  LeCORRE, Y., *J. Phys. radium* **18**, 629 (1957).

(M 1) MATTHIAS, B. T. and HULM, J. K., *Phys. Rev.* **82**, 108 (1951).
(M 2) MERZ, W. J., *Phys. Rev.* **82**, 562 (1951).
(M 3) MASON, W. P., *Piezoelectric Crystals and Their Application to Ultrasonics*, p. 232, Van Nostrand, New York (1950).
(M 4) MATTHIAS, B. T. and REMEIKA, J. P., *Phys. Rev.* **107**, 1727 (1957).
(M 5) MIEKK-OJA, H., *Ann. Acad. Sci. Fennicae*, Ser. A, **1**, No. 7 (1941).
(N 1) NITTA, I. and WATANABE, T., *Sci. Papers Inst. Phys. Chem. Research (Tokyo)* **26**, 164 (1935).
(N 2) NICOLINI, L., *Nuovo cimento* **13**, 257 (1959); 14, 905 (1959).
(O 1) ORIOLI, P. and PIERONI, M., *Ricerca sci.* **29**, 295 (1959).
(P 1) PEPINSKY, R. and ROCK, E. J., Unpublished; cited by SHIRANE, G., JONA, F. and PEPINSKY, R., *Proc. I.R.E.* **43**, 1776 (1955).
(P 2) PEPINSKY, R., VEDAM, K., HOSHINO, S. and OKAYA, Y., *Phys. Rev.* **111**, 430 (1958).
(P 3) PEPINSKY, R., OKAYA, Y., EASTMAN, D. P. and MITSUI, T., *Phys. Rev.* **107**, 1538 (1957).
(P 4) PEPINSKY, R., VEDAM, K. and OKAYA, Y., *Phys. Rev.* **110**, 1308 (1958).
(P 5) PEPINSKY, R., OKAYA, Y. and MITSUI, T., *Acta Cryst.* **10**, 600 (1957).
(R 1) REMEIKA, J. P., Private communication (1960).
(S 1) SCHOLZ, H., Dissertation, Goettingen (1940).
(S 2) SHIRANE, G., JONA, F. and PEPINSKY, R., *Proc. I.R.E.* **43**, 1776 (1955).
(S 3) SEKI, S., MOMOTANI, M. and NAKATSU, K. *J. Chem. Phys.* **19**, 1061 (1951).
(S 4) SEKI, S., MOMOTANI, M., NAKATSU, K. and OSHIMA, T., *Bull. Chem. Soc. Japan* **28**, 411 (1955).
(S 5) SAWADA, S., NOMURA, S. and FUJII, S., *J. Phys. Soc. Japan* **13**, 1549 (1958).
(S 6) SAWADA, S., NOMURA, S., FUJII, S. and YOSHIDA,, I. *Phys. Rev. Letters* **1**, 320 (1958).
(S 7) SOLOMON, A. L., *Phys. Rev.* **104**, 1191 (1956).
(S 8) STEWART, J. E., *J. Chem. Phys.* **26**, 248 (1957).
(S 9) STRIJK, B. and MacGILLAVRY, C. H., *Rec. trav. chim.* **62**, 705 (1943); **65**, 127 (1946).
(T 1) TRUTER, M. R., *Nature* **168**, 344 (1951); *Acta Cryst.* **7**, 73 (1954).
(T 2) TOYODA, H., NIIZEKI, N. and WAKU, S., *J. Phys. Soc. Japan* **15**, 1831 (1960).
(V 1) VERNON, R. C. and PEPINSKY, R., Final Report, Contract DA-36-039-SC-21, Signal Corps Engineering Laboratories, January (1953).
(W 1) WAKU, S., HIRABAYASHI, H., TOYODA, H., IWASAKI, H. and KIRIYAMA, R., *J. Phys. Soc. Japan* **14**, 973 (1959).
(W 2) WIEDER, H. H., *J. Appl. Phys.* **30**, 1010 (1959).
(W 3) WIEDER, H. H., *J. Appl. Phys.* **31**, 180 (1960).
(W 4) WYCKOFF, R. W. G., *Crystal Structures*, vol. II, ch. VII, table page 17c, Interscience, New York.
(W 5) WYCKOFF, R. W. G., *Crystal Structures*, vol. II, ch. VI, text p. 10a, illustr. p. 5, Interscience, New York.
(W 6) WYCKOFF, R. W. G. and COREY, R. D., *Z. Krist.* **81**, 386 (1932).
(W 7) WAKU, S., MASUNO, K., TANAKA, T. and IWASAKI, H., *J. Phys. Soc. Japan* **15**, 1185 (1960).
(W 8) WAKU, S., MASUNO, K. and TANAKA, T. *J. Phys. Soc. Japan* **15**, 1698 (1960).
(Z 1) ZIEGLER, G. E., *Phys. Rev.* **38**, 1040 (1931).
(Z 2) ZVONKOVA, Z. V. and TASHPULATOV, IU., *Kristallografiya* **3**, 553 (1958).

# CHAPTER X

# STRUCTURAL PROBLEMS AND RECENT DEVELOPMENTS

## 1. Introduction

In the preceding chapters, we have described the physical properties of various ferroelectric crystals, and in so doing we have classified the compounds partially according to their structural and partially according to their chemical characteristics. We have seen that the physical properties of a number of ferroelectrics can be described satisfactorily, in the vicinity of their Curie points, by the phenomenological theory developed by Mueller, Cady and Devonshire, but we have also encountered examples of ferroelectric materials that can hardly be treated with such a theory in its simplest form. It is quite obvious, therefore, that this large variety of characteristics makes it impossible to develop a unified treatment of the phenomenon of ferroelectricity at the present stage. The main difficulty of this problem arises from the necessity of identifying the fundamental interactions that are at work in causing the onset of spontaneous polarization, the dielectric anomalies and the domain structures in a large variety of compounds. We will devote this final chapter to a review of some of the experimental and theoretical problems involved in the process of understanding the nature of these interactions. It may be instructive, in this respect, to start from a comparison between the basic concepts involved with ferroelectricity on the one side and ferro- and antiferro-magnetism on the other (see, e.g., Kittel (K 1)).

## 2. Ferroelectricity vs. Ferro- and Antiferro-magnetism

In the case of magnetic substances, the nature of the "unit moment" responsible for the observed phenomena has been established beyond doubt. This moment has its origin in the magnetic moment of the electron spin (or orbit), and is assigned to particular atoms or ions within a given crystal. Its magnitude, which is usually independent of temperature, can be determined by a number of measurements, such as those of the magnetic susceptibility. In the case of ferro- and antiferro-magnetic materials, the distribution and the orientation of the moments in the structure can be established by way of neutron diffraction analyses (B 1), (S1).

In the case of ferroelectric substances, the existence of permanent electric moments is not always certain. There are indeed a few crystals (such as tri-glycine sulfate, potassium dihydrogen phosphate, Rochelle salt and thiourea), in which permanent moments could be assigned to certain ions or group of ions, and these moments could be considered temperature independent. The identification of these moments can only be done by means of very careful structural investigations. In general, the phase changes occurring in these crystals can be regarded and treated as transitions between ordered and disordered states. But even in these cases, as was already pointed out previously, the contribution from, and the interaction with the "background" are by no means negligible. There are other crystals, on the other hand, such as barium titanate and lead titanate, in which permanent electric moments most probably do not exist. These are the crystals that undergo transitions of the displacive type, and it is again only through very careful structural studies that one may obtain, at best, a clue as to the role played by the individual ions in bringing about an ionic polarization.

Once the structure of the compound and the identity of the "dipoles" have been established with sufficient confidence, attention is devoted to the study of the interactions among such dipoles imbedded in highly polarizable media. In magnetic compounds, the nature of these interactions is well known, being of the exchange, super-exchange or double exchange type, depending upon the compound; these interactions generally have the character of short-range forces. In dielectric and ferroelectric materials, on the other hand, long-range forces, such as Coloumb interaction, play often an equally important role.

At the present stage, the problem of unraveling the various aspects of interaction forces among dielectric dipoles could be handled satisfactorily only in a very few favorable cases. Long-range interactions can occasionally be treated by means of a properly applied Lorentz correction of the internal field, and Slater's theory of $BaTiO_3$ (S 2) is the now-classical example of successful appplication of this method. But when the structure is more complicated than cubic or pseudo-cubic, this method may encounter prohibitive difficulties. As for the short-range interactions, their treatment poses even harder problems, and no general method seems to be available yet for this purpose. The only case in which the treatment of short-range coupling could be accomplished with successful results is that of $KH_2PO_4$, where Slater's theory (S 3) proved to be capable of explaining the ferroelectric transition via the statistical treatment of the hydrogen bonds. It is important to notice that long-range interactions play a very little role in this particular theory.

A new approach to the problem has been recently proposed by Cochran (C 1), who related the occurrence of ferroelectricity to a problem of lattice dynamics. This approach seems to provide a proper way for treating both long-range and short-range interactions correctly, and will be discussed in some detail in Section 5. Before this is done, however, it may be worth devoting some time to review a few of the problems involved in the determination of ferroelectric structures by means of X-rays. First, we will discuss the ambiguity of structural results which arises from the pseudo-symmetric character of the ferroelectric phases. Then we will discuss the problem of determining the absolute configuration of such crystalline phases.

## 3. Structure Analysis of Ferroelectric Crystals

The structure analysis of ferroelectric crystals involves generally two stages of research. The first stage may be called, for the purpose of the present discussion, a "conventional" structure analysis of the "reference" phase, namely of the phase of the crystal above its Curie temperature. In some special cases, knowledge of the reference structure may be sufficient for a successful speculation about the mechanism of the ferroelectric transition. However, the validity of the theoretical model need be confirmed by a structural study of the ferroelectric phase; Slater's theory of $KH_2PO_4$ (S 3) is the most typical of such cases.

The second stage of the analysis consists in establishing accurately the positions of the atoms in the ferroelectric phase. Generally, these atomic positions differ only little from those found in the reference phase, which has higher symmetry. This is the reason why ferroelectric structures are often called "pseudo-symmetric" structures. The difficulties encountered in the accurate determination of ferroelectric structures arise indeed from their pseudo-symmetric character and from the smallness of the atomic displacements involved.

The first problem is that of determining the space group. Starting from the higher symmetry of the reference phase, one generally has the choice of several space groups that the crystal can possibly assume below its Curie temperature. In a few cases (such as those of $BaTiO_3$, $KH_2PO_4$ and Rochelle salt), the space groups were determined without difficulty. In the case of antiferroelectric crystals, such as $NaNbO_3$ and $PbZrO_3$, the ambiguity in the choice of the proper space group is larger, because the variety of configurations that can be obtained from a centrosymmetric structure via antiparallel atomic displacements is larger than that which can be obtained via parallel shifts. A procedure to be followed in these cases has been proposed by Vousden (V 1) and by Megaw and Wells (M 1), and then successfully applied to the analysis of $NaNbO_3$ by the latter authors (M 1).

Once the space group has been established, the positions $x_j$, $y_j$, $z_j$ of every $j$th atom in the structure must be determined accurately. This goal is achieved when best agreement is obtained between the observed and the calculated structure factors. The magnitudes of the *observed* structure factors are obtained from the observed intensities $I(h, k, l)$ of all X-ray reflections $(h, k, l)$ as follows:

$$|F_{obs}(h, k, l)| = \text{constant} \times \sqrt{[L_p \cdot I(h, k, l)]},$$

where $L_p$ is the Lorentz-polarization factor. The calculated structure factors, on the other hand, are obtained from the relation

$$F_{calc}(h, k, l) = \sum_{j=1}^{N} f_j \exp[2\pi i(hx_j + ky_j + lz_j)], \qquad \text{(X-1)}$$

where the $f_j$'s are the scattering factors defined in Section IV-6 and $N$ is the number of atoms in the unit cell. The calculated structure factors can be rewritten as follows:

$$F_{calc}(h, k, l) = \sum_{j=1}^{N} f_j \cos 2\pi(hx_j + ky_j + lz_j)$$
$$+ i \sum_{j=1}^{N} f_j \sin 2\pi(hx_j + ky_j + lz_j) = A + iB. \qquad \text{(X-2)}$$

The difficulty encountered in the analysis of pseudo-symmetric structures was first demonstrated by Evans (E 1) in his X-ray study of BaTiO$_3$. Evans showed that in the process of matching the calculated with the observed structure factors there occurs an interaction between positional parameters and temperature factors of the lighter atoms in the structure. The reason for this interaction is that the scattering power of heavy atoms such as barium is so much more massive

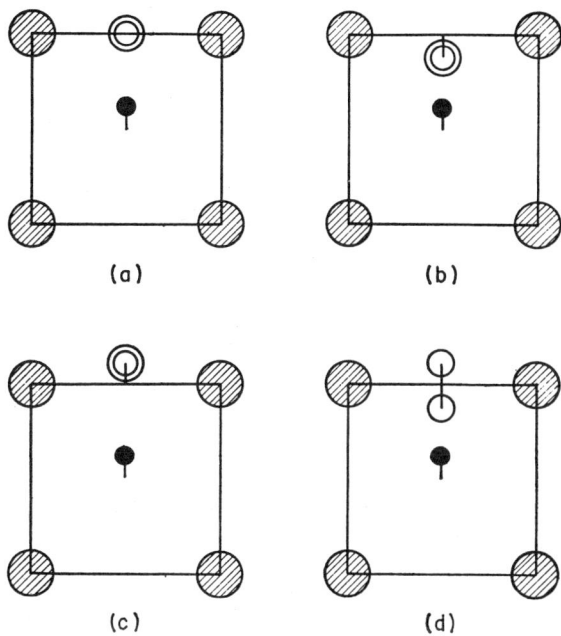

FIG. X-1. Schematic representation of possible configurations resulting from the shift of the oxygen atoms OI in BaTiO$_3$ (according to Shirane *et al.* (S 5)).

than that of oxygen that it is extremely difficult to distinguish between small co-ordinate shifts and temperature oscillations of the lighter oxygen atoms. This problem was discussed in detail by Shirane *et al.* (S 4) in the course of their structural analysis of PbTiO$_3$. The ambiguities encountered in a problem of this sort are evidenced by the schematic drawings of Fig. X-1. Each one of the four sketches depicted in this figure may represent the projection of a cell of BaTiO$_3$ on the (100) plane, the Ba atoms being at the corners of the squares, the Ti atom in the center, the OI atoms directly above Ti, and the OII atoms being omitted for convenience. In all four cases, the displacement of the Ti atom is assumed to be positive, but that of oxygen OI is assumed to be zero in case (a), negative in case (b), and positive in case (c). Using X-rays, it is possible to distinguish model (a) from models (b) or (c), but is difficult to distinguish between (b) and (c). This can be seen from the fact that the X-ray intensities are proportional to the square of the structural factor $F$, and from (X-2)

$$F^2 = A^2 + B^2 = \left(\sum_{j=1}^{N} A_j\right)^2 + \left(\sum_{j=1}^{N} B_j\right)^2 =$$

$$= (A_{\mathrm{Ba}} + A_{\mathrm{Ti}} + A_{\mathrm{OI}})^2 + (B_{\mathrm{Ti}} + B_{\mathrm{OI}})^2. \qquad \text{(X-3)}$$

The Ba atom does not contribute to the $B$ terms because it is located at the origin of the reference system of co-ordinates, so that $2\pi l \delta z_{\mathrm{Ba}} \equiv 0$. Now, it is clear that the three models (a), (b) and (c) of Fig. X-1 have identical values of $A_{\mathrm{Ti}}$ and $B_{\mathrm{Ti}}$ and differ only in the values of $A_{\mathrm{OI}}$ and $B_{\mathrm{OI}}$. Models (b) and (c) have identical values of $A_{\mathrm{OI}}$, and equal magnitudes of $B_{\mathrm{OI}}$ values but with opposite signs. The $A$ terms are relatively large, thus a distinction between (a) on one side and (b) or (c) on the other is possible, but the $B_{\mathrm{OI}}$ terms are much smaller, in comparison, and therefore a distinction between (b) and (c) is hardly possible within the experimental accuracy. If we now introduce a fictitious model such as (d) in Fig. X-1 where we assume that "half" of oxygen OI is displaced by a distance $+\delta$ and the other "half" by $-\delta$, we see that $B_{\mathrm{OI}}$ is equal to zero, but the value of $A_{\mathrm{OI}}$ is the same as for models (b) and (c). Consequently, the magnitude of $F^2$ is very nearly the same for all three models (b), (c) and (d). When the co-ordinate shift $\delta$ is small, the X-ray intensity obtained from model (d) cannot easily be distinguished from that obtained from a model in which OI has a large anisotropic thermal vibration along the vertical direction.

These qualitative arguments illustrate the difficulties encountered in the determination of accurate positional parameters in pseudosymmetric structures by means of X-rays. In structures more complicated than those of $BaTiO_3$ and $PbTiO_3$ (i.e. structures such as those of $KH_2PO_4$ and Rochelle salt), it is generally unlikely that the imaginary parts are smaller than the real parts for *all* structure factors. This means that the contribution of the sum of all $B_j$'s to the intensity (see Eq. X-3) is not negligible, although the values of the single $B_j$'s may be difficult to determine. On the other hand, the determination of these complex structures requires the knowledge of a larger number of reflection intensities than does that of simpler pseudo-cubic structures. In most cases, the ambiguities can be resolved (as mentioned in the previous chapters) by using neutron diffraction methods. In this case, the scattering factors are such that the $B$ terms have the same order of magnitude as the $A$ terms, and therefore a distinction between models such as (a), (b), (c) or (d) in Fig. X-1 is more easily attainable.

We have shown in the previous chapters that the structural characteristics of several ferroelectrics have been established to a rather high degree of accuracy. It should be emphasized, in this respect, that an accurate structure analysis is a very time-consuming and not always rewarding enterprise, because the atomistic mechanism leading to ferroelectricity cannot always be inferred uniquely from the structure. This aspect is again in contrast with the situation encountered in the neutron diffraction studies of magnetic materials, where the structural results usually provide unambiguous answers to the problem of ordering of magnetic dipoles.

## 4. The Absolute Configuration of Ferroelectric Structures

It is well known that when the unit cell of a given structure contains more than one kind of atom it is possible for the phase of the scattered X-ray radiation to be different for different atoms, thus introducing into the structure factor phase differences other than those due to the positions of the atoms. In ordinary X-ray diffraction analysis, it is generally assumed that the change of phase upon scattering is the same for all the constituent atoms in the structure. This is the basis for the so–called Friedel's law

$$| F(h, k, l) | \; = \; | F(\bar{h}, \bar{k}, \bar{l}) |, \qquad (X\text{-}4)$$

expressing the fact that the intensity of reflection from opposite sides of the same set of crystal planes is the same. Consequently, ordinary structure analysis cannot show the sense of any polar direction in the structure (J 1). When ferroelectric crystals are studied, it is usually necessary to apply a d.c. field in the direction of the polar axis in order to obtain a single domain, but identical intensities are obtained for any given reflection irrespective of whether the field is applied in the positive or in the negative direction.*

The extent of this limitation may become more significant if we consider some practical examples. Consider simple structures such as those of $BaTiO_3$ and $PbTiO_3$, and choose the origin in such a way that the oxygens OII remain undisplaced by the onset of the ferroelectric phase (see Sections IV-6 and V-5). It has been *assumed* that the shift of the Ti ion occurs in the same and that of the OI ions in the opposite direction as the polarizing field. Indeed, this assumption seems to be quite logical, but there is really no *a priori* reason why this should be so, especially if we consider that a large distortion of the electron clouds is likely to occur as a consequence of the spontaneous polarization. It was pointed out by Känzig (K 2) that this question may be particularly interesting for $PbTiO_3$, whose anomalies are generally more pronounced than those of $BaTiO_3$.

The problem of the sense of the spontaneous polarization becomes even more obscure when we deal with structures more complex than the perovskite structure. We have briefly discussed the situation in $KH_2PO_4$ in Section III-7. We may recall that Bacon and Pease (B 2) calculated the spontaneous polarization of $KH_2PO_4$ on the basis of charge assignments corresponding to $P^{3-}$ and $K^{1+}$ ions (the effective charge of the O and the H ions were not specified), obtaining $4.7 \times 10^{-6}$ C/cm². The data listed in Table X-1 indicate that in this case the

---

* This statement is only correct, of course, when reversal of the electric field does not produce an interchange of axes. It has been shown by Bacon and Pease (B 2) in the course of their neutron diffraction study of $KH_2PO_4$ that the intensities of the reflections (400) and (16, 0, 0) vary approximately by a factor of 10 when the direction of the biasing field is reversed. In this case, reversal of the field direction produces an interchange of the $a$ and $b$ axes of the orthorhombic unit cell, and this changes the (400) and (16, 0, 0) reflections into the (040) and (0, 16, 0) reflections, respectively, with consequent modification of the structure factors and intensities. However, the intensities of the (00$l$) reflections are unaltered by the reversal of the biasing field.

spontaneous polarization points in the direction opposite to that of the displacement of the phosphorous ion. This situation is depicted schematically in Fig. X-2 (b), which shows that the hydrogen atoms are close to the "upper" oxygens of any $PO_4$ group, and the phosphorous ion is displaced in the direction opposite to the

TABLE X-1. SPONTANEOUS POLARIZATION OF $KH_2PO_4$ AS CALCULATED FOR DIFFERENT EFFECTIVE CHARGES OF THE IONS
($\delta z$ = ionic shifts as obtained from the positional parameters reported by Bacon and Pease (B 2). Model (b) was proposed by Bacon and Pease.)

| Ion | $\delta z$ (Å) | Effective charges | |
|---|---|---|---|
| | | Model (a) | Model (b) |
| K | $-0.04$ | $+1$ | $+1$ |
| P | $+0.08$ | $+5$ | $-3$ |
| H | $-0.02$ | $+1$ | $-$ |
| Spontaneous polarization ($10^{-6}$ C/cm²) | | $+5.0$ | $-4.7$ |

polarization vector. In Fig. X-2, the polarization vector is assumed to point upwards for both case (a) and (b). On the other hand, if we compute the value of the spontaneous polarization on the basis of full ionic charges $K^{1+}$ and $P^{5+}$ we obtain $5.0 \times 10^6$ C/cm², but in this case the sense of the spontaneous polarization is the same as that of the phosphorous displacement.

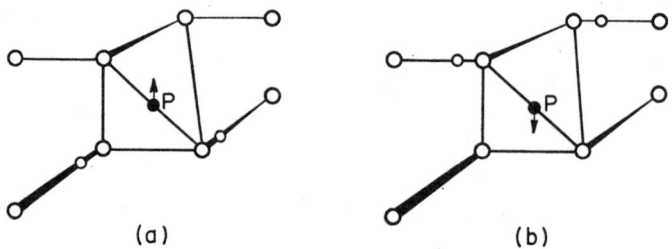

(a)                          (b)

FIG. X-2. Two models for the direction of ionic shifts in $KH_2PO_4$ (schematic). The figures represent $(H_2PO_4)^-$ groups with opposite polarities. The spontaneous polarization of the crystal is pointing "up" for both models. The ionic charge of the phosphorous ion is assumed to be $+5$ in (a) and $-3$ in (b) (see Table X-1).

These ambiguities can be resolved by resorting to the phenomenon of anomalous dispersion of X-rays. The assumption that the change of phase of the X-ray wave upon scattering is the same for all the constituent atoms of the structure is only valid when the wavelength of the incident radiation is much greater than any one of the wavelengths of the natural vibrations of the oscillators in the $K$, $L$ or $M$ groups of electrons. When this is not the case, i.e. the wavelength of the incident radiation is shorter than, say, that of the $K$ absorption, then the incident wavelength must always coincide with that of some oscillator of the $K$ continuum, and anomalous dispersion occurs (see, e.g. James (J 1)). In such a case, the atomic scattering factor is written as a complex number, thus

$$f = f_0 + \Delta f' + i f'' = f' + i f'' \qquad (\text{X-5})$$

where $f_0$ is the atomic scattering factor for wavelengths short in comparison with any atomic absorption, and is independent of the incident wavelength, while $\Delta f'$ and $f''$ are the real and imaginary parts of $f$ that depend on the wavelength. The effect of the imaginary part of $f$ is to add to the resultant scattered wave a component that at each point of the crystal is out of phase with the primary wave; there is a progressive diminution in the amplitude of the transmitted wave, or, in other words, an absorption. With the complex scattering factor, the square of the amplitude of the crystal structure factor becomes

$$|F(h, k, l)|^2 = \sum_{j, i} (f_j' f_i' + f_j'' f_i'') \cos 2\pi \left[ h(x_j - x_i) + k(y_j - y_i) + l(z_j - z_i) \right] +$$

$$+ \sum_{j, i} (f_j' f_i'' - f_j'' f_i') \sin 2\pi \left[ h(x_j - x_i) + k(y_j - y_i) + l(z_j - z_i) \right]. \qquad (\text{X-6})$$

Hence, we see that Friedel's law is no longer obeyed, i.e.

$$|F(h, k, l)| \neq |F(\bar{h}, \bar{k}, \bar{l})|.$$

This fact was confirmed in beautiful experiments with zincblende by Koster et al. (K 3) and later utilized by Bijvoet (B 3) to determine the absolute configuration of structures in enantiomorphous arrangements. More recently, Okaya, Saito and Pepinsky (O 1), (P 1) developed a method by which it is possible to obtain directly information about the structure as well as the sense of the non-centric distribution of non-anomalous scatterers about an anomalous scatterer, i.e. the absolute configuration. The latter goal can be achieved by studying the odd Fourier series that has the $|F(h, k, l)|^2$ as coefficients, i.e. the Patterson function

$$P_{\text{odd}}(u, v, w) = \sum_{\substack{h, k, l \\ -\infty}}^{+\infty} |F(h, k, l)|^2 \sin 2\pi (hu + kv + lw). \qquad (\text{X-7})$$

This odd function contains peaks of heights equal to $(f_j' f_i'' - f_j'' f_i')$ at the points $(u, v, w) = (x_j - x_i, y_j - y_i, z_j - z_i)$, together with peaks of height $- (f_j' f_i'' - f_j'' f_i')$ at the points $(-u, -v, -w) = (x_i - x_j, y_i - y_j, z_i - z_j)$.

If the $j$th and the $i$th atoms are of the same kind, the peaks at $\pm (x_j - x_i, y_j - y_i, z_j - z_i)$ in $P_{\text{odd}}(u, v, w)$ disappear. Thus, this function indicates only interactions between different kinds of atoms. If only one atom, say the $m$th, scatters X-rays with anomalous phase and amplitude, we have,

$$f_j' = f_j, f_j'' = 0 \text{ for } j \neq m,$$

and

$$f_j'' \neq 0 \text{ for } j = m.$$

Then the function $P_{\text{odd}}(u, v, w)$ has peaks of height $f_j f_m''$ at the points $(x_j - x_m, y_j - y_m, z_j - z_m)$, and of height $-f_j f_m''$ at the points $(x_m - x_j, y_m - y_j, z_m - z_j)$. We can distinguish between positive and negative peaks, and hence we can specify the magnitude and direction (including the sense) of the vector between the $m$th and $j$th atoms. When there is only one anomalous scatterer per cell, we can write

$$|F(h, k, l)|^2 - |F(\bar{h}, \bar{k}, \bar{l})|^2 =$$
$$2f''_m \sum_j f_j \sin 2\pi \left[h(x_j - x_m) + k(y_j - y_m) + l(z_j - z_m)\right]. \qquad (\text{X-8})$$

Thus the difference between the intensities $I(h, k, l)$ and $I(\bar{h}, \bar{k}, \bar{l})$ depends only upon the imaginary component of the anomalous scatterer and upon the atomic scattering factors and distribution of the normal scatterers around each anomalous scatterer. The larger the component $f''_m$, the more significant the difference $I(h, k, l) - I(\bar{h}, \bar{k}, \bar{l})$. This is expected to be the case if the value of $f''_m$ is larger than 3.0. Values between 3.0 and 3.5 can be achieved if the incident wavelength is $K\alpha$ radiation originating from an element with atomic number 2–3 times higher than the anomalous scatterer (P 1). In general, however, both the values of $f''$ and of $\Delta f'$ are comparatively small for $K$ electrons (J 1). In 1955, Dauben and Templeton (D 1) published the values of $\Delta f'$ and $f''$ for all atoms with atomic number larger than 20, as computed from the contributions of $K$, $L$ and $M$ electrons. These values are considerably larger than those calculated from the contribution of $K$ electrons alone. For example, Ba has $f'' = 8.9$ and $\Delta f' = -2.1$ for $CuK\alpha$ radiation.

In recent years, anomalous scattering of X-rays has been used extensively for the study of the absolute configuration of various ferroelectrics. Okaya et al. (O 2) have investigated $BaTiO_3$ with $FeK\alpha$ and $CoK\alpha$ radiations, with particular attention to the differences between $|F(h01)|^2$ and $|F(\bar{h}0\bar{1})|^2$. Although the ratio between the intensities of reflections from opposite sides of the same set of planes was found to vary only between 1.05 and 0.95, it was possible to confirm that the Ti ions are indeed displaced in the same sense as the spontaneous polarization, provided that the origin of the reference system is chosen at the level of the OII ions. This is indeed an expected result; nevertheless, it is important to have direct experimental evidence in its favor.

The absolute configuration of $KH_2PO_4$ was studied by Unterleitner et al. (U 1), actually not on $KH_2PO_4$ itself but rather on the isomorphous crystal $KD_2AsO_4$. The reason for this choice is that the value of $f''$ with respect to the employed radiation ($MoK\alpha$) is larger for As ($f'' = 2.2$) than for P. The Curie point of $KD_2AsO_4$ is at 162 °K (see Section III-6), but the investigations were carried out at room temperature on a strained crystal. A configuration similar to that of the ferroelectric phase was forced upon the crystal at room temperature by means of a compressive stress along [110] via the piezoelectric effect. The data show that the polarization has the same sense as the displacement of the P atoms within the $PO_4$ tetrahedra, as indicated schematically in Fig. X-2(a). This result is in contradiction with the assumptions made by Bacon and Pease (B 2) and discussed above, but appears to be more reasonable in view of the fact that it agrees with the model based on a more conventional charge assignment for the phosphorous ion (see Table X-1). It may be interesting to point out, however, that this is not the result that one would expect from a primitive picture of a $(H_2PO_4)^-$ dipole, because the positive hydrogen atoms turn out to be located on the negative end of the dipole.

The absolute configuration of the ferroelectric phase of tri-glycine sulfate was also studied by Unterleitner et al. (U 1) who used $CuK\alpha$ radiation to excite the S atoms in the $SO_4$ groups. The assumption of Hoshino et al. (H 1) that the $(NH)^{3+}$ group is displaced in the direction of the polarization was confirmed.

## 5. Lattice Vibrations and Ferroelectricity

The most promising method of dealing with the problem of ferroelectricity, at the present stage, seems to be that of treating it as a problem in lattice dynamics (A 1), (C 1), (L 1). The basic idea behind this approach is fairly simple, and we may first attempt to describe it with qualitative arguments. This idea is based on the probable existence of a low-frequency optical vibration just above the Curie temperature of those ferroelectrics which show a large anomaly of the dielectric constant. Consider the well-known dispersion formula for the dielectric constant in the hypothetical case of zero damping (see, e.g. Born and Huang (B 4)),

$$\varepsilon(\omega) = \varepsilon_\infty + \sum_i \frac{b_i^2}{\omega_i^2 - \omega^2}, \tag{X-9}$$

where $\varepsilon_\infty$ is the high frequency (or optical) dielectric constant, $\omega_i$ is a measure of the strength of the restoring force related to the $i$th optical mode, and $b_i$ the strength of the coupling of this mode to the applied field $E$. For a given mode, the frequency dependence of the dielectric constant has the well-known behavior depicted in Fig. X-3, becoming infinitely large at the frequency $\omega = \omega_i$ and equal to zero at the frequency $\omega = \omega_i(\varepsilon_0/\varepsilon_\infty)^{1/2}$. The static dielectric constant $\varepsilon_0$ is given by

$$\varepsilon_0 = \varepsilon_\infty + \sum_i \frac{b_i^2}{\omega_i^2}, \tag{X-10}$$

and is the one which becomes very large at the Curie point of a number of ferroelectrics. The optical constant $\varepsilon_\infty$, on the other hand, is very slightly dependent upon temperature (see, for example, Fig. IV-11), so that an anomaly in $\varepsilon_0$ can only obtain if the second term in (X-10) becomes very large. Assuming that the coupling constants $b_i$ remain essentially finite at all temperatures of interest, one must conclude that the restoring force of at least one of the vibrational modes considered here becomes very small (or zero) at a given temperature, the Curie temperature $T_c$. It is therefore logical to postulate that this particular restoring force is proportional to $(T - T_c)$, so that Eq. (X-10) assumes the form of a Curie–Weiss law. The particular mode for which this is the case may be called the ferroelectric mode (L 1), and it is the purpose of a proper theory to express above arguments in quantitative terms and provide an estimate for the frequency $\omega_i$ corresponding to the restoring force of the ferroelectric mode.

A suitable theory must necessarily be based on a very simple structure, since only in such a case is it possible to cope with the mathematical difficulties involved. It is therefore reasonable to try and use, at first, the fairly successful concepts developed in the theory of the dielectric behavior of alkali halides. In one of its simplest forms, such a theory was based on the assumption that a unique polarizability could be assigned to individual ions in different crystals. We have mentioned previously that this was done by Tessman *et al.* (T 1) for the electronic polarizability, and by Roberts (R 1) for the total (i.e. electronic plus

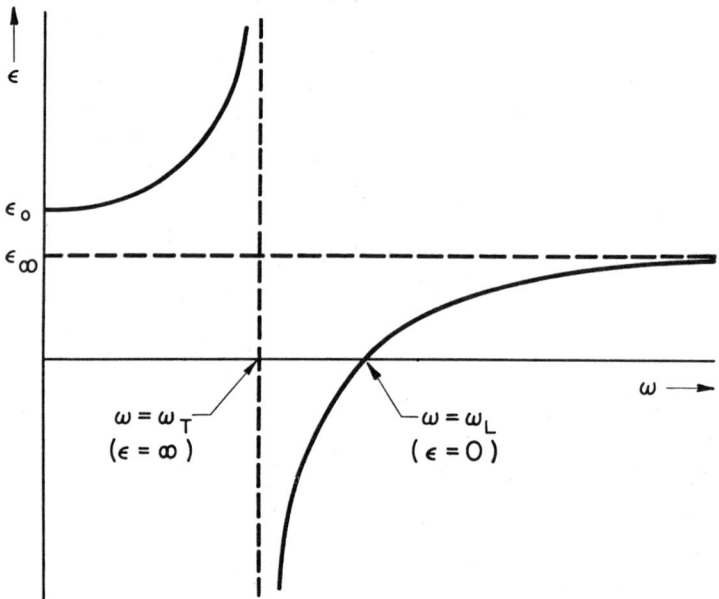

FIG. X-3. Frequency dependence of the real part of the dielectric constant (schematic).

ionic) polarizability. Roberts (R 2) also developed a classical model-theory for the dielectric constant of alkali halides. Such sets of polarizabilities are quite convenient for an estimate of the dielectric properties of various compounds, but it must be remembered that their general validity depends upon the assumption of their independence of volume, so that their values should not be altered, for example, when the substance is compressed. Szigeti (S 6) has shown, however, that this assumption is not justified; in an optical mode of simple ionic crystals, where all the positive ions are displaced against the negative, the overlap forces between nearest and next to nearest neighbors should be taken into account. Consideration of the overlap potential between nearest neighbors has led Szigeti to a relationship between the dispersion frequency and the compressibility, relationship which is only approximately fulfilled by the experimental data for alkali halides and simple oxides. In their discussion of the problem, Born and Huang (B 4) have pointed out a variety of causes that are probably responsible for the discrepancies between observed and calculated compressibilities, namely the fact that the theory does not take into account van der Waals forces and overlap potentials between second neighbors, and also some possible genuine deviations from the ideal ionic structure in some cases. It has also been pointed out by Szigeti that mutual distortion of neighboring ions must accompany the overlap between these ions, and these distortions may modify the strength of the coupling with the effective field acting in the ions. A phenomenological treatment of such distortions has been presented by Born and Huang (B 4), who did not investigate, however, the details of the origin of the distortion dipoles. The problem has also been treated by Yamashita and Kurosawa (Y 1), (Y 2) in terms of quantum-mechanical calculations, and later reconsidered by Dick and Over-hauser (D 2). Using the quantum-mechanical treatment of the short-range re-pulsive interactions between nearest neighbors, the latter authors have introduced a simple physical picture of these interactions, which allows a quantitative evaluation of the theory. The model of Dick and Overhauser is based upon the recognition that the outer electrons of an ion, being less tightly bound, are more profoundly affected by the application of an electric field than the inner elec-trons. Accordingly, the ions with rare-gas configurations are regarded as con-sisting of an outer spherical shell of $n$ electrons and a core constituted by the nucl-eus and the remaining electrons. In an electric field the shell retains its spherical charge distribution but moves bodily with respect to the core. The polarizability is made finite by a harmonic restoring force which acts between the core and the shell. The two unknown parameters ($n$ and the spring constant of the restoring force) are chosen by considering the polarizability and ultraviolet dispersion of a gas of such model ions. Using this model, Dick and Overhauser introduced two polarization mechanisms which were neglected in the simple dielectric theory; namely, the "short-range interaction" polarization and the "exchange charge" polarization. The short-range interaction arises when pairs of positive and nega-tive ions are moved toward each other because the shells of the ions repel one another and tend to become displaced with respect to the ion cores. This is equivalent to a polarization of the ions, which may have different values for nega-tive and positive ions, depending on the spring constants and the shell charges $ne$. The exchange charge polarization arises from the overlap of the ions and the resulting exchange charge. This alteration of the exchange-charge distribution is re-sponsible for the repulsive force between nearest neighbors and gives rise to a net dipole moment per unit volume. When applied to a crystal, the model of Dick and Overhauser has been found to clarify many of the discrepancies of the older theories.

It is on the model of Dick and Overhauser that Cochran (C 1) based his theory of ferroelectricity for the simple case of diatomic cubic crystal. The theory (W 1) gives the following expressions for the frequencies of the transverse optic (T.O.) and longitudinal optical (L.O.) modes of wave vector zero:

$$\overline{M}\,\omega_T^2 = R_0' - \frac{4\pi}{9v}(\varepsilon_\infty + 2)(Z'\,e)^2 \qquad \text{(X-11)}$$

$$\overline{M}\,\omega_L^2 = R_0' + \frac{8\pi}{9v\,\varepsilon_\infty}(\varepsilon_\infty + 2)(Z'\,e)^2. \qquad \text{(X-12)}$$

Here, $\overline{M}$ is the reduced mass of an ion pair defined by

$$\frac{1}{\overline{M}} = \frac{1}{M_1} + \frac{1}{M_2},$$

$M_1$ and $M_2$ being the masses of the two ions; $v$ is the volume of the unit cell, $Z'e$ the effective ionic charge, and $R_0'$ the "short-range" restoring force. The two latter quantities depend explicitly on the parameters of the shell model. In a crystal such as NaI (W 1) the short-range term $R_0'$ is about twice as great as the second term in (X-11), which arises from the Coulomb interaction. The case of interest here is that in which the two terms on the right hand side of (X-11) become equal to one another. The theory shows that the T.O. frequency $\omega_T$ may approach zero without the crystal becoming unstable against other vibration modes; indeed all T.O. modes for which the wave vector is not close to zero may retain quite usual frequencies. We may therefore postulate that

$$\frac{\overline{M}\,\omega_T^2}{R_0'} = 1 - \frac{4\pi}{9v R_0'}\,(\varepsilon_\infty + 2)(Z'e)^2 = \eta\,(T - T_c) \qquad \text{(X-13)}$$

and using the relation

$$\frac{\omega_T^2}{\omega_L^2} = \frac{\varepsilon_\infty}{\varepsilon_0}, \qquad\qquad \text{(X-14)}$$

we obtain a Curie–Weiss law

$$\varepsilon_0 - \varepsilon_\infty = \frac{4\pi}{9v R_0' \eta}\,(\varepsilon_\infty + 2)^2 (Z'e)^2 \,\frac{1}{T - T_c}. \qquad \text{(X-15)}$$

It should be pointed out that this equation is a modified form of one of the equations originally derived by Szigeti (S 6). The Curie constant $C$ is therefore given by

$$C = \frac{4\pi}{9v R_0' \eta}\,(\varepsilon_\infty + 2)^2 (Z'e)^2 \simeq \frac{\varepsilon_\infty + 2}{\eta}. \qquad \text{(X-16)}$$

Eq. (X-15) shows that when $T$ approaches $T_c$, $\varepsilon_0$ approaches $\infty$; Eq. (X-13) shows that when this occurs, $\omega_T^2$ approaches zero. We can gain insight into the meaning of this by considering that the ionic polarizability of one unit cell is given by

$$\alpha_{\text{ion}} = \frac{(Z'e)^2}{R_0'} \qquad\qquad \text{(X-17)}$$

and the electronic polarizability by

$$\frac{4\pi\,\alpha_{\text{el}}}{3v} = \frac{\varepsilon_\infty - 1}{\varepsilon_\infty + 2} \qquad\qquad \text{(X-18)}$$

so that the condition $\omega_T^2 = 0$ is equivalent to

$$\frac{4\pi}{3v}\,(\alpha_{\text{ion}} + \alpha_{\text{el}}) = 1, \qquad\qquad \text{(X-19)}$$

which is the condition for the $4\pi/3$ catastrophe (cf. Eq. IV-61).

An estimate of the order of magnitude for $\omega_T$ can be made on the basis of Eq. (X-14) by considering that typical values for the frequencies of longitudinal waves in ionic crystals (B 4) are about $10^{13}$ c/s, so that $\omega_L^2 \sim 4 \times 10^{27}$ c/s. The ratio $\varepsilon_\infty/\varepsilon_0$, on the other hand, is of the order of $10^{-4}$–$10^{-5}$ at the Curie temperature of typical ferroelectrics, and thus $\omega_T$ should be of the order of $10^{11}$ c/s. More accurate calculations for $BaTiO_3$ and similar perovskite ferroelectrics have led Cochran (C 1) to the value $\nu = 2$–$3 \times 10^{11}$ c/s for the frequency of the ferro-electric mode just above the Curie temperature. This frequency is abnormally low for optical modes but unfortunately lies in the range of millimeter waves which is experimentally difficult to handle. It may be recalled that the measurements of Benedict and Durand (B 5) on single crystal $BaTiO_3$ revealed no dispersion up to the frequency $2.4 \times 10^{10}$ c/s (see Section IV-2). Measurements of the dielectric properties in the millimeter wavelength range are not available at present.

Concluding, it seems reasonable to say that Cochran's theory may eventually provide the clearest picture of the balance between short- and long-range forces in ferroelectrics. What seems difficult to do, at the present stage, is to give physical meaning to the factors that cause the condition $\omega_T = 0$ in a given crystal, and to decide to which type of ferroelectric crystals the theory can be applied. Obviously, the first step is to give experimental support to the theory by means of measurements of lattice vibrations or dielectric dispersion. It was pointed out by Cochran (C 2) that the critical scattering experiments with tri-glycine sulfate of Shibuya and Mitsui (S 7) are closely related to the phenomenon discussed above, although the problem was approached in quite a different way. The origin of the critical scattering, in this case, is the correlation of atomic positions in neighboring unit cells and this, in turn, can be regarded as an optical mode with very low frequency. Further experimental and theoretical treatments are awaited with great interest.*

## BIBLIOGRAPHY

(A 1) ANDERSON, P. W., Preprint of paper presented at the Moscow Conference on Dielectrics, December (1958).
(B 1) BACON, G. E., *Neutron Diffraction*, Clarendon Press, Oxford (1955).
(B 2) BACON, G. E. and PEASE, R. S., *Proc. Roy. Soc. (London)* **A 220**, 397 (1953); **A 230**, 359 (1955).
(B 3) BIJVOET, J. M., *Nature* **173**, 888 (1954).
(B 4) BORN, M. and HUANG, K., *Dynamic Theory of Crystal Lattices*, Oxford University Press (1954).
(B 5) BENEDICT, T. S. and DURAND, J. L., *Phys. Rev.* **109**, 1091 (1958).
(C 1) COCHRAN, W., *Phys. Rev. Letters* **3**, 412 (1959).
(C 2) COCHRAN, W., Private communication (1960).

* After this text was written, a comprehensive description of the theory was published by Cochran (*Phil. Mag.* Suppl. 9, 387 (1960)). Critical X-ray scattering was also observed in $NaNO_2$ by Shibuya (*J. Phys. Soc. Japan* **16**, 490 (1961)).

(D 1) DAUBEN, C. H. and TEMPLETON, D. H., *Acta Cryst.* **8**, 841 (1955).

(D 2) DICK, B. G., JR. and OVERHAUSER, A. W., *Phys. Rev.* **112**, 90 (1958).

(E 1) EVANS, H. T., Technical Report No. 58, Laboratory for Insulation Research, Massachusetts Institute of Technology (1953).

(H 1) HOSHINO, S., OKAYA, Y. and PEPINSKY, R., *Phys. Rev.* **115**, 323 (1959).

(J 1) JAMES, R. W., *The Optical Principles of the Diffraction of X-Rays*, Bell, London (1954).

(K 1) KITTEL, C., *Introduction to Solid State Physics*, John Wiley, New York (1956).

(K 2) KÄNZIG, W., *Ferroelectrics and Antiferroelectrics, Solid State Physics*, vol. 4, p. 165, Academic Press, New York (1957).

(K 3) KOSTER, D., KNOL, K. S. and PRINS, J. A., *Z. Physik* **63**, 345 (1930).

(L 1) LANDAUER, R. and THOMAS, L. H., *Bull. Am. Phys. Soc.* Ser. II **4**, 424 (1959).

(M 1) MEGAW, H. D. and WELLS, M., *Acta Cryst.* **11**, 858 (1958).

(O 1) OKAYA, Y., SAITO, Y. and PEPINSKY, R., *Phys. Rev.* **98**, 1857 (1955).

(O 2) OKAYA,, Y. PEPINSKY, R. and UNTERLEITNER, F., *Bull. Am. Phys. Soc.* Ser. II **4**, 62 (1959).

(P 1) PEPINSKY, R. and OKAYA, Y., *Proc. Nat. Acad. Sci. U.S.* **42**, 286 (1956).

(R 1) ROBERTS, S., *Phys. Rev.* **76**, 1215 (1949); **81**, 161, 865 (1951).

(R 2) ROBERTS, S., *Phys. Rev.* **77**, 258 (1950).

(S 1) SHULL, C. G. and WOLLAN, E. O., *Application of Neutron Diffraction to Solid State Physics Problems, Solid State Physics*, vol. 2, pp. 137—217, Academic Press, New York (1956).

(S 2) SLATER, J. C., *Phys. Rev.* **78**, 748 (1950).

(S 3) SLATER, J. C., *J. Chem. Phys.* **9**, 16 (1941).

(S 4) SHIRANE, G., PEPINSKY, R. and FRAZER, B. C., *Acta Cryst.* **9**, 131 (1956).

(S 5) SHIRANE, G., JONA, F. and PEPINSKY, R., *Proc. I.R.E.* **43**, 1738 (1955).

(S 6) SZIGETI, B., *Trans. Faraday Soc.* **45**, 155 (1949); *Proc. Roy. Soc. (London)* **A 204**, 51 (1950).

(S 7) SHIBUYA, I. and MITSUI, T., *J. Phys. Soc. Japan* **16**, 479 (1961).

(T 1) TESSMAN, J. R., KAHN, A. H. and SHOCKLEY, W., *Phys. Rev.* **92**, 890 (1953).

(U 1) UNTERLEITNER, F., OKAYA, Y., VEDAM, K. and PEPINSKY, R., *Am. Cryst. Assoc., Abstracts Annual Meeting*, Ithaca, N.Y., July (1959).

(V 1) VOUSDEN, P., Acta Cryst. **7**, 321 (1954).

(W 1) WOOD, A. D. B., COCHRAN, W. and BROCKHOUSE, B. N., *Bull. Am. Phys. Soc.* Ser. II **4**, 246 (1959); *Phys. Rev.* **119**, 980 (1960).

(Y 1) YAMASHITA, J. and KUROSAWA, T., *J. Phys. Soc. Japan* **10**, 610 (1955).

(Y 2) YAMASHITA, J., *Progr. Theor. Phys. (Kyoto)* **8**, 280 (1953); **12**, 454 (1955).

# TABLE OF FERROELECTRIC CRYSTALS

The present Table lists the ferroelectric compounds known up to January 1960 (most solid solutions are *not* considered). $T_c$ = Curie temperature in (°C); $P_s$ (at $T$ °C) = spontaneous polarization $P_s$ (in units of $10^{-6}$ C/cm$^2$) measured at the temperature $T$ (in °C) listed in brackets. References to the original literature are not given here but can be found in the chapters and sections of this book that are listed in the last column.

| Name (abbreviation) | Formula | $T_c$ (°C) | $P_s$ (at $T$ °C) ($10^{-6}$ C/cm$^2$) | Chapter and Section |
|---|---|---|---|---|
| Potassium di-hydrogen phosphate (KDP) | $KH_2PO_4$ | − 150 | 4.75[− 177] | III |
| Potassium di-deuterium phosphate | $KD_2PO_4$ | − 60 | 4.83[− 93] | III-6 |
| Potassium di-hydrogen arsenate | $KH_2AsO_4$ | − 176 | 5.0[− 195] | III-6 |
| Potassium di-deuterium arsenate | $KD_2AsO_4$ | − 111 | ? | III-6 |
| Rubidium di-hydrogen phosphate | $RbH_2PO_4$ | − 126 | 5.6[− 183] | III-6 |
| Rubidium di-deuterium phosphate | $RbD_2PO_4$ | − 55 | ? | III-6 |
| Rubidium di-hydrogen arsenate | $RbH_2AsO_4$ | − 163 | ? | III-6 |
| Rubidium di-deuterium arsenate | $RbD_2AsO_4$ | − 95 | ? | III-6 |
| Cesium di-hydrogen phosphate | $CsH_2PO_4$ | − 114 | ? | III-6 |
| Cesium di-hydrogen arsenate | $CsH_2AsO_4$ | − 130 | ? | III-6 |
| Cesium di-deuterium arsenate | $CsD_2AsO_4$ | − 61 | ? | III-6 |
| Barium titanate | $BaTiO_3$ | 120 | 26.0 [23] | IV |
| Lead titanate | $PbTiO_3$ | 490 | >50 [23] | V-5 |
| Potassium niobate | $KNbO_3$ | 435 | 30.0 [250] | V-2 |
| Potassium tantalate | $KTaO_3$ | − 260 | ? | V-2 |
| Sodium niobate | $NaNbO_3$ | − 200 | 12.0 [− 200] | V-3 |
| Cadmium titanate | $CdTiO_3$ | ∼ − 218 | ? | V-8 |
| Tungsten trioxide | $WO_3$ | − 50 | ? | V-10 |
| Lead iron niobate | $Pb_2(FeNb)O_6$ | 112 | ? | V-9 |
| Lead magnesium niobate | $Pb_3(MgNb_2)O_9$ | − 10 | ? | V-9 |
| Cadium (pyro) niobate | $Cd_2Nb_2O_7$ | − 88 | ∼6.0 [− 185] | VI-2 |
| Strontium (pyro) tantalate | $Sr_2Ta_2O_7$ | − 80 | ? | VI-3 |

| Name (abbreviation) | Formula | $T_c$ (°C) | $P_s$ (at $T$ °C) ($10^{-6}$ C/cm²) | Chapter and Section |
|---|---|---|---|---|
| Lead (meta) niobate | $PbNb_2O_6$ | 570 | ? | VI-4 |
| Lead (meta) tantalate | $PbTa_2O_6$ | 260 | 10.0 [25] | VI-4 |
| Lead bismuth niobate | $PbBi_2Nb_2O_9$ | 526 | ? | VI-6 |
| Lithium tantalate (?) | $LiTaO_3$ | ? | 23.0 [450] | VI-7 |
| Sodium vanadate (?) | $NaVO_3$ | 380 | ? | VI-8 |
| Silver vanadate | $AgVO_3$ | 180 | ? | VI-8 |
| Barium lithium oxy-fluoaluminate | $Ba(Al_{1.4}Li_{0.6})(O_{2.8}F_{1.2})$ | 150 | 0.1 [23] | VI-9 |
| Guanidinium aluminum sulfate hexahydrate (GASH) | $C(NH_2)_3Al(SO_4)_2 \cdot 6H_2O$ | none | 0.35 [23] | VIII-2 |
| Guanidinium chromium sulfate hexahydrate (GCrSH) | $C(NH_2)_3Cr(SO_4)_2 \cdot 6H_2O$ | none | 0.36 [23] | VIII-2 |
| Guanidinium gallium sulfate hexahydrate (GGaSH) | $C(NH_2)_3Ga(SO_4)_2 \cdot 6H_2O$ | none | 0.36 [23] | VIII-2 |
| Guanidinium vanadium sulfate hexahydrate (GVSH) | $C(NH_2)_3V(SO_4)_2 \cdot 6H_2O$ | none | 0.36 [23] | VIII-2 |
| Guanidinium aluminum selenate hexahydrate (GASeH) | $C(NH_2)_3Al(SeO_4)_2 \cdot 6H_2O$ | none | 0.45 [23] | VIII-2 |
| Guanidinium chromium selenate hexahydrate (GCrSeH) | $C(NH_2)_3Cr(SeO_4)_2 \cdot 6H_2O$ | none | 0.47 [23] | VIII-2 |
| Guanidinium gallium selenate hexahydrate (GGaSeH) | $C(NH_2)_3Ga(SeO_4)_2 \cdot 6H_2O$ | none | 0.47 [23] | VIII-2 |
| Deuterated guanidinium aluminum sulfate | $C(ND_2)_3Al(SO_4)_2 \cdot 6D_2O$ | none | 0.35 [23] | VIII-2 |
| Ammonium iron alum | $NH_4Fe(SO_4)_2 \cdot 12H_2O$ | −185 | 0.40 [−187] | VIII-3 |
| Ammonium vanadium alum | $NH_4V(SO_4)_2 \cdot 12H_2O$ | −157 | 1.0 [−159] | VIII-3 |
| Ammonium indium alum | $NH_4In(SO_4)_2 \cdot 12H_2O$ | −146 | 1.2 [−148] | VIII-3 |
| Deuterated ammonium iron alum | $ND_4Fe(SO_4)_2 \cdot 12D_2O$ | −185 | 0.40 [−187] | VIII-3 |
| Methylammonium aluminum alum (MASD) | $CH_3NH_3Al(SO_4)_2 \cdot 12H_2O$ | − 96 | 1.0 [−98] | VIII-3 |
| Methylammonium gallium alum | $CH_3NH_3Ga(SO_4)_2 \cdot 12H_2O$ | −102 | ? | VIII-3 |
| Methylammonium chrome alum | $CH_3NH_3Cr(SO_4)_2 \cdot 12H_2O$ | −109 | 1.0 [−111] | VIII-3 |
| Methylammonium iron alum | $CH_3NH_3Fe(SO_4)_2 \cdot 12H_2O$ | −104 | 1.3 [−106] | VIII-3 |
| Methylammonium vanadium alum | $CH_3NH_3V(SO_4)_2 \cdot 12H_2O$ | −116 | 0.9 [−118] | VIII-3 |
| Methylammonium indium alum | $CH_3NH_3In(SO_4)_2 \cdot 12H_2O$ | −109 | 1.2 [−111] | VIII-3 |

| Name (abbreviation) | Formula | $T_c$ (°C) | $P_s$ (at $T$ °C) ($10^{-6}$ C/cm²) | Chapter and Section |
|---|---|---|---|---|
| Deuterated methylammonium aluminum alum | $CH_3ND_3Al(SO_4)_2 \cdot 12\,D_2O$ | − 96 | 1.0 [− 98] | VIII-3 |
| Methylammonium aluminum (selenate) alum | $CH_3NH_3Al(SeO_4)_2 \cdot 12\,H_2O$ | − 57 | 1.2 [− 59] | VIII-3 |
| Urea chrome alum | $CO(NH_2)_2HCr(SO_4)_2 \cdot 12\,H_2O$ | − 113 | 0.2 [− 115] | VIII-3 |
| Di-ammonium di-cadmium sulfate | $(NH_4)_2Cd_2(SO_4)_3$ | − 178 | 0.5 [− 180] | VIII-4 |
| Ammonium sulfate | $(NH_4)_2SO_4$ | − 50 | 0.45 [− 60] | VIII-5 |
| Ammonium fluoberyllate | $(NH_4)_2BeF_4$ | − 97 | 0.2 [− 110] | VIII-5 |
| Ammonium bisulfate | $(NH_4)HSO_4$ | upper−3 lower−119 | 0.8 [− 118] | VIII-6 |
| Rubidium bisulfate | $RbHSO_4$ | − 15 | 0.65 [− 170] | VIII-6 |
| Lithium hydrazinium sulfate | $Li(N_2H_5)SO_4$ | none | 0.3 [23] | VIII-7 |
| Lithium tri-hydrogen selenite | $LiH_3(SeO_3)_2$ | none | 15.0 [23] | VIII-8 |
| Sodium tri-hydrogen selenite | $NaH_3(SeO_3)_2$ | − 79 | ? | VIII-8 |
| Tri-glycine sulfate (TGS) | $(NH_2CH_2COOH)_3 \cdot H_2SO_4$ | 49 | 2.8 [20] | II |
| Tri-glycine selenate (TGSe) | $(NH_2CH_2COOH)_3 \cdot H_2SeO_4$ | 22 | 3.2 [0] | II-10 |
| Tri-glycine fluoberyllate (TGFB) | $(NH_2CH_2COOH)_3 \cdot H_2BeF_4$ | 70 | 3.2 [20] | II-10 |
| Di-glycine nitrate | $(NH_2CH_2COOH)_2 \cdot HNO_3$ | − 67 | 1.5 [− 190] | IX-3 |
| Glycine silver nitrate | $NH_2CH_2COOH \cdot AgNO_3$ | − 55 | 0.6 [− 195] | IX-4 |
| Di-glycine manganous chloride dihydrate | $(NH_2CH_2COOH)_2 \cdot MnCl_2 \cdot 2\,H_2O$ | none | 1.3 [25] | IX-5 |
| Ammonium monochloroacetate | $ClCH_2COONH_4$ | − 150 | 0.12 [− 170] | IX-6 |
| Di-calcium strontium propionate | $Ca_2Sr(CH_3CH_2COO)_6$ | 8.5 | 0.3 [− 45] | IX-7 |
| Sodium potassium tartrate tetrahydrate (Rochelle salt, RS) | $NaKC_4H_4O_6 \cdot 4\,H_2O$ | upper 24 lower −18 | 0.25 [5] | VII |
| Deuterated Rochelle salt | $NaKC_4H_2D_2O_6 \cdot 4\,D_2O$ | upper 35 lower −22 | 0.35 [6] | VII-7 |
| Lithium ammonium tartrate monohydrate (LAT) | $LiNH_4C_4H_4O_6 \cdot H_2O$ | − 175 | 0.22 [− 178] | IX-2 |
| Lithium thallium tartrate monohydrate (LTT) | $LiTlC_4H_4O_6 \cdot H_2O$ | − 263 | 0.14 [− 272] | IX-2 |
| Tetramethylammonium tri-chloro mercuriate | $[N(CH_3)_4] \cdot HgCl_3$ | none | 1.2 [23] | IX-8 |
| Potassium ferrocyanide trihydrate | $K_4Fe(CN)_6 \cdot 3\,H_2O$ | − 22 | 0.45 [− 50] | IX-9 |
| Potassium nitrate | $KNO_3$ | upper 124 lower 110 | 6.3 [121] | IX-10 |
| Sodium nitrite | $NaNO_2$ | 160 | 6.4 [143] | IX-11 |
| Colemanite | $Ca_2B_6O_{11} \cdot 5\,H_2O$ | − 7 | 0.45 [− 20] | IX-12 |
| Thiourea | $SC(NH_2)_2$ | − 104 | 3.0 [− 133] | IX-13 |

## APPENDIX II

Short analytical index for a few selected topics. Roman numerals refer to chapters, arabic numerals to sections. TGS = tri-glycine sulfate; KDP = potassium dihydrogen phosphate; BT = barium titanate; RS = Rochelle salt

| | II<br>TGS | III<br>KDP | IV<br>BT | VII<br>RS | Other chapters |
|---|---|---|---|---|---|
| **A. Phenomenological** | | | | | |
| Free energy expansion | 3 | | 4 | 9 | I-5 general theory<br><br>VIII-3 alums |
| Dielectric constant | | | | | |
|   non-linearity of | 3 | 2 | 4 | | I-5 |
|   adiabatic | 5 | 2 | | | |
|   high frequency | 2 | 2 | 2 | 2 | |
| Isotope effects | 10 | 6 | | 7 | VIII-3 alums<br>VIII-5 ammonium sulfate<br>IX-9 potassium ferrocyanide |
| Piezoelectricity and elasticity | 8 | 4 | 5 | 4 | VIII-2 GASH |
| Electrostriction | | | 5 | | |
| Electric-optic effect | | 5 | 2 | 2 | |
| Anomalies of | | | | | |
|   specific heat | 4 | 3 | 3 | 3 | |
|   spontaneous strain | | 3 | 3 | 3 | |
| Effect on Curie point by | | | | | |
|   electric fields | | | 4 | | IX-12 colemanite |
|   mechanical stresses | 6 | | 4 | 3 | |
| **B. Others** | | | | | |
| Domains | | | | | |
|   static | 7 | 9 | 7 | 6 | VIII-3 alums |
|   dynamic | 7 | | 8 | 6 | |
| Surface layer | 2 | | 9 | | |
| Crystal structure | 9 | 7 | 6 | 8 | X-3, 4 general |
| Model theories | | 8 | 11 | 9 | |

# SUBJECT INDEX

Absolute configuration of structure, 93, 375, 379

Aging effect, 45, 209

Aluminum niobate, $AlNb_3O_9$, 273

Alums, 324
  table of, 326

Ammonium bisulfate, $(NH_4)HSO_4$, 344

Ammonium dihydrogen phosphate (ADP), 25, 63, 89

Ammonium fluoberyllate, $(NH_4)_2BeF_4$, 341, 354

Ammonium monochloroacetate, $ClCH_2COONH_4$, 356

Ammonium Rochelle salt, 303

Ammonium sulfate, $(NH_4)_2SO_4$, 340, 354

Antiferroelectric
  definition, 25
  crystals, 226, 243

Antiferromagnetism, 374

Antipolar, 25

Aragonite, 360, 361


$Ba_2Al_2O_4$-type, 278

Barium niobate, $BaNb_2O_6$, 273

Barium titanate, $BaTiO_3$, 108
  absolute configuration, 379, 382
  ceramic, 203
  hexagonal phase, 200
  properties of ceramics, 203
  solid solutions of, 241, 248, 255
  conductivity of, 256

Barkhausen pulse, 177, 296, 347

Birefringence, 82, 121, 237, 246, 285, 288, 319

Boracite, $Mg_3B_7O_{12}Cl$, 371

Butterfly loop, 149, 289, 294


Ceramics, 203

Cadmium (pyro) niobate, $Cd_2Nb_2O_7$, 263

Cadmium titanate, $CdTiO_3$, 252

Calcite, 360, 361

Clausius–Clapeyron relation, 127, 135

Clausius–Mosotti equation, 9, 191

Cleavage, 28, 320, 344, 358, 363, 366

Coercive field
  definition, 4
  frequency dependence, 32, 117
  theoretical, 21
  thickness dependence, 118

Colemanite, $Ca_2B_6O_{11} \cdot 5H_2O$, 363

Convention for orthorhombic axes, 340, 367

Covalency, 199

Crystal classes, 10

Crystals growth, 338, 344, 346, 354, 355
  $BaTiO_3$, 199
  Guanidinium aluminum sulfate hexahydrate, 319
  potassium dihydrogen phosphate, 63
  Rochelle Salt, 281
  thiourea, 366
  tri-glycine sulfate, 28

Crystal structure, 392
  determination of, 376, 379

Curie constant
  definition, 7
  relation involving, 125
  significance of, 14

Curie point, see Curie temperature

Curie temperature, 6
  field dependence, 131, 364, 365
  pressure dependence, 41, 142, 290
  table of, 12, 389

Curie–Weiss law, 7, 30, 64, 112, 224, 226, 243, 264, 267, 283, 326, 352, 354, 360, 361, 363, 367

Curie–Weiss temperature, 7


Deliquescence, 281

Deuterium substitution, see Isotope effect

Devonshire theory of $BaTiO_3$, 132, 188

Di-ammonium di-cadmium sulfate, $(NH_4)_2Cd_2(SO_4)_3$, 338

Di-calcium strontium propionate, $Ca_2Sr(CH_3CH_2COO)_6$, 356, 357

Dielectric constant, 6, 383
  adiabatic and isothermal, 38, 65
  free and clamped, 64
  high frequency, 31, 64, 113, 283, 284
  non-linearity of, 35, 65, 131

Dielectric susceptibility, see Susceptibility

# AUTHOR INDEX